Novel Technologies in Food Science

Scrivener Publishing
100 Cummings Center, Suite 541J
Beverly, MA 01915-6106

Bioprocessing in Food Science

Series Editor: Anil Panghal, PhD

Scope: Bioprocessing in Food Science will comprise a series of volumes covering the entirety of food science, unit operations in food processing, nutrition, food chemistry, microbiology, biotechnology, physics and engineering during harvesting, processing, packaging, food safety, and storage and supply chain of food. The main objectives of this series are to disseminate knowledge pertaining to recent technologies developed in the field of food science and food process engineering to students, researchers and industry people. This will enable them to make crucial decisions regarding adoption, implementation, economics and constraints of the different technologies.

As the demand of healthy food is increasing in the current global scenario, so manufacturers are searching for new possibilities for occupying a major share in a rapidly changing food market. Compiled reports and knowledge on bioprocessing and food products is a must for industry people. In the current scenario, academia, researchers and food industries are working in a scattered manner and different technologies developed at each level are not implemented for the benefits of different stake holders. However, the advancements in bioprocesses are required at all levels for betterment of food industries and consumers.

The volumes in this series will be comprehensive compilations of all the research that has been carried out so far, their practical applications and the future scope of research and development in the food bioprocessing industry. The novel technologies employed for processing different types of foods, encompassing the background, principles, classification, applications, equipment, effect on foods, legislative issue, technology implementation, constraints, and food and human safety concerns will be covered in this series in an orderly fashion. These volumes will comprehensively meet the knowledge requirements for the curriculum of undergraduate, postgraduate and research students for learning the concepts of bioprocessing in food engineering. Undergraduate, post graduate students and academicians, researchers in academics and in the industry, large- and small-scale manufacturers, national research laboratories, all working in the field of food science, agri-processing and food biotechnology will benefit.

Publishers at Scrivener
Martin Scrivener (martin@scrivenerpublishing.com)
Phillip Carmical (pcarmical@scrivenerpublishing.com)

Novel Technologies in Food Science

Edited by
Navnidhi Chhikara
Anil Panghal
and
Gaurav Chaudhary

Scrivener Publishing

WILEY

This edition first published 2023 by John Wiley & Sons, Inc., 111 River Street, Hoboken, NJ 07030, USA and Scrivener Publishing LLC, 100 Cummings Center, Suite 541J, Beverly, MA 01915, USA

For more information about Scrivener publications please visit www.scrivenerpublishing.com.

Wiley Global Headquarters
111 River Street, Hoboken, NJ 07030, USA

For details of our global editorial offices, customer services, and more information about Wiley products visit us at www.wiley.com.

Library of Congress Cataloging-in-Publication Data

ISBN 9781119775577

Front cover images supplied by Pixabay.com
Cover design by Russell Richardson

Set in size of 11pt and Minion Pro by Manila Typesetting Company, Makati, Philippines

Printed in the USA

10 9 8 7 6 5 4 3 2 1

Contents

Preface

Novel Technologies in Food Science is gradually moving in a new direction, incorporating technologies from different fields and at the same time taking into account consumers' concerns about food safety, quality, and sensory attributes. The engineering and science aspects of these technologies are significant component of the undergraduate and postgraduate degree programs in Agriculture Engineering, Food Engineering, Food Science & Technology, and Nutrition & Health Science throughout the world. These novel technology operations in food processing are considered as one of the core competencies for these programs and in industries as well. Researchers around the globe will be able to use the information as a guide in establishing the direction of future research on nutritional properties during food processing. The main reason for writing this book now is to disseminate the wealth of knowledge on novel processing and its effects on food products. It is envisioned for scientists; technologists/engineers working in the area of food processing, process equipment design, and product development; and students of food science, technology, nutrition, health science, and engineering.

International peers having both the academic and professional expertise contributed the chapters in this book. The book illustrates a very clear and concise discussion on different novel technologies. The chapters included in this volume throw light on a number of research subjects, providing critical information on different processes, their impact on various food components, and their feasibility in food industry. The chapters' highlight different newer technologies which are making huge impact towards revolutionizing food industry growth and to the resulting finished products. The adoption of modern technologies has increased the efficiency and productivity within the factory. Most importantly, utilization of newer technologies in food processing operations has greatly improved product quality. Each of the chapter is supported with references owing to be an invaluable resource for the reader. Self-explanatory illustrations and tables of each chapter are an added advantage to understand the technology process and its outcome easily.

We thank all the contributing authors for their quality chapters and helping us to complete the book. We also thank the authorities of Chaudhary Charan Singh Haryana Agricultural University, Hisar and Guru Jambheshwar University of Science and Technology for their support. Finally, we also express indebtedness and thankfulness to Scrivener Publishing and John Wiley Sons Inc., USA for their unfailing guidance and assistance provided in finalization and publishing the book.

Navnidhi Chhikara
Anil Panghal
Gaurav Chaudhary

1

Ultrasound

Hugo Scudino[1], Jonas Toledo Guimarães[1], Angela Suárez-Jacobo[2], Hilda María Hernández-Hernández[3], Tatiana Colombo Pimentel[4], Socorro Josefina Villanueva Rodríguez[5], Vitoria Hagemann Cauduro[6], Erick Almeida Esmerino[1], Erico Marlon Moraes Flores[6], and Adriano Gomes da Cruz[7]*

[1]Fluminense Federal University, Faculty of Veterinary, Niteroi, RJ, Brazil
[2]Industrial Biotechnology Unit, Centro de Investigación y Asistencia en Tecnología y Diseño del Estado de Jalisco A.C., Zapopan, Jalisco, México
[3]Catedratica CONACYT in Food Technology Unit, Centro de Investigación y Asistencia en Tecnología y Diseño del Estado de Jalisco A.C., Guadalajara, Jalisco, México
[4]Federal Institute of Paraná, Paranavaí, Paraná, Brazil
[5]Food Technology Unit, Centro de Investigación y Asistencia en Tecnología y Diseño del Estado de Jalisco A.C., Guadalajara, Jalisco, México
[6]Federal University of Santa Maria, Department of Chemistry, Santa Maria, Brazil
[7]Federal Institute of Food Science and Technology (IFRJ), Department of Food, Rio de Janeiro, Brazil

Abstract
Ultrasound is an non-thermal technology which present enormous versatility at the food industry, with positive impact on the intrisic quality parameters of different food matrices. In this chapter, the definitons and the applications of ultrasound are presented as well as the challenges involved in their use.

***Keywords*:** Ultrasound, technology, food matrices

**Corresponding author*: adriano.cruz@ifrj.edu.br

Navnidhi Chhikara, Anil Panghal and Gaurav Chaudhary (eds.) Novel Technologies in Food Science, (1–38) © 2023 Scrivener Publishing LLC

1.1 Introduction

Conventional thermal technologies, such as sterilization, pasteurization, evaporation, and drying, are commonly used by the food industry. They can be used to inactivate enzymes and microorganisms, ensuring a safe product [1]. However, their effectiveness depends on the temperature and time of treatment, which may lead to nutrient loss, undesirable flavor and color development, and impairment of the sensory properties of foods [2–4]. Therefore, there is a considerable demand for new technologies to replace conventional thermal processing, providing microbiologically safe food without nutritional losses that are sensory pleasing to the consumer [5].

Among the various emerging technologies studied, high-intensity ultrasound (HIUS) technology has stood out. HIUS is a non-thermal emerging technology that can be used in food processing because of the ability to cause chemical and physical changes in the medium at mild temperatures. Ultrasound in a liquid medium can cause many changes, and the primary phenomenon that provides these changes is unstable acoustic cavitation [6]. The micro implosions generated during the collapse of the microbubbles result in high pressures and temperatures in the environment. Hence, they may induce chemical, physical, and microbiological changes in food [7].

Ultrasound is gaining prominence for presenting significant advantages compared to conventional thermal processes, such as lower production costs, preservation of the nutritional characteristics of food, improvement of sensory attributes, and improvements of product properties [8, 9]. In addition, it causes a low impact on the environment, and HIUS is known as a green technology [10].

There are currently many applications for HIUS in food processing. Ultrasound can be used individually in the processing or in combination with other methods, such as pressure (manosonication) and temperature (thermosonication), to produce a synergistic effect, increasing the effectiveness [11]. One of the main applications of ultrasound is microbial inactivation, which is widely studied in dairy and vegetable products. The impact on physicochemical properties is also widely explored. HIUS has wide application uses in defoaming, emulsification, texture modification of fatty products, as well as regulation of microstructures, functional properties of food proteins, and sonocrystallization. It can also be applied in various processing such as drying, freezing, concentration, tenderization, and thawing [12–15].

This chapter presents the applications of HIUS in food processing, with a focus on its use in food processing and preservation. The objective is to discuss what makes ultrasound one of the most promising research areas in food science.

1.2 Basic Principles of Ultrasound

Ultrasound research began in the 1920s with the researchers Harvey and Loomis and later in the 1960s. They verified that the sound waves produced by war submarine sonars could inactivate microorganisms [16, 17]. Then, other studies using ultrasound technology were developed using high frequencies and, then, high intensities, in parallel with the development of small equipment that would allow testing in these conditions [18].

Acoustic waves are classified considering the frequency audible to the human ear. Infrasound encompasses frequencies below the human hearing range (0 - 20 Hz), sound refers to the human audible range (20 Hz - 20 kHz), and ultrasound corresponds to the wave range above 20 kHz [19].

The ultrasound applications in food analysis, processing, and quality control can be divided based on the sound wave intensity in low and high-intensity ultrasound (Figure 1.1). Low-intensity ultrasound is related to intensities below 1 W/cm^2 and high frequencies (> 1 MHz), resulting in no changes in the treated material. Therefore, it is generally used for analytical applications, such as determining the structures and composition of food [20]. High-intensity ultrasound is related to intensities greater than 1 W/cm^2 and low frequencies (20 - 100 kHz). It can result in alterations

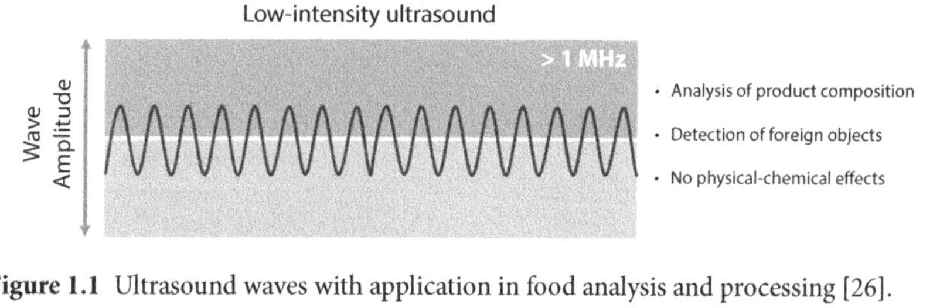

Figure 1.1 Ultrasound waves with application in food analysis and processing [26].

in the microbiological and physicochemical characteristics of the food. It is considered a non-thermal technology, and it has been used as an alternative to conventional food processing [7]. It is widely studied to replace thermal processes for microbial inactivation to reduce the adverse effects caused by heat [21]. Furthermore, it is used in the development of new products due to its ability to increase the activities of bioactive compounds [22], change viscosity [23], improve food stability by decreasing fat globules [24], alter the crystallization of dairy components, and water [25], among other applications.

1.2.1 Generation of the Ultrasonic Wave

Ultrasound is a form of vibrational energy produced by an ultrasound transducer that converts electrical energy into acoustic energy. A reliable and stable ultrasound is created through the electrical energy produced by a power generator (in general, using a piezoelectric device). This energy is transformed into mechanical energy by ultrasonic vibrations and then applied indirectly or directly to food [27]. When passing through food, the ultrasound will find resistance, and the waves will be partially dissipated. The resistance is dependent on the materials used. Therefore, an increase in the temperature may occur due to the transducer energy dissipation loss in the form of heat [7].

The most used ultrasound systems in food processing are probes and baths (Figure 1.2). The most commonly used is the probe system. An acoustic signal is directed to food in an amplified form and through a metal rod

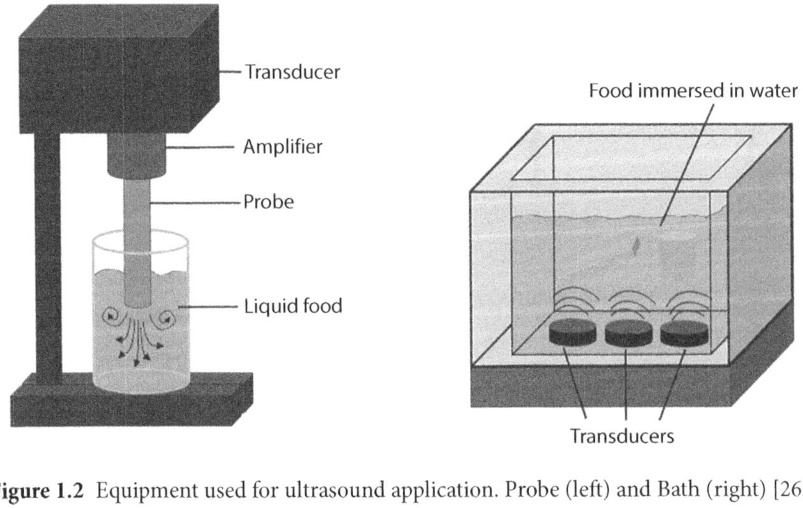

Figure 1.2 Equipment used for ultrasound application. Probe (left) and Bath (right) [26].

in this system. In this way, a liquid medium between probe and food is not needed. Therefore, energy is applied directly to the medium. In this way, Due to its direct action, there is greater cavitation intensity. The baths are considered an indirect system. In this system, a transducer is attached to the walls or base of the tank. The ultrasonic energy is transmitted to the liquid (generally water) and is then transferred to the food, resulting in greater dissipation of the waves. In this way, the ultrasound intensity is lower than expected [28, 29].

1.2.2 Principles of Acoustic Cavitation

Stable cavitation and transient or unstable cavitation are the two forms of acoustic cavitation. Stable cavitation is generally formed at low or moderate acoustic intensities. A micro agitation in the medium is observed because the bubbles oscillate regularly and long cycles around the equilibrium size. However, there is no implosion of the bubbles. The unstable cavitation is generally formed at high-intensity sound fields. The bubbles oscillate in short cycles, and they gradually grow. Contractions and expansions of the bubbles are observed several times, resulting in an unstable size. Then, the bubbles violently collapse, and there is the formation of shock waves and jets [30]. The unstable cavitation is responsible for chemical, physical, and microbiological alterations in food [31] (Figure 1.3).

The bubble size depends on the sound frequency. Other parameters also influence the cavitation, e.g., the viscosity, temperature, surface tension, vapor pressure of the liquid medium. Larger bubbles are produced at lower frequencies (20 - 40 kHz). In a general view, greater ultrasound intensities

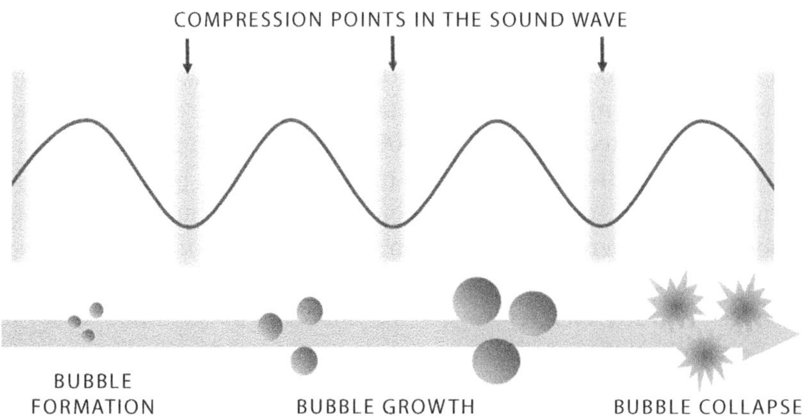

Figure 1.3 Acoustic cavitation in liquids [26].

result in a higher number of bubbles, as the number of bubbles is dependent on the ultrasound amplitude and energy supplied [32].

Once formed, a rapid translational motion of the bubbles relative to the liquid can occur. The origin for bubble translational motion in an acoustic field (called Bjerknes forces) is related to the oscillation period of the generalized buoyancy force exerting on anybody in an accelerating liquid. This phenomenon results from the effect known as "microstreaming," which is characterized by the streaming flow of fluid around the oscillating cavity [33, 34]. Another effect observed during acoustic cavitation is bubble expansion and compression caused by rectified diffusion. It will depend on the liquid medium saturation with dissolved gases and the sound field application at suitable levels [35]. When the specific size is achieved in the bubbles, they collapse and generate intense localized micro implosions, resulting in increased temperatures and pressures [7]. Temperatures of thousands of degrees Kelvin [36] and pressures of hundreds of bars are possible in the vicinity of the bubble [37]. As the temperature and pressure rise sharply at the bubble collapse, many highly reactive radicals are formed. They are formed from the decomposition of the molecules due to extreme conditions [31, 37].

1.3 Mechanisms of Microbial Inactivation

Several research works have evaluated the use of ultrasound to inactivate microorganisms and overcome undesirable results such as the formation of off-flavors and loss of nutrients. Ultrasound treatments are mainly applied at frequencies from 20 kHz to 100 kHz, and its effects in liquid medium are attributed to the cavitation phenomenon [38].

When the liquid medium is subjected to a sonic wave during the sonication process, longitudinal waves are formed, and expansion and compression regions may be observed [39]. The bubble collapse creates localized transitory areas with elevated temperature and pressure. These increases have been estimated to be up to 5000 K and 1000 atm. They accelerate the chemical reactivity dramatically into the medium [40]. The pressure changes resulting from the implosion of cavitation bubbles are mainly responsible for microbial inactivation [41]. In food processing, it is mandatory to reduce the microbial load and reach acceptable levels, as the presence of undesired microorganisms can result in economic losses and consumer rejection.

Butz and Tauscher [42] and Dolas, Saravanan [43] explained that ultrasound microbial inactivation is proportionated by intracellular cavitation. In this way, micro-mechanical shocks disrupt the membrane structure and cell wall, causing functional activity losses and DNA damage. For example, the generated OH• can react with the sugar-phosphate backbone of the DNA chain and drive the secession of the phosphate-ester bonds, breaking the double-strand microbial DNA. Thus, the effectiveness of bacterial reduction comes from the cavitation bubbles that disrupt the cell, modifying cellular activity from the outside to the inside, and from the high pressure that increases temperature weakening the bacteria membrane [44, 45]. These phenomena resulted in the inactivation of the microbial cells.

The ultrasound process parameters necessary for microbial inactivation are exposure time, ultrasonic wave intensity, type of microorganism, volume, food composition, and processing temperature. Different inactivation rates are observed for different types of microorganisms from literature data, with highly variable results according to the evaluated ultrasound conditions. Piyasena, Mohareb [41] emphasized that microbial inactivation by ultrasound is dependent on microbial species. Ultrasound effectively decreases bacterial growth and increases the kill rate of microbes at a wide range of frequencies (20 – 1000 kHz), being very effective at frequencies from 370 to 850 kHz. However, at these high frequencies, there is a high level of oxidative radical formation. Therefore, the lowest frequencies are still preferable to avoid interference with food compounds [46].

In general, gram-negative bacteria are less resistant to ultrasound than gram-positive [47, 48]. The lower resistance is associated with the outer membrane and thinner cell wall of the Gram-negative bacteria. On the other hand, Gram-positive bacteria have thicker cell walls with several peptidoglycan layers. Additionally, Specifically, *E. coli* was less resistant to ultrasound treatment than the other gram-negative bacteria.

Villamiel and Jong [45] studied the *Pseudomonas fluorescens* and *Streptococcus thermophilus* inactivation in Trypticase Soy Broth (TSB) using ultrasound. They reported that gram-negative bacteria were less resistant than gram-positive bacteria. Kang, Jiang [49] examined the effects of ultrasound on the number of *Escherichia coli* O157:H7 and vegetative cells of *Bacillus cereus* in brining and beef during the curing process. *B. cereus* had a higher survival to (30-min treatment). This was thought to be due to the formation of hydrogen peroxide, which can act in the sterilization process. It is formed from the recombination of OH• radicals caused by sonication utilization. However, Scherba, Weigel [50] have reported no differences between Gram-negative (*E. coli*) and Gram-positive (*Staphylococcus aureus* and *Bacillus subtilis*) bacteria. The differences may be associated with the different processing parameters.

Yeasts are generally more susceptible to ultrasound treatments due to the large cells (5-20 μm), resulting in the larger surface area [50]. Kapturowska,

Table 1.1 Evaluation of different ultrasound treatments on microbial populations.

A) Ultrasound (US)				
Microorganism	**Food matrix**	**Treatment conditions**	**Lethality/% reduced counts**	**References**
Total mesophilic aerobic counts *Total fungi (mold and yeast)*	Orange juice	23 and 500 kHz, 120-600 W Time: 15 min	0.8–1.5 log reduction Counts not affected.	[57]
Escherichia Coli 057 H7	Apple cider	150 W, 20 kHz, <57 °C, and 118 W/cm², Time: 6 min,	6-log reduction	[58]
Escherichia coli 25922	Model juice	85 W/cm², 24 kHz; 100 μm, Malic and citric acid Soluble solids: 0-16 g/100 mL Time: 9-11 min	>5 log reduction	[59]
Streptococcus mutants JCM 5175 *Escherichia coli IAM 12058*	Microbial suspension (0.89 wt% physiological salt solution)	1.7–12.4 W 20-500 kHz Time: 5, 10, 15 min	0.3- 2.0 log reduction	[60]

(*Continued*)

Table 1.1 Evaluation of different ultrasound treatments on microbial populations. (*Continued*)

A) Ultrasound (US)				
Microorganism	**Food matrix**	**Treatment conditions**	**Lethality/% reduced counts**	**References**
Enterobacter aerogenes *Aureobasidium pullulans*	Physiological salt solution (0.9% NaCl)	20 kHz Time: 5, 10, 20, 40, 60 min Bacterial suspensions: 4, 8, 12 W Yeast suspensions: 4, 8, 10, 12 W	1.0 up to 5.0 log reduction 0.1 up to 2.7 log	[61]
Lactobacillus acidophilus, *Escherichia coli* *Saccharomyces cerevisiae*	Saline salt solutions/ UHT milk	Sonication: 20 kHz, 750 W 100%/124 µm	0.55/0.82 log reduction 3.88/4.42 log reduction 3.62/2.10 log reduction	[62]
Mycobacterium sp (6PY1)	Aqueous suspensions	20 kHz (58.2, 114, 244, and 411 W L^{-1}) 612 kHz (13–114 W L^{-1})	93% 43%	[63]

(*Continued*)

Table 1.1 Evaluation of different ultrasound treatments on microbial populations. (*Continued*)

A) Ultrasound (US)				
Microorganism	**Food matrix**	**Treatment conditions**	**Lethality/% reduced counts**	**References**
Bacillus subtilis	Nutrient broth	Low Frequencies: 20 and 38 kHz High Frequencies: 512 and 850 kHz. Time: 1, 2, 5, 10, and 15 min	No effect in viability 40% increase	[64]
Bacillus licheniformis ATCC 6634	Non-fat milk	20 kHz, 76.3 W 3 s on/ 3 s off, 10 min, Ice bath	0.2 log reduction	[65]
Bacillus cereus ATCC 14579	Distilled water	20 kHz, 500 W, <20 W/mL 30 min, 20 °C	0.9 log reduction	[66]

(*Continued*)

Table 1.1 Evaluation of different ultrasound treatments on microbial populations. (*Continued*)

B) TS, MS or MTS					
Microorganism	**Food matrix**	**Treatment conditions**		**Lethality (log reductions)/% Reduced counts**	**References**
Total viable count	Raw milk	TS	20 kHz 48.6, 62 and 76 °C 11 mL/min	From 0.2 up to 2.9 log	[45]
Escherichia coli K12	Buffer solutions	TS, MTS MS	100 kPa/40 °C, (MS) at 400 kPa/40 °C, (TS) at 100 kPa/61 °C, (MTS) at 400 kPa/61 °C pH: 3, 4, 5, 7	5.0 log reached	[67]
Serovar Enteritidis Salmonella enterica	McIlvaine citrate phosphate buffer	MTS	117-µm; 20 kHz, 175 kPa; Different Temp and water activities	Up to 3.0 log	[68]

(*Continued*)

Table 1.1 Evaluation of different ultrasound treatments on microbial populations. (*Continued*)

B) TS, MS or MTS					
Microorganism	**Food matrix**	**Treatment conditions**		**Lethality (log reductions)/% Reduced counts**	**References**
Salmonella Senftenberg 775W	Media of varying water activities	MS MTS	20 kHz, 117 μm, 175 kPa 57 up to 80 °C	From 1.0 up to 2.0 log	[69]
Cronobacter sakazakii	Citrate–phosphate buffer, pH 7.0/ Rehydrated powdered milk	MTS	20 kHz, 200 kPa; 117 μm 35–68 °C 0-30 min	3.5 log / 3.2 log	[70]
	Apple juice	MTS	200 kPa; 20 kHz, 5 W/mL. 35 – 64 °C, 117 μm;	Up to 4.8 log	[71]
Listeria Monocytogenes Scott A	0.85% NaCl saline solution, pH 7.0	TS	22.3kHz; 1.43, 0.85, 0.49 W/mL, 0-20 min. 35°C, 50°C, 65°C.	Up to 2.7 log	[72]

(*Continued*)

Table 1.1 Evaluation of different ultrasound treatments on microbial populations. (*Continued*)

B) TS, MS or MTS					
Microorganism	**Food matrix**	**Treatment conditions**		**Lethality (log reductions)/% Reduced counts**	**References**
Listeria Monocytogenes 10403S	Apple cider, pH 3.4	TS	20 kHz, 457 mW/mL 0-20 min 20, 30, and 40°C 50, 55, and 60°C	Up to 2.5 log Up to 7.8 log	[73]
Listeria innocua ATCC 51742	Raw whole milk fat (free, 1%, 2%)	TS	24 kHz, 120 µm, 400 W, 30 min, 63 °C	Up to 4.9 logs in fat-free milk	[74]
Saccharomyces cerevisiae	Pineapple juice, Grape juice, Cranberry juice.	TS	24 kHz, 400 W, 120 µm; 0-10 min 40°C, 50°C, 60°C, Continuous pulses	6.4 log 5.0 log 7.0 log	[75]

(*Continued*)

Table 1.1 Evaluation of different ultrasound treatments on microbial populations. (*Continued*)

B) TS, MS or MTS					
Microorganism	**Food matrix**	**Treatment conditions**		**Lethality (log reductions)/% Reduced counts**	**References**
Alicyclobacillus acidoterrestris ATCC 49025 (spores) *Saccharomyces cerevisiae KE162*	Natural squeezed and commercial apple juice, pH 3.4	TS	20 kHz, 600 W, 95.2 µm 0- 30 min. 20, 30 y 40 °C	No reduction Up to 2.8 log reduction	[76]
Clostridium perfringes NZRM 2621 and NZRM 898 (spores)	Beef slurry	TS	Heat shock (80 °C for 10 min) followed by 1-min US treatment. 460 W/cm^2, 24 kHz, 0.33 W/g, 33W 0-60 min	1.0 to 1.5 log r	[77]

(*Continued*)

Table 1.1 Evaluation of different ultrasound treatments on microbial populations. (*Continued*)

B) TS, MS or MTS					
Microorganism	**Food matrix**	**Treatment conditions**		**Lethality (log reductions)/% Reduced counts**	**References**
Brettanomyces yeasts *Lactic acid bacteria*	Wine samples	TS	24 kHz, 400 W, 100 μm 30° and 40° C	1.9-2.1 log (89.1–99.7%) 1.8-2.6 log (71.8–99.3%)	[52]
Pseudomonas fluorescens *Streptococcus thermophilus*	3% Trypticase Soy Broth	TS	20 kHz, 150 W 45.2, 62.3, and 75.9 °C Continuos-flow (11 to 174 ml/min)	≤ 2.7 log reduction	[45]
Escherichia coli *Pseudomonas fluorescens* *Staphylococcus aureus*	Milk	MTS	20 kHz, 700W, 60-120 W/cm²; 225 kPa 40-240s 20- 52°C	Up to 1.6 log 1.05 log 1.58 log	[78]

(*Continued*)

Table 1.1 Evaluation of different ultrasound treatments on microbial populations. (*Continued*)

B) TS, MS or MTS					
Microorganism	**Food matrix**	**Treatment conditions**		**Lethality (log reductions)/% Reduced counts**	**References**
Escherichia coli *Staphylococcus aureus*	Milk	TS	20 kHz, 600 W, 60-120 mm 20°; 40°, 60° C 6-12 min	1.37 log 3.07 log	[79]
Yersinia enterocolitica ATCC 9610	Suspension	MS	20 kHz, 21-150 μm, 200 kPa, 30 °C	Up to 3.5 log	[80]
Bacillus subtilis (spores)	Spore suspension	MS MTS	20 kHz, 500 kPa, 117 μm, 12 min 20 kHz, 300 kPa, 117 μm, 6 m, 70–90 °C	99.0% 99.9 %	[81]

Stolarzewicz [51] observed that ultrasound treatment decreased the yeast numbers (100-1000 times). Gracin, Jambrak [52] observed satisfactory *Brettanomyces* yeasts reduction (89.1–99.7%) using high power ultrasound continuous flow treatment (400 W, 24 kHz, 100 μm amplitude) in wine (< 1 min). On the other hand, *Aureobasidium* was very resistant to ultrasound treatment [53].

Typically, cocci are more resistant than bacilli because of volume and cell surface. From published data, scattered information is available about the influence of stresses on microbial resistance to ultrasound technology. For example, vegetative cells in the growth phase are less resistant than cells in the stationary phase [54], and aerobes are more resistant than anaerobes [55]. In addition, the effectiveness of ultrasound on microbial control is highly dependent on three factors: the medium characteristics (volume, viscosity, initial microbial number, and pH), cavitation process (amplitude, intensity, temperature, frequency, and external pressure), and type of microorganism (size and shape, cell wall, growth phases, and spores or vegetative cells) which could present highly complex interactions among them [56].

Bacteria, yeast, and mold spores are more resistant to ultrasonic cavitation effects than vegetative cells. Ultrasound technology does not inactivate microbial spores. Greater efficiency is obtained by the combination of ultrasound with heat treatment (Thermosonication, TS) or external pressure applied (Manosonication, MS), or both (Manothermosonication, MTS), to get a complete sterile product and a most significant potential for industry application. Furthermore, ultrasound may also be combined with natural antimicrobials, pulsed electric fields (PEF), and high hydrostatic pressure (HPP) to improve the inactivation capacity and extend the food shelf life. Table 1.1 presents the effect of ultrasound on microbial safety.

1.4 Ultrasound Application in the Food Industry

This technology has many applications in the food industry and its subsidiaries, such as in the meat industry. It increases the water-holding capacity, enhances color, and improves tenderness. For fruits and vegetables, it can enhance firmness, reduces microbial load, causes color changes, enzyme inactivation, and enhances drying characteristics. In cereal products, ultrasound allows better mixing of doughs and improves textural, sensory, and visual aspects. In the dairy industry, ultrasound has been used to inactivate microorganisms, enhance the homogenization step, reduce fat globules size, and improve sensory and nutritional quality [11, 82–86].

Europe and the USA have adopted ultrasound technology for commercial applications; the food and drug administration approved the technology as a potential alternative to the traditional thermal preservation approach [87]. Ultrasound application in liquid medium is the most used. Thus, it is commonly used for beverages. The food matrix characteristics and the application define the choice of the equipment architecture and the processing conditions. Large-sized companies have designed ultrasound equipment for large-scale extractions, such as REUS (France) and Hielscher (Germany). Hielscher Company commercializes a wide power range for analytical (50 to 400 W) and industrial (500 to 16 000 W) scales. REUS Company has developed a wide range of reactors from the pilot (30 to 50 L) to industrial (500 to 1000 L) scale [86, 88, 89].

The major challenge for the industrial application of ultrasound has been defining the optimal processing parameters related to the study of nutritional, quality, and sensory properties aiming to scale up the process [83, 85, 86]. Ultrasound technology has been proposed as an extraction alternative to overcome the technological and scientific bottlenecks presented by conventional technologies that involve: time-consuming extractions, high organic solvent, and energy consumption, and higher costs [89–93].

The bioactive compounds from plant and animal materials can be extracted by ultrasound. Many authors report the economic feasibility of ultrasound bio compounds extraction from different materials. Vieira, Cavalcanti [94] utilized an agitated bed assisted by ultrasound to produce natural antioxidant extracts from jussara pulp (*Euterpe edulis*). The costs of ultrasound extract (90.57–165.34 USD/kg) were slightly higher than those obtained using a conventional extraction (87.32–167.48 USD/kg).

Another study examined the economics of ultrasound-assisted extraction of phenolic compounds from *Iraca infructescence* (Carludovica palmata, Ruiz & Pav.). The cost of manufacturing changed depending on the capabilities of the equipment. By scaling the volume from 5 L to 500 L, the price decreased from 456.04 USD/kg to 93.43 USD/kg [95]. The anthocyanin extraction from purple yam was economically feasible with a selling price above 170 USD/kg [96]. Then, ultrasound extraction is, in general, a more efficient and rapid method in comparison with conventional processes. However, the manufacturing costs will depend on the manufacturing capacity.

An innovation patent search related to ultrasound in food processing was carried out in the CLARIVATE ANALYTICS database. Figure 1.4 shows the results between 2015-2020, focusing on designing and constructing ultrasound equipment. The ultrasound is primarily applied for microbiological and enzymatic inactivation (41 registered patents), followed by

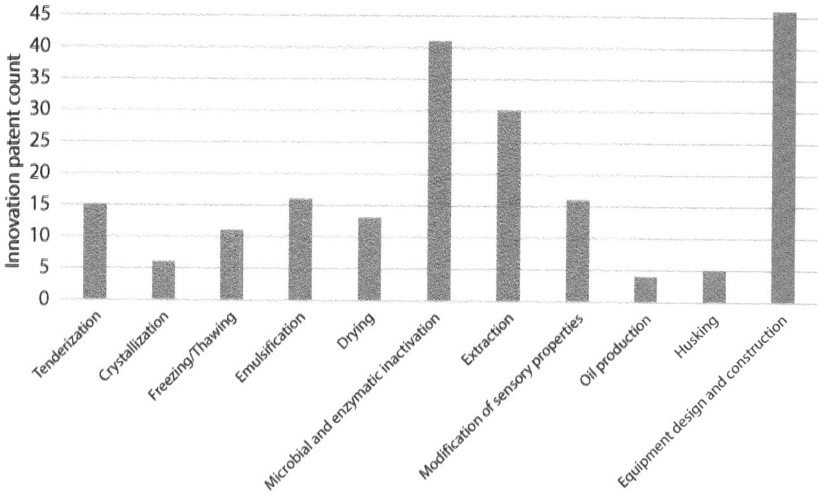

Figure 1.4 Some patents of ultrasound for food processing.

extraction. The top five countries with registered patents are China, Korea, the USA, Colombia, and Ukraine. The registration of patents has increased by more than 150%, reporting 12 patents for 2015. For 2020, the number increased to 57; however, in 2019, the highest number was reported in the analyzed period with 77 patents (Figure 1.5).

Through the years, ultrasound has proven its effectiveness in replacing and improving several traditional processing technologies. However, ultrasound has shown to be more potent combined with other techniques,

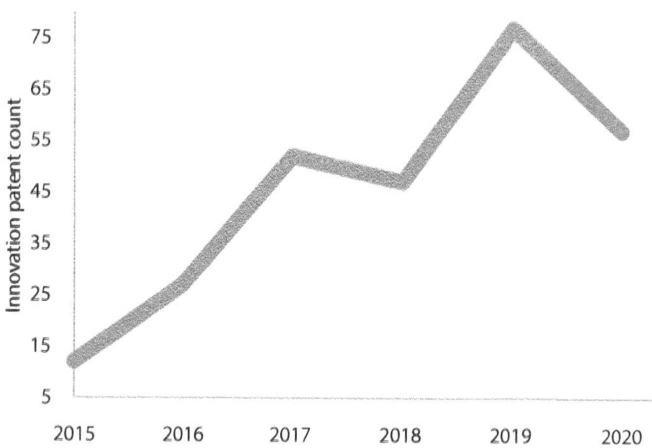

Figure 1.5 Patent publishing trends from 2015-2020.

resulting in better overall quality in the products. Ultrasound applications at the industrial level should consider improvements such as energy efficiency, high-power process design, competitive energy consumption, easy installation, and low maintenance cost.

1.4.1 Impact of Ultrasound on Physicochemical Quality Indicators of Food

The microbial inactivation efficacy using ultrasound treatment is well documented. However, the effect on other quality attributes as color, texture, and sensory characteristics have not been extensively studied [97]. Ultrasound treatment has been reported in the processing of various food products. The most relevant data in this area is given below.

1.4.1.1 Meat Products

The application of ultrasound in the meat industry began around the 1950s by evaluating fat and muscle on live cattle. Several studies reported the different uses of ultrasound in meat with attractive advantages in tenderizing, meat brining, cooking, thawing, freezing, and bacterial inhibition [11, 98].

Meat color is a primary quality factor because it may be associated with aging time, shelf life, hardness, and juiciness. In red meat, a bright red color is related to freshness and therefore has higher consumer acceptance. Many studies suggest that the color is not modified by ultrasound treatment due to the heat generated by ultrasound not being enough for protein or pigment denaturation [99, 100]. However, some studies suggest that ultrasound influence the color changes in meat due to the reduction in oxymyoglobin and metmyoglobin formation [11, 98].

One of the indicators of meat quality is tenderness. The texture of the meat can be improved using ultrasound, and there are applications with poultry meat, rabbit, fish, veal, and beef. Ultrasound acts by modifying muscle fiber properties, myofibrillar protein properties, and intramuscular connective tissue properties. Hence, ultrasound helps improve the physical properties of meat products, which includes tenderness, pH, water-binding capacity, and cohesiveness [84, 98, 101]. González-González evaluated the impact of ultrasound on the bovine *cleidooccipitalis, longissimus lumborum,* and *infraspinatus* muscles and aging for 14 days. Their results showed that the ultrasound did not affect pH, affected the color negatively, and improved water-holding capacity. Ultrasound helps in tenderization and accelerates the maturation of the meat [102]. Another study evaluated the effects of ultrasound, applied after storage, on beef's physical, microstructural, and sensory characteristics. The treated meat showed increased pH and luminosity, reduced redness and saturation value, higher hue angle, and variable changes in the water holding capacity. Overall, the application of ultrasound on meat during storage affects textural and microstructural properties and accelerates aging without negative impacts on other technological and sensory attributes at seven days of storage [103].

The use of ultrasound has been reported as an accelerator in compound impregnation and improved meat product quality. In fresh rabbit meat, the effect of acid marination assisted by ultrasound was evaluated. The results showed that marinating time, NaCl concentration, and ultrasound increased weight losses, meat hardness, and luminosity but reduced water content, water-holding capacity, and pH [98, 104].

1.4.1.2 Fruits and Vegetables

Fruits and vegetables are highly perishable, which makes them a challenge for the food industry. Ultrasound has been applied in fruits and vegetables to modify firmness and inactivate phenol peroxidase and polyphenol oxidase enzymes. In certain conditions, higher rates were observed when combining ultrasound technology with heat (40–60°C). Ultrasound waves can provide information about the composition, structure, and physical properties of fruits. The challenge of ultrasound application in fruit and vegetables is the pores and voids present in them since these voids limit or attenuate ultrasound dispersion [11, 105].

One important fruit or vegetable sensory attribute is color, an indicator of freshness and flavor quality. Inappropriate color affects consumers' buying decisions as it could be a sign of loss of freshness or lack of ripeness [106].

Firmness and juiciness are two attributes that consumers look for in fruits or vegetables. Most of the studies reported that ultrasound processing could affect fruit or vegetable firmness. However, this is a controversial point due to some studies not reporting significant differences in textural parameters. Thus, the ultrasound effect on fruit or vegetable texture depends on several parameters, such as food matrix, variety, maturity stage, and intensity or time processing. These parameters should be assessed independently for each fruit or vegetable [11, 105, 106].

Another aspect that has not been extensively studied is the effect of ultrasound on the flavor of fruits or vegetables. Thus more investigation may be necessary to elucidate these properties. A study on the impact of ultrasound combined with the controlled atmosphere on the quality of cucumbers showed that the combination of technologies reduced the loss of mass and firmness, inhibited the color change, maintained soluble solids content and cell wall integrity, and limited the redistribution of water in cucumbers. Regarding the taste, it reported no differences in astringency, aftertaste, bitterness, umami, richness, and saltiness between ultrasound processed cucumber and controls [106, 107].

Ultrasound has been widely used to treat fruit juices, demonstrating its effectiveness in improving their properties such as apparent viscosity, color, cloudy stability, and sensory quality (appearance, texture, taste, and aroma) [43]. In combination with other preservation technologies or alone, ultrasound can be suitable for preserving liquid foods. Ultrasound was proven effective for reducing polyphenol oxidase, and peroxidase activity, which are the enzymes commonly involved in fruit juices browning reactions [105].

The combined effect of ultrasound (40 kHz, 200W, 21 min) and pulsed electric field (1kHz, 9 kV/cm, 335 μs) technologies were studied to improve the quality and microbial safety of spinach juice [106]. After treatment, the spinach juice showed a significant reduction in microbial inactivation. However, juice rheology was changed by the non-thermal technologies, the apparent viscosity and the consistency index were changed in spinach juice [108].

Another work focused on examining the effect of chemical treatment (ascorbic acid and calcium chloride), physical treatment (water bath at 65 °C), and ultrasound (Bath type, 28 kHz, 100 kW/m^3 intensity, 50 W, 15 min) to prevent enzymatic browning of fresh-cut quince slices during 14 days of storage. The results showed that ultrasound treatment inhibited bacteria, mold, and yeast growth in fresh-cut quince slices at 4 °C storage. The ultrasound-treated samples showed improvements in bioactive compounds, and these samples showed better color values showing less enzymatic browning [109].

1.4.1.3 Dairy Industry

Ultrasound is a novel technology with significant potential in the dairy industry for commercialization. The ultrasound applications in the dairy industry include microbial inactivation, homogenization, skimming, and emulsification, and it can accelerate milk fermentation. Other applications, such as defoaming, defouling, denaturation, functional improvements to proteins, and lactose crystallization, have been explored. The ultrasound treatment of milk produces homogeneous and evenly distributed fat globules, which facilitates this milk in various dairy products. For example, milk treated by ultrasound will have better fat dispersion, making yogurt more homogeneous and dense [11, 105, 110, 111]. Intensities higher than 1 W/cm^2 have been used for emulsification and homogenization. Intensities below 1 W/cm^2 were used for non-invasive monitoring and non-destructive separation of mixtures [111].

Scudino, Silva [112] evaluated the ultrasound effect on the enzymatic and microbial inactivations, kinetic stability, and physicochemical properties of raw milk. With the application of ultrasound processing (20 kHz, 475 W, 3 kJ/mL), raw milk obtained similar microbial and enzymatic standards to conventional thermal processes. Furthermore, the size of the fat globules was reduced (< 1 μm), the color parameters were improved, and kinetic stability was observed during storage. However, studies of this technology still need to be carried out in dairy products to evaluate nutritional changes and the sensory acceptance of these products [112].

The application of ultrasound in cheese elaboration, coagulation processes, and fermentation was commonly studied [113]. Ultrasound and texture analysis has been used for quality control purposes to assess the texture and the maturation degree in manchego cheese during storage [114]. Ultrasound has been used for fractionating the milk liquid and removing fat layers [6].

1.4.2 Effects of Ultrasound Treatment on Sensory Characteristics of Foods

It is essential to underline that shear forces, turbulence, and high localized pressures and temperatures generated when applying sound waves with frequencies higher than 20 kHz inactivates microorganisms and enzymes. However, they produce physical changes capable of modifying the mechanical and rheological properties of the food matrices. These forces can cause ruptures of the microstructure and destabilization of the colloidal systems and the balances of food interfaces. In the same way, they weaken the cell walls [115] of plant products or cellular membranes of animal products which, in turn, can promote chemical changes by enabling the contact of different substances or incorporation of oxygen into the matrix. All this cascade of changes, positive or negative for the product, contribute to the different sensory perceptible changes in foods treated by ultrasound. Evaluating these impacts is decisive in the successful application of said technology to preserve food and beverages since the consumer carries out the last quality control faced by a food or a drink with their senses and their criteria of preference.

There are few studies aimed at evaluating the impact of ultrasound technology on sensory quality. Table 1.2 shows a summary of research aimed at

Table 1.2 Impact of ultrasound on sensory characteristics of foods.

Product	Applied process	Sensory effects	Reference
Beef m. *Longissimus dorsi* muscle	40 kHz bath, 11 W cm-2, 60 min. 0, 7, and 14 days of storage at 4 °C	Twelve training panels evaluated the product. The meat treated with ultrasound was perceived to be softer, less elastic, and more fibrous. In terms of color, ultrasound caused a slight decrease in the fresh reddish color, and in terms of flavor, the meat treated with ultrasound developed a smell and taste of fat and a few metallic tastes. These characteristics were accentuated as a function of the storage time of the meat.	[103]
Orange juice	20 kHz probe, 100 W, 15 min, the controlled temperature of 40 ± 1 °C,	The sensory properties of the ultrasound-treated juices were most acceptable to the consumers and evaluated as similar using analytical sensory tests.	[116]
Apple juice and nectar.	20 kHz probe, the amplitude of 60, 90, and 120 μm, 20, 40, and 60 °C, 3, 6, and 9 min.	Ultrasonic treatment changed the volatile compounds profile. The smell, color, and taste of the nectars and juices treated with ultrasound decreased compared to the non-treated products, but the preference was maintained.	[117]

(Continued)

Table 1.2 Impact of ultrasound on sensory characteristics of foods. (*Continued*)

Product	Applied process	Sensory effects	Reference
Tea sample mixed with 300 ml of distilled water.	40 kHz bath, 250 W, 60 °C, 40 min. compared to conventional extraction at 85 °C for 15 min.	The product subjected to ultrasonic-assisted extraction showed a higher fresh taste and flavor. The conventional extraction resulted in products with plain taste and flavor properties, astringency, and slight bitterness.	[118]
Restructured cooked ham	20 kHz probe, 600 W cm^{-2} for 10 min	A study was carried out with one hundred fifteen consumers. The ultrasound-treated product showed improved sensory acceptance in texture, taste, and global evaluation. Furthermore, the purchase intent of the product with 50% salt reduction increased. Ultrasound probably promoted a better salt diffusion.	[119]
Strawberry juice	40 kHz bath, 10 and 30 min, compared to pasteurized juice (90 °C for 60 s).	The application of ultrasound treatments did not affect color and off-odor scores compared to untreated or thermally treated samples. However, higher acidity and sweetness scores were found on both sonicated samples compared to untreated.	[120]

(*Continued*)

Table 1.2 Impact of ultrasound on sensory characteristics of foods. (*Continued*)

Product	Applied process	Sensory effects	Reference
Apple dried slices	21 and 35 kHz bath, 30 min, 3 and 4 W/ cm^2, then drying in a convective oven at 70 °C and air velocity of 2 m/s.	Ultrasound treatment resulted in red color discoloration, yellow color increase, and higher chromaticity and lightness.	[121]
Natural fish flavor production	37 kHz probe, 112 – 337 W, 25 °C for 30 min,	Ultrasound treatment resulted in products with increases toasty and fish-like aroma, mouthiness, and umami. Furthermore, it reduced the bitter taste.	[122]
Passion fruit juice	20 kHz 13 mm probe, 375 W, 89.25 μm for 8 min.	The panel scored sonication passion fruit juice significantly ($P<0.05$) lower in color, flavor, and aroma scores, but the global acceptance was similar ($P>0.05$) than that of non-sonicated juice. Therefore, sonication does not impact the global sensory quality of PFJ.	[123]

(*Continued*)

Table 1.2 Impact of ultrasound on sensory characteristics of foods. (*Continued*)

Product	Applied process	Sensory effects	Reference
Olive oil extraction	26 kHz probe, 200 W, 5 - 40 min. 3 L tank volume.	Ultrasound application for 15 min provided the optimum conditions to extract olive oil with the highest yield and lowest free acidity. Furthermore, oil with positive attributes (fruity, spicy, and bitterness) was obtained, and no defects (heated-muddy residue, musty-moist-earthy, winey-vinegary-acidic, wet-woody, rancid-stale, metallic, and black water) were observed.	[124]
Dried apples	37 kHz bath, 200 W, 0.02 –0.03 W/ cm^3, during 0, 9, 22.5, 45 and 54 min.	Although pre-treatment with ultrasound improved drying, the sensory characteristics were significantly decreased, so more research is required to obtain competitive sensory characteristics.	[125]
Milk to produce Yogurt	20 kHz probe, 150–750 W, for 10 min.	Results proved that milk treated by high-intensity ultrasounds led to acceptable yogurt with superior texture characteristics (cohesiveness, hardness, firmness) but bad taste and flavor (off-flavors).	[126]

(*Continued*)

Table 1.2 Impact of ultrasound on sensory characteristics of foods. (*Continued*)

Product	Applied process	Sensory effects	Reference
Iranian ultrafiltered feta-type cheese	20, 40, and 60 kHz probe, 160 W, 20 min.	Texture scores were similar for the sample sonicated at 60 kHz and the control. The highest texture scores were observed in samples sonicated at 20 and 40 kHz. Ultrasound increased taste and flavor scores due to higher proteolysis and lipolysis. The highest overall acceptance scores were kept for UF cheese sonicated at 60 kHz, and the lowest scores for the untreated control sample.	[127]

assessing the effect of ultrasound on the sensory quality of different foods or beverages and the production of natural flavoring of fish.

No single aspect describes the impact of ultrasound on sensory characteristics (Table 1.2). This impact depends on the composition of the food, its physical state, the intensity of the treatment, the sensory analysis methodology used, the type and quantity of participating judges, and the quality of their training. Some studies use analytical tests and descriptive tests to identify perceived characteristics, intensity or kind of smell and taste, changes in mechanical or rheological properties; others are based mainly on acceptance.

1.5 Conclusion

Ultrasound is an emerging technology that has been studied due to its beneficial effects on food preservation and processing. Furthermore, it can improve product yield, and it has shorter operating times and lower maintenance costs. Lower microorganism survival rate at low temperatures and better textural properties have also been reported. In addition, this technology is gaining prominence. It is not limited to improving quality or increasing preservation effects and offers the possibility of developing new products with unique functional attributes.

Combining ultrasound with other technologies generates better results in the overall quality of the final product, as it can extend its advantages over other technologies and help face or overcome its deficiencies. However, there is a need for studies to enable the use of ultrasound at the industrial level. The need to scale up on an industrial level would require adequate modeling and analysis to improve the design variables and optimize the parameters needed to offer quality and efficient food processing.

Therefore, ultrasound is a technology of great importance in food science and needs to be further developed by performing optimization for each specific process, assessing nutritional changes and sensory acceptance of these products by consumers, in addition to studies on how to scale processing without changing the energy performance of the ultrasound.

References

1. Fox, P.F., *et al.*, *Heat-Induced Changes in Milk*, in *Dairy Chemistry and Biochemistry*, P.F. Fox, *et al.*, Editors. 2015, Springer International Publishing: Cham. p. 345-375.
2. Livney, Y.D., M. Corredig, and D.G. Dalgleish, *Influence of thermal processing on the properties of dairy colloids.* Current Opinion in Colloid & Interface Science, 2003. **8**(4): p. 359-364.
3. Awuah, G.B., H.S. Ramaswamy, and A. Economides, *Thermal processing and quality: Principles and overview.* Chemical Engineering and Processing: Process Intensification, 2007. **46**(6): p. 584-602.
4. Shanmugam, A., J. Chandrapala, and M. Ashokkumar, *The effect of ultrasound on the physical and functional properties of skim milk.* Innovative Food Science & Emerging Technologies, 2012. **16**: p. 251-258.
5. Alegbeleye, O.O., *et al.*, *Hazards of a 'healthy' trend? An appraisal of the risks of raw milk consumption and the potential of novel treatment technologies to serve as alternatives to pasteurization.* Trends in Food Science & Technology, 2018. **82**: p. 148-166.
6. Paniwnyk, L., *Applications of ultrasound in processing of liquid foods: A review.* Ultrasonics Sonochemistry, 2017. **38**: p. 794-806.
7. Kentish, S. and H. Feng, *Applications of power ultrasound in food processing.* Annual review of food science and technology, 2014. **5**: p. 263-284.
8. Balthazar, C.F., *et al.*, *Ultrasound processing of fresh and frozen semi-skimmed sheep milk and its effects on microbiological and physical-chemical quality.* Ultrasonics Sonochemistry, 2019. **51**: p. 241-248.

9. Chouliara, E., *et al.*, *Effect of ultrasonication on microbiological, chemical and sensory properties of raw, thermized and pasteurized milk.* International Dairy Journal, 2010. **20**(5): p. 307-313.
10. Téllez-Morales, J.A., B. Hernández-Santo, and J. Rodríguez-Miranda, *Effect of ultrasound on the techno-functional properties of food components/ingredients: A review.* Ultrasonics Sonochemistry, 2020. **61**: p. 104787.
11. Bhargava, N., *et al.*, *Advances in application of ultrasound in food processing: A review.* Ultrasonics Sonochemistry, 2021. **70**: p. 105293.
12. Arvanitoyannis, I.S., K.V. Kotsanopoulos, and A.G. Savva, *Use of ultrasounds in the food industry–Methods and effects on quality, safety, and organoleptic characteristics of foods: A review.* Critical Reviews in Food Science and Nutrition, 2017. **57**(1): p. 109-128.
13. Buckin, V., E. Kudryashov, and B. O'Driscoll, *High-resolution ultrasonic spectroscopy for material analysis.* American laboratory, 2002. **34**(5): p. 28-31.
14. Dukhin, A.S., P.J. Goetz, and B. Travers, *Use of Ultrasound for Characterizing Dairy Products.* Journal of Dairy Science, 2005. **88**(4): p. 1320-1334.
15. Fu, X., *et al.*, *Sono-physical and sono-chemical effects of ultrasound: Primary applications in extraction and freezing operations and influence on food components.* Ultrasonics Sonochemistry, 2020. **60**: p. 104726.
16. Harvey, E.N. and A.L. Loomis, *The destruction of luminous bacteria by high frequency sound waves.* Journal of Bacteriology, 1929. **17**(5): p. 373.
17. Mason, T.J., *et al.*, *Potential uses of ultrasound in the biological decontamination of water.* Ultrasonics Sonochemistry, 2003. **10**(6): p. 319-323.
18. Zisu, B. and J. Chandrapala, *High power ultrasound processing in milk and dairy products*, in *Emerging Dairy Processing Technologies - Opportunities for the Dairy Industry,* N. Datta and P.M. Tomasula, Editors. 2015, John Wiley & Sons. p. 149 -180.
19. Kentish, S. and M. Ashokkumar, *The physical and chemical effects of ultrasound*, in *Ultrasound technologies for food and bioprocessing*, H. Feng, G. Barbosa-Canovas, and J. Weiss, Editors. 2011, Springer: New York, NY. p. 1-12.
20. Mohammadi, V., M. Ghasemi-Varnamkhasti, and L.A. González, *Analytical measurements of ultrasound propagation in dairy products: A review.* Trends in Food Science & Technology, 2017. **61**: p. 38-48.
21. Marchesini, G., *et al.*, *Ultrasonic inactivation of microorganisms: a compromise between lethal capacity and sensory quality of milk.* Innovative Food Science & Emerging Technologies, 2015. **29**: p. 215-221.
22. Guimaraes, J.T., *et al.*, *Physicochemical changes and microbial inactivation after high-intensity ultrasound processing of prebiotic whey beverage applying different ultrasonic power levels.* Ultrasonics Sonochemistry, 2018. **44**: p. 251-260.
23. Sfakianakis, P., E. Topakas, and C. Tzia, *Comparative Study on High-Intensity Ultrasound and Pressure Milk Homogenization: Effect on the Kinetics of*

Yogurt Fermentation Process. Food and Bioprocess Technology, 2014. **8**(3): p. 548-557.

24. Ertugay, M.F., M. Şengül, and M. Şengül, *Effect of ultrasound treatment on milk homogenisation and particle size distribution of fat.* Turkish Journal of Veterinary Animal Sciences, 2004. **28**(2): p. 303-308.
25. Chandrapala, J. and T. Leong, *Ultrasonic Processing for Dairy Applications: Recent Advances.* Food Engineering Reviews, 2015. **7**(2): p. 143-158.
26. Scudino, H., Guimarães, J. T., Lino, D. L., Duarte, M. C. K. H., Esmerino, E. A., Freitas, M. Q., Cruz, A. G. Chapter 16 - Ultrasound for Probiotic and Prebiotic Foods, Editor(s): A,G. Cruz, C. Senaka Ranadheera, Filomena Nazzaro, Amir Mortazavian, Probiotics and Prebiotics in Foods, Academic Press, 2021, pp. 293-307,
27. Arvanitoyannis, I.S., K.V. Kotsanopoulos, and A.G. Savva, *Use of ultrasounds in the food industry-Methods and effects on quality, safety, and organoleptic characteristics of foods: A review.* Critical Reviews in Food Science and Nutrition, 2017. **57**(1): p. 109 -128.
28. Jambrak, A.R., *et al.*, *Effect of ultrasound treatment on particle size and molecular weight of whey proteins.* Journal of Food Engineering, 2014. **121**: p. 15-23.
29. Capelo-Martínez, J.-L., *Ultrasound in chemistry: analytical applications.* 2009: John Wiley & Sons.
30. Villamiel, M., *et al.*, *Ultrasound in food processing: Recent advances.* 2017: John Wiley & Sons. 515.
31. Leong, T., M. Ashokkumar, and S. Kentish, *The fundamentals of power ultrasound - A review.* Acoustics Australia, 2011. **39**: p. 43-52.
32. Bermúdez-Aguirre, D., *Ultrasound: Advances in Food Processing and Preservation.* 2017: Academic Press.
33. Jalal, J. and T.S. Leong, *Microstreaming and its role in applications: a mini-review.* Fluids, 2018. **3**(4): p. 93.
34. Leighton, T.G., A.J. Walton, and M.J.W. Pickworth, *Primary bjerknes forces.* European Journal of Physics, 1990. **11**(1): p. 47.
35. Crum, L.A., *Acoustic cavitation series: part five rectified diffusion.* Ultrasonics, 1984. **22**(5): p. 215-223.
36. Suslick, K.S., *et al.*, *Acoustic cavitation and its chemical consequences.* Philosophical Transactions of the Royal Society of London. Series A: Mathematical, Physical and Engineering Sciences, 1999. **357**(1751): p. 335-353.
37. Ashokkumar, M. and F. Grieser, *Ultrasound assisted chemical processes.* Reviews in Chemical Engineering, 1999. **15**(1): p. 41-83.
38. McClements, D.J., *Advances in the application of ultrasound in food analysis and processing.* Trends in Food Science & Technology, 1995. **6**(9): p. 293-299.
39. Sala, F.J., *et al.*, *Effect of heat and ultrasound on microorganisms and enzymes*, in *New Methods of Food Preservation*, G.W. Gould, Editor. 1995, Springer US: Boston, MA. p. 176-204.

40. Suslick, K.S. and D.J. Flannigan, *Inside a Collapsing Bubble: Sonoluminescence and the Conditions During Cavitation.* Annual Review of Physical Chemistry, 2008. **59**(1): p. 659-683.
41. Piyasena, P., E. Mohareb, and R.C. McKellar, *Inactivation of microbes using ultrasound: a review.* International Journal of Food Microbiology, 2003. **87**(3): p. 207-216.
42. Butz, P. and B. Tauscher, *Emerging technologies: chemical aspects.* Food Research International, 2002. **35**(2): p. 279-284.
43. Dolas, R., C. Saravanan, and B.P. Kaur, *Emergence and era of ultrasonic's in fruit juice preservation: A review.* Ultrasonics Sonochemistry, 2019. **58**: p. 104609.
44. Earnshaw, R.G., J. Appleyard, and R.M. Hurst, *Understanding physical inactivation processes: combined preservation opportunities using heat, ultrasound and pressure.* International Journal of Food Microbiology, 1995. **28**(2): p. 197-219.
45. Villamiel, M. and P. Jong, *Influence of High-Intensity Ultrasound and Heat Treatment in Continuous Flow on Fat, Proteins, and Native Enzymes of Milk.* Journal of Agricultural and Food Chemistry, 2000. **48**(2): p. 472-478.
46. Režek Jambrak, A., *et al., Three Pillars of Novel Nonthermal Food Technologies: Food Safety, Quality, and Environment.* Journal of Food Quality, 2018. **2018**: p. 8619707.
47. Hulsen, U., *Alternative heat treatment processes.* European Dairy Magazine, 1999. **3**: p. 20-24.
48. Alliger, H., *Ultrasonic disruption.* American Laboratory, 1975. **10**(10): p. 75-85.
49. Kang, D., *et al., Inactivation of Escherichia coli O157:H7 and Bacillus cereus by power ultrasound during the curing processing in brining liquid and beef.* Food Research International, 2017. **102**: p. 717-727.
50. Scherba, G., R.M. Weigel, and W.D. Brien, *Quantitative assessment of the germicidal efficacy of ultrasonic energy.* Applied and Environmental Microbiology, 1991. **57**(7): p. 2079.
51. Kapturowska, A., *et al., Ultrasounds-a tool to inactivate yeast and to extract intracellular protein.* Zywnosc Nauka Technologia Jakosc, 2011. **18**: p. 160-171.
52. Gracin, L., *et al., Influence of high power ultrasound on Brettanomyces and lactic acid bacteria in wine in continuous flow treatment.* Applied Acoustics, 2016. **103**: p. 143-147.
53. Gao, S., *et al., Inactivation of bacteria and yeast using high-frequency ultrasound treatment.* Water Research, 2014. **60**: p. 93-104.
54. Mañas, P. and R. Pagán, *Microbial inactivation by new technologies of food preservation.* Journal of Applied Microbiology, 2005. **98**(6): p. 1387-1399.
55. Chandrapala, J., *et al., Ultrasonics in food processing – Food quality assurance and food safety.* Trends in Food Science & Technology, 2012. **26**(2): p. 88-98.
56. Yang, G., *et al., Handbook of ultrasonics and sonochemistry.* 2016: Springer.

57. Valero, M., *et al.*, *Effects of ultrasonic treatments in orange juice processing.* Journal of Food Engineering, 2007. **80**(2): p. 509-516.
58. D'Amico, D.J., *et al.*, *Inactivation of Microorganisms in Milk and Apple Cider Treated with Ultrasound.* Journal of Food Protection, 2006. **69**(3): p. 556-563.
59. Salleh-Mack, S.Z. and J.S. Roberts, *Ultrasound pasteurization: The effects of temperature, soluble solids, organic acids and pH on the inactivation of Escherichia coli ATCC 25922.* Ultrasonics Sonochemistry, 2007. **14**(3): p. 323-329.
60. Koda, S., *et al.*, *Inactivation of Escherichia coli and Streptococcus mutans by ultrasound at 500kHz.* Ultrasonics Sonochemistry, 2009. **16**(5): p. 655-659.
61. Gao, S., *et al.*, *Inactivation of microorganisms by low-frequency high-power ultrasound: 2. A simple model for the inactivation mechanism.* Ultrasonics Sonochemistry, 2014. **21**(1): p. 454-460.
62. Cameron, M., L.D. McMaster, and T.J. Britz, *Electron microscopic analysis of dairy microbes inactivated by ultrasound.* Ultrasonics Sonochemistry, 2008. **15**(6): p. 960-964.
63. Al Bsoul, A., *et al.*, *Effectiveness of ultrasound for the destruction of Mycobacterium sp. strain (6PY1).* Ultrasonics Sonochemistry, 2010. **17**(1): p. 106-110.
64. Joyce, E., *et al.*, *The development and evaluation of ultrasound for the treatment of bacterial suspensions. A study of frequency, power and sonication time on cultured Bacillus species.* Ultrasonics Sonochemistry, 2003. **10**(6): p. 315-318.
65. Khanal, S.N., S. Anand, and K. Muthukumarappan, *Evaluation of high-intensity ultrasonication for the inactivation of endospores of 3 bacillus species in non-fat milk.* Journal of Dairy Science, 2014. **97**(10): p. 5952-5963.
66. Lv, R., *et al.*, *Hurdle enhancement of acidic electrolyzed water antimicrobial efficacy on Bacillus cereus spores using ultrasonication.* Applied Microbiology and Biotechnology, 2020. **104**(10): p. 4505-4513.
67. Lee, H., *et al.*, *Effect of pH on Inactivation of Escherichia coli K12 by Sonication, Manosonication, Thermosonication, and Manothermosonication.* Journal of Food Science, 2009. **74**(4): p. E191-E198.
68. Álvarez, I., *et al.*, *Inactivation of <em>Salmonella enterica</em> Serovar Enteritidis by Ultrasonic Waves under Pressure at Different Water Activities.* Applied and Environmental Microbiology, 2003. **69**(1): p. 668.
69. Álvarez, I., *et al.*, *Inactivation of Salmonella Senftenberg 775W by ultrasonic waves under pressure at different water activities.* International Journal of Food Microbiology, 2006. **108**(2): p. 218-225.
70. Arroyo, C., *et al.*, *Inactivation of Cronobacter sakazakii by manothermosonication in buffer and milk.* International Journal of Food Microbiology, 2011. **151**(1): p. 21-28.
71. Arroyo, C., *et al.*, *Synergistic combination of heat and ultrasonic waves under pressure for Cronobacter sakazakii inactivation in apple juice.* Food Control, 2012. **25**(1): p. 342-348.

72. Ugarte-Romero, E., H. Feng, and S.E. Martin, *Inactivation of Shigella boydii 18 IDPH and Listeria monocytogenes Scott A with Power Ultrasound at Different Acoustic Energy Densities and Temperatures.* Journal of Food Science, 2007. **72**(4): p. M103-M107.
73. Baumann, A.R., S.E. Martin, and H.A.O. Feng, *Power Ultrasound Treatment of Listeria monocytogenes in Apple Cider.* Journal of Food Protection, 2005. **68**(11): p. 2333-2340.
74. Bermúdez-Aguirre, D. and G.V. Barbosa-Cánovas, *Study of butter fat content in milk on the inactivation of Listeria innocua ATCC 51742 by thermo-sonication.* Innovative Food Science & Emerging Technologies, 2008. **9**(2): p. 176-185.
75. Bermúdez-Aguirre, D. and G.V. Barbosa-Cánovas, *Inactivation of Saccharomyces cerevisiae in pineapple, grape and cranberry juices under pulsed and continuous thermo-sonication treatments.* Journal of Food Engineering, 2012. **108**(3): p. 383-392.
76. Ferrario, M., S.M. Alzamora, and S. Guerrero, *Study of the inactivation of spoilage microorganisms in apple juice by pulsed light and ultrasound.* Food Microbiology, 2015. **46**: p. 635-642.
77. Evelyn and F.V.M. Silva, *Use of power ultrasound to enhance the thermal inactivation of Clostridium perfringens spores in beef slurry.* International Journal of Food Microbiology, 2015. **206**: p. 17-23.
78. Cregenzán-Alberti, O., *et al.*, *Suitability of ccRSM as a tool to predict inactivation and its kinetics for Escherichia coli, Staphylococcus aureus and Pseudomonas fluorescens in homogenized milk treated by manothermosonication (MTS).* Food Control, 2014. **39**: p. 41-48.
79. Herceg, Z., *et al.*, *The effect of high intensity ultrasound treatment on the amount of Staphylococcus aureus and Escherichia coli in milk.* Food Technology Biotechnology 2012. **50**(1): p. 46-52.
80. Raso, J., *et al.*, *Influence of Temperature and Pressure on the Lethality of Ultrasound.* Applied and Environmental Microbiology, 1998. **64**(2): p. 465.
81. Raso, J., *et al.*, *Inactivation of Bacillus subtilis spores by combining ultrasonic waves under pressure and mild heat treatment.* Journal of Applied Microbiology, 1998. **85**(5): p. 849-854.
82. Welti-Chanes, J., *et al.*, *Chapter 19 - Opportunities and Challenges of Ultrasound for Food Processing: An Industry Point of View*, in *Ultrasound: Advances for Food Processing and Preservation*, D. Bermudez-Aguirre, Editor. 2017, Academic Press. p. 457-497.
83. Hernández-Hernández, H.M., L. Moreno-Vilet, and S.J. Villanueva-Rodríguez, *Current status of emerging food processing technologies in Latin America: Novel non-thermal processing.* Innovative Food Science & Emerging Technologies, 2019. **58**: p. 102233.
84. Rana, A., Meena, and Shweta, *Ultrasonic processing and its use in food industry : A review.* Journal of Chemical Studies, 2017. **5**(6): p. 1961-1968.

85. Astráin-Redín, L., *et al.*, *Application of High-Power Ultrasound in the Food Industry*, in *Sonochemical Reactions* S. Karakuş, Editor. 2019, IntechOpen: Rijeka. p. 1–24.
86. Chemat, F., H. Zill e, and M.K. Khan, *Applications of ultrasound in food technology: Processing, preservation and extraction.* Ultrasonics Sonochemistry, 2011. **18**(4): p. 813-835.
87. Priyadarshini, A., *et al.*, *Emerging food processing technologies and factors impacting their industrial adoption.* Critical Reviews in Food Science and Nutrition, 2019. **59**(19): p. 3082-3101.
88. Wen, C., *et al.*, *Advances in ultrasound assisted extraction of bioactive compounds from cash crops – A review.* Ultrasonics Sonochemistry, 2018. **48**: p. 538-549.
89. Chemat, F., *et al.*, *Ultrasound assisted extraction of food and natural products. Mechanisms, techniques, combinations, protocols and applications. A review.* Ultrasonics Sonochemistry, 2017. **34**: p. 540-560.
90. Rajha, H.N., *et al.*, *Comparison of aqueous extraction efficiency and biological activities of polyphenols from pomegranate peels assisted by infrared, ultrasound, pulsed electric fields and high-voltage electrical discharges.* Innovative Food Science & Emerging Technologies, 2019. **58**: p. 102212.
91. Xu, D.-P., *et al.*, *Ultrasound-assisted extraction of natural antioxidants from the flower of Limonium sinuatum: Optimization and comparison with conventional methods.* Food Chemistry, 2017. **217**: p. 552-559.
92. Viganó, J., *et al.*, *Extraction of bioactive compounds from defatted passion fruit bagasse (Passiflora edulis sp.) applying pressurized liquids assisted by ultrasound.* Ultrasonics Sonochemistry, 2020. **64**: p. 104999.
93. Plazzotta, S., *et al.*, *Optimizing the antioxidant biocompound recovery from peach waste extraction assisted by ultrasounds or microwaves.* Ultrasonics Sonochemistry, 2020. **63**: p. 104954.
94. Vieira, G.S., *et al.*, *Chemical and economic evaluation of natural antioxidant extracts obtained by ultrasound-assisted and agitated bed extraction from jussara pulp (Euterpe edulis).* Journal of Food Engineering, 2013. **119**(2): p. 196-204.
95. Galviz-Quezada, A., *et al.*, *Valorization of iraca (Carludovica palmata, Ruiz & Pav.) infructescence by ultrasound-assisted extraction: An economic evaluation.* Food and Bioproducts Processing, 2019. **118**: p. 91-102.
96. Ochoa, S., M.M. Durango-Zuleta, and J. Felipe Osorio-Tobón, *Techno-economic evaluation of the extraction of anthocyanins from purple yam (Dioscorea alata) using ultrasound-assisted extraction and conventional extraction processes.* Food and Bioproducts Processing, 2020. **122**: p. 111-123.
97. Guerrero, S.N., *et al.*, *Chapter 3 - Hurdle Technology Using Ultrasound for Food Preservation*, in *Ultrasound: Advances for Food Processing and Preservation*, D. Bermudez-Aguirre, Editor. 2017, Academic Press. p. 39-99.

98. Alarcon-Rojo, A.D., *et al.*, *Ultrasound and meat quality: A review.* Ultrasonics Sonochemistry, 2019. **55**: p. 369-382.
99. Alves, L.d.L., *et al.*, *Effect of ultrasound on the physicochemical and microbiological characteristics of Italian salami.* Food Research International, 2018. **106**: p. 363-373.
100. Caraveo, O., *et al.*, *Physicochemical and microbiological characteristics of beef treated with high-intensity ultrasound and stored at 4 °C.* Journal of the Science of Food and Agriculture, 2015. **95**(12): p. 2487-2493.
101. Chen, F., M. Zhang, and C.-h. Yang, *Application of ultrasound technology in processing of ready-to-eat fresh food: A review.* Ultrasonics Sonochemistry, 2020. **63**: p. 104953.
102. Gonzalez-Gonzalez, L., *et al.*, *Does ultrasound equally improve the quality of beef? An insight into longissimus lumborum, infraspinatus and cleidooccipitalis.* Meat Science, 2020. **160**: p. 107963.
103. Peña-Gonzalez, E., *et al.*, *Ultrasound as a potential process to tenderize beef: Sensory and technological parameters.* Ultrasonics Sonochemistry, 2019. **53**: p. 134-141.
104. Gómez-Salazar, J.A., *et al.*, *Effect of Acid Marination Assisted by Power Ultrasound on the Quality of Rabbit Meat.* Journal of Food Quality, 2018. **2018**: p. 5754930.
105. Soltani Firouz, M., A. Farahmandi, and S. Hosseinpour, *Recent advances in ultrasound application as a novel technique in analysis, processing and quality control of fruits, juices and dairy products industries: A review.* Ultrasonics Sonochemistry, 2019. **57**: p. 73-88.
106. Nicolau-Lapeña, I., *et al.*, *Ultrasound Processing Alone or in Combination with Other Chemical or Physical Treatments as a Safety and Quality Preservation Strategy of Fresh and Processed Fruits and Vegetables: A Review.* Food and Bioprocess Technology, 2019. **12**(9): p. 1452-1471.
107. Feng, L., *et al.*, *Effect of Ultrasound Combined with Controlled Atmosphere on Postharvest Storage Quality of Cucumbers (Cucumis sativus L.).* Food and Bioprocess Technology, 2018. **11**(7): p. 1328-1338.
108. Manzoor, M.F., *et al.*, *Novel processing techniques and spinach juice: Quality and safety improvements.* Journal of Food Science, 2020. **85**(4): p. 1018-1026.
109. Yildiz, G., G. Izli, and R.M. Aadil, *Comparison of chemical, physical, and ultrasound treatments on the shelf life of fresh-cut quince fruit (Cydonia oblonga Mill.).* Journal of Food Processing and Preservation, 2020. **44**(3): p. e14366.
110. Gallo, M., L. Ferrara, and D. Naviglio, *Application of Ultrasound in Food Science and Technology: A Perspective.* Foods, 2018. 7(10).
111. Mohan, M.S., *et al.*, *Milk fat: opportunities, challenges and innovation.* Critical Reviews in Food Science and Nutrition, 2020: p. 1-33.
112. Scudino, H., *et al.*, *Ultrasound stabilization of raw milk: Microbial and enzymatic inactivation, physicochemical properties and kinetic stability.* Ultrasonics Sonochemistry, 2020. **67**: p. 105185.

113. Chávez-Martínez, A., *et al.*, *Low and High-Intensity Ultrasound in Dairy Products: Applications and Effects on Physicochemical and Microbiological Quality.* Foods, 2020. **9**(11).
114. Benedito, J., *et al.*, *Manchego cheese texture evaluation by ultrasonics and surface probes.* International Dairy Journal, 2006. **16**(5): p. 431-438.
115. Qin, P., *et al.*, *Ultrasound-Microbubble Mediated Cavitation of Plant Cells: Effects on Morphology and Viability.* Ultrasound in Medicine & Biology, 2012. **38**(6): p. 1085-1096.
116. Khandpur, P. and P.R. Gogate, *Understanding the effect of novel approaches based on ultrasound on sensory profile of orange juice.* Ultrasonics Sonochemistry, 2015. **27**: p. 87-95.
117. Šimunek, M., *et al.*, *Aroma profile and sensory properties of ultrasound-treated apple juice and nectar.* Food Technology Biotechnology, 2013. **51**(1): p. 101-111.
118. Xia, T., S. Shi, and X. Wan, *Impact of ultrasonic-assisted extraction on the chemical and sensory quality of tea infusion.* Journal of Food Engineering, 2006. **74**(4): p. 557-560.
119. Barretto, T.L., *et al.*, *Improving sensory acceptance and physicochemical properties by ultrasound application to restructured cooked ham with salt (NaCl) reduction.* Meat Science, 2018. **145**: p. 55-62.
120. Tomadoni, B., *et al.*, *Effect of ultrasound and storage time on quality attributes of strawberry juice.* Journal of Food Process Engineering, 2017. **40**(5): p. e12533.
121. Fijalkowska, A., *et al.*, *Ultrasound as a Pre-treatment Method to Improve Drying Kinetics and Sensory Properties of Dried Apple.* Journal of Food Process Engineering, 2016. **39**(3): p. 256-265.
122. Siewe, F.B., *et al.*, *Effects of ultrasound-assisted heating on aroma profile, peptide structure, peptide molecular weight, antioxidant activities and sensory characteristics of natural fish flavouring.* Ultrasonics Sonochemistry, 2020. **65**: p. 105055.
123. Gómez-López, V.M., M.E. Buitrago, and A. Martínez-Yepez, *Effect of ultrasonication on sensory and chemical stability of passion fruit juice during refrigerated storage.* Emirates Journal of Food Agriculture, 2018. **30**: p. 85-89.
124. Sari, H.A. and R. Ekinci, *The effect of ultrasound application and addition of leaves in the malaxation of olive oil extraction on the olive oil yield, oxidative stability and organoleptic quality.* Food Science and Technology, 2017. **37**: p. 493-499.
125. Opalić, M., *et al.*, *The effect of ultrasound pre-treatment and air-drying on the quality of dried apples.* Czech Journal of Food Sciences, 2009. **27**(Special Issue 1): p. 297-300.
126. Sfakianakis, P. and C. Tzia, *Flavor and Sensory Characteristics of Yogurt Derived from Milk Treated by High Intensity Ultrasound*, in *Nutrition,*

Functional and Sensory Properties of Foods. 2013, The Royal Society of Chemistry. p. 92-97.

127. Jalilzadeh, A., *et al.*, *The effect of ultrasound treatment on microbial and physicochemical properties of Iranian ultrafiltered feta-type cheese.* Journal of Dairy Science, 2018. **101**(7): p. 5809-5820.

2

Pulse Electric Field: Novel Technology in Food Processing

Navnidhi Chhikara[1], Anil Panghal[2*], D.N. Yadav[3], Sandeep Mann[3] and Priya Bishnoi[1]

[1]Department of Food Technology, Guru Jambheshwar University of Science & Technology, Hisar, India
[2]Department of Processing & Food Engineering, AICRP-PHET, Chaudhary Charan Singh Haryana Agricultural University, Hisar, India
[3]Central Institute of Post-Harvest Engineering & Technology, Ludhiana, India

Abstract
Growing consumer demand and improvement in lifestyle and consumer awareness, demands natural food of high quality, which stimulated the quest for novel techniques. PEF (Pulse electric field) is one such non-thermal technique attaining heights in food processing and technology. Applicability of PEF is not only restricted to the inactivation of microorganisms and extraction of active constituents but it also helps in the enhancing chemical reactions, accelerates the aging of fermented foods and in modifying bio-macromolecules. Advancements are going on in making PEF a more sustainable approach by combining it with different techniques (Hurdle concept) such as Ultrasound, manothermosonication, HILP (High Intensity Light Pulse) etc. Pulse electric field therefore, has come a long way through research and advancements adding to its overall development and applications in the industry. The chapter deals in the latest advancements in PEF and its overall development.

Keywords: Non-thermal processing, food preservation, high voltage

2.1 Introduction

Competitiveness of the food industry is on rise as consumer demands for fresh, minimally processed and high quality food. Furthermore, the

**Corresponding author*: anilpanghal@gmail.com

Navnidhi Chhikara, Anil Panghal and Gaurav Chaudhary (eds.) Novel Technologies in Food Science, (39–64) © 2023 Scrivener Publishing LLC

concepts of "fresh," "natural," or "organic" foods has shifted towards environmentally safe and sustainable food production systems. Earlier, the use of thermal processing techniques like Evaporation, Drying, High processing temperature (> 60°C) takes longer time and often degrades the qualitative and quantitative attributes of the food and adds to the financial cost [1]. The consumer awareness about quality and safety of foods has led to the development of advanced technologies using radiations, pressure, magnetic fields, nanotechnology and pulse electric field. These technologies have the potential for better nutrient retention and sensory characteristics [2]. Pulse electric field is novel technology using short, high voltage pulse to obtain fresh-like food products with quality and safety [3, 4]. PEF is targeted, instant, flexible, energy efficient and helps in the longer storagability of the product as compared to other conventional techniques. The pulse induces poration of microbial, plant and animal cell leading to microbial inactivation and cell disintegration. Applying electric energy is novel approach for non-thermal food preservation began in the early 1900s. The electric energy was passed through food sample and electrical resistance alleviated the temperature of food resulting in microbial inactivation, milk being the first electrically pasteurized product. Further electric field was applied to increase the cell permeability and thus efficient juice extraction was achieved in late 1940s [100]. The potential application of PEF have been investigated worldwide for treating a wide range of commodities [5–8]. Study shows that PEF speeds up the extraction rates of juices, coloring agents and other valuable components as well as drying of foods thus extending the shelf life [9, 10]. Despite scientific researches and publications and the knowledge about its application and principle, it is still considered to be developing for industries.

2.2 Principle

The basic principle of PEF deals in applying electrical waves with high voltage magnitude to food with the help of controllable system. High Voltage short electrical impulses ranging (10-80 kV/cm) are applied to the product holded in between the electrodes in the chamber for duration of microseconds to milliseconds [11]. The force per charge experienced by food in between the gap of electrodes is called electric field. The electrodes inside a chamber are capable of passing electric energy because of the presence of several ions and electrical conductivity of food. This would retain the overall quality of food including nutrients, bioactive and sensory attributes with lesser energy consumption than other conventional preservation techniques. The treatment time t (pulse duration * no. of pulses), electric field strength are the centrally processing factors for enhancing the efficiency of PEF and by optimization of these factors, overall efficiency of the process in terms of yield, quality and energy used can be increased [6]. Treatment chambers generally used are batch type and continuous type as per product nature (solid, semi-solid, liquid, semi-liquid). The electric field may be applied at ambient, sub-ambient or above ambient temperature in the form of exponentially decaying, square, wave, bipolar or oscillatory pulses. Electrically sensitive components show changes in their electro chemical and physiochemical properties on getting exposed to electric field [12–14]. The food packaged aseptically after PEF processing and then can be stored in refrigeration for long time [15].

The primary purpose of physical and chemical methods of food preservation is to protect the food from spoilage causing microorganisms and pathogens for food quality and safety. These methods primarily focused on pasteurization of food and the microorganisms are viable and start multiplying on availing favorable conditions. Pulse electric field technology is based on the mechanism of electroporation and electrical breakdown of microbial membrane.

Electroporation is applying the electric field of particular strength for particular duration to increase the permeability of membrane due to compression and poration, of which will lead to osmotic imbalance in cell followed by cell inactivation. High electric field results in dielectric breakdown of the microflora cell membranes and interaction with the charged molecules of food [16]. PEF efficacy is based on a phenomenon of electroporation initially followed by electrical breakdown, which leads to cell wall perforations and leak out of cytoplasmic fluid resulting in cell death. High intensity pulse for microseconds to milliseconds causes temporary or permanent pore formation in the cell membrane and thus cell death. However, experimental studies indicate that when long and intense electrical impulses are applied rearrangement of lipids and water occur and leads to the formation of aqueous hydrophilic pores [17]. So, computation methods are used to observe the cell membrane effect caused by electric field [18]. PEF technology has been suggested for the sterilization of foods such as juices, milk, yogurt, soups, and liquid eggs [19–21].

2.3 Electroporation

Electroporation is the phenomenon leading to temporary destabilization of lipid bilayer and proteinaceous components of cell membranes due to cell exposure to high voltage electric field [22]. Electroporation takes place in two stages as suggested by [23], water molecules organized in single file penetrate the hydrophobic core of the bilayer and then it grow into length and expand in water filled pores which can be stabilized by reorganization of lipid molecules. The altered permeability of cell membrane leads to permeation, cell swelling and thus cell membrane ruptures. Occurrence of Induced trans-membrane voltage is the result of exposing a biological cell (cytoplasm surrounded by plasma membrane) to an external electric field [24]. By altering strength of electric field, duration and medium strength, pores formation can be controlled.

Electrical breakdown of microbial cell membranes results in inactivation of microorganisms. The resistance of membranes due to charge separation across the membranes results in difference in membrane potential (V) and thus cell membrane thickness decreases. This membrane potential is dependent on the strength of electric field applied and cell dimensions. This may further result in development of minute pores affecting the ionic stability of cell and thus cell membrane breakdown occurs. This pore formation is reversible when small pores are developed as compared to total surface of cell membrane. This reversible process drift towards irreversible when the size and number of pores increases on application of high field strength and longer exposure time causing in complete cell destruction [25]. To avoid di-electric breakdown applications of PEF are limited at low conductivity. When the transmembrane potential difference is exceeded beyond 250mV it results in permeabilization of cell membrane. It is dependent on the strength induced by field along with pulse parameters (amplitude, repetition, pulse number, duration), cell size, surrounding media, temperature, orientation of electric field and shape [26]. The corresponding electric field is $E_{critical} = V_{critical}/f_a$, where a is the radius of the cell and f is a form dependent on cell dimensions [27].

2.4 PEF System

The PEF equipment system consists of a high voltage pulse generator, current limiting resistor, treatment chamber with sample handling system, energy storage capacitor bank and requisite monitoring and controlling devices (Figure 2.1). The pulse generator provides pulse electric field of particular voltage, type for specific time. The treatment chamber has two electrodes connected with each other through non-conductive material to prevent electric field flow in between and these electrodes pass high intensity pulse to the food sample holded in chamber [28]. This arrangement of electric circuit, capacitors, resistors, power supplies and treatment chambers form pulse forming network (PFN).

The purpose of treatment chambers, designed in parallel plate and coaxial geometry, is to hold the food product during processing. The parallel plate treatment chamber is preferably used for batch processing and coaxial chamber for continuous processing. However, coaxial chambers ensures consistent electric field distribution in the food and thus results in increased rate of microbial inactivation [29]. The dielectric breakdown of food occurs on increasing the strength of electric field above the electric field potential of food. The food resistance, uniformity in electric field application, pulse movement strongly influences the design of PEF system in terms of economic and environmental impact [30]. The electric field homogeneity, electrical resistance, and generation of field areas are few important criteria for designing efficient PEF system for lesser heat generation in heat and

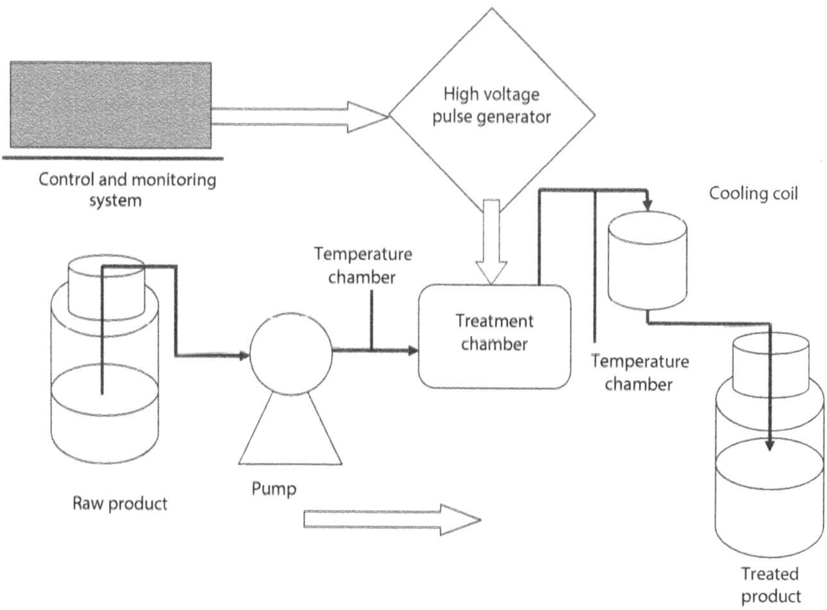

Figure 2.1 Flow chart of a PEF food processing system with basic component.

better energy utilization. In case of cold sterilization electric field intensity is achieved by placing multiple PEF treatment chambers in series [31]. Commonly used material for treatment chambers are corrosion resistant stainless steel electrodes and PTFE (Polytetrafluoroethylene) shells [5].

2.5 Factors Affecting PEF

The efficiency of pulse electric field treatment is dependent on pulse shape, strength, frequency, intensity, duration and type of food matrix [32]. The vegetative cells are killed at about 150,000 V/cm and microbes are at 35,000 V/cm. In general, the microbial inactivation increases with an increasing number of pulses, but the significant heating of products is likely to occur.

2.5.1 Process Factors

The main processing parameters which may affect the microbial resistance towards PEF are electric field strength, treatment time, frequency, pulse shape and specific energy [33, 34]. The strength of electric field applied on food matrix significantly influences the rate of microbial inactivation. Pataro *et al.* [35] observed the effect of process parameters; electric field (13–30 kV/cm), specific energy input (20–110 J/mL), number of passes of sample (1-5) and flow rate (1–4 L/h) on the process efficiency of pulse electric field on the basis of destruction of *Saccharomyces cerevisiae* and *Escherichia coli* and energy efficiency. The study stated that on increasing the electric field and energy at flow rate of (2 L/h), the inactivation of both microbial strains was increased. The difference in cell structure, dimension, composition, tendency of microorganisms for cluster formation are the predominant factors in showing different response to microbial strains for particular process conditions. Sharma *et al.* [36] reported that Gram negative bacteria are more sensitive towards PEF treatment in comparison to Gram positive bacteria.

The exponentially decaying and square wave pulses are used in pulse electric field processing of food. Square waves are more lethal and energy efficient than exponentially decaying waves. Square wave forms are recommended over exponential waves as these are more consistent and maintained uniform intensity for complete treatment duration [20]. Bipolar square pulse causes more structural alteration to membrane induced by stress generated by rapid reversal of pulses and thus showed enhanced permeability in cellular structure [14]. Along with this, bipolar wave's treatment showcased minimal deposition of solids on the surface of electrodes, minimum energy usage and decreased electrolysis of food.

Treatment time is the product of the number of pulses applied and the duration of each pulse. The treatment time and electric field strength are the centrally processing factor for enhancing the efficiency of PEF. By increasing the intensity of these factors PEF efficiency increases [6]. When exponential decay pulse are used it is tough to determine the treatment duration so instead specific energy is used as a process criterion. Energy requirement and cost of different preservation system is determined by this parameter [37].

The temperature during PEF processing also controls the microbial destruction rate. Sensoy *et al.* [38] suggested the first-order kinetic model to describe the effect of treatment chamber on microbial inactivation following the Arrhenius model. The impact of temperature on microbial inactivation has been analyzed in continuous flow treatment chamber [37, 39, 40]. Due to non-homogenous pulse distribution during continuous processing a wide range of temperature is reported in particularly for collinear treatment chambers [41, 42]. Therefore, in order to get a clear perspective of the effect of temperature on microbial inactivation a static treatment chamber is used, which helps in obtaining data at different temperature under uniform electric strength and quasi-isothermal conditions [43]. Reina *et al.*, [44] also observed that an increase in treatment temperature (50°C) leads to higher effectiveness in the inactivation of *Listeria monocytogenes.* Pressure is useful to inhibit the formation of air bubbles in which electrical arcing could occur with fields above 20,000 V/cm. During the treatment, a constant pressure is applied to each sample corresponding to the weight of the upper system, assuring the uniformity of treatment conditions. Moreover, polished surfaces of electrodes are used in order to reduce the risk of dielectric breakdown of the food during pulses generation.

2.5.2 Food Matrix

Food matrix composition, concentration, constituents, pH, ionic strength, conductivity (σ, Siems/m) and resistance (r, ohm-meters) affects the efficiency of PEF process. The reduced microbial inactivation rate was observed for *Lactobacillus brevis* in phosphate buffer on increasing the conductivity as resistance of medium was reduced along with pulse width [45]. The differential potential created due to difference in conductivity of food matrix and microorganism cell fluid results in weakening of membrane structural components. Dunn [101] reported that resistivity had no impact on PEF effectiveness on *E. coli* and *L. innocua*. These apparent controversial results may be due to the microorganisms or media used. Acids play important role as antimicrobial agents in food industry and acids used along with PEF have synergistic effect on microbial inactivation. Liu [46] reported that electric filed (12.5 kV/cm) and pH 3.4 by using benzoic/sorbic acid significantly inactivated *E. coli*. This clearly recommended that various preservation methods can be applied along with PEF and this hurdle design will be more effective in extension of shelf life of food samples.

2.5.3 Microbial Factors

The microorganism type, composition, structure, its life phase and concentration also play important role in defining the efficiency of PEF. The large surface area of yeasts makes them more sensitive towards PEF in comparison to bacteria [47]. Gram-positive bacteria membrane consists of firm peptidoglycan layer making them resistant towards PEF treatment as compared with Gram-negative bacteria [48]. The lifecycle of a microorganism passes through lag, exponential, stationery and death phase. The log phase of life is more sensitive to external stress as the cell itself performing functional, growth and multiplication activities and cell membrane becomes very sensitive in response to electric field [49].

2.6 Benefits and Shortcomings of PEF

PEF processing is quite successful in comparisons to conventional thermal treatments due to minimal changes in food quality and safety characteristics. PEF technology has been presented as an energy efficient, green, non-thermal technology that kills microorganisms in seconds while retaining color, flavor, texture, and nutritional value of the food. PEF inactivates vegetative micro-organisms including yeasts, spoilage micro-organisms and pathogens; furthermore, it can be used to pasteurize fluids such as juices, milk and soups. The pores formed through electroporation results in faster drying, more juice yield, extraction of valuable functional components, and rehydration properties of dried products [50, 51]. It can assist conventional processing techniques such as enzymatic maceration and mechanical disintegration. Considerable progress in continuous flow treatment will be effective in scaling up the industrial progress [52]. Recently,

PEF has found its application in solid food too mainly, in potato processing industry. Also, the whole treatment of tomatoes using PEF helps in easy peeling and less usage of natural gas and steam [53].

However, the food after PEF treatment requires proper storage conditions. The food is packaged after processing, so the chances of contamination are there. The PEF processing is preferably used for variety of liquid foods, recently it has been applied to potato and tomato products. The effectiveness of PEF is also limited by the presence of different types of microroagnisms in food as the sensitivity of yeast, bacteria and spores is different. Studies indicate that electrode material constituents (Cr, Fe, Mn, Ni) are leached out in to the liquid food samples due to corrosion [54]. Technically, there are some concern in industrial application of pulse electric field e.g. designing of industry scale treatment chambers, specific pulse generator and efficiency of system.

2.7 Application in Food Industry

Pulse electric field (PEF) technology is suitable approach for inactivation of microorganisms over thermal technologies in liquid foods. PEF technology is now exploring it usage in drying, extraction of valuables, preservation of solid and semisolid food products, and waste water treatment (Table 2.1).

2.7.1 Drying

Drying is the process of moisture removal from food for preservation and value addition. The conventional, mechanical or in conjugation drying techniques are often associated with nutrient losses as well as reduction in appealability of product. PEF can be used as supportive treatment for better quality of dried products, faster drying rates and better rehydration profile [59, 62, 63] and is energy efficient. Telfser and Gómez [64] observed the effect on basil leaves (*Ocimum bascilicum L.*). Application of PEF reduced the drying time 25% for freeze drying followed by 33% for vacuum drying and 57% for air drying. The drying time was reduced to 21% at 60° and 28% at 70°C for sliced carrot and parsnip in a experiment carried out to study pre-treatment of drying with PEF [60]. Ostermeier in [65] observed the effect of pre-treatment of PEF on onion. An increase in cell disintegration was observed on increasing the electric field strength

Table 2.1 Applications of high pulsed electric fields in food processing.

Product	Treatment	Impact	Ref.
Milk	25.7 kV/cm, 34 μs	Inactivation of alkaline phosphatase. Reduced xanthine (30%) and plasmin oxidase (7%) activity	[55]
Cloudy red apple juice	PEF (35 kV/cm for 258 μs) with mild temperature (50 °C)	*E. coli, S. enteritidis, and S. cerevisiae* shows 5.21, 6.02, and 5.49 log reduction, respectively	[56]
Orange juice	33.6-35.7kV/cm; 35 pulses; 1-100 μs; 45-55 °C	5D Inactivation of Natural microflora, Shelflife increased form 3d to 1 w	[102]
Milk	36.7 kV/cm; 40 pulses; 100 μs; 63 °C	3D reduction in *Salmonella dublin*	[103]
Skim Milk	45 kV/cm; 64 pulses; 1.8-6 μs, 35°C	2D reduction in *Escherichia coli*	[104]
Yoghurt (Starwberry, Grapes & Blueberry)	30 kV/cm electric field strength for 32 μs	Significant reduction in total viable count of bacteria, yeast and mold	[57]
Liquid Whole egg	11 kV/cm; 40 pulses; 60 °C	4 D reduction in *Escherichia coli*	[58]

(Continued)

Table 2.1 Applications of high pulsed electric fields in food processing. (*Continued*)

Product	Treatment	Impact	Ref.
Carrot	Pulse number 10, 50 and 100; 1,85 and 5 kV/cm; 5,63, 8 and 80 kJ/kg	Drying reduction time by 8.2%. Decrease of sample lightness up to 25.3%	[59]
Parsnip and carrot	20 μs, 50 Hz, 0.9 kV/cm, after 1000 pulses	Drying reduction time by 28% at 70 °C and to 21% at 60 °C	[60]
Citrus fruits and peel (orange, pomelo and lemon) Extraction	10 kV/cm—peel; 3 kV/cm—fruits.	Polyphenol extraction increased by 50% Increased yield of juice by 25% for oranges, 37% for pomelos and 59% for lemon	[61]

up to 1.07 kV/cm which facilitated the release of moisture content on the products surface. The PEF treated dried sample at 45°C led to a decrease in time by 30%.

2.7.2 Food Preservation

The shelf life of food is dependent on the microbial load and is must to control for food safety and quality. PEF processing has been successful in a variety of fruit juices with low viscosity and electrical conductivity such as orange, apple, cranberry, strawberry, watermelon juice [56, 66–72] (Table 2.1). MTS (Mano-thermosonication) along with PEF resulted in average reduction of 5.6 log cfu/ml of *L. innocua* in milk based smoothie [73]. PEF and PAA (60 ppm, peracetic acid) in conjugation was applied on blueberries

Table 2.2 PEF treatment for optimal reduction of enzymes in food.

Enzyme	Food	Treatment conditions	Impact	Reference
Amylase activity	-	15 kV/cm electric field intensity and 100 mL/ min flow	Icreased α amylase activity	[78]
Peroxidase	Tomato	35 kV/cm, 200 Hz, >5.5-/ μs, 2000 μs, bipolar pulses	Complete Inactivation	[105]
	Orange	35 kV/cm, 250 Hz, 7- μs, 1500 μs, bipolar pulses	Complete Inactivation	[69]
	Watermelon	35 kV/cm, 2000 μs, > 114 Hz, 7.0 -s bipolar pulses	Complete Inactivation	[106]
Polyphenol-oxidase	Apple juice	40 kV/cm, ≥100 pulses	Complete inactivation	[107]
	Grape	30 kV/cm, 3000 μs, 600 Hz, 3.23-4.00- μs bipolar pulses	Complete Inactivation	[108]
	Apple juice	31kV/cm, 1000 μs	<10% Inactivation	[109]
Lipoxygenase	Tomato	35 kV/cm, 250 Hz, 7.0 μs, bipolar pulses	20% Inactivation	[110]

(Continued)

Table 2.2 PEF treatment for optimal reduction of enzymes in food. (*Continued*)

Enzyme	Food	Treatment conditions	Impact	Reference
	Watermelon	35 kV/cm, 220-250 Hz,50 μs, bipolar pulses	50% Inactivation	[106]
Hydroperoxide liase	Tomato	35 kV/cm, 250 Hz, 50 μs, bipolar pulses	93 % Inactivation	[110]
Phosphatase	Skim milk, 2% fat milk, Whole milk	18.8 and 22 kV/cm, up to 70 pulses	65 % and 59% inactivation in skim milk, 2%fat milk and whole milk	[22]
	Raw milk	6.7-20 kV/cm, up to 200 pulses	74% inactivation	[109]
Peroxidase	Potassium phosphate	40--73.3 kV/cm, 30 and 100 pulses	30% inactivation	[111]
	Raw milk	13-19 kV/cm, up to 200 pulses	2-13% inactivation	[109]
Proteases from P. fluorescens	Casein solution Skim milk (SM)	14-15 kV/cm, up to 98 pulses	60% inactivation	[112]
	Whole milk	35.5kV/cm, 866 μs	81.1% Inactivation	[19]
Pectin methyl esterase	Orange juice	35kV/cm, 1500 μs	80% Inactivation	[113]
Lipase	Whole milk	35.5kV/cm, 866 μs	57.1% Inactivation	[19]

with artificially grown population of *E. coli* and *Listeria innocua*. The treatment doesn't affect the color and appearance of blueberries rather reduced the population of *Listeria innocua* and *E. coli*. Phenolic compounds and anthocyanins increased by 25% and 10% respectively. The only adverse effect shown was the softening of the berry structure as studied by Jin *et al.* [74]. PEF on combination with other technology can be proven effective in inactivation of endospores. A score of 4.67 log 10 was achieved by Reineke *et al.* [75] for inactivation of *B. subtilis* spores in saline water with 70°C, with a product flow rate of 5 l/h, 150 Hz pulse frequency with 226.5 kJ/Kg energy. PEF at 3 kV/cm fruits significantly increased juice yield by 25%, 37% and 59% in orange, pomelos and lemon juice as well as polyphenols extraction was increased to 50% with 10 kV/cm electric field [61].

PEF facilitates the permeability in cells through electroporation or electro-permeability, this increased permeability results in faster absorption of curing agents, spices, flavor enhancer etc in meat products. The retaining of freshness, sound nutritional quality with extended shelf life in meat products upon application of PEF for faction of second have been reported by various researchers [22, 76, 77]. Studies recommended that PEF is a feasible technique for producing safe and fresh-like food products.

Enzymes are also responsible for various biochemical degradative changes in food. PEF have been applied for numerous enzymes e.g. b-glucosidase (b-GLUC), peroxidase (POD), lipoxygenase (LOX), pectin methylesterase (PME), polyphenol oxidase (PPO) inactivation in different food samples (Table 2.2).

2.7.3 Improvement of Extraction of Intracellular Compounds

Different techniques based on mechanical and thermal energy are used for extraction of valuable compounds, however, activated endogenous enzymes and heat may results in losses of functional components. Increased extract, oil, pigment, juice, pectin and other valuables yield have been observed by various researchers on treatment of horticultural produce with PEF [58, 79–83, 114]. Bozinou *et al.* [84] compared the extraction efficiency and antioxidant potential of extract by water maceration, boiling water extraction, microwave assisted, ultrasound assisted and pulse electric field extraction methods from *Moringa oleifera* Dry Leaves. The study showed that PEF extraction was highly efficient in terms of yield and antioxidant activity.

Zhao *et al.*, [85] compared polysaccharide extraction yield from corn silk by PEF, hot water and microwave assisted extraction. Maximum yield of 7.31%, 1.95% and 1.13% was obtained by PEF (30 kV/cm, 6 μs), microwave assisted (560 W, 3 min) and hot water (100 °C, 60 min) respectively. PEF-assisted enzymatic extraction (20 kV/cm, 600 μs) reported a higher yield (10.10 mg/mL) of abalone viscera protein having ratio material to solvent of 4:1. compared to purely enzymatic extraction. The isolated protein had better hydrolysis, emulsifying properties with reduced foamability and viscosity behavior [86].

2.8 Effect of PEF on Food Components

Pulsed electric field (PEF) processing is an innovative, emerging and high yielding non-thermal technology for pasteurization of pumpable foods. The various phyto-constituents are present in food accounting for health promoting and disease preventing characteristics. These bioactive compounds are synthesized primarily for self defense and also contribute for specific color, flavour and functional value in food. Evidences of the benefits to human-health associated with the consumption of plant-derived phytochemicals have caused an increase in the demand for fresh-like fruits and vegetables., where are present in different forms as alkaloids (eg., caffeine and threbromine), carotenoids (e.g. lycopene), flavonoids (e.g., flavon-3-ols), isoflavones (e.g. genistein), phenolic acids (e.g., capsaicin, gallic acid and tannic acid), etc., depending on plant species [87].

2.8.1 Proximate Composition

PEF is novel technology for modifying the functionality of starch for requisite functionality in food industry. Han *et al.* [88] evaluated the impact of PEF on corn starch thermal, structural and pasting properties. The study revealed that on increasing the electric field strength, enthalpy, crystallinity, viscosity and gelatinization temperature of starch granules were decreased. PEF treatment induced electrolysis and polarization causing replacement of hydroxyl groups in starch chains that further reduce inter- and intra-molecular hydrogen bonds formation and damages carbonyl groups [7, 89].

High-intensity PEF treatment caused minimal losses in vitamins than conventional methods of processing. Slight or negligible reduction in vitamins content of orange, tomato, strawberry, carrot, broccoli juice, other fruit beverages and milk [90–93, 113]. It was observed that content of Vitamin B in a fruit juice (Orange, Mango, Kiwi & Pineapple) and skin, whole milk thiamin and niacin content in food beverage observed no change on PEF processing with conditions pulse frequency 200Hz, electric field strength of 35kV/cm for 1800μs and 4μs bipolar pulse [92].

Fats are highly prone to degradative changes in presence of light, oxygen, metal ions and heat and results in oxidation and rancidity of fats and fats based products. Zeng *et al.* [94] studied the storage life of pulse electric field treated peanut oil and concluded that PEF treated oil has better stability evaluated in terms of acid value, peroxide value and carbonyl group value over untreated oil sample.

2.8.2 Other Components

The bioactive compounds accounts for taste, color and flavour of food products. Phenolic compounds and anthocyanin showed an increase of 10% and 25% respectively when PEF and PAA (60ppm, peracetic acid) were used in treatment of blueberries [74]. Yin *et al.*, [82] found that electric field (60 kV/cm) resulted in inactivation of micro-organisms and enzymes responsible for chlorophyll degradation in spinach puree, but on further increasing electric field color of puree was detrimentally affected. Carotenoids are pigments present in different food source with huge antioxidant potential. López-Gámez *et al.*, [95] conducted a study to observe effect of PEF on carrots. It was reported that PEF treatment modifiy the perameability of cell membrane and thus enhanced the bioaccessibility of carotenoids without any impact on carotenoids amount. Similar findings indicating non-significant destruction of carotenoids have been reported by others [57, 113]. On the contrary, Vallverdú-Queralt *et al.*, [96]concluded that the content of phenolic compounds and carotenoids was increased in PEF treated tomato samples.

2.8.3 Sensory Attributes

Color is strong determinant for food quality and consumer acceptance. Color undergoes changes due to physiological changes and oxidation. The color retention or enhancing can be achieved by treatment of PEF along with cryoprotectants and vacuum infusion in case of freezing foods [97]. Potato chips maintained their color, firmness and structural integrity after thawing when $CaCl_2$ and Trehalose were used in combination with PEF [98]. The losses of volatile compounds in PEF-treated products are usually dependent on the type of compound. Aguilar-Rosas *et al.* [99] compared the effect of PEF and conventional thermal treatment adopting high temperature-short time process on apple juice quality. The study revealed that there was significant difference in quality attributes of juice and PEF treatment showed lesser impact on physicochemical and flavour characteristics of apple juice. However, the effect of PEF on flavoring compounds release and accessibility is still need to be explored for formulation of flavour rich foods.

2.9 Conclusion

PEF is energy efficient, faster process which provides quality and safe food to consumer. The technology is nowadays preferred over conventional thermal processing due to minimal degradation of physical and chemical quality attributes of food. The alteration of protein functionality and starch modification by PEF has opened new applications of modified components in food industry. The high cost, scale up problems and applicability to pumpable food limits its application at commercial level. However, the effect of PEF on microstructure of food, electrochemical reactions, enzymes and configuration of different constituents is still unexplored. So, researchers from academia and industry should come forward to understand the PEF impact at structural level and a techno-economic robust system of PFE needs to be designed for industrial scale.

References

1. Zhang, L., Li, C. Q., Jiang, W., Wu, M., Rao, S. Q., & Qian, J. Y. (2019). Pulsed electric field as a means to elevate activity and expression of α-amylase in barley (Hordeum vulgare L.) malting. Food and Bioprocess Technology, 12(6), 1010–1020.
2. Saidatul, S. W., Noriham, A., Zainal, S., Khairusy, S. Z., & Nurain, A. (2013). Impact of Non-Thermal Processing on Antioxidant Activity, Phenolic content, Ascorbic Acid content and Color of Winter Melon Puree. *International Food Research Journal, 20*(2).
3. Evans G, Cox DN. (2006). Australian consumers' antecedents of attitudes towards foods produced by novel technologies. Br Food J 108:916–930.
4. Soliva-Fortuny R, Balasa A, Knorr D, Martín-Belloso O. (2009). Effects of pulsed electric field son bioactive compounds in foods: a review. Trends Food Sci Technol 20: 544-556.

5. Zhang, R. B., Chen, J., Xiao, J. F., Liao, X. J., Wang, L. M., & Guan, Z. C. (2011). Pulsed electric fields system and its application in non-thermal food processing. Gaodianya Jishu/ High Voltage Engineering, 37(3), 777–786.
6. Gabrić, D., Barba, F., Roohinejad, S., Gharibzahedi, S. M. T., Radojčin, M., Putnik, P., *et al.* (2018). Pulsed electric fields as an alternative to thermal processing for preservation of nutritive and physicochemical properties of beverages: A review. Journal of Food Process Engineering, 41(1), 1–14.
7. Giteru, S. G., Oey, I., & Ali, M. A. (2018). Feasibility of using pulsed electric fields to modify biomacromolecules: A review. Trends in Food Science & Technology, 72, 91–113.
8. Bhat, Z. F., Morton, J. D., Mason, S. L., & Bekhit, A. E. D. A. (2019). Current and future prospects for the use of pulsed electric field in the meat industry. Critical Reviews in Food Science and Nutrition, 59(10), 1660–1674.
9. Neri, L., Giancaterino, M., Rocchi, R., Tylewicz, U., Valbonetti, L., Faieta, M., & Pittia, P. (2021). Pulsed electric fields (PEF) as hot air drying pre-treatment: Effect on quality and functional properties of saffron (Crocus sativus L.). *Innovative Food Science & Emerging Technologies*, *67*, 102592.
10. Rábago-Panduro, L. M., Morales-de la Peña, M., Martín-Belloso, O., & Welti-Chanes, J. (2021). Application of Pulsed Electric Fields PEF on Pecan Nuts Carya illinoinensis Wangenh. K. Koch: Oil Extraction Yield and Compositional Characteristics of the Oil and Its By-product. *Food Engineering Reviews*, 1-10.
11. Deeth HC, Datta N, Ross AIV, Dam XT. (2008). Pulsed electric field technology: effect on milk and fruit juices. In: Tewari Gaurav, Juneja VK (eds) Advances in thermal and non thermal food preservation. Wiley-Blackwell, Hoboken.
12. Li, Y., Chen, Z., & Mo, H. (2007). Effects of pulsed electric fields on physicochemical properties of soybean protein isolates. LWT-Food Science and Technology, 40(7), 1167–1175.
13. Zhao, W., Yang, R., Liang, Q., Zhang, W., Hua, X., & Tang, Y. (2012). Electrochemical reaction and oxidation of lecithin under pulsed electric fields (PEF) processing. Journal of Agricultural and Food Chemistry, 60(49), 12204–12209.
14. Wang, Q., Li, Y., Sun, D. W., & Zhu, Z. (2018). Enhancing food processing by pulsed and high voltage electric fields: Principles and applications. Critical Reviews in Food Science and Nutrition, 58(13), 2285–2298.
15. Guerrero-Beltran, J. A., & Welti-Chanes, J. (2016). Pulsed electric fields. Encyclopedia of Food and Health. 561-65.
16. Syed, Q. A., Ishaq, A., Rahman, U. U., Aslam, S., & Shukat, R. (2017). Pulsed electric field technology in food preservation: A review. *Journal of Nutritional Health & Food Engineering*, *6*(6), 168-172.
17. Weaver JC (2003) Electroporation of biological membranes from multicellular to nanoscales. Dielectrics and electrical insulation. IEEE Trans Dielectr Electr Insul 10(5):754–768.

18. Leach AR (2001) Molecular modelling: principles and applications, 8th edn. Prentice Hall, Upper Saddle River.
19. Bendicho S, Barbosa-Canovas GV, Martın O (2003) Reduction of protease activity in milk by continuous flow high-intensity pulsed electric field treatments. J Dairy Sci 86: 697-703.
20. Wan, J., Coventry, J., Swiergon, P., Sanguansri, P., & Versteeg, C. (2009). Advances in innovative processing technologies for microbial inactivation and enhancement of food safety–pulsed electric field and low-temperature plasma. Trends in Food Science & Technology, 20(9), 414–424.
21. Kebede, B., Lee, P. Y., Leong, S. Y., Kethireddy, V., Ma, Q., Aganovic, K., ... & Oey, I. (2018). A chemometrics approach comparing volatile changes during the shelf life of apple juice processed by pulsed electric fields, high pressure and thermal pasteurization. *Foods*, *7*(10), 169.
22. Castro AJ, Swanson BG, Barbosa-Canovas GV, Zhang QH. (2001). Pulsed electric field modification of milk alkaline phosphatase activity. In Barbosa-Canovas G. V. & Zhang Q. H. (ed.) Pulsed electric fields in food processing. Lancaster: Technomic Publishing. pp. 65-82.
23. Tieleman, D. P. (2004). The molecular basis of electroporation. *BMC biochemistry*, *5*(1), 1-12.
24. Kotnik T, Pucihar G, Miklavcic D (2010) Induced transmembrane voltage and its correlation with electroporation-mediated molecular transport. J Membr Biol 236:3–13.
25. Golzio MJ, Teissié J, Rols MP (2002) Direct visualization at the single-cell level of electrically mediated gene delivery. Proc Natl Acad Sci USA 99:1292–1297.
26. Valič B, Golzio M, Pavlin M, Schatz A, Faurie C, Gabriel B, Teissié J, Rols MP, Miklavčič D (2003) Effect of electric field induced transmembrane potential on spheroidal cells: theory and experiment. Eur Biophys J 32:519.
27. Schoenbach KH, Peterkin FE, Alden RW, Beebe SJ. (1997). The effect of pulsed electric fields on biological cells: Experiments and applications, IEEE Transactions Plasma Sci 25: 284-292.
28. Oyetunji, O., Adeagbo, O., Quadri, F., Ishola, F., Mamudu, A., & Akinbode, E. (2020). Development of a Pilot-scale Pulse Electric Field system for Processing Liquid Foods. *International Journal*, *9*(4).
29. Arshad, R. N., Abdul-Malek, Z., Ahmad, M. H., Buntat, Z., Nawawi, Z., Kumara, C. P., ... & Sidik, M. A. B. (2020). Coaxial treatment chamber for liquid food treatment through pulsed electric field. *Indonesian Journal of Electrical Engineering and Computer Science*, *19*(3), 1169-1176.
30. Barbosa-Canovas, G. V., & Sepúlveda, D. (2005). Present status and the future of PEF technology. *Novel food processing technologies*, 1-44.
31. Buckow R, Ng S, Toepfl S. (2013). Pulsed electric field processing of orange juice: a review on microbial, enzymatic, nutritional, and sensory quality and stability. Compr Rev Food Sci Food Saf 12:455–467.

32. Buchmann, L., & Mathys, A. (2019). Perspective on pulsed electric field treatment in the bio-based industry. *Frontiers in bioengineering and biotechnology, 7*, 265.
33. Ortega-Rivas, O. (2011). Critical issues pertaining to application of pulsed electric fields in microbial control and quality of processed fruit juices. Food Bioprocess Technol. 4:631–645.
34. Feng, X., Jiang, Q., Zhu, M., Jiang, J., Yin, W., & Li, H. (2019). Damaging Effects of Pulsed Electric Field Process Parameters on Rhizoctonia solani Cells Using Response Surface Methodology. *Applied Sciences, 9*(21), 4706.
35. Pataro, G., Senatore, B., Donsì, G., & Ferrari, G. (2011). Effect of electric and flow parameters on PEF treatment efficiency. *Journal of Food Engineering, 105*(1), 79-88.
36. Sharma, P, Bremer, P, Oey, D.W.E. Bacterial inactivation in whole milk using pulsed electric field processing. *Int. Dairy J.* **2014**, *35*, 49–56.
37. Heinz V, Toepfl S, Knorr D (2003) Impact of temperature on lethality and energy efficiency of apple juice pasteurization by pulsed electric fields treatment. Inno Food Sci Emerg Technol 4 (2): 167-175.
38. Sensoy I, Zhang Q, Sastry SK. (1997). Inactivation kinetics of Salmonella dublin by pulsed electric fields. J Food Proces Engin 20: 367-381.
39. Sepulveda-Ahumada DR, Gongora-Nieto MM, Guerrero JA, Barbosa-Canovas GV. (2008). Production of extended-shelf life milk by processing pasteurized milk with pulsed electric fields. J Food Engin 67: 81-86.
40. Amiali M, Ngadi MO, Raghavan GSV, Smith JP (2006) Inactivation of Escherichia coli O157:H7 and Salmonella enteritidis in liquid egg white using pulsed electric field. J Food Sci 71: 88-94.
41. Buckow, R., Baumann, P., Schroeder, S. and Knoerzer, K. (2011). Effect of dimensions and geometry of co-field and co-linear pulsed electric field treat_ ment chambers on electric field strength and energy utilization. J. Food Eng. 105:545–556.
42. Buckow, R., Schroeder, S., Berres, P., Baumann, P. and Knoerzer, K. (2010). Simulation and evaluation of pilot-scale pulsed electric field (PEF) process_ ing. J. Food Eng. 100:67–77.
43. Saldana, G., Puértolas, E., Álvarez, I., Meneses, N., Knorr, D. and Raso, J. (2010). Evaluation of a static treatment chamber to investigate kinetics of microbial inactivation by pulsed electric fields at different temperatures at quasi-isothermal conditions. J. Food Eng. 100:349–356.
44. Reina LD, Jin ZT, Zhang QH, Yousef AE. (1998). Inactivation of *Listeria monocytogenes* in milk by pulsed electric field. J Food Prot 61: 1203-1206.
45. Jayaram S, Castle GSP, Margaritis A (1992) Kinetics of sterilization of Lactobacillus brevis cells by the application of high voltage pulses. Biotechnol Bioengin 40: 1412-1420.
46. Liu X, Yousef AE, Chism GW. (1997). Inactivation of *Escherichia coli* O157:H7 by the combination of organic acids and pulsed electric fields. J Food Safety 16: 287-299.

47. Marsellés-Fontanet, À. R., Puig, A., Olmos, P., Mínguez-Sanz, S., & Martín-Belloso, O. (2009). Optimising the inactivation of grape juice spoilage organisms by pulse electric fields. *International Journal of Food Microbiology, 130*(3), 159-165.
48. Barbosa-Cánovas GV, Pierson MD, Zhang QH, Schaffner DW (2000) Pulsed electric fields. J Food Sci Suppl 65(8): 65-79.
49. Pothakamury UR, Vega H, Zhang Q, Barbosa-Canovas GV, Swanson BG (1996) Effect of growth stage and processing temperature on the inactivation of *Escherichia coli* by pulsed electric fields. J Food Prot 59: 1167-1171.
50. Onwude D, *et al.* (2017). Non-thermal hybrid drying of fruits and vegetables: A review of current technologies. *Innovative Food Science & Emerging Technologies* 43:223-238.
51. Fauster T, Schlossnikl D, Rath F, Ostermeier R, Teufel F, Toepfl S, Jaeger H (2018) Impact of pulsed electric field (PEF) pretreatment on process performance of industrial French fries production. J Food Eng 235:16–22.
52. Huang, K. and Wang, J. (2009). Designs of pulsed electric fields treatment chambers for liquid foods pasteurization process: A review. J. Food Eng. 95:227–239.
53. Arnal, Á. J., Royo, P., Pataro, G., Ferrari, G., Ferreira, V. J., López-Sabirón, A. M., & Ferreira, G. A. (2018). Implementation of PEF treatment at real-scale tomatoes processing considering LCA methodology as an innovation strategy in the agri-food sector. *Sustainability, 10*(4), 979.
54. Pataro G, Falcone M, Donsì G, Ferrari G. (2014). Metal release from stainless steel electrodes of a PEF treatment chamber: effects of electrical parameters and food composition. Innov Food Sci Emerg Technol 21:58–65.
55. Sharma P, Oey I, Bremer P, Everett D. (2017). Microbiological and enzymatic activity of bovine whole milk treated by pulsed electric fields. Int J Dairy Technol 71(1):10–19.
56. Katiyo, W., Yang, R., & Zhao, W. (2017). Effects of combined pulsed electric fields and mild temperature pasteurization on microbial inactivation and physicochemical properties of cloudy red apple juice (Malus pumila Niedzwetzkyana (Dieck)). Journal of Food Safety, 37(4), Article e12369.
57. Yeom, H. W., EVRENDILEK, G. A., JIN, Z. T., & ZHANG, Q. H. (2004). Processing of yogurt-based products with pulsed electric fields: microbial, sensory and physical evaluations. *Journal of food processing and preservation, 28*(3), 161-178.
58. Bazhal M, Lebovka NI, Vorobiev E (2006). Optimisation of pulsed electric field strength for electroplasmolysis of vegetable tissues. Biosys Engin 86(3): 339-345.
59. Wiktor A, Nowacka M, Dadan M, Rybak K, Lojkowski W, Chaduba T, Witrowa-Rajchert D (2016) The effect of pulsed electric field on drying kinetics, color, and microstructure of carrot. Dry Technol 34(11): 1286–1296.

60. Alam M, Lyng J, Frontuto D, Marra F, Cinquanta L (2018) Effect of pulsed electric field pretreatment on drying kinetics, color, and texture of parsnip and carrot. J Food Sci 83:2159–2166.
61. El Kantar S, Boussetta N, Lebovka N, Foucart F, Rajha H, Maroun H, Louka N, Vorobiev E. (2018). Pulsed electric field treatment of citrus fruits: improvement of juice and polyphenols extraction. Innov Food Sci Emerg Technol 46:153–161.
62. Taiwo KA, Angersbach A, Knorr D (2002) Influence of high intensity electric field pulses and osmotic dehydration on the rehydration characteristics of apple slices at different temperatures. J Food Engin 52: 185-192.
63. Angersbach A, Knorr D (1997) High intensity electric field pulses as pre-treatment for affecting dehydration characteristics and rehydration properties of potato cubes. Nahrung 55: 143-146.
64. Telfser A, Gómez Galindo F (2019) Effect of reversible permeabilization in combination with different drying methods on the structure and sensorial quality of dried basil (Ocimum basilicum L.) leaves. LWT 99:148–155.
65. Ostermeier R, Giersemehl P, Siemer C, Töpfl S, Jäger H (2018) Influence of pulsed electric field (PEF) pre-treatment on the convective drying kinetics of onions. J Food Eng 237:110–117.
66. Dziadek, K., Kopeć, A., Dróżdż, T., Kiełbasa, P., Ostafin, M., Bulski, K., & Oziembłowski, M. (2019). Effect of pulsed electric field treatment on shelf life and nutritional value of apple juice. *Journal of food science and technology*, *56*(3), 1184-1191.
67. Timmermans, R. A. H., Mastwijk, H. C., Berendsen, L. B. J. M., Nederhoff, A. L., Matser, A. M., Van Boekel, M. A. J. S., & Groot, M. N. (2019). Moderate intensity Pulsed Electric Fields (PEF) as alternative mild preservation technology for fruit juice. *International journal of food microbiology*, *298*, 63-73.
68. Charles-Rodríguez AV, Nevárez-Moorillón GV, Zhang QH, Ortega-Rivas E. (2007). Comparison of thermal processing and pulsed electric fields treatment in pasteurization of apple juice. Food Bioprod Process 85(2C): 93-97.
69. Elez-Martínez P, Martín-Belloso O. (2007). Effects of high intensity pulsed electric field processing conditions on vitamin C and antioxidant capacity of orange juice and gazpacho, a cold vegetable soup. Food Chem 102: 201-209.
70. Torregrosa F, Esteve MD, Frigola A, Cortes C (2006) Ascorbic acid stability during refrigerated storage of orange–carrot juice treated by high pulsed electric field and comparison with pasteurized juice. J Food Eng 73: 339-345.
71. Oms-Oliu G, Odriozola-Serrano I, Soliva-Fortuny R, Martín-Belloso O. (2009). Effects of high-intensity pulsed electric field processing conditions on lycopene, vitamin C and antioxidant capacity of watermelon juice. Food Chem 115(4): 1312-1319.
72. Lee, H., Choi, S., Kim, E., Kim, Y. N., Lee, J., & Lee, D. U. (2021). Effects of Pulsed Electric Field and Thermal Treatments on Microbial Reduction, Volatile Composition, and Sensory Properties of Orange Juice, and Their

Characterization by a Principal Component Analysis. *Applied Sciences, 11*(1), 186.
73. Palgan I, Muñoz A, Noci F, Whyte P, Morgan DJ, Cronin DA, Lyng JG. (2012). Effectiveness of combined pulsed electric field (PEF) and manothermosonication (MTS) for the control of Listeria innocua in a smoothie type beverage. Food Control 25:621–625.
74. Jin TZ, Yu Y, Gurtler JB (2017) Effects of pulsed electric field processing on microbial survival, quality change and nutritional characteristics of blueberries. LWT Food Sci Technol 77:517–524
75. Reineke K, Schottroff F, Meneses N, Knorr D. (2015). Sterilization of liquid foods by pulsed electric fields—an innovative ultra-high temperature process. Front Microbiol 6:400.
76. Gudmundsson M, Hafsteinsson H (2001) Effect of electric field pulses on microstructure of muscle foods and roes. Trends Food Sci Technol 12(3-4): 122-128.
77. O'Dowd LP, Arimi JM, Noci F, Cronin DA, Lyng JG (2013) An assessment of the effect of pulsed electrical fields on tenderness and selected quality attributes of post rigour beef muscle. Meat Sci 93:303-309.
78. Tian, M. L., Fang, T., Du, M. Y., & Zhang, F. S. (2016). Effects of pulsed electric field (PEF) treatment on enhancing activity and conformation of α-amylase. *The protein journal, 35*(2), 154-162.
79. Guderjan M, Töpfl S, Angersbach A, Knorr D (2005) Impact of pulsed electric field treatment on the recovery and quality of plant cells. J Food Engin 67: 281-287.
80. Fincan M, DeVito F, Dejmek P (2004) Pulsed electric field treatment for solid-liquid extraction of red beetroot pigment. J. Food Engin 64: 381-388.
81. Guderjan M, Elez-Martínez P, Knorr D (2007) Application of pulsed electric fields at oil yield and content of functional food ingredients at the production of rapeseed oil. Innov Food Sci Emerg Technol 8: 55-62.
82. Yin, Y. G., Fan, X. D., Liu, F. X., Yu, Q. Y., & He, G. D. (2009). Fast extraction of pectin from apple pomace by high intensity pulsed electric field. *Journal of Jilin University (Engineering and Technology Edition), 5*, 021-029.
83. Luengo, E., Álvarez, I., & Raso, J. (2014). Improving carotenoid extraction from tomato waste by pulsed electric fields. *Frontiers in nutrition, 1*, 12.
84. Bozinou, E., Karageorgou, I., Batra, G., G Dourtoglou, V., & I Lalas, S. (2019). Pulsed electric field extraction and antioxidant activity determination of Moringa oleifera dry leaves: A comparative study with other extraction techniques. *Beverages, 5*(1), 8.
85. Zhao, W., Yu, Z., Liu, J., Yu, Y., Yin, Y., Lin, S., *et al.* (2011). Optimized extraction of polysaccharides from corn silk by pulsed electric field and response surface quadratic design. Journal of the Science of Food and Agriculture, 91(12), 2201–2209.

86. Li, M., Lin, J., Chen, J., & Fang, T. (2016). Pulsed Electric Field-Assisted Enzymatic Extraction of Protein from Abalone (Haliotis Discus Hannai Ino) Viscera. *Journal of Food Process Engineering, 39*(6), 702-710.
87. Chhikara, N., Kour, R., Jaglan, S., Gupta, P., Gat, Y., & Panghal, A. (2018). Citrus medica: nutritional, phytochemical composition and health benefits–a review. *Food & function, 9*(4), 1978-1992.
88. Han Z, Zeng XA, Zhang BS, Yu SJ (2009) Effects of pulsed electric fields (PEF) treatment on the properties of corn starch. J Food Engin 93: 318-323.
89. Hong, J., Zeng, X. A., Han, Z., & Brennan, C. S. (2018). Effect of pulsed electric fields treatment on the nanostructure of esterified potato starch and their potential glycemic digestibility. Innovative Food Science & Emerging Technologies, 45, 438–446.
90. Odriozola-Serrano, I., Soliva-Fortuny, R., & Martín-Belloso, O. (2008). Changes of health-related compounds throughout cold storage of tomato juice stabilized by thermal or high intensity pulsed electric field treatments. *Innovative Food Science & Emerging Technologies, 9*(3), 272-279.
91. Quitão-Teixeira, L. J., Odriozola-Serrano, I., Soliva-Fortuny, R., Mota-Ramos, A., & Martín-Belloso, O. (2009). Comparative study on antioxidant properties of carrot juice stabilised by high-intensity pulsed electric fields or heat treatments. *Journal of the Science of Food and Agriculture, 89*(15), 2636-2642.
92. Salvia-Trujillo, L., Morales-de la Pena, M., Rojas-Graü, A., & Martin-Belloso, O. (2011). Changes in water-soluble vitamins and antioxidant capacity of fruit juice–milk beverages as affected by high-intensity pulsed electric fields (HIPEF) or heat during chilled storage. *Journal of agricultural and food chemistry, 59*(18), 10034-10043.
93. Sánchez-Vega, R., Elez-Martínez, P., & Martín-Belloso, O. (2015). Influence of high-intensity pulsed electric field processing parameters on antioxidant compounds of broccoli juice. *Innovative Food Science & Emerging Technologies, 29*, 70-77.
94. Zeng, X. A., Han, Z., & Zi, Z. H. (2010). Effects of pulsed electric field treatments on quality of peanut oil. *Food Control, 21*(5), 611-614.
95. López-Gámez, G., Elez-Martínez, P., Martín-Belloso, O., & Soliva-Fortuny, R. (2021). Changes of carotenoid content in carrots after application of pulsed electric field treatments. *LWT, 147*, 111408.
96. Vallverdú-Queralt, A., Oms-Oliu, G., Odriozola-Serrano, I., Lamuela-Raventós, R. M., Martín-Belloso, O., & Elez-Martínez, P. (2013). Metabolite profiling of phenolic and carotenoid contents in tomatoes after moderate-intensity pulsed electric field treatments. *Food chemistry, 136*(1), 199-205.
97. Velickova E, Tylewicz U, Dalla Rosa M, Winkelhausen E, Kuzmanova S, Romani S (2018) Effect of pulsed electric field coupled with vacuum infusion on quality parameters of frozen/ thawed strawberries. J Food Eng 233:57–64.

98. Shayanfar S, Chauhan OP, Toepfl S, Heinz V. (2013). The interaction of pulsed electric fields and texturizing-antifreezing agents in quality retention of defrosted potato strips. Int J Food Sci Technol 48:1289–1295.
99. Aguilar-Rosas, S. F., Ballinas-Casarrubias, M. L., Nevarez-Moorillon, G. V., Martin-Belloso, O., & Ortega-Rivas, E. (2007). Thermal and pulsed electric fields pasteurization of apple juice: Effects on physicochemical properties and flavour compounds. *Journal of Food Engineering*, *83*(1), 41–46.
100. Heinz V, Alvarez I, Angersbach A, Knorr D (2001) Preservation of liquid foods by high intensity pulsed electric fields-basic concepts for process design. Trends Food Sci Technol 12: 103-111.
101. Dunn, J. (1996). Pulsed light and pulsed electric field for foods and eggs. *Poultry Science*, *75*(9), 1133-1136.
102. Dunn JE, Pearlman JS (1987) Methods and apparatus for extending the shelf life of fluid food products. U.S. patent 4,695,472.
103. Vega-Mercado, H., Martin-Belloso, O., Qin, B.L., Chang, F.J., Góngora-Nieto, M.M., Barbosa-Cánovas, G.V. and Swanson, B.G. (1997). Non-thermal food preservation: pulsed electric fields. Trends in Food Science & Technology, 8(5), pp.151-157.
104. Martín O, Qin BL, Chang FJ, Barbosa-Cánovas GV, Swanson BG (1997) Inactivation of *Escherichia coli* in skimmed milk by high intensity pulsed electric fields. Journal of Food Process and Engineering, 20, 317–336.
105. Aguilo-Aguayo, I., Odriozola-Serrano, I., Quintao-Teixeira, L. J., & Martın-Belloso, O. (2008). Inactivation of tomato juice peroxidase by high-intensity pulsed electric fields as affected by process conditions. Food Chemistry, 107(2), 949-955.
106. Aguilo-Aguayo, I., Soliva-Fortuny, R., & Martın-Belloso, O. (2010). Impact of high-intensity pulsed electric field variables affecting peroxidase and lipoxygenase activities of watermelon juice. LWT-Food Science and Technology, 43(6), 897-902.
107. Ertugay, M.F., Başlar, M. and Ortakci, F., 2013. Effect of pulsed electric field treatment on polyphenol oxidase, total phenolic compounds, and microbial growth of apple juice. Turkish Journal of Agriculture and Forestry, 37(6), pp.772-780.
108. Marsellés-Fontanet, Á.R. and Martin-Belloso, O., 2007. Optimization and validation of PEF processing conditions to inactivate oxidative enzymes of grape juice. Journal of food engineering, 83(3), pp.452-462.
109. Van Loey, A., Verachtert, B. and Hendrickx, M., 2001. Effects of high electric field pulses on enzymes. Trends in Food Science & Technology, 12(3-4), pp.94-102.
110. Aguilo-Aguayo, I., Soliva-Fortuny, R., & Martın-Belloso, O. (2009). Effects of high-intensity pulsed electric fields on lipoxygenase and hydroperoxide lyase activities in tomato juice. Journal of Food Science, 74(8), C595-C601.

111. Ho, S. Y., Mittal, G. S., & Cross, J. D. (1997). Effects of high field electric pulses on the activity of selected enzymes. Journal of Food Engineering, 31, 69–84.
112. Vega-Mercado, H., Gongora-Nieto, M.M., Barbosa-Canovas, G.V. and Swanson, B.G. (2007). Pulsed electric fields in food preservation. In Handbook of Food Preservation (pp. 801-832). CRC Press.
113. Elez-Martinez, P., Suárez-Recio, M. and Martin-Belloso, O., 2007. Modeling the reduction of pectin methyl esterase activity in orange juice by high intensity pulsed electric fields. Journal of Food Engineering, 78(1), pp.184-193
114. Eshtiaghi MN, Knorr D. (2002). High electric field pulse pretreatment: potential for sugar beet processing. J Food Engin 52: 265-272.

3

An Overview of Membrane Technology in Dairy & Food Industry

Sunil Kumar Khatkar[1]*, Kuldeep Dudi[1], Shubham Arjun Lonkar[1], Kiranpreet Singh Sidhu[1], Anju Boora Khatkar[2], Narender Kumar Chandla[3] and Anil Panghal[4]

[1]*By-products Utilization Lab; Department of Dairy Technology, College of Dairy Science & Technology, Guru Angad Dev Veterinary & Animal Sciences University, Ludhiana, Punjab, India*
[2]*Department of Food Technology, Punjab Agricultural University, Ludhiana, Punjab, India*
[3]*Department of Dairy Engineering, College of Dairy Science & Technology, Guru Angad Dev Veterinary & Animal Sciences University, Ludhiana, Punjab, India*
[4]*Department of Processing & Food Engineering, AICRP-PHET, CCS Haryana Agricultural University, Hisar, Haryana, India*

Abstract
Membrane technology is widely used in dairy and food industry. The membrane processes like microfiltration (0.1 to 10 μm); ultrafiltration (0.01 to 0.1 μm); nano-filtration (nano-voids or less than 1 nm) & reverse osmosis (0.001 – 0.0001 μm) are generally being used for pre-treatment, separation, fractionation, purification, recovery, standardization, clarification, concentration, partial or full demineralization, cold pasteurization, etc. which are owing to their property to produce high-value biological products, low temperature processing, retention & rejection selectivity, economy, and their environment friendly operation in comparison to conventional methods. Additionally, the knowledge of various factors related to the membrane operation and the feed like the nature of feed & foulants, configuration, requirement & material of membrane, density & size of pores, operational conditions like pressure, temperature, velocity & concentration factor, retention & rejection phenomenon, etc. are also the key components for the optimum application and diversification of membrane technology.

**Corresponding author*: absuneelkhatkar@gmail.com

Navnidhi Chhikara, Anil Panghal and Gaurav Chaudhary (eds.) Novel Technologies in Food Science, (65–108) © 2023 Scrivener Publishing LLC

Keywords: Membrane, ultrafiltration, microfiltration, fouling, concentration polarization, demineralization, milk proteins

List of Abbreviations

CP	Concentration polarization
CF	Concentration Factor
MWCO	Molecular weight cut off
kDa	Kilo Dalton
IUPAC	International Union of Pure and Applied Chemistry
ED	Electro-dialysis
MF	Micro-filtration
Nm	Nanometer
Mm	Micrometer
UF	Ultrafiltration
m^2	Meter square
m^3	Meter cube
A	Area
NF	Nanofiltration
%	Percentage
<	Less than
>	More than
RO	Reverse osmosis
TMP	Transmembrane pressure
EPS	Extracellular polymeric substances
°C	Degree centigrade
μ	Dynamic viscosity
ΔP	Pressure drop

S	Specific area per unit volume
T	Tortuosity of capillaries
Mg	milli gram
L	Litre
IgG	Immunoglobulin
N	Netwon
MMV	Maubois, Macquot and Vassal
NaCl	Sodium chloride
KCl	Potassium chloride
PPO	Polyphenol oxidase
°Brix	Degree brix
kWh	Kilo watt hour
HCl	Hydro chloric acid
NaOH	Sodium hydroxide
K	Potassium
Na	Sodium
Ca	Calcium
Mg	Magnesium
CM	Concentrated must
RCM	Rectified concentrated must
KHT	Potassium hydrogen tartrate
PAA	Polyacrylic acid
PEI	Polyethyleneimine
MS	Mass Spectroscopy
L	Leavo
D	Dextro
MD	Membrane Distillation
LEP	Liquid entry pressure

3.1 Introduction

Historical aspects of membrane technology are very vast. Membrane technology placed an important role to produce novel and functional food ingredients. A lot of laboratory work had been carried out on separation technology before its implementation at industrial scale in 1960s. Various studies were carried out to understand the separation phenomenon and its effectiveness on the industrial level process, started in the 18th century. Implementation of synthetic membrane at industrial level was started in 1960s and effective implementation was carried out in mid of 18th century. With the development of technology, various types of membrane were used for different application of purification and separation. Membrane technology is widely used in dairy and food applications since 1990s and it grows with a rate of 6-7% per year since last 5-7 years. Various membrane processes are based on requirements like ultrafiltration (for fractionation, partial demineralization), reverse osmosis (for concentration), nanofiltration (for demineralization), microfiltration (for cold pasteurization, separation of large molecules) are used. Membrane processing delivers numerous advantages in the processing of high-value biological products, like processing at desirable temperature and lower pressure, retains the nutritional quality of materials, economical operation, better yield, and environment friendly in comparison to conventional methods. From the last decades, the requirement of foods comprises of three main keys, like nutrition, dietetic, and health. Therefore, to formulate novel foods, membrane technology has an important role. Abbe Nolet coined the term osmosis in 1748, and described the permeation phenomenon of water [1]. Based on requirement, different types of materials were used to design different types of membrane modules. Bechhold prepared the membrane with different pore size by using nitrocellulose in 1907 [2]. The first significant application of the membrane for desalination of water was used at the end of World War 2. Loeb–Sourirajan transformed the use of the membrane process for desalination of water at industrial scale and developed first spiral wound modules from cellulose acetate based asymmetric/anisotropic reverse osmosis membranes having high-flux and defect-free [3]. From the last 2-3 decades, membrane technology is widely used for various applications, but its use in food industry increased significantly. A 25-30% of membrane technology turnover is covered only by food sector. Application of membrane technology in dairy sector is around 40% for different applications like, purification, separation, fractionation etc. for the production of novel food ingredients. Out of various dairy processes, milk protein standardization placed its contribution of around 10%. Other applications of membrane techniques are production and clarification of beer, wine, and fruit juices to produce quality product. Several other applications of membrane techniques in dairy and food industry are discussed in detail in this chapter.

3.2 Terminology in Membrane Processing

3.2.1 Membrane

Membrane is a sieving process, which allows the selective permeation of solute molecules through it, based on its pore size and molecular weight cut off.

3.2.2 Permeate

The part of the feed that passes through the membrane and collected separately is called permeate. In some of the membrane processes permeate is the main desired product. For E.g. Water that is permeating from the reverse osmosis plant which meant for the purification of the water.

3.2.3 Retentive/Retentate

The part of the solution or feed that is not able to pass through the pores of membrane and obtained in the concentrated form is called the retentive/ retentate. The term concentrate is also used in some of the literature.

3.2.4 Fouling

Fouling is deposition/accumulation of the feed material on the surface or in the pores of membrane, which results in decrease the overall performance of membrane during processing.

3.2.5 Concentration Polarization

It is the change in the concentration (either decrease or increase) of specific component from the feed at the boundary layer close to the surface of the membrane due to the gathering of solutes, which results in the formation layer of solutes adjacent to the surface of membrane. Due to the formation of the layer, a concentration gradient is set up b/w the solute layer and bulk solution. Due to setting up of steep concentration gradient, back transport of solutes to the bulk solution takes place.

3.2.6 Concentration Factor

It is the number of times the component of the feed gets increased in the final product or the retentate.

$$CF = \frac{Cr}{Cf}$$

Where: C_r is the final concentration of the component in the feed (retentate) and C_f is the initial concentration of the component in the feed.

3.2.7 Feed

It is the fluid or the solution which enters the membrane module or the plant for the separation or concentration or purification or fractionation.

3.2.8 Flux

Flux is the quantity of the permeate or the quantity of the component of the permeate that passes through per unit of the membrane area per unit of time.

3.2.9 Pore Size

It is an opening in the membrane which can of various sizes.

3.2.10 Molecular Weight Cut-Off

It is the lowest molecular weight at which minimum 90% of solute molecules of given molecular weight is retained by the membrane.

3.3 Types of Membrane

Morphology of the membrane is of two types, one is an anisotropic or asymmetric membrane and the other is an anisotropic or asymmetric membrane. Symmetrical membranes can be of the following types.

3.3.1 Microporous Membrane

The solid material which is having pores or voids is considered as the porous membrane. It is well known that all solid materials are capable of providing a porous medium. According to the International Union of Pure and Applied Chemistry (IUPAC) classification of the membrane material is done based on their pore size. Membrane having pore size under 2 nm, falls under the category of microporous membrane. Pore size falls in the range of 2-50 nm comes under the category of mesoporous membrane, and from 50-1000 nm comes under the category of macroporous membrane [4]. By managing the pore size, pore volume can be maintained. There are two categories of porous solids, one is crystalline and another is amorphous. By estimating the chemical composition and behavior, physio-chemical properties of that specific material can be determined, thus the membrane properties largely depend upon several factors like, chemical and structural characteristics of that material and their pore characteristics. The microporous membrane has a similar function to the conventional filter. Particles with the largest size than the largest pores are not allowed to pass through it and are completely rejected. Particles larger than the smallest pore and smaller than the larger pore are partially rejected.

3.3.2 Nonporous, Dense Membrane

This membrane works on the principle of diffusion phenomenon mainly driven due to concentration, or electrical potential gradient. Based on the comparative (relative) transport rate, components can be separated. Effectiveness of separation largely depends on diffusion rate (diffusivity) and interaction with membrane material. Hence, on the basis of concentration difference, permeate can be separated using nonporous membrane.

3.3.3 Electrically Charged Membranes

The electrically charged membrane could be either be microporous or dense, but in most cases, they are very finely microporous, contains fixed positive or negative charge on the surface. Anion exchange membrane contains positive charge on the surface, while cation exchange membrane contains negative charge on the surface. When fluid passes through the membrane, opposite charge present in fluid binds to the membrane, and fluid of single charge can be obtained. In this case, separation takes place as the ions with a similar charge as that of the fixed ions are excluded by the membrane. Membrane efficiency largely affected by the concentration of ions and charge density. These membranes are used for processing electrolytic solutions in electro-dialysis.

3.3.4 Anisotropic Membranes (Asymmetrical)

In this type of membrane there is a difference in layered structures, porosity, and pore size over the whole membrane cross-section i.e. it is unsymmetrical [5]. Asymmetric membranes contains a thin surface layer supported with a thick microporous layer. The thin layer performs the work of the separation and used as the selective layer. High permeate flux will be obtained in this type of membranes due to low thickness. Asymmetrical membranes, composite membranes and supported liquid membranes falls under the category of anisotropic membranes.

3.3.5 Ceramic, Metal and Liquid Membranes

With the development in membrane technology, use of non-conventional methods for membrane manufacturing and polymers as base materials increases significantly. To achieve the higher thermal stability and solvent resistance during process like ultrafiltration and microfiltration, ceramic membranes are used. For separation of gases like hydrogen, non-porous metal membranes (especially Palladium membranes) are used for effective separation of hydrogen from gaseous mixture. To support the liquid transport process, liquid films are comes under application.

3.4 Processes in Membrane Technology

3.4.1 Microfiltration (MF)

It is basically a filtration process which is driven by pressure for the separation of the fine particles, microorganisms and the emulsion droplets. The diameter of the separated particle ranges from 0.1 to 10 μm [6–8]. It is done under the pressure gradient of 1-3 bars. The wide range of the pore allows the separation of the versatile fields like virus, bacteria, aerosol and innumerable macromolecules from fluids. The membrane in MF is permeable to the fluid but retain the particles, which cause the separation. Microfiltration can also be used as pretreatment for Nano filtration or reverse osmosis [9].

3.4.2 Ultrafiltration (UF)

It is one of the membrane separation process in which the pore size of the membrane ranges from 1 to 100 nm (0.01 - 0.1 μm). Ultrafiltration has pore size between the Nano filtration and the Microfiltration. MF and UF membranes both have porous nature, but the structure of the UF membrane is always asymmetrical, as it contains small pore size and lower surface porosity, which results in development of higher hydrodynamic resistance.

Permeate flux of UF membranes can be calculated using Darcy's law and is given as

$$J = A(\Delta P)$$

Where:
A = Membrane permeability constant
ΔP = Trans membrane pressure

The value of the A depends upon porosity of membrane, thickness, and pore size. Calculated values of A for open and dense types of membranes are5 m^3/m^2.day.bar and 0.5 m^3/m^2.day.bar, respectively [10].

At lower trans-membrane pressure, flux varies linearly with trans-membrane pressure, and at higher values it is independent of pressure after achievement of threshold value [11].

3.4.3 Nano-Filtration (NF)

It is comparatively newly originated pressure driven separation process. NF falls between the UF and RO, in terms of rejection of molecular and ionic components on the basis of MWCO and pore size of membrane. NF membranes contains small pore size (nano-voids or 0.01 - 0.001 μm) with larger free space [12–14] which are available for transportation of feed material. Modelling of nano-filtration process is hard to describe due to its complicated processing conditions and different separation parameters as compared to UF and RO [15, 16]. NF rejects multivalent ions (>99% rejection rate) and high molecular weight organic compoundswith the lower rejection rate of monovalent ions (<70%).

3.4.4 Reverse Osmosis (RO)

Osmosis is a natural phenomenon, which is originated as early as 1748. According to this movement of the solvent takes place from lower to higher concentration, until achievement of equilibrium condition, as separated via semi permeable membrane [17]. Chemical potential difference is the major driving force for the transportation of water. The reverse osmosis (0.001 – 0.0001 μm) can be achieved by application of the pressure more than the osmotic pressure, which causes the flow of the water in the reverse direction i.e. from the solution with the higher solute concentration to the solution with the lower solute concentration.

3.5 Membrane Modules

Productivity of the membranes can be exploited by assembling it into adequate modules by the optimization of the process variables. The main objective of the membrane design is to realize the higher value for specific membrane area per unit volume, which results in higher flow rate. Geometry of membrane module largely affects the parameters like concentration polarization, fluid dynamics phenomenon, membrane performance, and thermal conditions during processing.

The design of the module should provide high rate of heat and mass transfer, it should minimize the concentration and polarization and polarization due to the temperature. It should ensure low fouling density. To overcome the problem of concentration polarization, higher feed velocity is desired, so that higher turbulence created which helps in the reduction of concentration polarization.

Different types of membrane modules have been suggested according to the requirements but among these following are the main modules [18] and their schematic configuration and mode of operation is presented in Figure 3.1.

- Shell-and-tube type modules for hollow fiber or tubular membranes;
- Spiral wound type for flat-sheet membranes; and
- Plate and frame modules type for flat-sheet membranes

Amongst above shell-and – tubuler model (A) is widely studied and used because of its simplicity and easy manufacturing, and high surface area to volume ratio The tubular and the hollow fiber membrane can operate in two types of the configuration. If the flow of the feed is from outside the fiber it is called as the outside-in configuration where permeate is received from the inside of the fiber. If the feed flows from the inside of the hollow fiber and the permeate is received from the outside of the fiber then it is called as the inside-out configuration. Due to the wider variability in pore size and channel diameter, different types of feed materials can be processed, even feed containing suspended particles (using Outside-in configuration). Main drawback of this type of modules is the lower flux rate and low processing life of membrane fibers [18, 19].

In case of the plate and frame (C) and spiral wound modules (B) flat sheet membranes can be assembled. These types of membranes are comparatively thin than the spiral wound membrane and that's why can provide

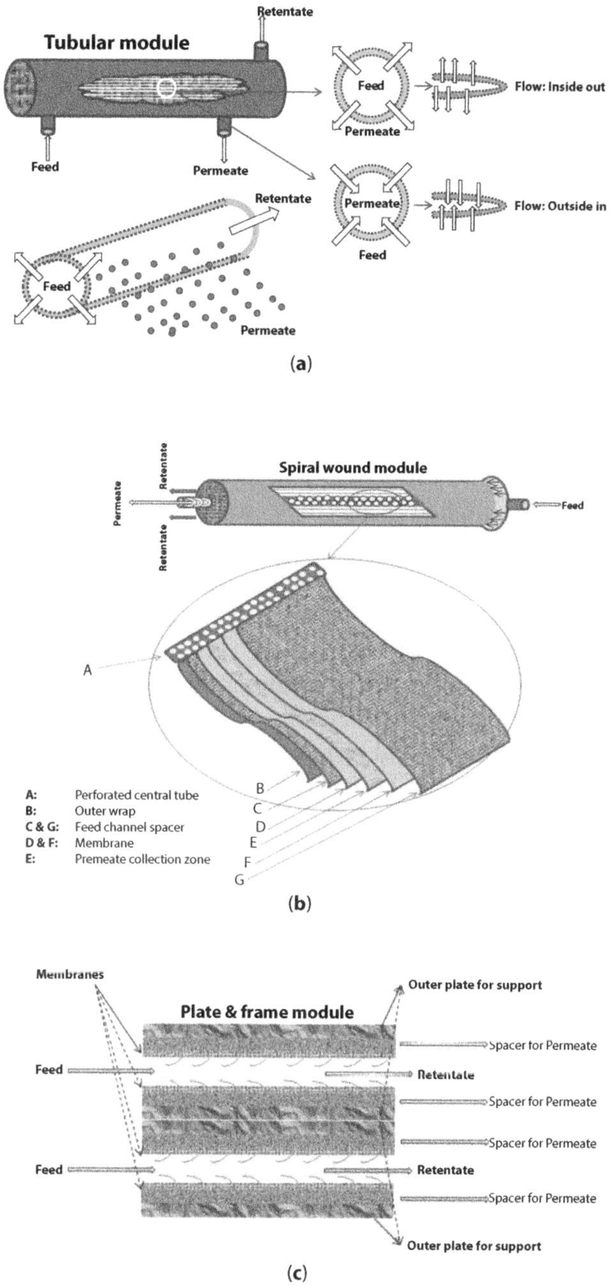

Figure 3.1 Schematic configuration and mode of operation of various type of membrane module (a) tubular module, (b) spiral wound module & (c) plate & frame module.

higher permeability with the requirement of porous support which provides the mechanical strength.

Plate and frame membrane has advantages such as it can provide higher tangential flow rate which in turn lowers the concentration and polarization and fouling effect and this type of membrane can be easily replaced. Low packing density and poor energy efficiency are some disadvantages of the plate and frame type of membrane [20–22].

3.6 Mechanism of Mass Transfer in Membrane Separation

In the field of novel technology, membrane processing has its place. The membrane has a significant means to uplift the growth of processing sectors due to its simplicity, efficiency, and economy. The membrane is a pressure-driven process, has potential application in clarification, concentration, fractionation, and molecular separation. Reduced permeate flux with processing time is one of the major problems which restricts the extensive use of membrane technology. This major occurrence is well-known as membrane fouling. Membrane fouling is the most serious and major problem in membrane processing, and also it is the key challenge for membrane manufacturers and researchers to overcome the fouling of the membrane. Fouling of membrane creates the obstacles like reduced mass transfer, higher processing time and cost, short life of the membrane, and increased cleaning time.

The reduced permeate flux corresponds to the pure water flux/pure solvent flux is due to two main reasons. First is the phenomenon of concentration polarization and another is membrane fouling.

3.6.1 Concentration Polarization (CP)

During membrane processing, due to the hydraulic transport fluid feed comes towards the membrane surface. On the basis of pore size and molecular weight cut off (MWCO), a part of the solvent goes to permeate along with or without some solutes. Along with permeate, a part of solutes moves towards the surface of membrane. Due to the gathering of solutes, a layer of solutes adjacent to the surface of membrane is formed. Due to the formation of the layer, a concentration gradient is set up b/w the solute layer and bulk solution. Due to setting up of steep concentration gradient, back transport of solutes to the bulk solution takes place. At the condition of steady-state where the convective (towards the membrane) and diffusive (away from the membrane) transport balance each other, the concentration of solutes in the concentration polarization layer attained to the maximum. This maximum concentration is known as gel concentration. After the attainment of gel concentration permeate flux starts decreasing and becomes independent of transmembrane pressure (TMP) [23]. Concentration polarization largely depends upon operating conditions like local hydrodynamic and mass transport conditions [24]. Based on several assumptions like uniform CP layer and constant permeate flux, classical film theory (stagnant film model) explains the phenomenon of CP [25]. This theory is not applicable for cross-flow membrane filtration due to the reason that in this type there is a gradual increase in the CP layer with filtration channel. As in concentration polarization flux gradually decreases, due to this reason constant flux assumption also arises some significant errors in the modeling [26]. Due to these reasons, classical film theory fails to explain the CP phenomenon. Some efforts were also made by researchers to present the model using different numerical methods [27], but those were not so effective. The mass transport model will be discussed later in this chapter.

3.6.2 Membrane Fouling

Fouling of membrane is the major factor that affects the efficiency of membrane processing [28]. Concentration polarization is a time-dependent reversible phenomenon, while fouling is an irreversible phenomenon. As a result of fouling, flux decreased greatly in the early stage and slowly in the posterior stage. Fouling mainly arises due to the accumulation or adsorption of macromolecules like proteins, fat, colloidal particles, fibers, enzymatic residues, microorganisms, salts of inorganic origin, etc. on the surface of the membrane, or due to the precipitation of permeable/soluble solutes within the pores of the membrane due to overcrowding (as shown in Figure 3.2). As a result of precipitation and overcrowding resistance to the permeate flow increases. Due to increasing hydraulic resistance, a decrease in flux rate occurs at the same trans-membrane pressure (TMP). By increasing TMP, permeate flux may be invariable [29], but increases energy requirement. Trans-membrane pressure and permeate flux are the two major indicators for membrane fouling. There is a direct correlation between operation time and irreversible fouling fraction. To obtain the maximum permeate flux repeated chemical cleaning is required. Repeated chemical cleaning ultimately increases the processing cost and decreases the life of the membrane.

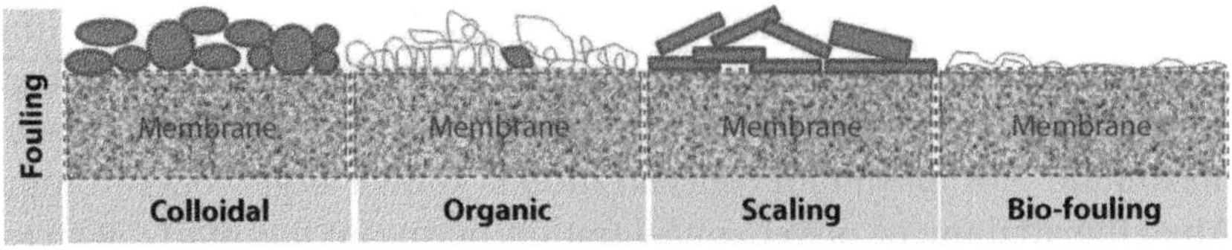

Figure 3.2 Types of membrane fouling (Reproduced from [45]).

Fouling is majorly categorized into back washable or non-back washable and reversible or irreversible. Removal of back-washable fouling takes place by employing the permeate flow in the reverse direction; but non-back washable fouling can't be removed by this technique. It required a chemical cleaning procedure to attain efficient cleaning and better efficiency of the process. Irreversible fouling cannot be removed by employing backwashing, flushing, chemical, or any other methods. Irreversible fouled membrane unable to restore their original flux.

3.6.3 Major Categories of Fouling

3.6.3.1 Inorganic Fouling

Inorganic fouling is majorly triggered due to the over-crowding of the inorganic materials. As a result of the gathering of inorganic materials, after attainment of the saturation concentration precipitation of the compounds occurs and the formation of scaling takes place. Ca^{2+} and Mg^{2+} are the major ions responsible for inorganic fouling [30–32].

3.6.3.2 Organic Fouling

Due to the accumulation of low molecular weight organic materials, clogging of membrane takes place. Accumulation of organic materials generally takes place on the internal portion of the membrane. By adopting a normal cleaning schedule, organic foulants can be removed. Mo and Huanga [33] analyzed the chemical solution after cleaning. He found that the major components which are responsible for organic fouling are low molecular weight organic compounds and calcium.

3.6.3.3 Colloidal Fouling

The major limitation of pressure-driven membranes like, microfiltration, ultrafiltration, and reverse osmosis is fouling. Feed generally treated with these membranes contains colloids. Species having a particle size less than

5 μm, which remains stable in suspension or having a charge density like in proteins comes under the category of colloids. In the case of colloids, electro-static interaction plays an important role, which controls the rate of coagulation and diffusivity variation with concentration during diffusion of particles [34, 35]. The deposition of colloidal particles and adsorption of proteins on the surface depends on the interaction forces besides diffusion and hydrodynamics [36]. The rate of deposition of colloidal suspension on the membrane surface largely depends on the surface interaction. Variation in pH during UF treatment of colloidal suspension highly affect the efficiency of the UF process, due to the reason that variation in pH affect the charge density of protein and thus enhances the fouling phenomenon [37, 38]. In the UF of colloidal clay, deposition kinetic largely depends on the suspension salinity [39].

3.6.3.4 Biological Fouling

The development of undesirable microbial on the membrane surface is known as biofouling [40]. Raw water sources contain several microorganisms like bacteria, algae, and fungi. Due to the presence of appropriate nutrient growth of these occurs. In the RO system, the most severe problem is biofouling [41, 42]. In simple words, biofilm is the accumulation of microbial aggregates having attachment to the surface. Bacteria produce extracellular polymeric substances (EPS) and form a network of EPS, which results in the formation of biofilm. EPS generally contains polysaccharides, proteins, and nucleic acids [43]. By the application of hydrogen bonds, electrostatic forces, and London dispersion forces, EPS increases the adhesiveness and structural integrity of biofilm. Due to the excellent adhesive and elastic properties, it is difficult to remove the biofilms from the surface of membranes. The presence of Ca^{2+} and Mg^{2+} ions promotes the formation of bridges b/w negatively charged bacteria and the surface of the membrane [44].

To control the formation of biofilms, pretreatments of feed are necessary. Pre-treatments like application of MF to feed so that most microorganisms can be removed from the feed. Another method is the disinfection of feed using biocides or ultra-violet treatment.

3.7 Mechanism of Membrane Fouling

As shown in the flux - time curve (Figure 3.3) for the UF process. In portion I, flux rapidly drops, then followed by a gradual decrease in portion II and

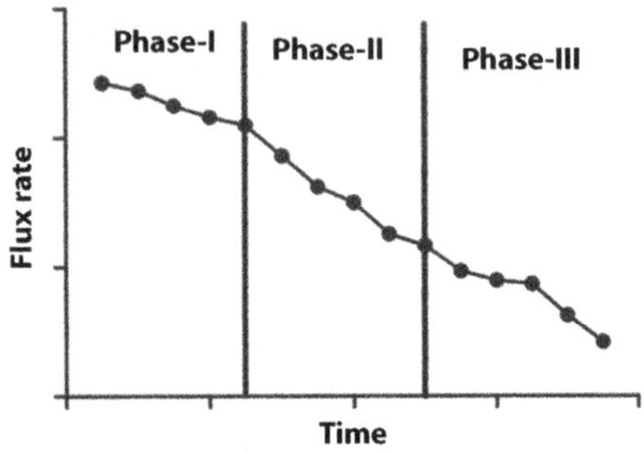

Figure 3.3 Presentation of three-stage flux decline (Reproduced from [50]).

then steady state is achieved in ending in portion III. A decrease in permeate flux of membrane is mainly due to two principal phenomena. The first phenomenon is the blockage of the pore, which increases the resistance towards the permeate flow and the second is the formation of the cake layer after the blockage of pores. The cake layer formation neighboring to the surface of membrane creates the hurdles in the flow of permeate.

At the starting of the membrane process, high permeate flux is obtained because membrane pores are open and clean. As processing time increases, blockage of pores starts due to the retention of particles in the pores. The retention of particles in the pores mainly depends on the size of particles, pore size, and shape of particles [46, 47]. Due to blockage of pores, flux declines rapidly because pore blockage is a fast process and less than one particle can contribute to the full blockage of the pore [48, 49]. Due to increasing resistance to permeate flow by the blockage of pores, particles start accumulates near the membrane surface, which results in the formation of the cake layer. Depending upon the thickness of the cake layer, the rate of permeate flux decreases, and with time decrease continuously.

3.8 Factors Influencing Fouling of Membrane

(a) Properties of membrane: pore size, surface hydrophobicity, distribution of pore size, and type of membrane material.
(b) Feed properties: Concentration and nature of feed, feed particle size.
(c) Operational parameters: operating temperature and pressure, feed rate.

3.8.1 Properties of Membrane

Interaction of water with membrane largely depends on the composition of material and its surface chemistry. Based on the surface chemistry of membrane material, the wetting behavior of the membrane is decided. The presence of active groups enhances the hydrophilicity and charge density of the membrane thus, increases the wettability of the membrane. As a reason for high hydrophilicity, the membrane allows the formation of the water layer on it, so decreasing the probability of membrane fouling. While the absence of active groups enhances the hydrophobicity of the membrane, thus increases the wetting time of the membrane. Due to the lower wettability of the hydrophobic membrane, chances of fouling increase because particles present in an aqueous medium that are responsible for fouling are hydrophobic. Due to low interfacial free energy, these particles tend to form aggregates and enhance the fouling phenomenon. Transmembrane transport of particles and fouling of membrane largely depends upon the morphology of the membrane. Membrane morphology (surface roughness) depends on the factors like size, shape, and surface protuberance. It was observed that higher roughness provides the larger surface area to the foulants for the accumulation thus enhances the fouling, even in the presence of larger turbulence but lower effective shear rate [51, 52]. Permeate flux decreases with increasing the surface roughness [53]. The electrostatic charge of opposite natures of the membrane surface and colloidal particles enhances the membrane fouling. Due to higher magnitude of electrostatic interactions, the accumulation of colloidal particles takes place on the surface of the membrane thus enhances the rate of fouling. To reduce the fouling rate, membranes of opposite charges as of feed solution, so the generation of repulsive forces takes place which results in the decrease in membrane fouling [54].

3.8.2 Feed Properties

Properties of feed solution also affect the membrane performance by inducing fouling. Important feed properties which greatly affect the membrane performance are feed concentration, properties of particles, pH of the solution, the temperature of feed, ionic strength, and feed rate. By increasing the feed concentration fouling rate also increases due to greater accumulation of particles, as discussed in the pore block phenomenon. Larger particle size decreases the deposition but increases the cake formation results in the greater permeate flux. Particle geometry also influences the porosity of cake formation; particles with lower spheroucity cake with higher porosity will be obtained. The pH of the feed solution is an important factor that determines the performance of the membrane. Lower pH results in the conformational change in the protein, also the change in charge density; thus, increases the extent of fouling.

3.8.3 Operating Parameters

Temperature is an important parameter in membrane processing. Vyas, Bennett, and Marshall [55] investigated the effect of temperature on the membrane permeate flux. They found that higher permeate flux was obtained with lower fouling at higher operating temperatures. This might be due to the increasing diffusion rate of feed material. Cross-flow velocity is the apparent velocity of stream, feed corresponding to the surface of the membrane. An increase in crossflow velocity results in an increase of mass transfer rate increases in permeates flux with a lower fouling rate [56]. By increasing the cross-flow velocity, a higher mixing rate will be obtained this will results in the lowering of accumulation of feed particles and hence decreasing the effect of concentration polarization.

Transmembrane pressure (TMP) greatly affects the permeate flux. Higher TMP delivers higher permeate flux. But as processing time increases, concentration polarization increases. At the initial stage of concentration polarization, by increasing TMP, permeate flux increases. But when the gel concentration is attained, the increase in TMP has no positive effect on the permeate flux. At this stage, TMP has attained the value of critical pressure, and permeate flux is denoted with critical flux. It was observed that, problem of fouling can be solved when membrane operated under critical flux condition [57].

3.9 Prevention of Membrane Fouling

3.9.1 Type of Feed and Pre-Treatment

Interaction b/w solutes and membrane significantly contribute to reduction in flux due to the precipitation of colloidal calcium phosphate and adsorption of milk protein on the surface of UF membrane. By changing the pH of milk to lower side converts the colloidal form into soluble form, thus most part of calcium phosphate goes to permeate. On the other hand, increase in pH of milk made the colloidal calcium phosphate insoluble and gets precipitated on the membrane surface, thus contribute to fouling. Protein also carries the high charge density. By changing the pH of feed containing protein away from their isoelectric points increases the dispersion of proteins and thus lower chances of fouling [58–60].

3.9.2 Operating Parameters

For efficient membrane operation, maximum permeate flux should be obtained from the given area of membrane. Permeate flux depends upon several parameters like feed temperature, feed composition, operating pressure, pH and ionic strength of feed.

3.9.2.1 *Operating Pressure*

As discussed earlier in concentration polarization, as pressure increases permeate flux increases until gel concentration achieved. At the stage of gel concentration flux becomes independent of applied pressure. This pressure is known as critical pressure and flux is known as critical flux [61, 62]. At this stage, increase in flux velocity is more effective than increasing pressure.

3.9.2.2 *Operating Temperature*

In UF processing of milk, flux increases with increase in feed temperature. It was observed that by increasing temperature of milk from 10 to 50 °C, flux increases by 3.5 times. Increase in permeate flux is due to higher diffusivity and lower viscosity of milk at higher temperature [63]. UF operation at higher temperature enhances the rate of fouling due to the precipitation of colloidal calcium phosphate [64].

3.9.2.3 *Feed Velocity*

Rate of fouling decreases as flow rate increases. By increasing flow rate higher turbulence occurs, as a result of which deposited materials flows continuously into permeate, hydraulic resistance created by fouling layer is reduced. Increasing flow velocity is more effective than increasing pressure [65].

3.10 Mass Transfer Model for Filtration Process in Absence of Fouling

Mass transfer through membrane can be classified under two categories in the situation, where fouling is absent. First type is diffusion through dense membrane which is independent of membrane structure and obey Fick's Law of Diffusion and 2nd is the diffusion through porous membrane (Figure 3.4).

3.10.1 Diffusion Theory Through Dense Membrane

It consists of five process, which depends on each other, these are as -

(a) Diffusion of the solute molecules through the liquid layer of the feed phase on feed side of membrane.
(b) Solute molecules of feed solution reach to the surface of membrane from feed side (i.e. upstream).
(c) Diffusion of soluble particles through the membrane matrix occurs.
(d) On downstream of membrane, desorption phenomenon occurs. Solute molecules get desorbed from the membrane and come on the downstream surface.
(e) From the boundary layer, diffusion of soluble molecules on permeate side occurs.

Under steady state condition, diffusive flux can be explained by Fick's first law. According to Fick's first law, flux is directly proportional to the concentration gradient [66].

i.e. $J \alpha (dC/dZ)$

$$J = -D\,(dC/dZ)$$

Where, J = Rate of transfer of permeate flux per unit area of membrane; C = Concentration of diffusing materials&Z = Space co-ordinate normal to surface.

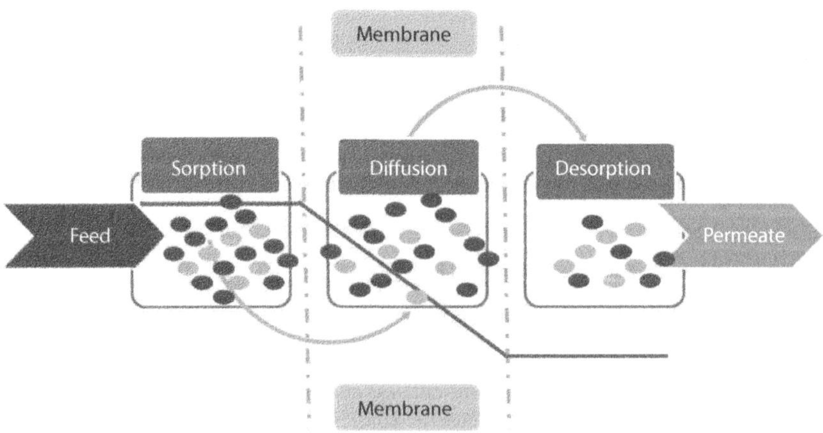

Figure 3.4 Mass transfer through dense membrane (Reproduced from [67]).

3.10.2 Transfer Through Porous Membrane - Convective Transfer - Pore Flow Model

Flow through capillary tubes can be effectively described using pore flow model in which pressure assisted convective flow occurs. In membranes like UF and MF, selective rejection of molecules occurs based on the size of molecules and pore size of membrane. Permeability of molecules through porous membrane can be described in two ways-

(a) Arrangement of particles in the membrane close to spherical, in this situation Carman - Kozney equation can be applied [68].
(b) In another approach, differential momentum equation is effectively used to describe the laminar fluid flow through capillaries and Hagen - Poiseuille equation is used to describe the average convective velocity through porous membrane [69, 70].

3.11 Application of the Membrane Technology in Dairy Industry

Dairy industry is the important part of the food industry, which provides the number of nutritionally rich products for the human consumption. Until now various technological advances have simplified the manufacturing of the various dairy products improving its overall quality, acceptability and economic output. Membrane technology is one of that and due to it various processes in the dairy industry are now simplified. It focuses on the environment friendly manner and the cost effectiveness which is must inn the dairy industry for various useful processes. Here we can divide the application of the Membrane technology under its four main processes, which are microfiltration, ultrafiltration, nano filtration and finally reverse osmosis.

3.11.1 Microfiltration

3.11.1.1 Waste Water Processing

MF is used in the Dairy industry for the purification of the wastewater so that the load of the polluting components in the wastewater can be decreased up to the permissible level and also the valuable components from it can be also recovered. We can take the example of the fat rich dairy production where the presence of the fats and oil in the wastewater creates several problems in its biological treatments. Their presence in the wastewater increases the overall processing cost of effluent treatment plant like for aeration and the sludge disposal. In contrary in the case of the MF treatment of this wastewater with 0.2 μm ceramic membrane have seen to reduce the initial chemical oxygen demand from 5000 to 10000 mg/L to less than 250 mg/L. The obtained permeate from this separation can be mixed with the water having low to medium contamination and which later can be subjected to the biological treatment. Retentate produced from MF can be further processed for the production of the soap [71].

3.11.1.2 Production of the Protein Concentrate

Due to the precise pore size of the MF membrane and processing at lower temperature, fat and bacteria can be efficiently removed from the feed, which can be further used for production of the high protein products like, protein concentrates and isolates without affecting the desirable properties of final products [72]. Pre-treatment of whey for the production of whey protein concentrate can efficiently remove the residual fat present in whey, which further helps in the enhancement of foaming properties [73].

3.11.1.3 Isolation

Immunoglobulin (IgG) can also be isolated from the colostrum whey with the help of the MF. Treatment of the colostrum with (0.1 μm) MF membranes, obtained from the different type of the species can be efficiently used to recover IgG upto 80%. The obtained micro filtered liquid (permeate) named serocolostrum which is free from fat and casein proteins [74].

3.11.1.4 Separation of Micellar Casein from the Milk

Concentrated micellar caseins can obtain from both full fat milk and skimmed milk, which can account up to 95% of the total protein. This type of the high casein retentate can be obtained with the help of the microfiltration process along with the diafiltration. When these micellar casein are concentrated, permeate obtained from this process contains native whey proteins. The obtained native whey is microbiologically sterile in nature and almost resembles to the sweet cheese whey [75].

3.11.1.5 Pretreatment of the Cheese Milk

MF has shown increase in the cheese firmness and decrease in the coagulation time [76, 77]. The reduction of need to add additives like calcium chloride have also noticed with enabling the stronger heat treatment of the cheese milk [78]. With the standardization of the milk protein level with the MF, quality attributes of cheese can be enhanced [79]. Due to standardized protein level, rennet coagulation time can be optimized and the obtained coagulum will be rigid in nature. MF pre-treatment increased the hardness of cheese (around 40%) as measured using instron universal testing machine [80]. The recoveries of casein and fat in cheese are observed to be increased reducing their loss in the whey. The yield of cheese shown to be increased by 2-4% as a result of increase in the casein content [81].

3.11.2 Ultrafiltration

3.11.2.1 Enzyme Recovery and Concentration

Ultrafiltration is widely used for the recovery of the enzymes [82]. It is less expensive than the other processes and it also known to prevent the loss of the enzyme activity. It is widely used for the concentration and purification of the protease enzyme [83] showed the possibility of the *Pseudomonas fluorescens* lipase purification by using UF capillary membranes. Separation of the species from the aqueous solution on the basis of size shape and the charge is possible with the UF [84].

3.11.2.2 Cheese Manufacturing

UF is used to pre-concentrate the milk before cheese making which provide the benefits such as increased yield of cheese, low energy consumption and volume reduction of the whey [85]. Maubois, Macquot and Vassal in 1969 [86] developed the use of UF for the manufacturing of the cheese. The method is therefore recognized as MMV process. For the manufacture of the lactic acid cheese varieties likeQuarg, cream cheese, mascarpone, and *fromages frais*in many plants UF has replaced filtration cloths or centrifugal seperators [87]. Cheese manufactured by the UF from fully concentrated milk with the by using mesophilic starter cultures have shown retardation in ripening as compared to the cheese manufactured by the traditional method [88].

3.11.3 Nanofiltration

Demineralization of the whey increased its nutritional value [89]. Nanofiltration is alternative to evaporation and electro-dialysis, which is meant for the demineralization of the whey [90]. NF membrane gives high permeability to the monovalent salts (NaCl, KCl) along with very low permeability to the organic compounds.

NF provides advantage over evaporation and electro dialysis that it preform the concentration and demineralization simultaneously. NF results in reduction in cost of energy, wastewater treatment and total cost. The sweet Gouda whey demineralized by NF in respect of the monovalent ions is comparable with the whey demineralized by employing Electro dialysis.

3.11.4 Reverse Osmosis

Concentration of the milk is done in many cases for increasing the shelf life and also concentration results in reduction in the cost of packaging, storage and transportation of milk and milk products. For these purposes, various types of heat exchangers and falling film evaporators are used. These techniques of concentration result in the change in the composition, nutritional properties, rheological properties and heat stability due to the high processing temperature [91]. Increase in the output of the cheese has observed by using UF and RO as a method for pre-concentration [92].

In the dairy industry, RO membrane is used mostly as the concentration technique [93–96]. The main advantage offered by the RO in the dairy industry is reduced energy consumption than the conventional techniques.

3.12 Application of Membrane Technology in Food Industry

As the development in the field of novel technology increased, the use of membrane technology also increased in the dairy and food sectors. Various membrane processes are based on requirements like Ultrafiltration (for fractionation, partial demineralization), Reverse osmosis (for concentration), Nano-filtration (for demineralization), Micro-filtration (for cold pasteurization, separation of large molecules) are used.

Membrane processing delivers numerous advantages in the processing of high-value biological products, like processing at desirable temperature and lower pressure, retains the nutritional quality of materials, economical operation, better yield, and environment friendly in comparison to conventional methods [97]. Because from the last decades, the requirement of foods comprises of three main keys, like nutrition, dietetic, and health. Therefore, to formulate novel foods, membrane technology has an important role. Every method also has its drawbacks, like limited concentration of products, problems like fouling, and concentration polarization which leads to the inefficient separation of molecules and decrease in permeate flux.

3.12.1 Beverages

Membrane operation is generally used in two or more combinations like sequential filtration of liquids using membranes of varying molecular weight cut off (MWCO), pre-treatments like heat treatment, enzymes, and post-treatments like evaporation, adjustment of pH in case of fruit juices, vegetable juices, and sugar solutions [98–100].

3.12.2 Clarification, Concentration, and Sterilization of Fruit Juices

Several variables like the maturity of the fruit, ripening index, climate, and natural variations affects the composition and processing of fruit juices [101]. Clarification of fruit juices is one of the major challenges. In traditional methods, enzymes are used for clarification purposes; but time and labor requirement is high in this process. Membrane technology using ultrafiltration membrane has several advantages over traditional method; as it retains the enzymes based on their pore size like depectinization enzymes, as well as polyphenol oxidase (PPO). Due to the retention of the enzyme, mainly pectinase less amount of enzyme is required in the process; ultimately reducing the production cost. Preservation of juices is also a major challenge; because preservation by heat treatment causes the problem of haziness in the juices. To avoid such types of problems microfiltration for cold sterilization of juices can be used.

Juices produced by traditional methods and membranes are almost similar in composition. Heatherbell *et al.* [102] observed that the unpasteurized bottled juice from UF had a superior quality to traditional and UF canned juice; but the problem of sedimentation during storage. Rwabahizi and Wrolstad, [103] found a 55% loss of anthocyanin in UF strawberry clarified juice using 10 kDa membrane; which was quite higher than the traditional filtration method. Drake and Nelson [104] observed lower turbidity and 5% higher total soluble solids in UF-treated apple juices than the conventional filtration method.

3.12.3 Concentration, De-Acidification, and Demineralization of Juices

To reduce the cost of transportation and also to increase the shelf life of juices; concentration is done. The most common method is used for concentration is vacuum evaporation. But this can cause several changes in the fruit juices like loss of aroma, decrease in the nutritional quality of juice. To avoid such kinds of problems concentration using reverse osmosis (RO) is commercially used. Bengtsson [105] reported the high retention of aroma compounds in polyamide RO membrane. The concentration of apple juice can be achieved by 25° Brix employing reverse osmosis.

Juices of single strength extracted from orange, pineapple, lemon, and grapes are highly acidic with an acid concentration ranges from 1.0-1.2% [106, 107]. Due to high acidity, it is hard to use these juices in single strength or concentrated form for canning purposes [108, 109]. Sourness in juices depends on the ratio of total soluble sugar to the acid content. The concentration of soluble sugar remains almost the same in fruits but citric content varies with the variation in season. The ratio of total soluble sugar to the acid concentration is known as Brix/acid ratio. Brix/acid ratio for acceptable quality of grapes and orange juices are 10 to 11 and 13.5-14.5 respectively [110]. One method to adjust the Brix/acid ratio is the addition of natural sweet juice into sour juice, but this is not much effective method. Therefore, for effective de-acidification of juices, the electro-dialysis technique is generally used. Two stacks containing cation and anion exchangers (bipolar) are used with periodic reversal of voltage to prevent fouling problems. Total energy requirement varies from 0.02-0.1 kWh/equivalent with a current efficiency of 52 to 90%. The efficiency of electro-dialysis mainly depends upon the quality of juice [111, 112]. Due to the high fiber and nutrient content in native/original apple juice, demand is also high, but it is very challenging to produce high-quality cloudy juice. Due to the high fiber content in unclarified juice, a higher concentration of polyphenol oxidase and polyphenol is present. These cause undesirable changes, like enzymatic browning of the apple juice. The activity of polyphenol oxidase can be permanently suppressed by changing the pH of apple juice to 2 and then readjusted to the original pH [113, 114]. Earlier this was done using salts like hydrochloric acid (HCl) and caustic soda (NaOH). But the combination of acid and base leads to the formation of salts in the product and contributes inferior flavors to the product. To avoid this kind of problem electro-dialysis technique is used nowadays. A bipolar electro-dialysis system with potassium hydroxide is used for the adjustment of pH of unclarified/cloudy apple juice. By using electro-dialysis, pH of juice can be attained up to 2-3.5 so that the activity of polyphenol oxidase can be suppressed and then readjusted to the original pH [115, 116].

3.12.4 Demineralization of Sugar Syrup

Alkali and alkaline earth metals are highly melassigenic in nature. Due to this property, they prevent the crystallization of sugars by holding the sugar in molasses. Alkali metals exhibit a higher melassigenic effect than alkaline earth metals (i.e. K > Na > Ca > Mg). Ionized metal content in raw sugar beet or sugar cane juice is up to 3.5% [117–120]. The presence of these metal impurities inhibits the crystallization of sugar and causes fouling in the evaporator. To overcome this problem technologies like ion-exchange resins, synthetic resins, adsorbents, and membranes have been employed, but due to the short life cycle and higher processing cost, these technologies are not much used. Electro-dialysis an emerging technology for reducing ion concentration is generally employed for the demineralization of sugar syrups. A bipolar module is used for this process. The major problem of using electro-dialysis is the higher fouling of membrane modules caused by anionic organic materials which results in the un-appropriate cleaning of assembly, decreasing efficiency, and increasing the resistance of the membrane. The smooth operation of electro-dialysis requires energy about 1.10 kWh/kg of salt removed from the juice [121, 122]. The efficiency of this process varies from 40-45%, which largely depends on the initial quality of juice [123].

3.12.5 Manufacturing of Beverages Using Vegetable Proteins

Several vegetables and their derivatives are widely used for the manufacturing of beverages. Soybean extract called soya milk contains proteins, fat, and glucides as 47%, 24%, and 15% respectively. By employing UF and diafiltration, retentate can be obtained which is a concentrated source of protein (47-60%) and fat (24-34%). This type of retentate can be further processed for the manufacturing of protein hydrolysates; useful for sportsmen and also for the preparation of protein-enriched diets for the peoples having allergy from the parent or native proteins [124]. By employing a combination of membranes and different processing conditions, it is quite easy to separate/fractionate the desirable components from the raw materials. Peptides from wheat gluten hydrolysate can be fractionated based on their charge using acidic conditions and zirconium oxide UF membrane [125].

3.12.6 Rough Beer Clarification

Clarification is one of the most important steps in the manufacturing of beer. Filtration of rough beer is done to eliminate the residue of yeast and colloidal particles, which are mainly responsible for the haziness in beer. Clarification is generally done to provide biological stability and the reformation of the previous form into bright and clear form. By using standard filtration techniques, we can't achieve the superior quality of beer. To produce the superior quality of beer with the economic operation; MF is generally used at a lower temperature almost 0°C to obtain the optimum permeate flux rate [126, 127]. Rough beer clarification has an emerging market for the application of membrane technology. With the opportunities, it is also a challenge for the engineers to avoid the problems of fouling and improvement in the quality of permeate [128, 129], as the rate of permeate flux in MF is quite lower than that of dead-end filtration and also complicated cleaning schedule like enzymatic, alkaline, and oxidative cleaning; consume a lot of time for cleaning.

3.12.7 Preservation of Beer

To maintain the quality and microbiological stability of beer heat treatment is generally done. Flash pasteurization is generally done for the preservation of clarified beer. Some heat-induced flavors may produce during this process, which can be easily eliminated by the application of sterile filters [130]. Nowadays there is a huge demand for cold sterile beer (draught beer). So to meet the quality and market demand requirement, cold sterilization of beer employed MF is done. MF technique eliminates the process like dead-end filtration with cartridges and heat treatment [131, 132] due to some advantages like -

a) Low operational temperature (0°C)
b) Maintains the biological, colloidal, and microbiological quality of beer
c) Economic operation
d) Maintains the color, flavor, and aroma of beer.

3.12.8 Membrane Processing in the Wine Industry

In the wine manufacturing process, grapes must have their place to produce superior quality wine. Enrichment of must is done before the fermentation process to overcome the reduced level of sugar. Traditionally this level is makeup by adding sucrose from different sources like beet and cane sugar, concentrated must (CM) of grapes, or rectified CM. Vacuum evaporation is generally used to produce concentrated must; but it has some drawbacks like loss of aroma and production of off-flavors [133]. To counter these types of phenomenon reverse osmosis is used for the concentration of must [134, 135]. If must rectification is considered there is the requirement of ion-exchange assembly, and this can be completely fulfilled by nano-filtration as proposed by Santos *et al.* [136] because nano-filtration can retain most of the sugar in the retentate and reject the acids in permeate from the grapes must. After the must concentration and rectification process, clarification of wine is an important step. Earlier wine clarification was done using diatomaceous-earth filtration, which is a discontinuous and time-consuming process. As there is the valorization of novel technology increases, this is replaced by tangential MF and UF. It has several advantages over conventional filtration in terms of process economy, time consumption, and process efficiency. Yeasts and suspended particles with colloidal nature are the major cause of turbidity in wine [137]. Serrano *et al.* [138] compared the efficiency of the clarification process by tangential MF and the traditional method. They found that tangential MF employed clarification wine had a lower amount of polysaccharide content. Polyphenols and polysaccharides are the major constituents that cause the fouling of organic membranes. Clarification of white wine using tangential MF retains higher amounts of polysaccharides and polyphenols in clarified white wine in comparison to clarified red wine [139, 140].

Another application of membrane technology in winemaking is tartaric stabilization using electro-dialysis. A natural constituent called Potassium hydrogen tartrate (KHT) present in grapes. As alcohol content increases, the solubility of KHT decreases. Due to super-saturation, precipitation of KHT occurs in bottles. To overcome this problem, earlier cold tartaric stabilization was done. This process consists of cooling, seeding, crystallization, and filtration using diatomaceous-earth. The main drawback of this process is that there is no control on the concentration of KHT and also the undesirable precipitation of polyphenols and polysaccharides occurs with the crystallization process. To overcome these types of problems electro-dialysis is done. The main advantage of this process is that there is precise control on the concentration of KHT and no unwanted precipitation of polysaccharides and polyphenols [141].

13.12.9 Membrane Processing in Fish, Poultry, and Gelatin Industry

Ultrafiltration (UF) technique is mainly used in fish industry to fractionate low molecular weight components. UF was applied to improve the bioactivity and fractionation of small peptides having molecular weight (<7kDa) of saithe protein hydrolysate by Chbeaud *et al.* [142]. Waste water originated from fish and poultry industries contains large amounts of small molecular weight compounds. To minimize the organic load, and increasing the recovery of by-products UF is used nowadays. Afonso and Borquez [143] used ceramic membrane of molecular weight cut off 15kDa to reduce the load (organic) of fish meal waste water. Results revealed that organic load was sufficiently reduced and higher recovery of valuable components from waste water.

3.13 Uses of Membrane Technology in Biotechnology

3.13.1 Purification of Proteins

Membrane processes based on poly-electrolytic films and polymer brushes as a support material are widely used in the separation and purification of biotechnological products. Based on the surface hydrophobicity, surface charge, pore size of membrane adsorption of proteins takes place [144]. The advantage of the membrane purification process is that it is a less time-consuming process. Besides this, the surface area used for the separation process is quite lower than that of the separation bed process; due to this reason, this process has a lower binding capacity than beds containing nanoparticles. To improve the separation process using membrane technology, modification based on polymeric films with the brush is done. The principle behind this is that in the presence of water polymers absorb water and by swelling increase in thickness takes place. Due to the increase in thickness binding capacity of the membrane increases, so the efficiency of purification and separation increases.

3.13.2 Purification of Antibody

Columns containing the costing of immobilized protein A and G are generally used for the purification of antibodies. Modified microporous membranes containing PAA (polyacrylic acid)/PEI (poly-ethyleneimine) films, small length peptides with antibodies were used by [145]. Antibodies placed an important role in biotechnological therapy. The cost of separation and purification of antibodies is quite high. To optimize the processing time and cost membrane processing is used instead of column separation because it took a long time for processing. In the protein purification process immobilization of antibodies in membrane pores may affect the binding capacity and thus ultimately rate of reaction. Electrostatic immobilization of antibodies is unstable but provides higher binding capacity; whereas higher stability and lower binding capacity were seen in the case of covalent immobilization. To improve the efficiency of the process, the two-step immobilization method involves both electrostatic and covalent immobilization to produce a higher binding capacity and stability.

3.13.3 Controlled Protein Digestion - A Substrate for Mass Spectroscopy

Mass Spectroscopy (MS) is a potent technique to detect the proteins and their post-translational modifications. Peptides are the most important substrate for MS in comparison to proteins. The conversion of proteins into peptides is the most crucial step in MS analysis. Earlier, digestion employed enzymes like protease (trypsin) was generally used. The main drawback of this process is the uncontrolled digestion and due to the low concentration of enzymes, it restricts the self -digestion of protease [146]. Researchers made several efforts to develop the digestion process like polymeric microfluidic channels [147, 148] reactors with solid supports like monolith containing enzymes [149, 150] resins [151] and membranes [152]. Membranes thickness ranges from 10-200 μm provides an excellent surface for controlled protein digestion. By varying the residence time in membrane peptides with desirable size can be obtained by lessening or extending the residence time [153]. By reducing the residence time, better results of post-translational modification can be obtained using MS and better flexibility in protein regions can be obtained.

3.13.4 Enantiomer Isolation from Racemic Mixtures

A racemic mixture is the mixture of two enantiomers having a mirror image of each other. Most of the drugs contain enantiomers. One enantiomer may have the same and positive effect as of other enantiomer or have a different and negative effect. Therefore, it is important to separate, isolate, and identify the enantiomers. For separation of enantiomers from the racemic mixture, membranes of regenerated cellulose with chiral

L- proline - copper complexes were used by [154]. The chiral compound can form coordination compounds with different enantiomers based on their stereo-chemistry. In this ligands exchange, the proposed mechanism is generally used based on their different functional group having a varying affinity towards the formation of coordination compounds with the ligands and effective separation of the racemic mixture takes place. This method was applied on D - or L - Phe using the membrane separation technique. The results revealed that the membrane allows a higher permeation rate to D - Phe than L - Phe.

3.14 Membrane Distillation

Among the membrane processes, membrane distillation (MD) is an emerging and popular technology that is widely used as a separation technique nowadays. As distillation also explain the use of thermal conditions. Besides non-thermal and pressure-driven processes, membrane distillation is a thermally pressure-driven process generally used for the distillation of feed solution. In membrane, distillation water is the main component as a feed solution. In MD, one side of the membrane which is microporous and hydrophobic remain in contact with the feed solution. As a result of the temperature gradient, partial pressure gets generated and mass transfer occurs. Permeation of liquid molecules is restricted in MD. Due to the difference in vapor pressure vaporization of liquid molecules take place and the volatile components through micro-porous structure go into permeate side and collected in condensed form. MD is a major alternative to reverse osmosis and has potent application in the desalination of water. MD is more economical and simpler than the traditional desalination process. Due to lower operating pressure than reverse osmosis and lower energy requirements, energy sources like solar and medium mechanical energy sources can be used [155–158]. Majorly four configurations of MD are generally used (as shown in Figure 3.5) based on the recovery of permeate, these are (a) Direct contact MD (b) Air gap MD (c) Sweeping gas MD (d) Vacuum MD. There is the requirement of special types of membranes having optimum surface hydrophobicity, high thermal resistance, high porosity, and adequate liquid entry pressure (LEP). MD technology undergoes some issues like module design optimization, requisite of energy, and need for superior membrane [159–162].

MD has some emerging application like desalination of seawater, recovery of volatile acids from industrial wastes, manufacturing of (milli-Q,

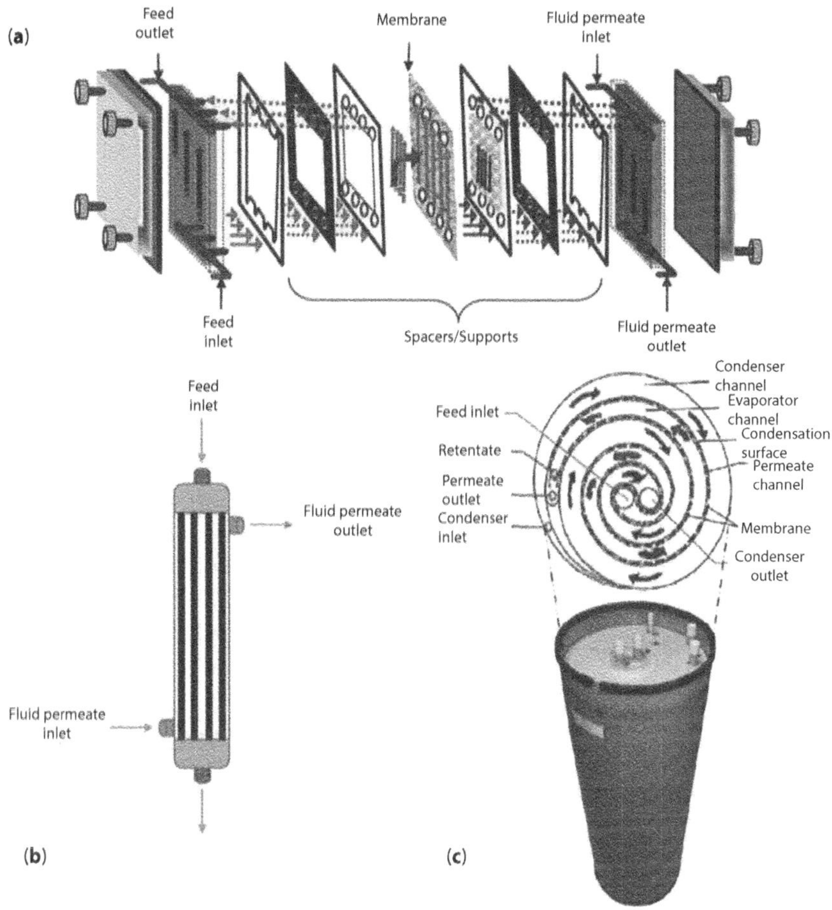

Figure 3.5 Different module of membrane distillation (Adopted from [163]), where a - Plate and Frame type, b - Capillary type, & c - Spiral wound type.

double distilled water), liquid food concentrate preparation (mandarin juice, sugar solution, paneer and cheese whey, apple juice concentration), alcoholic solution treatment, recovery of aroma compounds from (fruit juice and alcohol) fruit juice concentration, separation of a zeotropic mixtures, alcohol removal from fermented beverages, increasing concentration of alcohol, etc. Based on the treatment requirement different MD modules are used. Generally used MD modules are plate and frame, capillary, and spiral wound types.

References

1. Singh, S., Henderson, R. K., Baker, A., Stuetz, R. M., & Khan, S. J. (2012). Characterisation of reverse osmosis permeates from municipal recycled water systems using fluorescence spectroscopy: Implications for integrity monitoring. *Journal of membrane science, 421*, 180-189.
2. Baker, R. W. (2002). Membrane technology. *Encyclopedia of polymer science and technology, 3.*
3. Loeb, S., & Sourirajan, S. (1962). Sea water demineralization by means of an osmotic membrane.
4. IUPAC (1972). *Pure Appl. Chem.* 31: 577–638.
5. Baker, R. W. (2012). *Membrane technology and applications.* John Wiley & Sons.
6. Hwang, K. J., & Lin, T. T. (2002). Effect of morphology of polymeric membrane on the performance of cross-flow microfiltration. *Journal of Membrane Science, 199*(1-2), 41-52.
7. Chen, J. C., Li, Q., & Elimelech, M. (2004). *In situ* monitoring techniques for concentration polarization and fouling phenomena in membrane filtration. *Advances in Colloid and Interface Science, 107*(2-3), 83-108.
8. Saha, N. K., Balakrishnan, M., & Ulbricht, M. (2006). Polymeric membrane fouling in sugarcane juice ultrafiltration: role of juice polysaccharides. *Desalination, 189*(1-3), 59-70.
9. Ghayeni, S. S., Beatson, P. J., Schneider, R. P., & Fane, A. G. (1998). Water reclamation from municipal wastewater using combined microfiltration-reverse osmosis (ME-RO): preliminary performance data and microbiological aspects of system operation. *Desalination, 116*(1), 65-80.
10. Porter, M. C. (1972). Concentration polarization with membrane ultrafiltration. *Industrial & Engineering Chemistry Product Research and Development, 11*(3), 234-248.
11. S.B. Applebaum, Demineralization by Ion Exchange, Academic Press, New York, NY, 1968.
12. Van der Bruggen, B. (2009). Chemical modification of polyethersulfone nanofiltration membranes: a review. *Journal of applied polymer science, 114*(1), 630-642.
13. Crespo, J. G., & Brazinha, C. (2010). Membrane processing: Natural antioxidants from winemaking by-products. *Filtration & Separation, 47*(2), 32-35.
14. Bowen, W. R., & Welfoot, J. S. (2002). Modelling the performance of membrane nanofiltration—critical assessment and model development. *Chemical engineering science, 57*(7), 1121-1137.
15. Van der Bruggen, B., Mänttäri, M., & Nyström, M. (2008). Drawbacks of applying nanofiltration and how to avoid them: a review. *Separation and purification technology, 63*(2), 251-263.

16. Geens, J., Van der Bruggen, B., & Vandecasteele, C. (2006). Transport model for solvent permeation through nanofiltration membranes. *Separation and purification technology, 48*(3), 255-263.
17. Wang, Y. N., & Tang, C. Y. (2011). Protein fouling of nanofiltration, reverse osmosis, and ultrafiltration membranes—the role of hydrodynamic conditions, solution chemistry, and membrane properties. *Journal of Membrane Science, 376*(1-2), 275-282.
18. Winter, D., Koschikowski, J., Gross, F., Maucher, D., Düver, D., Jositz, M., ... & Hagedorn, A. (2017). Comparative analysis of full-scale membrane distillation contactors-methods and modules. *Journal of membrane science, 524*, 758-771.
19. Teoh, M. M., Chung, T. S., & Yeo, Y. S. (2011). Dual-layer PVDF/PTFE composite hollow fibers with a thin macrovoid-free selective layer for water production via membrane distillation. *Chemical engineering journal, 171*(2), 684-691.
20. Eykens, L., De Sitter, K., Dotremont, C., Pinoy, L., & Van der Bruggen, B. (2017). Membrane synthesis for membrane distillation: A review. *Separation and Purification Technology, 182*, 36-51.
21. Guillen-Burrieza, E., Servi, A., Lalia, B. S., & Arafat, H. A. (2015). Membrane structure and surface morphology impact on the wetting of MD membranes. *Journal of Membrane Science, 483*, 94-103.
22. Chmiel, H., Kaschek, M., Blöcher, C., Noronha, M., & Mavrov, V. (2003). Concepts for the treatment of spent process water in the food and beverage industries. *Desalination, 152*(1-3), 307-314.
23. Gekas, V., & Hallström, B. (1987). Mass transfer in the membrane concentration polarization layer under turbulent cross flow: I. Critical literature review and adaptation of existing sherwood correlations to membrane operations. Journal of Membrane Science, 30(2), 153-170.
24. Bouchard, C. R., Carreau, P. J., Matsuura, T., & Sourirajan, S. (1994). Modeling of ultrafiltration: predictions of concentration polarization effects. Journal of Membrane science, 97, 215-229.
25. Zydney, A. L. (1997). Stagnant film model for concentration polarization in membrane systems. Journal of Membrane Science, 130(1-2), 275-281.
26. Brian, P. L. T. (1965). Concentration polar zation in reverse osmosis desalination with variable flux and incomplete salt rejection. Industrial & Engineering Chemistry Fundamentals, 4(4), 439-445.
27. Bhattacharyya, D., Back, S. L., Kermode, R. I., & Roco, M. C. (1990). Prediction of concentration polarization and flux behavior in reverse osmosis by numerical analysis. Journal of Membrane Science, 48(2-3), 231-262.
28. Makardij, A., Chen, X. D., & Farid, M. M. (1999). Microfiltration and ultrafiltration of milk: some aspects of fouling and cleaning. Food and bioproducts processing, 77(2), 107-113.

29. Abdelrasoul, A., Doan, H., & Lohi, A. (2013). Fouling in membrane filtration and remediation methods. Mass transfer-advances in sustainable energy and environment oriented numerical modeling, 195.
30. Hong, S., & Elimelech, M. (1997). Chemical and physical aspects of natural organic matter (NOM) fouling of nanofiltration membranes. Journal of membrane science, 132(2), 159-181.
31. Quintanilla, V. Y. (2005). Colloidal and Non-colloidal NOM Fouling of Ultrafiltration Membranes: Analyses of Membrane Fouling and Cleaning. Unesco-IHE.
32. Lee, S., Cho, J., & Elimelech, M. (2005). Combined influence of natural organic matter (NOM) and colloidal particles on nanofiltration membrane fouling. Journal of Membrane Science, 262(1-2), 27-41.
33. Mo, L., & Huanga, X. (2003). Fouling characteristics and cleaning strategies in a coagulation-microfiltration combination process for water purification. Desalination, 159(1), 1-9.
34. Anderson, J. L., Rauh, F., & Morales, A. (1978). Particle diffusion as a function of concentration and ionic strength. The Journal of Physical Chemistry, 82(5), 608-616.
35. Verwey, E. J. W. (1962). Theory of the stability of lyophobic colloids: the interaction of sol particles having an electric double layer (No. 541.3453 V4).
36. Rajagopalan, R., & Kim, J. S. (1981). Adsorption of Brownian particles in the presence of potential barriers: effect of different modes of double-layer interaction. Journal of Colloid and Interface Science, 83(2), 428-448.
37. Fane, A. G., Fell, C. J. D., & Suki, A. (1983). The effect of pH and ionic environment on the ultrafiltration of protein solutions with retentive membranes. Journal of membrane science, 16, 195-210.
38. Mcdonogh, R. M., Fane, A. G., & Fell, C. J. (1989). Charge effects in the cross-flow filtration of colloids and particulates. Journal of Membrane Science, 43(1), 69-85.
39. Bacchin, P., Aimar, P., & Sanchez, V. (1995). Model for colloidal fouling of membranes. AIChE journal, 41(2), 368-376.
40. Flemming, H. C. (1997). Reverse osmosis membrane biofouling. Experimental thermal and fluid science, 14(4), 382-391.
41. Baker, J. S., & Dudley, L. Y. (1998). Biofouling in membrane systems—a review. Desalination, 118(1-3), 81-89.
42. Al-Ahmad, M., Aleem, F. A., Mutiri, A., & Ubaisy, A. (2000). Biofuoling in RO membrane systems Part 1: Fundamentals and control. Desalination, 132(1-3), 173-179.
43. Liu, H., & Fang, H. H. (2002). Extraction of extracellular polymeric substances (EPS) of sludges. Journal of biotechnology, 95(3), 249-256.
44. Flemming, H. C., & Wingender, J. (2001). Relevance of microbial extracellular polymeric substances (EPSs)-Part I: Structural and ecological aspects. Water science and technology, 43(6), 1-8.

45. Tiraferri, A. (2014). Membrane-based water treatment to increase water supply.
46. Belfort, G., Davis, R. H., & Zydney, A. L. (1994). The behavior of suspensions and macromolecular solutions in crossflow microfiltration. Journal of membrane science, 96(1-2), 1-58.
47. Hermia, J. (1982). Constant pressure blocking filtration laws: application to power-law non-Newtonian fluids.
48. Hlavacek, M., & Bouchet, F. (1993). Constant flowrate blocking laws and an example of their application to dead-end microfiltration of protein solutions. Journal of Membrane Science, 82(3), 285-295.
49. Granger, J., Leclerc, D., & Dodds, J. A. (1985). Filtration of dilute suspensions of latexes. Filtration and Separation, 22, 58-60.
50. Li, N. N., Fane, A. G., Ho, W. W., & Matsuura, T. (Eds.). (2011). Advanced membrane technology and applications. John Wiley & Sons.
51. Rana, D., & Matsuura, T. (2010). Surface modifications for antifouling membranes. Chemical reviews, 110(4), 2448-2471.
52. Kochkodan, V., Johnson, D. J., & Hilal, N. (2014). Polymeric membranes: Surface modification for minimizing (bio) colloidal fouling. Advances in colloid and interface science, 206, 116-140.
53. Vrijenhoek, E. M., Hong, S., & Elimelech, M. (2001). Influence of membrane surface properties on initial rate of colloidal fouling of reverse osmosis and nanofiltration membranes. Journal of membrane science, 188(1), 115-128.
54. Kang, G. D., & Cao, Y. M. (2012). Development of antifouling reverse osmosis membranes for water treatment: a review. Water research, 46(3), 584-600.
55. Vyas, H. K., Bennett, R. J., & Marshall, A. D. (2000). Influence of feed properties on membrane fouling in crossflow microfiltration of particulate suspensions. International Dairy Journal, 10(12), 855-861.
56. Salahi, A., Mohammadi, T., Rahmat Pour, A., & Rekabdar, F. (2009). Oily wastewater treatment using ultrafiltration. Desalination and Water Treatment, 6(1-3), 289-298.
57. Li, N. N., Fane, A. G., Ho, W. W., & Matsuura, T. (Eds.). (2011). Advanced membrane technology and applications. John Wiley & Sons.
58. Hanemaaijer, J. H., Robbertsen, T., Van den Boomgaard, T., & Gunnink, J. W. (1989). Fouling of ultrafiltration membranes. The role of protein adsorption and salt precipitation. Journal of Membrane Science, 40(2), 199-217.
59. Cheryan, M., & Merin, U. (1981). The effect of protein-mineral-membrane interactions on the fouling of ultrafiltration membranes. Fundamentals and Applications of Surface Phenomenon Associated with Fouling and Cleaning in Food Processing, 204-214.
60. Grandison, A. S., & Glover, F. A. (1994). Membrane processing of milk. In Robinson: Modern Dairy Technology (pp. 273-311). Springer, Boston, MA.
61. Cheryan, M. (1986). Ultrafiltration handbook. Technomic Publishing Co. Inc.

62. Renner, E., & Abd El-Salam, M. H. (1991). Ultrafiltration of milk. Application of Ultrafiltration in the Dairy Industry, 112-152.
63. CHERYAN, M. (1977). Mass transfer characteristics of hollow fiber ultrafiltration of soy protein systems. Journal of Food Process Engineering, 1(3), 269-287.
64. Maubois, J. L. (1980). Ultrafiltration of whey. International Journal of Dairy Technology, 33(2), 55-58.
65. Renner, E., & Abd El-Salam, M. H. (1991). Ultrafiltration of milk. Application of Ultrafiltration in the Dairy Industry, 112-152.
66. Nagy, E. (2018). Basic equations of mass transport through a membrane layer. Elsevier.
67. Abdelrasoul, A., Doan, H., Lohi, A., & Cheng, C. H. (2015). Mass transfer mechanisms and transport resistances in membrane separation process. Chapter, 2, 15-40.
68. Nagy, E. (2018). Basic equations of mass transport through a membrane layer. Elsevier
69. Bird, R. B., Lightfoot, E. N., & Stewart, W. E. (1960). Notes on Transport Phenomena. Transport Phenomena. By RB Bird, Warren E. Stewart, Edwin N. Lightfoot. A Rewritten and Enlarged Edition of" Notes on Transport Phenomena". John Wiley & Sons.
70. Geankoplis, C. J. (2003). Transport processes and separation process principles:(includes unit operations). Prentice Hall Professional Technical Reference.
71. Chmiel, H., Kaschek, M., Blöcher, C., Noronha, M., & Mavrov, V. (2003). Concepts for the treatment of spent process water in the food and beverage industries. *Desalination, 152*(1-3), 307-314.
72. Lipnizki, F. (2010). Cross-flow membrane applications in the food industry. *Membrane technology: membranes for food applications*, *3*, 1-24.
73. Pearce, R. J. (1992). Whey protein recovery and whey protein fractionation. In *Whey and lactose processing* (pp. 271-316). Springer, Dordrecht.
74. Piot, M., Fauquant, J., Madec, M. N., & Maubois, J. L. (2004). Preparation of serocolostrum by membrane microfiltration. *Le Lait, 84*(4), 333-341.
75. Fauquant, J., Maubois, J. L., & Pierre, A. (1988). Microfiltration du lait sur membrane minerale. *Technique laitière et marketing*, (1028), 21-23.
76. Caron, A., St-Gelais, D., & Pouliot, Y. (1997). Coagulation of milk enriched with ultrafiltered or diafiltered microfiltered milk retentate powders. *International Dairy Journal, 7*(6-7), 445-451.
77. Maubois, J. L. (2002). Membrane microfiltration: a tool for a new approach in dairy technology. *Australian Journal of Dairy Technology, 57*(2), 92.
78. Samuelsson, G., Dejmek, P., Trägårdh, G., & Paulsson, M. (1997). Minimizing whey protein retention in cross-flow microfiltration of skim milk. *International Dairy Journal, 7*(4), 237-242.
79. Klein, N., & Lortal, S. (1999). Attenuated starters: an efficient means to influence cheese ripening—a review. *International Dairy Journal, 9*(11), 751-762.

80. Neocleous, M., Barbano, D. M., & Rudan, M. A. (2002). Impact of low concentration factor microfiltration on the composition and aging of Cheddar cheese. *Journal of dairy science, 85*(10), 2425-2437.
81. Daviau, C., Pierre, A., Famelart, M. H., Goudédranche, H., Jacob, D., Garnier, M., & Maubois, J. L. (2000). Characterisation of whey drainage kinetics during soft cheese manufacture in relation with the physicochemical and technological factors, pH at renneting, casein concentration and ionic strength of milk. *Le Lait, 80*(4), 417-432.
82. Bohdziewicz, J. (1996). Ultrafiltration of technical amylolytic enzymes. *Process biochemistry, 31*(2), 185-191.
83. Sztajer, H., & Bryjak, M. (1989). Capillar membranes for purification of Pseudomonas fluorescens lipase. *Bioprocess Engineering, 4*(6), 257-259.
84. Schratter, P. (2004). Purification and concentration by ultrafiltration. In *Protein Purification Protocols* (pp. 101-116). Humana Press.
85. Cheryan, M. (1986). *Ultrafiltration handbook.* Technomic Publishing Co. Inc.
86. Maubois, J.-L., Mocquot, G., Vassal, L. (1969). Procede de traitement du lait et de sousproduits laitiers French Patent no. 2,052,121.
87. Mistry, V. V., & Maubois, J. L. (2004). Application of membrane separation technology to cheese production. *Cheese: Chemistry, physics and microbiology, 1*, 261-285.
88. Hannon, J. A., Lopez, C., Madec, M. N., & Lortal, S. (2006). Altering renneting pH changes microstructure, cell distribution, and lysis of Lactococcus lactis AM2 in cheese made from ultrafiltered milk. *Journal of dairy science, 89*(3), 812-823.
89. Kelly, P., Horton, B.S., Burling, H. (1992). Partial demineralization of whey by nanofiltration in new applications of membrane processes. *IDF* special issue no. 9201, pp. 130–140.
90. Gregory, A. G. (1987). Desalination of sweet-type whey: salt drippings for whey solids recovery. *International Dairy Federation.*.G
91. Kiesner, C., & Eggers, R. (1994). Concept of a sterile concentration process for milk by multistage flash evaporation. *Chemical Engineering & Technology: Industrial Chemistry-Plant Equipment-Process Engineering-Biotechnology, 17*(6), 374-381.
92. Kelly, P. M. (1987). Application of membranes to whole milk separations: a review. *Irish Journal of food science and technology, 11*(2), 153-161.
93. Eriksson, P. (1977). Concentration of whey by reverse osmosis-inventory and operating experiences. *Nordeuropaeisk Mejeri-Tidskrift, 43*(7).
94. Glover, F. A., P. J. Skudder, P. H. Stothart, and E. W. Evans. 1978. Reviews of the progress of dairy science: reverse osmosis and ultrafiltration in dairying. J. Dairy Res. 45:291.
95. Johnston, A. N. *1977.* How European cheese plants save costs in whey processing through reverse osmosis system. Dairy and Ice Cream Field 160:1160.

96. Short, J. L., & Hughes, I. R. (1978). The concentration of separated milk by reverse osmosis. *New Zealand Journal of Dairy Science and Technology*.
97. Zydney, A. L. (2016). Continuous downstream processing for high value biological products: a review. *Biotechnology and bioengineering, 113*(3), 465-475.
98. Cheryan, M. (1998). Ultrafiltration and Microfiltration Handbook (Technomic, Lancaster, USA).
99. Decloux, M., & Prothon, F. (1998). Fruit juices, vegetable juices and sugar juices. *Membrane Separations in the Processes of Food Industry*, 473-501.
100. Girard, B., & Fukumoto, L. R. (2000). Membrane processing of fruit juices and beverages: a review. *Critical Reviews in Food Science Nutrition, 40*(2), 91-157.
101. Hicks, D. (1990). *Production and packaging of non-carbonated fruit juices and fruit beverages* (No. 663.63 H5).
102. Heatherbell, D. A., Short, J. L., & Strubi, P. (1977). Apple juice clarification by ultrafiltration. *Confructa, 22*(5/6), 157.
103. Rwabahizi, S., & Wrolstad, R. E. (1988). Effects of mold contamination and ultrafiltration on the color stability of strawberry juice and concentrate. *Journal of Food Science, 53*(3), 857-861.
104. Drake, S. R. and Nelson, J. W., Apple juice quality as influenced by ultrafiltration. J. Food Quality, 9, 399, 1986.
105. Bengtsson, E., Trägårdh, G., & Hallström, B. (1989). Recovery and concentration of apple juice aroma compounds by pervaporation. *Journal of food engineering, 10*(1), 65-71.
106. Iglinski, B., & Buczkowski, R. (2008). Uses of electrodialysis through a bipolar membranes in industry and environmental protection. *Przemysl Chemiczny, 87*(1), 43-47.
107. Vera, E., Sandeaux, J., Persin, F., Pourcelly, G., Dornier, M., & Ruales, J. (2009). Modeling of clarified tropical fruit juice deacidification by electrodialysis. *Journal of membrane science, 326*(2), 472-483.
108. Lopez Leiva, M. H. (1988). The use of electrodialysis in food processing. Part 2: Review of practical applications. *Lebensmittel-Wissenschaft+ Technologie, 21*(4), 177-182.
109. Adhikary, S. K., Harkare, W. P., Govindan, K. P., Chikkappaji, K. C., Saroja, S., & Nanjundaswamy, A. M. (1987). Deacidification Of Fruit Juices By Electrodialysis. 2. *Indian Journal of Technology, 25*(1), 24-27.
110. Trägårdh, G. (1991). Membrane applications in the food industry. *Polymer Journal, 23*(5), 521-529.
111. Kothapalli, A., Hayes, K., Sadler, G., & Morgan, M. (2007). Comparison of Kinetic Profile of Free and Immobilized Glucose Oxidase, Immobilized on Low-Density Polyethylene Using UV Polymerization. *Journal of food science, 72*(9), C478-C482.
112. Vera, E., Sandeaux, J., Persin, F., Pourcelly, G., Dornier, M., Piombo, G., & Ruales, J. (2007). Deacidification of clarified tropical fruit juices by

electrodialysis. Part II. Characteristics of the deacidified juices. *Journal of food engineering, 78*(4), 1439-1445.
113. Quoc, A. L., Mondor, M., Lamarche, F., Ippersiel, D., Bazinet, L., & Makhlouf, J. (2006). Effect of a combination of electrodialysis with bipolar membranes and mild heat treatment on the browning and opalescence stability of cloudy apple juice. *Food research international, 39*(7), 755-760.
114. Mondor, M., Quoc, A. L., Lamarche, F., Ippersiel, D., & Makhlouf, J. (2005). Assessing ionic transport during apple juice electro-acidification: influence on system efficiency. *Journal of membrane science, 246*(2), 217-226.
115. Wucherpfennig, K., & Keding, K. (1982). Deacidification and acidification of stone, pome and berry juices and grape must by electrodialysis. *Flussiges Obst, 49*, 590-601.
116. Adhikary, S. K., Harkare, W. P., Govindan, K. P., & Nanjundaswamy, A. M. (1983). Deacidification of fruit juices by electrodialysis. *Indian Journal of Technology, 21*(3), 120-123.
117. Lutin, F. (2000). Electrodialysis as a purification technology in the sugar industry. *Zuckerindustrie, 125*(12), 982-984.
118. El Khattabi, M. O., Hafidi, M. R. A., & El Midaoui, A. (1996). Reduction of melassigenic ions in cane sugar juice by electrodialysis. *Desalination, 107*(2), 149-157.
119. Lutin, F., Bailly, M., & Bar, D. (2002). Process improvements with innovative technologies in the starch and sugar industries. *Desalination, 148*(1-3), 121-124.
120. Thampy, S. K., Narayanan, P. K., Trivedi, G. S., Gohil, D. K., & Indushekhar, V. K. (1999). Demineralization of sugar cane juice by electrodialysis. *International sugar journal, 101*(1207), 365-366.
121. Elmidaoui, A., Chay, L., Tahaikt, M., Sahli, M. M., Taky, M., Tiyal, F., ... & Hafidi, M. R. A. (2006). Demineralisation for beet sugar solutions using an electrodialysis pilot plant to reduce melassigenic ions. *Desalination, 189*(1-3), 209-214.
122. Yasinzai, M. (2007). Immobilization of oxidases and their analytical applications. *Pakistan Journal of Analytical & Environmental Chemistry, 8*(2), 5.
123. Daufin, G., Escudier, J., Carrere, H., Berot, S., Fillaudeau, L., and Decloux, M. (2001) Food Bioprod. Process., 79, 89–102.
124. DEESLIE, W. D., & CHERYAN, M. (1992). Fractionation of soy protein hydrolysates using ultrafiltration membranes. *Journal of food science, 57*(2), 411-413.
125. Bérot, S., Chaufer, B., Basso, Y., Legay, C., & Popineau, Y. (1999). Fractionation of gliadin hydrolysates in water-ethanol by ultrafiltration with modified or unmodified membranes. *Biotechnology and bioengineering, 62*(6), 649-658.
126. Fillaudeau, L., Boissier, B., Moreau, A., Blanpain-avet, P., Ermolaev, S., Jitariouk, N., & Gourdon, A. (2007). Investigation of rotating and vibrating filtration for clarification of rough beer. *Journal of Food Engineering, 1*(80), 206-217.

127. Hansen, N. L. (2001). From filter presses over centrifuges to cross flow and vibrating membrane filtration: a review of recovery of beer from surplus yeast. *Technical quarterly*.
128. Burrell, K. J., & Reed, R. J. R. (1994). Crossflow microfiltration of beer: Laboratory-scale studies on the effect of pore size. *Filtration & separation*, *31*(4), 399-386.
129. Bühler, T. M., McKechnie, M. T., & Wakeman, R. J. (1996). Temperature induced particle aggregation in mashing and its effect on filtration performance. *Food and Bioproducts Processing*, *74*(4), 207-211.
130. Leeder, G. (1993). Cold sterilization of beer. *Brauwelt international (Germany)*.
131. Blanpain-Avet, P., Fillaudeau, L., & Lalande, M. (1999). Investigation of mechanisms governing membrane fouling and protein rejection in the sterile microfiltration of beer with an organic membrane. *Food and bioproducts processing*, *77*(2), 75-89.
132. Stkwart, D. C., Hawthorne, D., & Evans, D. E. (1998). Cold sterile filtration: a small scale filtration test and investigation of membrane plugging. *Journal of the Institute of Brewing*, *104*(6), 321-326.
133. Jiao, B., Cassano, A., & Drioli, E. (2004). Recent advances on membrane processes for the concentration of fruit juices: a review. *Journal of food engineering*, *63*(3), 303-324.
134. Mietton-Peuchot, M., Milisic, V., & Noilet, P. (2002). Grape must concentration by using reverse osmosis. Comparison with chaptalization. *Desalination*, *148*(1-3), 125-129.
135. Ferrarini, R., Versari, A., & Galassi, S. (2001). A preliminary comparison between nanofiltration and reverse osmosis membranes for grape juice treatment. *Journal of Food Engineering*, *50*(2), 113-116.
136. Santos, F. R., Catarino, I., Geraldes, V., & Pinho, M. N. (2008). Concentration and rectification of grape must by nanofiltration. *American journal of enology and viticulture*, *59*(4), 446-450.
137. Feuillat, M. (1987). Stabilisation et clarification des vins: aspects colloidaux. Rev. Oenol, 45(8).
138. Serrano, M., Pontens, B., & Ribéreau-Gayon, P. (1992). Study of different tangential microfiltration membranes. Comparison with precoat filtration of diatoms. *OENO One*.
139. Ferreira Cameira Dos Santos, PJ (1995). Clogging by tangential microfiltration: demonstration of interactions between polysaccharides and polyphenols in wine and polymer membranes.
140. Vernhet, A., Bellon-Fontaine, M. N., Brillouet, J. M., Roesink, E., & Moutounet, M. (1997). Wettigg properties of microfiltration membrane: determination by means of the capillary rise technique and incidence on the adsorption of wine polysaccharide and tannins. *Journal of membrane science*, *128*(2), 163-174.

141. Gonçalves, F., Fernandes, C., dos Santos, P. C., & de Pinho, M. N. (2003). Wine tartaric stabilization by electrodialysis and its assessment by the saturation temperature. *Journal of Food Engineering*, *59*(2-3), 229-235.
142. Chabeaud, A., Vandanjon, L., Bourseau, P., Jaouen, P., Chaplain-Derouiniot, M., & Guérard, F. (2009). Performances of ultrafiltration membranes for fractionating a fish protein hydrolysate: Application to the refining of bioactive peptidic fractions. *Separation and Purification Technology*, *66*(3), 463-471.
143. Afonso, M. D., & Borquez, R. (2002). Review of the treatment of seafood processing wastewaters and recovery of proteins therein by membrane separation processes—prospects of the ultrafiltration of wastewaters from the fish meal industry. *Desalination*, *142*(1), 29-45.
144. Salomäki, M., Tervasmäki, P., Areva, S., & Kankare, J. (2004). The Hofmeister anion effect and the growth of polyelectrolyte multilayers. *Langmuir*, *20*(9), 3679-3683.
145. Bennett, A. L. (2016). *Manipulating antibody-antigen interactions in microporous membranes for selective antibody and protein purification*. Michigan State University. Chemistry.
146. Ji, J., Zhang, Y., Zhou, X., Kong, J., Tang, Y., & Liu, B. (2008). Enhanced protein digestion through the confinement of nanozeolite-assembled microchip reactors. *Analytical chemistry*, *80*(7), 2457-2463.
147. Gao, J., Xu, J., Locascio, L. E., & Lee, C. S. (2001). Integrated microfluidic system enabling protein digestion, peptide separation, and protein identification. *Analytical Chemistry*, *73*(11), 2648-2655.
148. Liuni, P., Rob, T., & Wilson, D. J. (2010). A microfluidic reactor for rapid, low-pressure proteolysis with on-chip electrospray ionization. *Rapid Communications in Mass Spectrometry*, *24*(3), 315-320.
149. Wang, G., Chen, X., Liu, S., Wong, C., & Chu, S. (2016). Mechanical chameleon through dynamic real-time plasmonic tuning. *Acs Nano*, *10*(2), 1788-1794.
150. Sproß, J., & Sinz, A. (2010). A capillary monolithic trypsin reactor for efficient protein digestion in online and offline coupling to ESI and MALDI mass spectrometry. *Analytical chemistry*, *82*(4), 1434-1443.
151. Slysz, G. W., & Schriemer, D. C. (2005). Blending protein separation and peptide analysis through real-time proteolytic digestion. *Analytical chemistry*, *77*(6), 1572-1579.
152. Cooper, J. W., Chen, J., Li, Y., & Lee, C. S. (2003). Membrane-based nanoscale proteolytic reactor enabling protein digestion, peptide separation, and protein identification using mass spectrometry. Analytical chemistry, 75(5), 1067-1074.
153. Dong, J., & Bruening, M. L. (2015). Functionalizing microporous membranes for protein purification and protein digestion. Annual Review of Analytical Chemistry, 8, 81-100.

154. Keating IV, J. J., Bhattacharya, S., & Belfort, G. (2018). Separation of D, L-amino acids using ligand exchange membranes. Journal of Membrane Science, 555, 30-37.
155. Gude, V. G. (2015). Energy storage for desalination processes powered by renewable energy and waste heat sources. Applied Energy, 137, 877-898.
156. Khayet, M. (2013). Solar desalination by membrane distillation: Dispersion in energy consumption analysis and water production costs (a review). Desalination, 308, 89-101.
157. Kullab, A., & Martin, A. (2011). Membrane distillation and applications for water purification in thermal cogeneration plants. Separation and Purification Technology, 76(3), 231-237.
158. Qtaishat, M. R., & Banat, F. (2013). Desalination by solar powered membrane distillation systems. Desalination, 308, 186-197.
159. Camacho, L. M., Dumée, L., Zhang, J., Li, J. D., Duke, M., Gomez, J., & Gray, S. (2013). Advances in membrane distillation for water desalination and purification applications. Water, 5(1), 94-196.
160. Tijing, L. D., Choi, J. S., Lee, S., Kim, S. H., & Shon, H. K. (2014). Recent progress of membrane distillation using electrospun nanofibrous membrane. Journal of Membrane Science, 453, 435-462.
161. Woo, Y. C., Tijing, L. D., Park, M. J., Yao, M., Choi, J. S., Lee, S.,& Shon, H. K. (2017). Electrospun dual-layer nonwoven membrane for desalination by air gap membrane distillation. Desalination, 403, 187-198.
162. Wang, P., & Chung, T. S. (2015). Recent advances in membrane distillation processes: Membrane development, configuration design and application exploring. Journal of membrane science, 474, 39-56.
163. Kebria, M. R. S., & Rahimpour, A. (2020). Membrane Distillation: Basics, Advances, and Applications. Advances in Membrane Technologies, 67.

4
Cold Plasma

Rodrigo Nunes Cavalcanti[1], Tatiana Colombo Pimentel[2], Erick Almeida Esmerino[3], Monica Queiroz de Freitas[3], Silvani Verruck[4], Marcia Cristina Silva[1] and Adriano Gomes da Cruz[1]*

[1]Federal Institute of Education, Food Science and Technology of Rio de Janeiro (IFRJ), Department of Food, Rio de Janeiro, Brazil
[2]Federal Institute of Paraná (IFPR), Paranavaí, Paraná, Brazil
[3]Federal Fluminense University (UFF), Faculty of Veterinary, Niterói, Brazil
[4]Federal University of Santa Catarina (UFSC), Department of Food Engineering, Florianópolis, Brazil

Abstract
In the last decades, consumers have shown great interest in products that provide convenience, diversity, long shelflife, low cost and are produced in an environmentally correct way. Cold plasma technology has been use as an alternative for thermal treatments by the food industry. In this chapter we will cover the principles and methods of plasma generation and the cold plasma applications in food systems. Cold plasma technology can modify food components' functionality (such as protein and carbohydrates), can be used for decontamination purposes and also seems promising on enzyme inactivation. Also, cold plasma has been used in the sterilization and modification of polymers in food packaging and the in-package treatment has gained great interest from industries. The biofilm control and treatment of waste, processing surfaces, and effluent is also under investigation. This chapter reviews the mechanisms associated with cold plasma in these applications and discuss future perspectives of its utilization by food industry.

***Keywords*:** Non-thermal technology, food preservation

4.1 Introduction

Traditionally, thermal treatments for pasteurization and sterilization are the methods used to destroy microorganisms and enzymes in food. However,

**Corresponding author*: adriano.cruz@ifrj.edu.br

Navnidhi Chhikara, Anil Panghal and Gaurav Chaudhary (eds.) Novel Technologies in Food Science, (109–170) © 2023 Scrivener Publishing LLC

these technologies are dependent on the processing time and temperature, and they can promote nutrient losses, develop off-flavors, change the color, and deteriorate the sensory characteristic of the food products [1]. In the last decades, consumers have shown great interest in products that provide convenience, diversity, long shelflife, low cost and are produced in an environmentally correct way [2]. As a result, new non-thermal technologies are receiving great interest. Cold plasma technology has been use as an alternative for thermal treatments by the food industry [3–6].

Plasma is considered the fourth stage of matter and is produced by passing a process gas through an electric field at atmospheric pressure [7]. When a sufficient amount of energy achieve the elements dispersed in the phase of gas, they become plasma, an ionized gas [8]. This occurs because the ionization processes can accelerate the electrons by impact. Plasma is composed of both positively and negatively charged ions in almost the same proportion, in addition to contain electrons, atoms, molecules, free radicals, and UV photons [7]. Plasma components and their reactive substances promote interactions with the food surface, and the high energy can dissociate covalent bonds of the organic molecules [8, 9].

The application of cold plasma in food systems can happen, basically, in two different ways. The first with the directly application on the food surface, and the second using plasma-activated water (PAW) as an agent that involves the food to be treated [10]. The applicability of cold plasma treatment in the food chain is high (Figure 4.1). Cold plasma can be used in the cultivation and production of seeds and grains by promoting decontamination and seeds germination and ground [11]. Furthermore, it can be used for decontamination of raw products (almond, lettuce, apple, melon, mangoes, and egg surface) or food products (cheese and cooked meat) [12]. The bacterial inactivation mechanism includes damage caused by electroporation, damage to DNA, lipid peroxidation, protein modulation and induction of apoptosis. The microbial inactivation is associated with the production of free radicals, which attack the components of the cells, and the thinning of the membranes of the cells [13].

Cold plasma has been used in the sterilization and modification of polymers in food packaging and the in-package treatment has gained great interest from industries [14]. The biofilm control and treatment of waste, processing surfaces, and effluent is also under investigation [15]. In addition, cold plasma treatment seems promising on enzyme inactivation and in the modification of food components' functionality (such as protein and carbohydrates) [11].

CP equipment is scalable, but, it has not been commonly used by the food industry because of many issues concerning the process parameters.

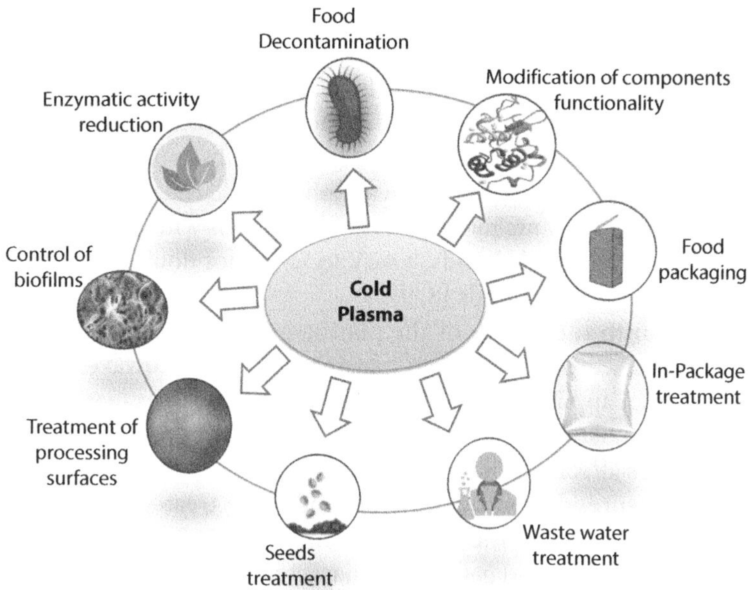

Figure 4.1 Applications of cold plasma in food systems.

Thus, this chapter reviews the mechanisms associated with cold plasma applications and discuss future perspectives of its utilization by food industry.

4.2 Principles and Methods of Plasma Generation

Plasma is defined as the fourth matter stage, while the other three are solid, liquid, and gaseous [7]. When a sufficient amount of energy is applied to the solid matter, it becomes liquid, which in turn, when applied with a greater amount of energy, becomes gas. Thus, it is known that when more energy is additionally applied to the elements dispersed in the gas phase, they become plasma, an ionized gas [8]. Plasma is composed of both positively and negatively charged ions in almost the same proportion, in addition to contain electrons, atoms, molecules, free radicals, and UV photons [7]. There is a differentiation between the plasma that is in its thermodynamic balance and the one that is usually applied for food treatment, cold plasma. Plasma generation via thermal heating requires a large amount of energy and high pressures to maintain thermodynamic balance with neutrons, negative ions, positive ions, and free electrons, and the temperature reached is too high for food processing [8, 9]. The infeasibility of using thermal plasma in the food industry means that studies are turning to using cold plasam, as it is a non-thermal plasma. An electric field is used to create the cold plasma field and it promotes the free electrons acceleration and, therefore, their collision with target chemical species causing the fission of bonds, as well as the formation of activated chemical radicals [9]. The electron energy is can dissociate the organic molecules covalent bonds with an application of 1.5 eV to 6.2 eV for single bonds, 4.4 eV to 7.4 eV for double bonds and 8.5 eV to 11.2 eV for triple bonds [8, 9].

In addition to the choice of the plasma generator (eg, dielectric barrier discharge, corona discharge, plasma jet, or microwave discharge) and the chosen application mode (in-field, off-field, direct, and indirect) other factors must be considered when choosing equipment to use [1]. Several factors affect the plasma formation, such as the choice of gas, gas flow and pressure, power, quantity of food to be treated, characteristics of microorganisms and the surface of the food to be treated, in addition to the packaging and geometric characteristics should be considered (Figure 4.2) [1]. The choice of gas to be transformed into plasma determines the effectiveness of sterilization, as it defines the type of active species formed, in addition to dictating the intensity and wavelengths of radiation emitted. That is, the free radicals type are dependent on the molecules present in the gas. Common gases to be used for this purpose are CO_2, O_2, O_2/Ar, O_2/H_2, and O_2/CF_4 [16]. In addition, the flow rate and gas pressure are other factors that greatly influence the treatment.

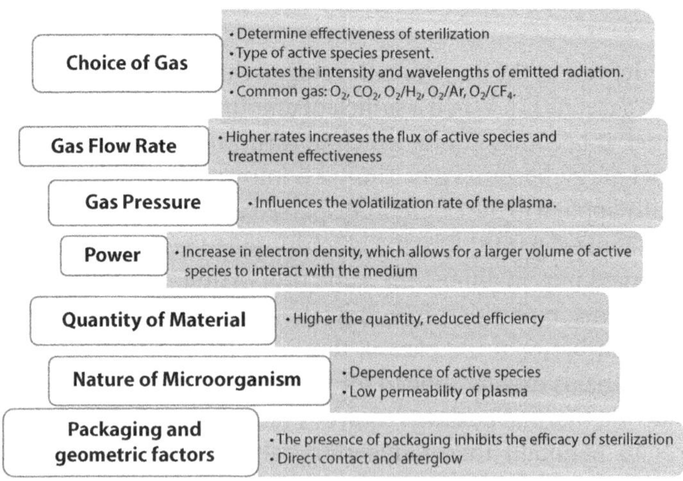

Figure 4.2 Factors that affect plasma treatment.

The treatment effectiveness can be increase by increasing the gas flow rate, and consequently, the number of environment active species [1]. Meanwhile, the pressure of the gas in the system influences the plasma volatilization rate. Another factor that influences the process is the voltage, current, frequency, or wavelength used, because an increase in power increased the density of electrons, and, then, a greater number of active species interact with food [1]. In addition, the amount of material to be sterilized, the microorganism nature and density, and the formation of the surface layer of the food to be treated must be considered [17]. Finally, since it is a superficial treatment, the packaging inhibits the efficacy of sterilization due to the low permeability of the plasma, however, the treatment can be used to direct sterilize packaging. In addition, it should not be forgotten that the geometric factors of the product must be considered since the design of the reactor strongly influences the concentrations of the active species [1].

The source of cold plasma generation can occur with the use of different methods. The most used for food applications are dielectric barrier discharge (DBD), glow discharge, pulse discharge, corona discharge, radiofrequency discharge, and microwave-powered plasma [8, 18].

When plasma is generated through dielectric barrier discharges (DBD), alternating high voltage pulses are applied through two electrodes. The application frequency varies from 10 Hz to several kHz [19]. In order not to form a thermal plasma, the electrodes are covered with some insulating material, which can be glass, ceramic, plastic, or silicone [18]. This dielectric insulation interrupts the current flow generated after the electrical breakdown, thus suppressing the arc discharge (thermal plasma) and allowing uniform distribution of discharges throughout the electrode area [20]. Generally, a post-luminescence discharge is generated with DBD, where the generated plasma is expelled from inside the chamber [8]. A method known as a plasma jet is also being tested (Figure 4.3). In this method, gases (helium or argon) are utilized to expel the plasma from the tube through a high gas flow. On the other hand, gases are generally easily flammable and can also make the use of the technology more expensive [18].

The generation of plasma through a glow discharge is used as a method of generating large volumes of plasma, using low temperatures [20]. Spatially, the luminescent discharge generates very homogeneous plasma when low pressures (1 to 1000 Pa) and voltages ranging from 100 V to several kV are applied [8]. Another form of cold plasma generation is the use of pulse discharges, where electrical pulses of short duration are applied under voltage of 100 V/ns or more [8]. As with the use of glow discharge,

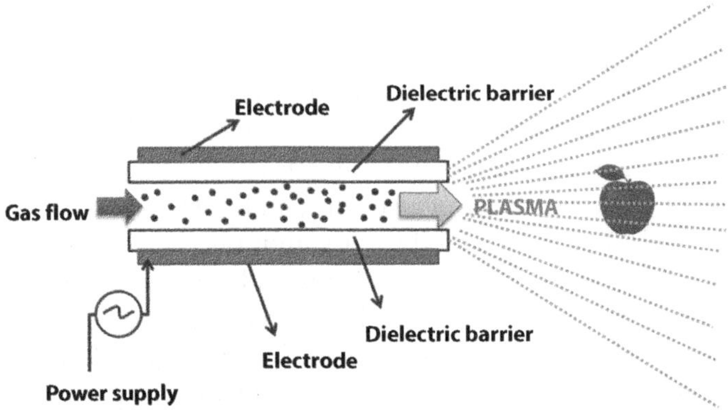

Figure 4.3 Cold plasma generation by dielectric barrier discharge. Adapted from [8].

the energy of electrons is highen compared to that of neutral gases and ions, since pulse discharge acts mainly on them [20].

The generation of cold plasma through corona discharge is a well-described method. Corona discharge is generated from electrical discharges of relatively low power, which occur at or near atmospheric pressure [21]. In such cases, the electrode composed of small diameter wires, needles, or sharp edges when associated with strong electric fields generates a crown of filamentary discharges that radiate out of the discharge electrode [8]. In addition, it is possible to use radiofrequency discharge (usually 13.56 MHz) to generate cold plasma. The shape of the electrodes or antenna allows a large amount of highly dense plasma to be generated when a high-frequency discharge is used, under atmospheric pressure conditions. The use of this method is recommended for products that do not want the surface to be damaged, in addition to being able to be customized according to the size of the food [8].

When plasma is generated using microwaves, the electrodes commonly used by other methods are replaced by antennas or waveguides [8]. Generally, microwave-generated plasma has the same mechanism as the plasma jets discussed earlier (with the high gas flow). Discharges are then generated by electromagnetic radiation, preferably at the frequency of 2.45 GHz (which can vary between 300 MHz and 300 GHz) and in the power range of 10 W-6 kW [18].

4.3 Cold Plasma Applied in Food Systems

4.3.1 Modification of Food Components Functionality

The quantity of both charged ions (negatively and positively), in addition to atoms, electrons, free radicals, molecules, and UV photons formed in plasma treatment and the reactions formed between them is quite complex [7]. Additionally, when all of these plasma components come into contact with a food matrix, many other interactions can occur. Modifications in only one compound of the plasma generation system or a food system can generate a very different behavior in comparison to previous studies [11]. Interactions of plasma components with food matrices are being studied and their effects on carbohydrates, proteins, lipids, water, vitamins, and minerals are being reported [22]. Figure 4.4 represents the main interactions between the components of the plasma on a model food surface. The application of plasma to a food produces reactive species and they come into contact with the molecules present in that environment, which generates a high amount of ROS and RNS. When they reach the surface of the food, they can be absorbed, and chemical reactions with the matrix are initiated or even show a slight migration into the matrix. In addition, the formed molecules can be reabsorbed into the gas phase, even causing the occurrence of new reactive species [11].

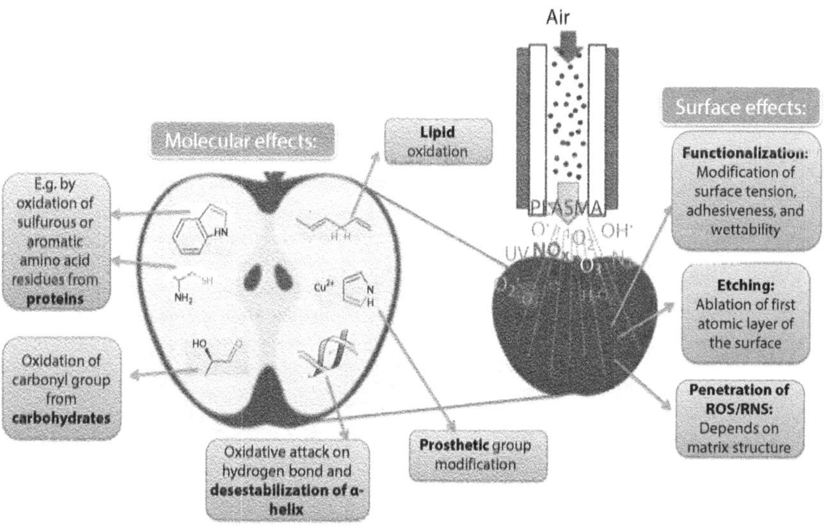

Figure 4.4 Main interactions between the components of the plasma and a food model. Adapted from [11].

When plasma is applied directly to the food, interaction phenomena that occur are limited to a few μm under the food surface, depending mainly on its porosity, water activity, and constituents [22]. Most of the effects on food are observed only on the contact surface, since, for example, ROS such as hydroxyl radicals and singlet oxygen have a half-life of only 1 ns and 1 μs, respectively. This also causes the depth of penetration of the radicals formed to be limited because of the half-life time [23]. In addition, cross-reactions between molecules can be formed, while the degradation of molecules and their by-products can trigger chain reactions on the surface of the food [11].

On the other hand, when cold plasma treatment is used in liquid foods or foods treated with plasma-activated water (PAW) the reactions that occur can be intensified [11]. Due to the interaction of plasma components with water, a series of reactions and formation of active species can occur (Figure 4.5) [7].

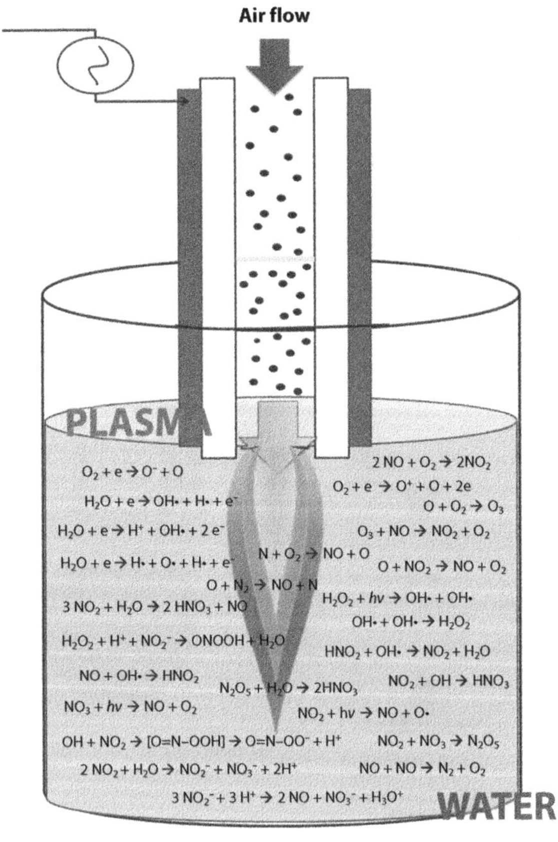

Figure 4.5 Reactions and formation of active species in plasma-activated water (PAW). Adapted from [7].

Recently, indirect plasma exposure (PAW) has been used to replace traditional methods of chlorinating water to sanitize fresh food without leaving residues in products [24]. In addition, PAW has been used to increase germination and plant development since reactive species formed from chemical donors seem to have a beneficial influence on the plant growth process [25]. This large number of active species in a aquous environment results in greater penetrability of the active species in food, as well as substantially modify the functionality of carbohydrates, proteins, and lipids in food. With the use of PAW, it is possible to visualize amino acids oxidation, hydrogen bonds rupture, and modifications of prosthetic groups with greater intensity [11] (Figure R).

Considering the diversity of reactions that can occur between plasma and food components, Table 4.1 provides a summary of recent studies on the functional changes that can occur when cold plasma is used in food components or complex matrices. In carbohydrates, degradation of reducing and non-reducing sugars by cold plasma has been reported, as well as changes in high molecular weight oligosaccharides [18]. One of the components that have been investigated is starch since in its native form this ingredient has some limitations to be used [26]. Generally, native starches have limited solubility at low temperatures, in addition to having high viscosity and undesirable retrogradation effects in most products in which they are applied [27]. After modifying its functionalities, starch becomes a highly valued ingredient and has a wide capacity for use in various food matrices. The cold plasma application to native starches has been studied, and active plasma species interact with amylose and amylopectin causing desirable configuration changes in these molecules [26]. Configuration changes can increase its expansion power, its hydrophilic and hydrophobic characteristics, its solubility, adhesion characteristics, its susceptibility to enzymes, in addition to causing structural and thermal changes [27]. The ozone present in the plasma is one of the agents responsible for the depolymerization of the molecules since it cleaves glycosidic bonds. In addition, as with hydroxyl radicals, it has the characteristic of oxidation of carbonyl functional groups and probable incorporation of other functional groups [18]. In addition, oxidation of aldoses and ketosis was observed, leading to the formation of formic acid. The addition of a hydroperoxide anion oxidizes aldohexose (for example glucose) leading to the formation of aldopentose and formic acid, which modifies the functional characteristics of the matrix [11].

Considering the above, the main starch products studied with cold plasma are legumes and cereals [18]. Among the observed effects the cooking time and cooking properties can be improved, in addition to the cold plasma processing can improve the grain surface characteristics [12].

Table 4.1 Summary of recent studies on the functional changes that can occur when cold plasma is used in food components or complex matrices. Adapted from [12, 22].

Matrix	Power source	Feed gas	Treatment parameters	Main findings	References
Banana starch	Corona discharge	Air	Power = 50 kV/cm Exposure time = 3 min	Pasting temperatures were increased and viscosity was decreased	[26]
Rice starch	Dielectric barrier discharge	Air	Power = 60 W Exposure time = 10 min	Lower tendency to starch gels degradation and higher pasting and final viscosities	[27]
Parboiled rice flour	Radiofrequency	Air	Power = 50 W Exposure time = 15 min	Lower stickiness, hardness, and endothermic enthalpy	[28]
Basmati rice flour	Radiofrequency	Air	Power = 40 W Exposure time = 10 min	Higher flour and gel hydration properties	[29]
Brown rice	Dielectric barrier discharge	Air	Power = 250 W Exposure time = 20 min	Improvements in α-amylase inhibitory activity. Decreases in hardness.	[30]
Brown rice	Radiofrequency	Air	Power = 50 W Exposure time = 10 min	Cooking time was reduced (29.1 min to 21.1 min). Increase in water absorption.	[29]

(Continued)

Table 4.1 Summary of recent studies on the functional changes that can occur when cold plasma is used in food components or complex matrices. Adapted from [12, 22]. (*Continued*)

Matrix	Power source	Feed gas	Treatment parameters	Main findings	References
Parboiled rice	Radiofrequency	Air	Power = 50 W Exposure time = 5 min	Cooking time and hardness were reduced. Water absorption was increased.	[31]
Basmati rice	Radiofrequency	Air	Power = 40 W Exposure time = 5 min	Cooking time and contact angle were reduced. Water absorption was increased.	[32]
Zein	Dielectric barrier discharge	Air	Power = 75 V Exposure time = 10 min	Solubility was improved. Surface hydrophilicity and tensile strength were reinforced.	[33]
Pea protein isolate	Dielectric barrier discharge	Air	Power = 8.8 kV *ww = 10 min	Water holding capacity, fat binding capacity and solubility were increased.	[34]
Wheat flour	Dielectric barrier discharge	Air	Power = 60 and 70 kV Exposure time = 5 and 10 min	Gluten protein secondary structure were modified. Increases in α-helix and β-turns and decreases in β-sheets were observed.	[35]

(*Continued*)

Table 4.1 Summary of recent studies on the functional changes that can occur when cold plasma is used in food components or complex matrices. Adapted from [12, 22]. (*Continued*)

Matrix	Power source	Feed gas	Treatment parameters	Main findings	References
Whey protein	Dielectric barrier discharge	Air	Power = 70 kV Exposure time = 1, 5, 10, 15, 30, 60 min	pH values were decreases. Protein oxidation and increases in surface hydrophobicity and carbonyl groups occurred after 15 min. Protein structure was chenages after 30 and 60 min.	[36]
Raw milk	Corona discharge	Air	Power = 9 kV Exposure time = 0, 3, 6, 9, 12, 15, 20 min	Increases in the total aldehyde contents after 20 min. No effect on total ketone, lipid composition, and alcohol.	[37]
Guava-flavored whey beverage	Glow discharge	Nitrogen	Power = 400 W and 50 kHz Exposure time = 5, 10, and 15 min	Increased concentration of bioactive and volatile compounds. The effect on the fatty acid profile and angiotensin-converting-enzyme inhibitory activity is dependent on the process parameters	[38]

(*Continued*)

Table 4.1 Summary of recent studies on the functional changes that can occur when cold plasma is used in food components or complex matrices. Adapted from [12, 22]. (*Continued*)

Matrix	Power source	Feed gas	Treatment parameters	Main findings	References
Guava-flavored whey beverage	Glow discharge	Nitrogen	Power = 400 W and 50 kHz Exposure time = 5, 10, and 15 min	Higher pH values, lower consistency and lower viscosity, and a flow behavior index similar to Newtonian fluids. Thermal properties were not changed.	[39]
Chocolate milk drink	Glow discharge	Nitrogen	Power = 400 W and 50 kHz Exposure time = 5, 10, and 15 min	Cold plasta at intermediate processing conditions kept the fatty acid profile, bioactive compounds concentration, and health indices, and improved the volatile compounds profile (increases in esters, lactones, and ketones).	[40]
Chocolate milk drink	Glow discharge	Nitrogen	Power = 400 W and 50 kHz Exposure time = 5, 10, and 15 min	Cold plasma increased the partile size and consistency. The melting profile was altered (higher enthalpy and lower bound water and temperature).	[41]

(*Continued*)

Table 4.1 Summary of recent studies on the functional changes that can occur when cold plasma is used in food components or complex matrices. Adapted from [12, 22]. (*Continued*)

Matrix	Power source	Feed gas	Treatment parameters	Main findings	References
Whey dairy beverage with xylooligosaccharide	Glow discharge	Nitrogen	Power = 15 kV Exposure time = 5, 10, and 15 min	Xylooligosaccharide and sugars were maintained. Decreased color intensity and consistency of whey beverages. Higher concentrations of bioactive compounds.	[42]
Pork jerky	Dielectric barrier discharge	Air	Power = 3.8 kV. Exposure time = 0, 20, 40, 60 min	Higher treatment times altered the red color and contents of nitrosoheme pigment and residual nitrite, and decreased lipid oxidation.	[43]
Canned ground ham	Dielectric barrier discharge	Air	Power = 7 kV; Frequency: 25 kHz Exposure time = 10, 20, 30, 40, 50, 60 min;	No difference in nitrosyl hemochrome, color, residual nitrite, texture, lipid oxidation, and protein oxidation compared to sodium nitrite treatment	[44]

(*Continued*)

Table 4.1 Summary of recent studies on the functional changes that can occur when cold plasma is used in food components or complex matrices. Adapted from [12, 22]. (*Continued*)

Matrix	Power source	Feed gas	Treatment parameters	Main findings	References
Acerola juice	Glow discharge	Nitrogen	Power = 80 kHz Exposure time = 5, 10, and 15 min	Increased the content of vitamins A (retention higher than 95%) and the carotenoid content and the color of the juice has improved.	[45]
Orange juice	Dielectric barrier discharge	Air	Power = 70 kV Exposure time = 15, 30, 45, 60 s	No changes in organic acids, amino acids, and sugar contents.	[46]
Orange juice	Dielectric barrier discharge	Air	Power = 70 kV; Frequency: 25 kHz Exposure time = 15, 30, 45, 60 s	Degradations of some compounds (p-cymene and limonene) and formation of others (terpinene-4-lol, linalool, and α-terpineol)	[47]
Sucrose, fructose, ketose, nystose, and fructofuranosyl nystose	Dielectric barrier discharge	Air	Power = 70 kV Exposure time = 15, 30, 45, 60 s	The contents of fructooligosaccharides, fructose, and sucrose were not altered.	[48]

(*Continued*)

Table 4.1 Summary of recent studies on the functional changes that can occur when cold plasma is used in food components or complex matrices. Adapted from [12, 22]. (*Continued*)

Matrix	Power source	Feed gas	Treatment parameters	Main findings	References
Cashew apple juice	Radiofrequency	Nitrogen	Power = 80 kV Exposure time = 5, 10, 15 min	Lower gas flow rate application resulted in products with higher total polyphenol content, vitamin C, and antioxidant activity. Higher gas flow rate application resulted in products with higher flavonoids and polyphenols content, but a degradation of bioactive compounds occurred.	[49]
Blueberries	Dielectric barrier discharge	Air	Power = 60 and 80 kV Frequency: 50 kHz Exposure time = 0, 1, 5 min	Nutritional properties are not impaired using 60 kV, for 5 min or 80 kV for 1 min. Decreases in vitamin C content was observed for longer processing times.	[50]

The corrosion and depolymerization of starch by ROS that occurs on the surface of the food, as well as the consequent increase in the binding sites with water, proved to be the main mechanism that induced the black grass to absorb more water [51]. Additionally, decreases in cooking time (30.25 to 20.45 min) and grain hardness (22.50 to 12.36 N) were observed. In another study with brown rice, the same behavior was observed, and the decrease in cooking time related to the polar groups absorption and increased degree of starch gelatinization [29]. In addition, cold plasma processing changed the functional, structural and rheological characterisricts of rice starch, as it was observed that the amylose content decreased, as well as the paste and gelatinization temperature [27]. In addition, increased phenolic compounds content were also observed after treatment of parboiled rice flour with cold plasma [28]. In general, most of these studies also observed an increase in luminosity values, as well as few changes in the proximate composition of the analyzed matrices. These observations lead to the belief that few changes in the nutritional and sensory profile exist when cold plasma treatment was applied to grains [12].

The effect of cold plasma treatment is also being evaluated in relation to proteins and protein foods. The effect of cold plasma specifically on enzymes will be described in a separate section. In whey proteins, an improvement in the foaming and emulsification capacity was observed when cold plasma treatment was applied for 15 min [36]. These functional changes in proteins were correlated with a strong oxidative effect of ROS and RNS on cysteine. In addition, the interaction of proteins with ROS and RNS led to an increase in surface hydrophobicity and carbonyl groups, while free SH groups decreased, pointing to effects on amino acid residues [36]. Finally, the dynamic light scattering also showed differences, with color changes (higher yellow color intensity), and decreases in the pH values [36].

Another protein complex that has been studied is gluten and its functionality. The nature of proteins and their functionality has been changed for gluten processes with cold plasma. Wheat flour processing with cold plasma resulted in more stable gluten secondary structure, which promoted changes in dough rheology [35]. The treatment conditions, time of exposure, and applied voltage to plasma were factors that influenced the rheological characteristics of viscosity and elasticity of the analyzed wheat flour. In this work, the main component of plasma that affects strong and weak wheat flours was ozone, which was generated with DBD plasma and air as the working gas [35].

When protein-rich pea flour was subjected to cold plasma, it was observed that the technological and functional properties were modified [34].

With the use of plasma treatment, the water and fat binding capacity increased by 113% and 116%, respectively. In addition, an increase of 191% was observed in the solubility of the flour [34]. The intensity of fluorescence emission of proteins (328-355 nm) was dependent on the processing time. This effect was correlated with the conformational transformation of tryptophan present in proteins [34]. In addition, there was a decrease in the pH of the flour that can be attributed to acidification that occurs from NO via NO_2 to nitric and nitrous acid [52]. When cold plasma was applied to *Tenebrio molitor* flour and compared to heat treatment, it was observed that the treatment with cold plasma led to greater stability of the protein composition in pH 4, but decreased the capacity of binding with water and increased the capacity connection with oil [53]. Meanwhile, the increased free sulfhydryl group concentration was associated with a increase in the zein solubility after cold plasma processing. This occurred because of the interaction of the protein surface with the active species O_2^-, O_2^+, H_3O^+ and, consequently, the increase in its hydrophilicity [33].

Another important effect that is being investigated is the cold plasma effect on allergenic proteins. Promising studies have been carried out with two soy allergens, β-conglycinin glycinin (Gly m6) and (Gly m5) [54] and milk allergens (β-lactoglobulin casein, and α-lactalbumin) [55]. What happens in these situations is that some reactions that occur between proteins with the components of the plasma can modify linear epitopes by their fragmentation, while the conformational epitopes are modified by the formation of crosslinking and aggregates [12]. In addition, it is possible that peptide bonds break, side-chain oxidation and fragmentation as a result of the OH radicals action. The cleavage of disulfide bonds in peptides is caused by the production of RSH and RSO· through the inclusion of a hydroxyl radical [56]. In addition, reactive nitrogen species, ozone (O_3), superoxide radical (O_2^-), hydroxyl radical (OH·), hydrogen peroxide (H_2O_2), when introduced into a liquid environment, are able to destroy binding sites of antibodies as they attack easily oxidizable amino acids [54].

When it comes to the interaction of cold plasma with lipids, it can be said that it is almost always unwanted due to the oxidative reactions that occur in an accelerated way [11]. Cold plasma contains a high reactive species number, such as $^1O_2$ and $OH^{\cdot}$, whose primary targets are methyl groups of fatty acids. Even greater action is observed when the methyl group is located between double bonds, as in this case, the energy required to abstract the hydrogen atom is only 272 kJ/mol compared to 422 kJ/mol elsewhere in the molecules [57]. The phenomenon of lipid oxidation is generally divided between the stages of initiation, propagation and termination since there is the formation of peroxy radicals and their formation of hydroperoxide and another radical through the breaking of CH bonds [58]. When there is the presence of radical-eliminating antioxidants in the food or the recombination of radicals, this chain reaction is interrupted. However, if these substances do not exist, the chain reaction is continued and hydroperoxides lead to the breakdown of lipids or bonds between fatty acids [11]. In addition, along with lipid peroxidation, volatile carbonyl compounds were formed such as cis-4-heptanal pentanal, and trans-6-nonenal, which may be associated with negative sensory impacts on these products [11]. The negative effects of accelerated lipid oxidation by the use of cold plasma were recently reviewed in wheat flour, rice, beef, pork, seafood, cheese, sushi, olive oil, and milk [59]. In general, this is not a problem for the processing of fruits and vegetables, however, it can be inconvenient in products with a high lipid content, such as meat, dairy and oilseeds [18]. On the contrary, cold plasma application has already been used to simulate the lipid oxidation accelation in real food matrices, as a way of assessing its stability [60, 61]. In addition, cold plasma has already been used in soybean oil partial hydrogenation, resulting in no *trans* fatty acids formation, showing to be a promising treatment compared to the conventional hydrogenation process [62]. Finally, cold plasma can be used for producing biodiesel using frying oils because of its ability to rapidly esterify [63].

4.3.2 Cold Plasma Mechanisms Involved in Microbial Inactivation

Food microbial decontamination is a common problem for several products during the entire food chain. The high porosity, different formats and high resistance of some microorganisms are generally considered to be the main problems during decontamination. Furthermore, with conventional methods that use chlorinated compounds, the waste and toxic substances formed can be a problem [10, 64]. Along to that, heat treatments are not always indicated for some types of food. Considering the above, cold plasma has been used as a very efficient alternative in the decontamination of food [3–6]. As previously described, the plasma can be applied directly to the food, at the plasma-air-surface interface or also through the plasma-activated water (PAW) in the plasma-water-surface contact of the food [3]. Generally, when PAW is used the microbial inactivation capacity increases due to better permeation of ROS and RNS into the microbial cells [65]. The main antimicrobial effect of plasma on bacteria, fungi, and spores is related to the interaction of these ROS and RNS generated by the plasma with cell membranes or spore coatings [3, 4]. In addition, the diffusion of ROS in viral capsids has already been reported, leading to its inactivation due to RNA damage [10, 64]. The effects of microbial inactivation of *Salmonella* sp., *Escherichia coli*, *Staphylococcus aureus*, *Enterococcus faecalis*, *Listeria monocytogenes*, among others, by cold plasma has already been reported in several types of food and are summarized in Table 4.2.

Reactive oxygen species formed by cold plasma treatment (hydroxyl radicals (·OH), hydrogen peroxide (H_2O_2), atomic oxygen (O), ozone (O_3), and singlet oxygen ($^1O_2$)) can act on cell membranes in many ways, causing cell leakage, irreversible DNA damage, inducing lipid peroxidation, protein denaturation and also interfering with cell metabolism [3, 4, 13]. Likewise, UV radiation, RNS, charged particles, and energy ions also act on them [98, 99]. As a way of surviving external stresses, bacteria have developed a series of protective responses, such as adaptation to the environment, change to the form of endospores, activation of stress-resistant genes, adaptive mutations, production of biofilms, and induction of viable, but not cultivable (VBNC) form [100]. Biofilm formation serves as external protection for ROS attack and will be discussed in a topic below. In addition, several authors have reported that if the damage caused by the plasma bombing of bacteria is not strong enough, under developmental conditions, many of them are able to restore their mechanisms with the consequent ability to grow and reproduce, maintaining their pathogenicity [4]. This recovery behavior of VBNC cells, after treatment with cold plasma, has already been observed for *Staphylococcus aureus* [101, 102], *Escherichia coli* [103, 104], *Listeria monocytogenes* [102, 105], *Salmonella enterica, Pseudomonas fluorescens* [102], and *Brochothrix thermosphacta* [94]. This occurs because when plasma components come into contact with the cell membrane or its internal components, a complex antioxidant system inside the cell is activated, in addition to activating the regulation of gene expression to prevent further ROS damage to cells [4]. This occurs, for example, with the OxyR and SoxR genes that are stimulated when the cell comes in contact with hydrogen peroxide and superoxide anion, respectively [106–108]. In this same sense, the bacteria's SOS repair system is activated when damage is caused by ROS in cellular DNA [4].

On the other hand, when cold plasma treatment is more drastic and well controlled (usually seen with PAW), microbial inactivation is irreversible. In this case, there is a higher concentration of ROS, RNS and UV photons permeating the cell membrane, which strongly induces membrane rupture, lipid peroxidation, membrane leakage, alteration of morphology, damage to RNA/DNA and protein [3–6]. Figure 4.6 shows the cold plasma bacterial

Table 4.2 Cold plasma effects on food microbial inactivation. Adapted from [5, 12, 13].

Food matrix	Microorganisms	Plasma source	Treatment conditions	Observation	References
Mandarin	*Penicillium italicum*	Microwave discharge	P = 900 W, ET = 10 min	84% reduction in disease incidence	[66]
Radish sprouts	*Salmonella* Typhimurium	Microwave Discharge	P = 900 W, ET = 20 min	Reductions higher than 2.6 log CFU/g	[67]
Romaine lettuce	*Escherichia coli* O157:H7	Dielectric barrier discharge	V = 42.6 kV, ET = 10 min	*E.coli* counts reduction in 0.4–0.8 log CFU/g	[68]
Onion Powder	*Bacillus cereus*, *Aspergillus brasiliensis*, *Escherichia coli*	Microwave discharge	P = 400 W, ET = 40 min	Reductions from 1.6 to 2.1 log spores/cm^2	[69]
Corn salad, apples, cucumber, and tomatoes	*Escherichia coli*	Atmospheric pressure plasma jet	F = 1.1 MHz, V = 2 to 6 kV ET = 60 s	*E. coli* counts reduction up to 4 log units for 20 s of processing time	[70]

(Continued)

Table 4.2 Cold plasma effects on food microbial inactivation. Adapted from [5, 12, 13]. (*Continued*)

Food matrix	Microorganisms	Plasma source	Treatment conditions	Observation	References
Pomaine lettuce, cocktail tomatoes, and baby carrots	*Escherichia coli*	Atmospheric pressure plasma	F = 60 Hz V = 3.95 to 12.83 kV ET = 30 s to 10 min	The food surface had an impact on microbial inactivation. Lettuce was a easier matrix for disinfection	[71]
Golden Delicious apples	*Escherichia coli*	Gliding arc cold plasma	F = 60 Hz V = 15 kV ET = 1, 2, or 3 min	Decreases of 3.4-3.6 log CFU/ml	[72]
Strawberries	*S. aureus*	Plasma Activated Water	F = 10 kHz V = 18 kV ET = 10 and 20 min	Reduction of *S. aureus* for 1.6 to 2.3 log CFU/g	[73]
Lettuce and cabbage	*L. monocytogenes*	Cold oxygen plasma	P = 10, 15, and 30W ET = 5 min	Reduction of >1 log CFU/cm^2 on lettuces, and >2 log on cabbages.	[74]

(*Continued*)

Table 4.2 Cold plasma effects on food microbial inactivation. Adapted from [5, 12, 13]. (*Continued*)

Food matrix	Microorganisms	Plasma source	Treatment conditions	Observation	References
Cabbage, dried figs, and lettuce	*L. monocytogenes*	Microwave cold plasma	P = 400 to 900 W ET = 1 to 10 min	Maximum reduction of 1.8 ± 0.2 log CFU/g	[75]
Fresh-cut dragon fruit	*L. monocytogenes*	Atmospheric radiofrequency plasma	F = 20 to 600 kHz P = 40 W ET = 60 s	Reduction of about 5 to 6 log CFU/g.	[76]
Strawberries and cherry tomatoes	*E. coli, Salmonella sp., L. monocytogenes*	Dielectric barrier discharge	F = 50 Hz V = 120 kV ET = 0 to 300 s	Reduction of *Salmonella, E. coli*, and *L. monocytogenes* populations by 3.8 , 3.5, and 4.2 log CFU/sample, respectively	[77]

(*Continued*)

Table 4.2 Cold plasma effects on food microbial inactivation. Adapted from [5, 12, 13]. (*Continued*)

Food matrix	Microorganisms	Plasma source	Treatment conditions	Observation	References
Spinach	*Salmonella sp.*	Oxygen radiofrequency plasma	F = 13.56 MHz P = 600 W ET = 0 to 800 s	Oxygen plasma had a higher microbial inactivation capacity against *S. typhimurium* LT_2 than washing with 3% H_2O_2	[78]
Golden Delicious apples	*Salmonella sp.*	Gliding arc cold plasma	F = 560 Hz V = 15 kV ET = 1, 2, or 3 min	*Salmonella* Stanley inactivation was in the range of 2.9-3.7 log CFU/ml	[72]
Lettuce and strawberry	*Salmonella* Typhimurium	Nitrogen plasma jet	F = 1 kHz P = 1 W ET = 28 hr	Reductions of 2.71 log CFU/g were achieved with 2 min processing time	[79]

(*Continued*)

Table 4.2 Cold plasma effects on food microbial inactivation. Adapted from [5, 12, 13]. *(Continued)*

Food matrix	Microorganisms	Plasma source	Treatment conditions	Observation	References
Apple juice	*Escherichia coli* O157:H7	Dielectric barrier discharge	P = 50 W ET = 30 s	Maximum reduction of 4.3 log CFU/mL	[80]
Apple Juice	*Zygosaccharomyces rouxii*	Dielectric barrier discharge	P = 90 W ET = 140 s	Maximum reduction of 5 log CFU/mL	[81]
Apple Juice	*Escherichia coli* O157:H7	Dielectric barrier discharge	V = 9 kV F = 100 Hz ET = 40 s	Reductions of >5 log CFU/mL	[82]
Orange juice	*Staphylococcus aureus* *Escherichia coli* *Candida albicans*	Dielectric barrier discharge	P = 1.14 W/cm^2 ET = 12 s, 8 s, and 25 s	Maximum reduction of >5 log CFU/mL	[83]
Orange Juice	*Salmonella enterica*	Dielectric barrier discharge	V = 90 kV F = 60 Hz ET = 120 s	Maximum reduction of 4.7 log CFU/mL	[84]

(Continued)

Table 4.2 Cold plasma effects on food microbial inactivation. Adapted from [5, 12, 13]. (*Continued*)

Food matrix	Microorganisms	Plasma source	Treatment conditions	Observation	References
White grape juice	*Saccharomyces cerevisiae*	Dielectric barrier discharge	V = 80 kV F = 60 Hz ET = 4 min	Reductions of 7.4 log CFU/mL	[85]
Apple juice, tomato juice, orange juice, and sour cherry nectar	*Escherichia coli*	*Atmospheric Jet Plasma*	P = 650 W ET = 120 s	Reductions in the range of 1.4-4 log CFU/mL and dependent on the food matrix.	[86]
Blueberry juice	*Bacillus sp.*	*Atmospheric Jet Plasma*	F = 1000 Hz V = 11 kV ET = 6 min	Maximum reduction of 7.2 log CFU/mL	[87]
Fresh tomato juice	Total aerobic mesophilic bacteria, molds, and yeasts	*Gliding Discharge Plasma*	P = 40 W V = 3.8 kV F = 50 Hz ET = 5 min	Reductions in the range of 3.5- >5 CFU/mL and dependent on the microorganism type	[88]

(*Continued*)

Table 4.2 Cold plasma effects on food microbial inactivation. Adapted from [5, 12, 13]. (*Continued*)

Food matrix	Microorganisms	Plasma source	Treatment conditions	Observation	References
Pasteurized tomato juice	*Candida albicans Saccharomyces cerevisiae*	*Gliding Discharge Plasma*	P = 40 W V = 3.8 kV F = 50 Hz ET = 1 min, 5 min	Reductions in the range of 3.5-3.7 log CFU/mL and dependent on the microorganism type	[88]
Barley grains and wheat	*Escherichia coli*	High voltage dielectric barrier discharge plasma system	V = 80 kV F = 50 Hz ET = 20 min	Reductions in the range of 3.5-3.7 log CFU/mL and dependent on the food matrix	[89]
Groundnuts	*Aspergillus parasiticus, Aspergillus flavus*	Radiofrequency plasma	P = 60 W F = 13.56 MHz ET = 5, 10, 15 min	97.9% and 99.3% reduction, respectively	[90]
Chicken eggs	*Salmonella enteritidis*	High voltage atmospheric cold plasma	V = 85 kV ET = 15 min F = 60 Hz	Reductions up to 5.53 log CFU/egg	[91]

(*Continued*)

Table 4.2 Cold plasma effects on food microbial inactivation. Adapted from [5, 12, 13]. (*Continued*)

Food matrix	Microorganisms	Plasma source	Treatment conditions	Observation	References
Egg shell	*Salmonella* Typhimurium	Dielectric barrier discharge	V = 25–30 kV ET = 10 min and 25 min F = 10–12 kHz	Complete inactivation of the microorganism	[92]
Egg shells	*Salmonella* Enteritidis	Atmospheric pressure plasma jet	V = 2–3 kV ET = 5 min F = 1 MHz	Reductions in the range of 0.22-2.27 log CFU/egg	[93]
Lamb meat	*Brochothrix thermosphacta*	Dielectric barrier discharge	V = 80 kV ET = 5 min F = 50 Hz	Reductions of 2 log cycles	[94]
Vacuum packaged beef loin	*Staphylococcus aureus, Escherichia coli, Listeria monocytogenes*	Dielectric barrier discharge	F= 9 kHz P = 29.9 W ET = 1 min	≥2 log reduction	[95]

(*Continued*)

Table 4.2 Cold plasma effects on food microbial inactivation. Adapted from [5, 12, 13]. (*Continued*)

Food matrix	Microorganisms	Plasma source	Treatment conditions	Observation	References
Dried squid shreds	*Aerobic bacteria, Staphylococcus aureus, marine bacteria*	Corona Discharges	V = 20 kV F = 58 kHz ET = 3 min	Reductions in the range of 0.9-2.0 log units and dependent on the microorganism type	[96]
Fresh Cheese	*Listeria innocua* and *Escherichia coli* K-12	High voltage atmospheric cold plasma	V = 60 kV, 80 kV, and 100 kV ET = 5 min	At 100 kV (5min), a reduction of 1.4 log10 CFU/g and 3.5 log10 CFU/g was observed for *L. innocua* and *E. coli* K-12 after 24 h storage at 4 °C, respectively	[97]

P = Power, V = Voltage, F = frequency, ET = exposure time.

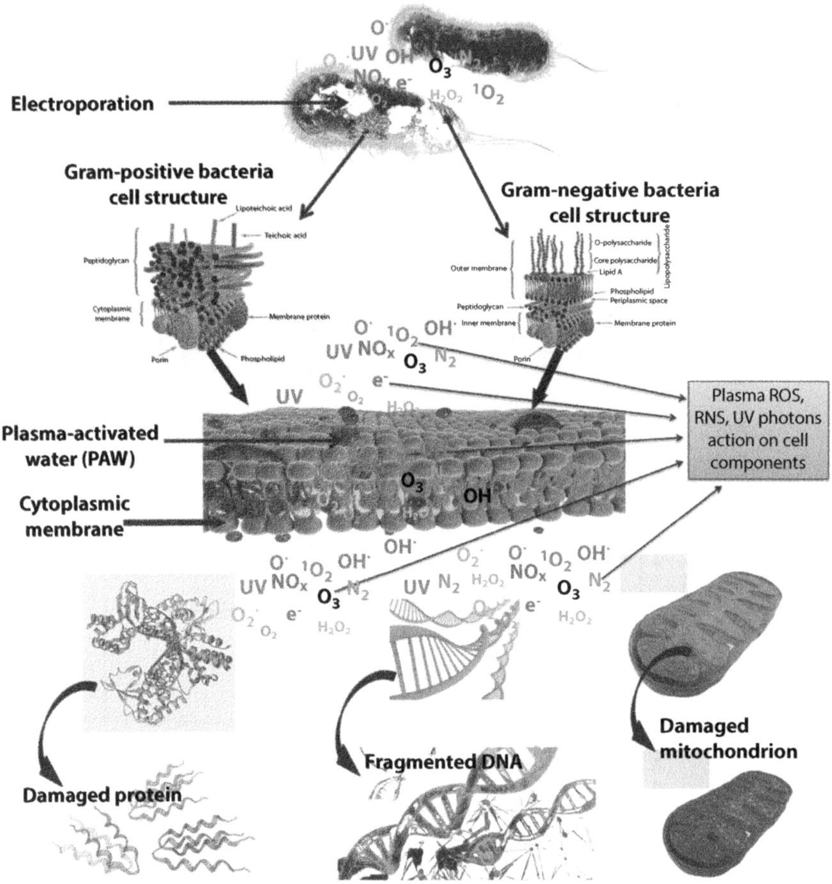

Figure 4.6 Cold plasma mechanism of microbial inactivation in the presence of water-activated plasma. Adapted from [3].

inactivation mechanism (in this case also in the presence of water-activated plasma). At the beginning, the effects of electroporation caused by the high-intensity field used and lipid peroxidation on the lipopolysaccharide wall present in the cell membrane are observed [3, 109]. For lipid peroxidation, the oxidation of lipid bilayers (especially those containing polyunsaturated fatty acids [PUFAs]) by the radicals •OH and CO_3• generated through OONO- [110, 111]. In addition, singlet oxygen ($^1O_2$) and H_2O_2 are also responsible for lipid peroxidation of the membrane [112]. Associated with these effects, the ROS and RNS that are formed also have the ability to cross the membrane of peptidoglycans acting on other components of cells [7]. Therefore, it is common to observe the cell membrane rupture and bacteria internal content leakage [3]. In this case, it was reported that smaller ions were first released from the cytoplasm, followed by DNA/RNA and then by proteins [110]. This occurs, for example, with intracellular inactivation caused when unloaded peroxynitric acid (ONOOH) enters the cell and decomposes to NO_2• and other radicals •OH which are highly toxic to cell components such as RNA/DNA, proteins, mitochondria, and ribosomes [109]. Deoxyguanosine and deoxyadenosine adducts are formed in the presence of malonaldehyde (at the end of the lipid peroxidation that occurs), which leads to cell death [113]. The effects on food proteins have been reported previously and occur in the same way here on cell proteins, with the destruction of hydrogen, sulfide bonds and peptide oxidation by ROS [114]. However, it is important to highlight that the sensitivity to cold plasma is different for Gram-negative and Gram-positive bacteria because of the thickness and composition of the cell membranes. The viable cell of Gram-positive bacteria has higher membrane thickness and because of that is less sensitive to cold plasma compounds while Gram-negative almost always has its membrane destroyed by the treatment [114]. In addition, plasma components can also inhibit mitochondrial respiration and promote genetic mutation, ribosomal damage, necrosis and cell apoptosis [109]. Finally, when PAW is used, it has been reported that the influx of water into the cell leads to cell swelling causing hydropic degeneration which can also be an event responsible for microbial inactivation [115].

4.3.3 Decontamination of Mycotoxins and Pesticides By Cold Plasma

Mycotoxins are produced by toxigenic fungi when they have the conditions, mainly humidity and temperature, to develop. These fungi strongly attack the production of grains and cereals worldwide, although these are not the only foods involved with mycotoxin contamination [116, 117]. Mycotoxins, when inhaled, ingested, or skin absorbed, cause illness, reduced performance, and/or death of mammals (humans and other animals) and poultry [118]. Depending on their classification and chemical structure, mycotoxins produce toxic reactions even in small concentrations, in addition to having a cumulative effect, with serious consequences such as being carcinogenic, teratogenic, mutagenic, hepatotoxic, among others, and can then cause long-term effects [119–121]. The high concentrations of these compounds found in a wide variety of foods make them of great interest in the area of food toxicology, and methods of detoxifying contaminated foods have been tested over the past few years [117].

Aflatoxin detoxification is one of the most studied when cold plasma treatment is applied, due to its high toxicity and wide distribution [118]. In addition, aflatoxin B1 is classified by the International Agency for Research on Cancer (IARC) as a group 1 carcinogen [122]. Aflatoxins (represented mainly by AFB1, AFB2, AFG1, AFG2 and AFM, the latter a hydroxylated metabolite of AFB1) have in their chemical structure, a difuran ring generally responsible for its basic toxicity, and coumarin, responsible for its carcinogenic effect [123]. Cold plasma was responsible for the C8 = C9 double bond loss in the furan ring, thus decreasing its toxicity. Therefore, mycotoxin is degraded by the plasma reactive species (O_3, O, NO, OH, and NO_2) in addition to the presence of UV photons that also act on them [124]. The efficiency of plasma on mycotoxin degradation varies according to the molecular structure of the toxin, the gas type and the interaction between mycotoxin molecules and reactive species [124, 125]. Table 4.3 shows the efficiency of inactivation of aflatoxins in some studies with cold plasma. The plasma efficiency on mycotoxins also depends on operational factors such as voltage, frequency, exposure time and interaction with the food matrix [125].

As occurs with mycotoxins, other residual molecules such as pesticides are taking cold plasma as an alternative for removing it from contaminated food. Pesticides and their various classes have been used extensively in primary

Table 4.3 Efficiency of inactivation of aflatoxins in some studies with cold plasma in food.

Aflatoxins	Food	Maximum degradation (%)	References
Aflatoxins	Peanut	23% for 2 min and 38% for 5 min	[126]
AFB1	Maize	62% for 1 min and 82% for 10 min	[127]
AFB1, AFB2, AFG1, AFG2	Hazelnut	More than 70% for 12 min	[128]
AFB1, AFB2, AFG1, AFG2	Hazelnut	72-73% for 30 min	[129]
AFB1	Rice and wheat	More than 95% for 30 min	[130]
AFB_1	Maize	65% for 10 min	[131]

food production [125]. However, the use of classes not recommended in certain foods or residual amounts above those legally permitted is a constant problem [132]. The consumption of high doses of these residues is harmful to health and, therefore, methods capable of degrading the residues of these compounds in food are of great interest [12]. Cold plasma can be used to degradate pesticides, such as imidacloprid and boscalid in blueberries [50], endosulfan, malathion, and dichlorvos in water [28], and fludioxonil, azoxystrobin, pyriproxyfen and cyprodinil in strawberries [133]. Table 4.4 shows the inactivation efficiency of some pesticides in studies with cold plasma. Studies have shown that reactive species generated by cold plasma, i.e. H^-, H^+,

Table 4.4 Inactivation efficiency of some pesticides in studies with cold plasma.

Pesticide	**Food**	**Maximum degradation**	**References**
Malathion Chlorpyrifos	Water	64.6% 62.7%	[135]
Carbaryl Methiocarb Aminocarb	Water	50.5% 99.6% 99.3%	[136]
Phoxim	Water	74%	[137]
Alachlor	Water	80%	[138]
Dichlorvos Malathion Endosulfan	Water	78% 69% 57%	[28]
Malathion Chlorpyrifos	Lettuce	59.0% 57.9%	[135]
Dichlorvos Omethoate	Wolfberry	96.8% 99.6%	[139]
Cypermethrin Chlorpyrifos	Mango	63% 74%	[140]
Boscalid Imidacloprid	Blueberries	75% 80%	[50]
Azoxystrobin Cyprodinil Fludioxonil Pyriproxyfen	Strawberries	69% 45% 71% 46%	[133]

O^-, O^+, OH^-, H_3O^+, N_2^+, H_2O_2, N_2, O_3, O_2, NO•, OH•, H•, O•, and UV photons, attacked regions of the pesticide molecules and modified or degraded them [28, 124]. For example, malathion was oxidized by plasma oxygen species, causing the P = S bond to be transformed into P = O, forming diethyl-2-(dimethoxyphosphorylthio) maleate (malaoxon) [134]. However, the cold plasma effectiveness on the pesticide degradation varies according to the gas type and composition and the pesticide type and structure [12].

4.3.4 Cold Plasma Mechanisms Involved in Enzyme Inactivation

The enzymatic browning of fresh-cut vegetables and fruits affects the quality of these products. In this sense, cold plasma could be a potential technology to inactivate the enzymes responsible for this problem, once cold plasma induces chemical and physical reactions and can be used in temperature-sensitive surfaces. Reviewed works clearly demonstrate that the generated reactive species of plasma can impact food enzymes and alter some of their conformations and structures which could lead to change in the function of the enzyme [141].

At present, plasma has already been used for inactivating food endogenous enzymes, such as peroxidase (POD), polyphenol oxidase (PPO), superoxide dismutase (SOD), pectin methylesterase (PME), lipase, and alkaline phosphatase in foods and model foods [98]. As can be seen in Table 4.5 POD and PPO are the enzymes more studied in different food matrixes. Alkaline phosphatase, lactate dehydrogenase were studied only in a enzyme solution model. In general, the inactivation mechanism by cold plasma is associated with amino acids oxidation (polypeptide chains or free aminoacids), polypeptide chain depolymerization, and protein secondary structural destruction. Cold plasma consists of ultraviolet (UV) photons, ions, free electrons, and reactive species (RNS and ROS). The presence of tyrosine as the free amino acid or in the polypeptide chain is the primary target for the ROS and RNS during the oxidative and nitrative stress conditions [142]. The applications of cold plasma in the field enzyme inactivation include aminoacids residues degradation and Wang *et al.* [143] observed using UV spectroscopy analysis that dielectric barrier discharge plasma treatment produced ROS and RNS and increased the acidity of the solution leading to the protein unfolding after treatment. Also, hydroxyl radicals, superoxide anion radicals, hydroperoxy radicals and nitric oxide generated can chemically modify the amino acids reactive side-chain causing loss of enzyme activity and damage to alfa-helix and beta-sheet structures [144]. Thus, amino acid side chains oxidation or hydrogen bonding disruption in the protein tertiary structure are needed to damage the active site of the enzyme native structure [145].

Furthermore, some cold plasma process parameters are important for inducing protein modification or denaturation and enzyme inactivation, such as the gas type, plasma source, protein/enzyme type, food matrix, and food volume. Furthermore, excitation frequency, voltage and time are also important parameters. Studies performed by Castro *et al.* [146] reported that the increase in excitations frequencies resulted in increases in endogenous enzyme inactivation (POD and PPO), with a consequent decrease in enzyme activity in juices. On the other hand, in studies performed by Farias *et al.* [147] reported that the increase in excitations frequencies resulted in decreases in endogenous enzyme inactivation (POD) in apple cubes. The higher POD activity was associared with the generation of hydrozyl radicals due to cold plasma application, and a consequent water dissociation.

In addition, the changes in the enzyme structure are dependent on processin time, and the higher the processing time the higher the enzyme structure modification. An increase in processing times decreased the residual POD in tomato samples [148]. Also, as the treatment time increased, the content of plasma active substances level increase and the inactivation effect gradually growths as occurred on the activity of litchi POD [143], PPO of cloudy apple juice [149], laccase [150] and orange juice [84].

Intrinsic factors also seems to affect enzymes stability. The free energy, which is a thermodynamic parameter Gibbs, may be used in quantitative description for protein conformation. Cold plasma application may decrease the enthalpy, the agreggation and prptides stretch of peanut protein isolate, and destruct the enzyme structure [151]. Furthermore, Choi *et al.* [152] observed that the thenodynamic lysozyme stability was impacted by cold plasma processing.

In summary, the cold plasma can be used to inactivate food enzymes and improve the food nutritional quality. However, it is limited to application to sliced samples or liquid foods, as the enzymes are associated with other compounds in the food products and the plasma penetrability is low. In this way, the equipment selection and optimization of the process parameters should be done for each food product to perform enzyme inactivation at the highest efficiency.

4.3.5 Cold Plasma for Food Packaging

Cold plasma was initially used to improve the printability and adhesion in food packaging. However, in the recent uears, it proved to be an important tool for surface decontamination of food packaging materials.

Table 4.5 Enzyme cold plasma inactivation in food.

Food matrix	Enzymes	Plasma source	Treatment conditions	Observation	References
Camu-camu juice	POD and PPO	dielectric barrier atmospheric	F = 200-960 Hz V = 24 kV	Residual activity of POD about 60 % and PPO about 70 %	[146]
Model food system	POD and PPO	dielectric barrier atmospheric (Argon)	F = 1.1 MHz V = 65 V ET = 0-360 s	Residual activity of 25 % and 30 % for PPO and POD, respectively, in 180 s	[56]
Tomato	POD	dielectric barrier discharge (He)	F = 20 kHz V = 14 kV	Residual activity about 3.5 % for POD in 5 s	[148]
Apple and potato	POD and PPO	plasma processed air	F = 2.45 GHz P = 1.2 kW	Apple = Residual activity about 45 % and 50 % for PPO and POD, respectively (5 min) Potato = Residual activity about 20 % and 40 % for PPO and POD, respectively (5 min)	[153]
Apple cubes and apple juice	POD and PPO	dielectric barrier discharge	F = 50, 200, 400, 600, and 900 Hz V = 20 kV ET = 15 min	Cold plasma partially inactivated POD in juice regardless of the plasma excitation frequencies, and in apple cubes when 50 or 200 Hz were used.	[147]

(Continued)

Table 4.5 Enzyme cold plasma inactivation in food. (*Continued*)

Food matrix	Enzymes	Plasma source	Treatment conditions	Observation	References
Cloudy apple juice	PPO	dielectric barrier discharge	F = 20-65 kHz V= 20 kV ET = 1-5 min	Residual activity about 55 and 16 % after glow and spark treatments, respectively.	[149]
Tender coconut water	POD and PPO	dielectric barrier discharge	F = 50 Hz V = 0–50 kV	Residual activity about 10 % for POD and PPO in 5 min	[154]
Green coconut water	POD	atmospheric cold plasm	F = 730 Hz V = 15-20 kV ET = 15 min	Residual activity about 28 % for POD and PPO in 5 min	[155]
Litchi	POD	dielectric barrier discharge	F = 50 kV, V = 10 kHz ET = 0-10 min	activity about 50 % for POD and PPO in 10 min	[143]
Banana	POD and PPO	dielectric barrier atmospheric	F = 10 kHz V = 40 V	Reduction in the enzimantic activitity about 38% for PPO and 35% for POD in 2 min	[156]

(*Continued*)

Table 4.5 Enzyme cold plasma inactivation in food. (*Continued*)

Food matrix	Enzymes	Plasma source	Treatment conditions	Observation	References
Wheat germ	Lipase and lipoxygenase	atmospheric cold plasma	F = 6 kHz V = 10–26 kV ET = 35 min	Residual activity about 66 % and 89 % for lipase using argon and nitrogen (5 min) and 42 % and 77 % for lipoxygenase using argon and nitrogen (30 min)	[157]
Wheat germ	Lipase and lipoxygenase	atmospheric cold plasma	F = 6 kHz V = 10–26 kV ET = 35 min	Residual activity about 42 % for lipase and 87 % for lipoxygenase using argon	[158]
Enzyme solution	Lactate dehydrogenase	atmospheric-pressure dielectric barrier discharge	F = 10 - 42kHz V = 0 - 50 kV	Decrease of nearly 36–37% in lactate dehydrogenase activity after the plasma treatment for 300 s	[159]
Fresh-cut melon pieces	Pectin methylesterase	atmospheric cold plasma	F = 12.5 kHz V = 15 kV ET = 60 min (30 min on each side)	Decreased the residual activity by 7%	[160]

(*Continued*)

Table 4.5 Enzyme cold plasma inactivation in food. (*Continued*)

Food matrix	Enzymes	Plasma source	Treatment conditions	Observation	References
Enzyme solution	Alkaline phosphatase	atmospheric cold plasma	F = 40, 50, 60 kHz ET =180 s	Residual activity below 10%	[144]
Enzyme solution	Laccase	corona plasma jet	F = 20 kHz V = 8 kV ET = 60 s	Residual activity about 8 %	[150]
Orange juice	Pectin methyl-esterase (PME)	dielectric barrier discharge	F = 60 Hz V = 90 kV ET = 30, 60, and 120s	Residual activity about 30 %. The increase in treatment time reduced PME activity.	[84]
Strawberry fruits	Pectin methyl-esterase (PME) and PPO	Cold atmospheric plasma	F = 42 kHz V = 6 kV ET = 10 min	During 17 days of storage at 1 °C, the lowest PME activities were observed in CAP-processed strawberries. Control samples presented higher PPO activities compared to nonthermal treated samples after 17 days of storage.	[161]

P = Power, V = Voltage, F = frequency, ET = exposure time.

In this context, there are several studies that report the CP potential in in this field. Plasma is suitable to polymers for surface cleaning, surface ablation or etching, crosslinking and modification of functional properties. Reseraches with protein films, for example, cold plasma increased surface roughness, ink adhesion, contact angle and biodegradability [162–164]. Another vision to improve bio-based polymer properties is the use of CP and the incorporation of hydrophobic components, as waxes or oils. In films, the use concomitant of CP and waxes reduced the water vapor permeability and solubility, which are important properties for food packaging materials [165, 166]. These effects could be explained by the energetic particles bombardment in CP, including ions, electrons, excited atoms/molecules, and radicals on the polymer surface, therefore altering the surface properties, as surface roughness. An equilibruim of the effects caused by plasma induction may be controled by regulating the type of plasma, determined by the type of gas (oxygen, nitrogen, air, or argon) and the conditions to generate plasma [163, 164]. Therefore, understanding the effects of plasma treatment and the parameters operation is essential to the satisfactory final result in the surface packing.

Non-thermal processing of foods in post-packaging is a new trend to perform the in-package decontamination. Meanwhile packaging generally represents the final stage of the production process, a subsequent decontamination phase is critical to eradicating any contaminants in the products or packaging materials. In this sense, CP sterilization proposals present some benefits, such as, shortened process time, low cost, elimination of toxic input chemicals and residues, and minimal effects on product quality [14, 136]. A summary of studies on the sterilization of packaging materials by cold plasma has been published by [167]. Also, several studies have been conducted on foods and model foods using different configurations of cold plasma in-package by Misra *et al.* [168].

Previous studies have reported in-package decontamination (yeast, bacteria, toxins, and spores) using several cold plasma sources. Furthermore, the *L. monocytogenes, E. coli,* and *Staphylococcus aureus* were reduced by 6-8 log on growth support media [94, 168, 169]. The effective inactivation of bacterial spores, bacteria, and biofilms using in-package surface and volumetric DBD plasma were reported in previous studies. Nevertheless, the studies need to be performed in real food systems (see below) to acquire more accurate data. Also, it seems that the CP exposure time influence the post-treatment storage extension, and it needed for maximum microbial inactivation [168]. Hence, it is recommended to optime the treatment time based on the maximum reactive species production.

Studies also was carried outo in fresh products. The in-package CP process seems to be imperative for vegetables, fruits, and ready-to-eat foods [168, 170], as ambient temperature are used to guaranteeing safety, with slight influence on the quality of the product. Reductions of microorganisms was observed and thein-package approach did not affect product quality. Noteworthy, the application of CP processing on strawberries reduced the *E. coli* and *L. innocua* counts by 2.0 and 1.3 log CFU/mL, while the reduction were of 2.2 and 1.7 log CFU/mL for spinach, respectively [171]. In mixed vegetables packed in plastic packages after CP (DBD) were observed inactivation rate of indigenous aerobes about 1.2 log CFU. An increase in the volume ratio of headspace to sample from 30:1 to 43:1 and the use of secondary packaging led to an increase in the *Salmonella* reduction, while tensile properties, transparency, glass transition temperature, and surface morphology of packed did not change [172].

CP technology have shown to be also important for meat safety in different meat matrices, such as chicken meat [136, 173], pork [43], beef [174] and lamb [94]. Moutiq *et al.* [173] and Lee *et al.* [30] reported that CP processing could reduce the counts of *Escherichia coli, Listeria monocytogenes,* aerobic bacteria, and *Salmonella typhimurium.* Furthermore, it controlled the growth of *psychrotrophs, mesophiles,* and *Enterobacteriaceae* during storage (24 days). Nevertheless, Wang *et al.* [174] and Huang *et al.* [175] demonstrated that when high DBD voltages were used, the pork color was impacted (decreases in redness), as well as, acceleration in lipid oxidation and myoglobin reactions, and discoloration of beef patties occurred. The increase in processing times (1-30 min) also increased the lipid oxidation on the surface of dried black mouth angler [176].

CP processing showed no significant impact on firmness, color, TSS, and pH values of vegetavles and fruits, and the products were stable durin shelf life period. The CP kept the quality characteristics of strawberries [171]. Similar results were observed in cherry tomatoes, with maintenance of the firmness when 100 kV and 2.5 min of CP processing were used [177]. On the other hand, in other products may have their quality attributes modified by CP. A voltage increase may accelerate the hydrogen peroxide formation and stimulate lipid oxidative reactions [174, 177, 178]. Therefore, microbial decontamination efficiency is dependent on the contamination nature, plasma arrangement, and food matrix. However, it is not known if these results could also be observed at industrial scale, as few studies have been conducted.

4.3.6 Cold Plasma in Biofilms and Surfaces Treatment

Biofilms are communities of microrganisms that stay attached to each other. Therefore, it could be formed by single or multiple species, and are established by the formation of EPS, and also because of the presence of DNA and proteins [179]. Tha inactivation of biofilms tooks longer time compared to planktonic cells because they are more resistant to inactivation methods. Jahid *et al.* [179] observed that CP processing for 5 s was able to inactivate higher than 5 log of planktonic *Aeromonas hydrophila,* while 5 min were required for the same efficiency when the microorganism was in a biofilm form.

Many researches are looking for strategies for food biofilm inactivation and CP results are promising. Patange *et al.* [104] studies the CP efficacy against several pathogens in fresh produce. The authors evaluated planktonic cultures, monoculture biofilms, and mixed culture biofilms of *Pseudomonas fluorescens, Escherichia coli, Listeria monocytogenes,* and *Salmonella enterica).* The utilization of 80 kV and 60 s of CP processing could reduce the biofilms, while the complete inactivation was achieved with 120 s. Han *et al.* [180] reported that the increase in voltage levels could produce a higher number of ROS and reduce the inactivation times.

El Kadri *et al.* [181] reported that the flow rate was an important parameter to be evaluated, and they observed that a higher inactivation rate of *L. innocua* and *E. coli* was achieved with higher air flow rates. The authors also reported that the microorganism in the biofilm state were more resistant than in planktonic form, and the CP was more efficienty in young biofilms. Similar results were observed by [182] and [105].

The efficienty of CP on biofilm inactivation is dependent on the active and lived reactive species [105, 183], as these species enter the biofilm and reacts with bacteria causing their death. Furthermore, they can prevent the formation of new biofilm by promoting the collapse of the biofilm structure [183, 184] .

In addition to biofilm inactivation, CP technologies have been studied discussed to process anti-biofilm contact materials for foods. In these way, different methodologies can be used such as the surface immobilization of the antimicrobial agents, the coating deposition (modifying the physicochemical characteristics), and the composite deposition with the embedded biocidal agent [185]. The second alternative is considered the most promising for food applications, as it can avoid the initial bacterial attachment and preclude the biofilm formation at early stages. Furthermore, this methodology can act as a permant anti-biofilm, as it dependents on the surface physicochemical characteristics and not on the release of agents.

Considering the production of antibacterial surfaces, the APPJ and DBD systems are the most recommended [185].

4.3.7 Cold Plasma in Wastewater Treatment

The significant impact of pollutants on the human health and environment is a huge concern, and legislative demands are becoming stricter worldwide. Therefore, some approaches have been proposed in the last decades to treat polluted effluents, as the cold plasmas that are useful to modify thermal-sensitive materials at operation temperatures lower than 40 °C or even at room temperature. In this sense, this technology has gained attention as a novel advanced oxidation processes (AOPs) method which brings considerable environmental and economical advantages for the degradation of organic contaminants.

In fact, cld plasma (CP) can bring about both physical and chemical effects over the performance of the treatment process. CP can produce diverse reactive species involving ions (H^+, H_3O^+, H^-, O^+, OH^-, N_2^+), molecular species (O_3, H_2O_2), and radicals ($O^\bullet$, $H^\bullet$, $OH^\bullet$, $NO^\bullet$) which contribute to efficient degradation of contaminants, known as chemical effects and ultraviolet (UV) radiation, shock wave, cavitation effects, and high temperature pyrolysis are some example of positive physical effects of non-thermal plasmas.

Researchers showed different approaches to wastewater treatment using CP. Rowan *et al.* [186] used a pulsed plasma gas-discharge system for decontamination of chilled poultry wash water and reported that at treatment times of 24 s and temperatures of 4 °C the microbial counts could be reduced to non-detectable levels. Kazemi *et al.* [187] studied the chemical reactive species involved in the decomposition of crystal violet (CV), a model organic compound, by a novel multi array dielectric barrier discharge plasma gas diffuser. Experimental results exhibited that for the best operating conditions decolororization, energy efficiency, and CV decolorization rate constant were 95%, 7.8 g/kWh, and 2.83 min. The same authors related that O_3 play the most key role in degradation of CV molecules in air plasma discharge. A CP treatment at continuous mode was used to treat meat and dairy effluents. A reduction in microbial counts was obsersed during 120 s of processing, but the complete inactivation was achieved with 300 s. Moreover, CP processing for 20 min realized reductions (≥ 2 Log) of *Bacillus megaterium endospores* counts in wastewater effluent [188].

4.4 Conclusions

Cold plasma technology has been applied in the food industry as a way to replace conventional thermal treatments and in the development of new products for the food indutry. In this chapter it was covered the principles and methods of plasma generation and the cold plasma applications in food systems. Cold plasma treatment is widely used in the modification of food components' functionality (such as protein and carbohydrates), while their functional compounds such as vitamins are preserved. The studies with decontamination purposes are one of the largest areas of application of cold plasma, having been tested on a wide range of fresh and processed products and also with various pathogenic microorganisms. The degradation of toxins and pesticides, control of microbial biofilms and treatment of waste, processing surfaces, and effluent is also under investigation. Treatment with cold plasma seems promising on enzyme inactivation in the sterilization and modification of polymers for food packages. In this way, the in-package cold plasma treatment has gained great interest from industries and consumers. Finally, this emerging non-thermal technology has shown potential applications at different interfaces of food science and technology. However, further work in scalability studies, industrial system design and mechanistic insights must be observed before a larger commercial scale utilization in the food industry.

References

1. Gómez-lópez, V.M., Pataro, G., Tiwari, B., Gozzi, M., Meireles, M.Á.A., Wang, S., Guamis, B., Pan, Z., Ramaswamy, H., Sastry, S., Kuntz, F., Cullen, P.J., Vidyarthi, K., Ling, B., Quevedo, J.M., Strasser, A., Vignali, G., Veggi, P.C., Gervilla, R., Kotilainen, H.M., Viganó, J., Morata, A., Meireles, Á.A., Wang, S., Guamis, B., Pan, Z., Ramaswamy, H., Sastry, S., Kuntz, F., Cullen, P.J., Vidyarthi, S.K., Ling, B., Quevedo, J.M., Strasser, A., Vignali, G., Veggi, P.C., Gervilla, R., Kotilainen, H.M., Pelacci, M., Viganó, J., Morata, A.: Guidelines on reporting treatment conditions for emerging technologies in food processing. Crit. Rev. Food Sci. Nutr. 0, 1–25 (2021). https://doi.org/10.1080/10408398.2021.1895058.
2. Priyadarshini, A., Rajauria, G., O'Donnell, C.P., Tiwari, B.K.: Emerging food processing technologies and factors impacting their industrial adoption. Crit. Rev. Food Sci. Nutr. 59, 3082–3101 (2019). https://doi.org/10.1080/10408398.2018.1483890.

3. Herianto, S., Hou, C.Y., Lin, C.M., Chen, H.L.: Nonthermal plasma-activated water: A comprehensive review of this new tool for enhanced food safety and quality, (2021).
4. Cheng, J.H., Lv, X., Pan, Y., Sun, D.W.: Foodborne bacterial stress responses to exogenous reactive oxygen species (ROS) induced by cold plasma treatments, (2020).
5. Ozen, E., Singh, R.K.: Atmospheric cold plasma treatment of fruit juices: A review, (2020).
6. Gavahian, M., Chu, Y.H., Jo, C.: Prospective Applications of Cold Plasma for Processing Poultry Products: Benefits, Effects on Quality Attributes, and Limitations. Compr. Rev. Food Sci. Food Saf. (2019). https://doi.org/10.1111/1541-4337.12460.
7. Thirumdas, R., Kothakota, A., Annapure, U., Siliveru, K., Blundell, R., Gatt, R., Valdramidis, V.P.: Plasma activated water (PAW): Chemistry, physico-chemical properties, applications in food and agriculture. Trends Food Sci. Technol. 77, 21–31 (2018). https://doi.org/10.1016/j.tifs.2018.05.007.
8. Sakudo, A., Misawa, T., Yagyu, Y.: Equipment design for cold plasma disinfection of food products. Elsevier Inc. (2019).
9. Pedrow, P., Hua, Z., Xie, S., Zhu, M.J.: Engineering principles of cold plasma. Elsevier Inc. (2019).
10. Bermudez-Aguirre, D.: Disinfection of high-moisture food using cold plasma. Elsevier Inc. (2019).
11. Surowsky, B., Bußler, S., Schlüter, O.K.: Cold Plasma Interactions With Food Constituents in Liquid and Solid Food Matrices. Cold Plasma Food Agric. Fundam. Appl. 179–203 (2016). https://doi.org/10.1016/B978-0-12-801365-6.00007-X.
12. Chizoba Ekezie, F.G., Sun, D.W., Cheng, J.H.: A review on recent advances in cold plasma technology for the food industry: Current applications and future trends. Trends Food Sci. Technol. 69, 46–58 (2017). https://doi.org/10.1016/j.tifs.2017.08.007.
13. Pan, Y., Cheng, J., Sun, D.: Cold Plasma-Mediated Treatments for Shelf Life Extension of Fresh Produce: A Review of Recent Research Developments. Compr. Rev. Food Sci. Food Saf. 18, 1312–1326 (2019). https://doi.org/10.1111/1541-4337.12474.
14. Peng, P., Chen, P., Zhou, N., Schiappacasse, C., Cheng, Y., Chen, D., Addy, M., Zhang, Y., Anderson, E., Fan, L., Hatzenbeller, R., Liu, Y., Ruan, R.: Packed food and packaging materials disinfected by cold plasma. Elsevier Inc. (2019).
15. Mai-Prochnow, A.: Cold plasma to control biofilms on food and in the food-processing environment. Elsevier Inc. (2019).
16. Girard, F., Peret, M., Dumont, N., Badets, V., Blanc, S., Gazeli, K., Noël, C., Belmonte, T., Marlin, L., Cambus, J.P., Simon, G., Sojic, N., Held, B., Arbault, S., Clément, F.: Correlations between gaseous and liquid phase chemistries

induced by cold atmospheric plasmas in a physiological buffer. Phys. Chem. Chem. Phys. 20, 9198–9210 (2018). https://doi.org/10.1039/c8cp00264a.
17. Smet, C., Govaert, M., Kyrylenko, A., Easdani, M., Walsh, J.L., Van Impe, J.F.: Inactivation of Single Strains of Listeria monocytogenes and Salmonella Typhimurium Planktonic Cells Biofilms With Plasma Activated Liquids. Front. Microbiol. 10, 1539 (2019). https://doi.org/10.3389/fmicb.2019.01539.
18. Ojha, S., Fröhling, A., Durek, J., Ehlbeck, J., Tiwari, B.K., Schlüter, O.K., Bußler, S.: Principles and Application of Cold Plasma in Food Processing. Innov. Food Process. Technol. 1857, 519–540 (2021). https://doi.org/10.1016/b978-0-08-100596-5.23033-3.
19. Ikmal Misnal, M.F., Redzuan, N., Firdaus Zainal, M.N., Raja Ibrahim, R.K., Ahmad, N., Agun, L.: Emerging cold plasma treatment on rice grains: A mini review. Chemosphere. 274, 129972 (2021). https://doi.org/10.1016/j.chemosphere.2021.129972.
20. Fridman, A.: Plasma Chemistry. Cambridge University Press, Philadelphia, PA (2008).
21. Alves Junior, C.: Plasma frio atmosférico – novas oportunidades de pesquisa numa plataforma versátil e portadora de futuro. Rev. Mater. 25, 1–7 (2020). https://doi.org/10.1590/S1517-707620200004.1212.
22. Alves Filho, E.G., de Brito, E.S., Rodrigues, S.: Effects of cold plasma processing in food components. Elsevier Inc. (2019).
23. Møller, I.M., Jensen, P.E., Hansson, A.: Oxidative modifications to cellular components in plants, (2007).
24. Scholtz, V., Pazlarova, J., Souskova, H., Khun, J., Julak, J.: Nonthermal plasma - A tool for decontamination and disinfection, (2015).
25. Ji, S.H., Kim, T., Panngom, K., Hong, Y.J., Pengkit, A., Park, D.H., Kang, M.H., Lee, S.H., Im, J.S., Kim, J.S., Uhm, H.S., Choi, E.H., Park, G.: Assessment of the Effects of Nitrogen Plasma and Plasma-Generated Nitric Oxide on Early Development of *Coriandum sativum*. Plasma Process. Polym. 12, 1164–1173 (2015). https://doi.org/10.1002/ppap.201500021.
26. Wu, T.Y., Sun, N.N., Chau, C.F.: Application of corona electrical discharge plasma on modifying the physicochemical properties of banana starch indigenous to Taiwan. J. Food Drug Anal. 26, 244–251 (2018). https://doi.org/10.1016/j.jfda.2017.03.005.
27. Thirumdas, R., Kadam, D., Annapure, U.S.: Cold Plasma: an Alternative Technology for the Starch Modification, (2017).
28. Sarangapani, C., Misra, N.N., Milosavljevic, V., Bourke, P., O'Regan, F., Cullen, P.J.: Pesticide degradation in water using atmospheric air cold plasma. J. Water Process Eng. 9, 225–232 (2016). https://doi.org/10.1016/j.jwpe.2016.01.003.
29. Thirumdas, R., Deshmukh, R.R., Annapure, U.S.: Effect of low temperature plasma on the functional properties of basmati rice flour. J. Food Sci. Technol. 53, 2742–2751 (2016). https://doi.org/10.1007/s13197-016-2246-4.

30. Lee, K.H., Kim, H.J., Woo, K.S., Jo, C., Kim, J.K., Kim, S.H., Park, H.Y., Oh, S.K., Kim, W.H.: Evaluation of cold plasma treatments for improved microbial and physicochemical qualities of brown rice. LWT - Food Sci. Technol. 73, 442–447 (2016). https://doi.org/10.1016/j.lwt.2016.06.055.
31. Sarangapani, C., Devi, Y., Thirundas, R., Annapure, U.S., Deshmukh, R.R.: Effect of low-pressure plasma on physico-chemical properties of parboiled rice. LWT - Food Sci. Technol. 63, 452–460 (2015). https://doi.org/10.1016/j.lwt.2015.03.026.
32. Thirumdas, R., Deshmukh, R.R., Annapure, U.S.: Effect of low temperature plasma processing on physicochemical properties and cooking quality of basmati rice. Innov. Food Sci. Emerg. Technol. 31, 83–90 (2015). https://doi.org/10.1016/j.ifset.2015.08.003.
33. Dong, S., Gao, A., Xu, H., Chen, Y.: Effects of Dielectric Barrier Discharges (DBD) Cold Plasma Treatment on Physicochemical and Structural Properties of Zein Powders. Food Bioprocess Technol. 10, 434–444 (2017). https://doi.org/10.1007/s11947-016-1814-y.
34. Bußler, S., Steins, V., Ehlbeck, J., Schlüter, O.: Impact of thermal treatment versus cold atmospheric plasma processing on the techno-functional protein properties from Pisum sativum "Salamanca." J. Food Eng. 167, 166–174 (2015). https://doi.org/10.1016/j.jfoodeng.2015.05.036.
35. Misra, N.N., Kaur, S., Tiwari, B.K., Kaur, A., Singh, N., Cullen, P.J.: Atmospheric pressure cold plasma (ACP) treatment of wheat flour. Food Hydrocoll. 44, 115–121 (2015). https://doi.org/10.1016/j.foodhyd.2014.08.019.
36. Segat, A., Misra, N.N., Cullen, P.J., Innocente, N.: Atmospheric pressure cold plasma (ACP) treatment of whey protein isolate model solution. Innov. Food Sci. Emerg. Technol. 29, 247–254 (2015). https://doi.org/10.1016/j.ifset.2015.03.014.
37. Korachi, M., Ozen, F., Aslan, N., Vannini, L., Guerzoni, M.E., Gottardi, D., Ekinci, F.Y.: Biochemical changes to milk following treatment by a novel, cold atmospheric plasma system. Int. Dairy J. 42, 64–69 (2015). https://doi.org/10.1016/j.idairyj.2014.10.006.
38. Silveira, M.R., Coutinho, N.M., Esmerino, E.A., Moraes, J., Fernandes, L.M., Pimentel, T.C., Freitas, M.Q., Silva, M.C., Raices, R.S.L., Senaka Ranadheera, C., Borges, F.O., Neto, R.P.C., Tavares, M.I.B., Fernandes, F.A.N., Fonteles, T. V., Nazzaro, F., Rodrigues, S., Cruz, A.G.: Guava-flavored whey beverage processed by cold plasma technology: Bioactive compounds, fatty acid profile and volatile compounds. Food Chem. 279, 120–127 (2019). https://doi.org/10.1016/j.foodchem.2018.11.128.
39. Silveira, M.R., Coutinho, N.M., Rocha, R.S., Moraes, J., Esmerino, E.A., Pimentel, T.C., Freitas, M.Q., Silva, M.C., Raices, R.S.L., Senaka Ranadheera, C., Borges, F.O., Fonteles, T. V., Neto, R.P.C., Tavares, M.I.B., Fernandes, F.A.N., Rodrigues, S., Cruz, A.G.: Guava flavored whey-beverage processed

by cold plasma: Physical characteristics, thermal behavior and microstructure. Food Res. Int. 119, 564–570 (2019). https://doi.org/10.1016/j.foodres.2018.10.033.

40. Coutinho, N.M., Silveira, M.R., Fernandes, L.M., Moraes, J., Pimentel, T.C., Freitas, M.Q., Silva, M.C., Raices, R.S.L., Ranadheera, C.S., Borges, F.O., Neto, R.P.C., Tavares, M.I.B., Fernandes, F.A.N., Fonteles, T. V., Nazzaro, F., Rodrigues, S., Cruz, A.G.: Processing chocolate milk drink by low-pressure cold plasma technology. Food Chem. 278, 276–283 (2019). https://doi.org/10.1016/j.foodchem.2018.11.061.
41. Coutinho, N.M., Silveira, M.R., Pimentel, T.C., Freitas, M.Q., Moraes, J., Fernandes, L.M., Silva, M.C., Raices, R.S.L., Ranadheera, C.S., Borges, F.O., Neto, R.P.C., Tavares, M.I.B., Fernandes, F.A.N., Nazzaro, F., Rodrigues, S., Cruz, A.G.: Chocolate milk drink processed by cold plasma technology: Physical characteristics, thermal behavior and microstructure. LWT. 102, 324–329 (2019). https://doi.org/10.1016/j.lwt.2018.12.055.
42. Ribeiro, K.C.S., Coutinho, N.M., Silveira, M.R., Rocha, R.S., Arruda, H.S., Pastore, G.M., Neto, R.P.C., Tavares, M.I.B., Pimentel, T.C., Silva, P.H.F., Freitas, M.Q., Esmerino, E.A., Silva, M.C., Duarte, M.C.K.H., Cruz, A.G.: Impact of cold plasma on the techno-functional and sensory properties of whey dairy beverage added with xylooligosaccharide. Food Res. Int. 142, 110232 (2021). https://doi.org/10.1016/j.foodres.2021.110232.
43. Yong, H.I., Lee, S.H., Kim, S.Y., Park, S., Park, J., Choe, W., Jo, C.: Color development, physiochemical properties, and microbiological safety of pork jerky processed with atmospheric pressure plasma. Innov. Food Sci. Emerg. Technol. 53, 78–84 (2019). https://doi.org/10.1016/j.ifset.2017.09.005.
44. Lee, J., Jo, K., Lim, Y., Jeon, H.J., Choe, J.H., Jo, C., Jung, S.: The use of atmospheric pressure plasma as a curing process for canned ground ham. Food Chem. 240,430–436(2018). https://doi.org/10.1016/j.foodchem.2017.07.148.
45. Fernandes, F.A.N., Santos, V.O., Rodrigues, S.: Effects of glow plasma technology on some bioactive compounds of acerola juice. Food Res. Int. 115, 16–22 (2019). https://doi.org/10.1016/j.foodres.2018.07.042.
46. Alves Filho, E.G., Almeida, F.D.L., Cavalcante, R.S., De Brito, E.S., Cullen, P.J., Frias, J.M., Bourke, P., Fernandes, F.A.N., Rodrigues, S.: 1H NMR spectroscopy and chemometrics evaluation of non-thermal processing of orange juice. Food Chem. 204, 102–107 (2016). https://doi.org/10.1016/j.foodchem.2016.02.121.
47. Alves Filho, E.G., Rodrigues, T.H.S., Fernandes, F.A.N., de Brito, E.S., Cullen, P.J., Frias, J.M., Bourke, P., Cavalcante, R.S., Almeida, F.D.L., Rodrigues, S.: An untargeted chemometric evaluation of plasma and ozone processing effect on volatile compounds in orange juice. Innov. Food Sci. Emerg. Technol. 53, 63–69 (2019). https://doi.org/10.1016/j.ifset.2017.10.001.
48. Alves Filho, E.G., Cullen, P.J., Frias, J.M., Bourke, P., Tiwari, B.K., Brito, E.S., Rodrigues, S., Fernandes, F.A.N.: Evaluation of plasma, high-pressure and

ultrasound processing on the stability of fructooligosaccharides. Int. J. Food Sci. Technol. 51, 2034–2040 (2016). https://doi.org/10.1111/ijfs.13175.

49. Rodríguez, Ó., Gomes, W.F., Rodrigues, S., Fernandes, F.A.N.: Effect of indirect cold plasma treatment on cashew apple juice (Anacardium occidentale L.). LWT - Food Sci. Technol. 84, 457–463 (2017). https://doi.org/10.1016/j.lwt.2017.06.010.
50. Sarangapani, C., O'Toole, G., Cullen, P.J., Bourke, P.: Atmospheric cold plasma dissipation efficiency of agrochemicals on blueberries. Innov. Food Sci. Emerg. Technol. 44, 235–241 (2017). https://doi.org/10.1016/j.ifset.2017.02.012.
51. Sarangapani, C., Yamuna Devi, R., Thirumdas, R., Trimukhe, A.M., Deshmukh, R.R., Annapure, U.S.: Physico-chemical properties of low-pressure plasma treated black gram. LWT - Food Sci. Technol. 79, 102–110 (2017). https://doi.org/10.1016/j.lwt.2017.01.017.
52. Sakiyama, Y., Graves, D.B.: Neutral gas flow and ring-shaped emission profile in non-thermal RF-excited plasma needle discharge at atmospheric pressure. Plasma Sources Sci. Technol. 18, 025022 (2009). https://doi.org/10.1088/0963-0252/18/2/025022.
53. Bußler, S., Rumpold, B.A., Fröhling, A., Jander, E., Rawel, H.M., Schlüter, O.K.: Cold atmospheric pressure plasma processing of insect flour from Tenebrio molitor: Impact on microbial load and quality attributes in comparison to dry heat treatment. Innov. Food Sci. Emerg. Technol. 36, 277–286 (2016). https://doi.org/10.1016/j.ifset.2016.07.002.
54. Meinlschmidt, P., Ueberham, E., Lehmann, J., Reineke, K., Schlüter, O., Schweiggert-Weisz, U., Eisner, P.: The effects of pulsed ultraviolet light, cold atmospheric pressure plasma, and gamma-irradiation on the immunoreactivity of soy protein isolate. Innov. Food Sci. Emerg. Technol. 38, 374–383 (2016). https://doi.org/10.1016/j.ifset.2016.06.007.
55. Tammineedi, C.V.R.K., Choudhary, R., Perez-Alvarado, G.C., Watson, D.G.: Determining the effect of UV-C, high intensity ultrasound and nonthermal atmospheric plasma treatments on reducing the allergenicity of α-casein and whey proteins. LWT - Food Sci. Technol. 54, 35–41 (2013). https://doi.org/10.1016/j.lwt.2013.05.020.
56. Surowsky, B., Fischer, A., Schlueter, O., Knorr, D.: Cold plasma effects on enzyme activity in a model food system. Innov. Food Sci. Emerg. Technol. 19, 146–152 (2013). https://doi.org/10.1016/j.ifset.2013.04.002.
57. Wollrab, A.: Organische Chemie. Springer Berlin Heidelberg, Berlin, Heidelberg (2014).
58. Vatansever, F., de Melo, W.C.M.A., Avci, P., Vecchio, D., Sadasivam, M., Gupta, A., Chandran, R., Karimi, M., Parizotto, N.A., Yin, R., Tegos, G.P., Hamblin, M.R.: Antimicrobial strategies centered around reactive oxygen species - bactericidal antibiotics, photodynamic therapy, and beyond, (2013).
59. Gavahian, M., Chu, Y.H., Mousavi Khaneghah, A., Barba, F.J., Misra, N.N.: A critical analysis of the cold plasma induced lipid oxidation in foods, (2018).

60. Van Durme, J., Vandamme, J.: Non-thermal plasma as preparative technique to evaluate olive oil adulteration. Food Chem. 208, 185–191 (2016). https://doi.org/10.1016/j.foodchem.2016.04.007.
61. Vandamme, J., Nikiforov, A., Dujardin, K., Leys, C., De Cooman, L., Van Durme, J.: Critical evaluation of non-thermal plasma as an innovative accelerated lipid oxidation technique in fish oil. Food Res. Int. 72, 115–125 (2015). https://doi.org/10.1016/j.foodres.2015.03.037.
62. Yepez, X. V., Keener, K.M.: High-voltage Atmospheric Cold Plasma (HVACP) hydrogenation of soybean oil without trans-fatty acids. Innov. Food Sci. Emerg. Technol. 38, 169–174 (2016). https://doi.org/10.1016/j.ifset.2016.09.001.
63. Cubas, A.L.V., Machado, M.M., Pinto, C.R.S.C., Moecke, E.H.S., Dutra, A.R.A.: Biodiesel production using fatty acids from food industry waste using corona discharge plasma technology. Waste Manag. 47, 149–154 (2016). https://doi.org/10.1016/j.wasman.2015.05.040.
64. Bermudez-Aguirre, D.: Advances in the inactivation of microorganisms and viruses in food and model systems using cold plasma. Elsevier Inc. (2019).
65. Zhao, Y.M., Patange, A., Sun, D.W., Tiwari, B.: Plasma-activated water: Physicochemical properties, microbial inactivation mechanisms, factors influencing antimicrobial effectiveness, and applications in the food industry. Compr. Rev. Food Sci. Food Saf. 19, 3951–3979 (2020). https://doi.org/10.1111/1541-4337.12644.
66. Won, M.Y., Lee, J., Min, S.C.: Mandarin preservation by microwave-powered cold plasma treatment. (2016). https://doi.org/10.1016/j.ifset.2016.10.021.
67. Oh, Y.J., Song, A.Y., Min, S.C.: Inhibition of Salmonella typhimurium on radish sprouts using nitrogen-cold plasma. (2017). https://doi.org/10.1016/j.ijfoodmicro.2017.03.005.
68. Min, S.C., Roh, S.H., Niemira, B.A., Boyd, G., Sites, J.E., Uknalis, J., Fan, X.: In-package inhibition of E. coli O157:H7 on bulk Romaine lettuce using cold plasma. (2017). https://doi.org/10.1016/j.fm.2017.01.010.
69. Kim, J.E., Oh, Y.J., Won, M.Y., Lee, K.-S., Min, S.C.: Microbial decontamination of onion powder using microwave-powered cold plasma treatments. (2016). https://doi.org/10.1016/j.fm.2016.10.006.
70. Baier, M., Görgen, M., Ehlbeck, J., Knorr, D., Herppich, W.B., Schlüter, O.: Non-thermal atmospheric pressure plasma: Screening for gentle process conditions and antibacterial efficiency on perishable fresh produce. Innov. Food Sci. Emerg. Technol. 22, 147–157 (2014). https://doi.org/10.1016/j.ifset.2014.01.011.
71. Bermúdez-Aguirre, D., Wemlinger, E., Pedrow, P., Barbosa-Cánovas, G., Garcia-Perez, M.: Effect of atmospheric pressure cold plasma (APCP) on the inactivation of Escherichia coli in fresh produce. Food Control. 34, 149–157 (2013). https://doi.org/10.1016/j.foodcont.2013.04.022.

72. Niemira, B.A., Sites, J.: Cold plasma inactivates Salmonella Stanley and Escherichia coli O157:H7 inoculated on golden delicious apples. J. Food Prot. 71, 1357–1365 (2008). https://doi.org/10.4315/0362-028X-71.7.1357.
73. Ma, R., Wang, G., Tian, Y., Wang, K., Zhang, J., Fang, J.: Non-thermal plasma-activated water inactivation of food-borne pathogen on fresh produce. J. Hazard. Mater. 300, 643–651 (2015). https://doi.org/10.1016/j.jhazmat.2015.07.061.
74. Srey, S., Park, S.Y., Jahid, I.K., Ha, S. Do: Reduction effect of the selected chemical and physical treatments to reduce L. monocytogenes biofilms formed on lettuce and cabbage. Food Res. Int. 62, 484–491 (2014). https://doi.org/10.1016/j.foodres.2014.03.067.
75. Lee, H., Kim, J.E., Chung, M.S., Min, S.C.: Cold plasma treatment for the microbiological safety of cabbage, lettuce, and dried figs. Food Microbiol. 51, 74–80 (2015). https://doi.org/10.1016/j.fm.2015.05.004.
76. Matan, N., Puangjinda, K., Phothisuwan, S., Nisoa, M.: Combined antibacterial activity of green tea extract with atmospheric radio-frequency plasma against pathogens on fresh-cut dragon fruit. Food Control. 50, 291–296 (2015). https://doi.org/10.1016/j.foodcont.2014.09.005.
77. Ziuzina, D., Patil, S., Cullen, P.J., Keener, K.M., Bourke, P.: Atmospheric cold plasma inactivation of Escherichia coli, Salmonella enterica serovar Typhimurium and Listeria monocytogenes inoculated on fresh produce. Food Microbiol. 42, 109–116 (2014). https://doi.org/10.1016/j.fm.2014.02.007.
78. Zhang, M., Oh, J.K., Cisneros-Zevallos, L., Akbulut, M.: Bactericidal effects of nonthermal low-pressure oxygen plasma on S. typhimurium LT2 attached to fresh produce surfaces. J. Food Eng. 119, 425–432 (2013). https://doi.org/10.1016/j.jfoodeng.2013.05.045.
79. Fernández, A., Noriega, E., Thompson, A.: Inactivation of Salmonella enterica serovar Typhimurium on fresh produce by cold atmospheric gas plasma technology. Food Microbiol. 33, 24–29 (2013). https://doi.org/10.1016/j.fm.2012.08.007.
80. Liao, L.K., Wei, X.Y., Gong, X., Li, J.H., Huang, T., Xiong, T.: Microencapsulation of Lactobacillus casei LK-1 by spray drying related to its stability and *in vitro* digestion. Lwt-Food Sci. Technol. 82, 82–89 (2017). https://doi.org/10.1016/j.lwt.2017.03.065.
81. Xiang, Q., Liu, X., Li, J., Liu, S., Zhang, H., Bai, Y.: Effects of dielectric barrier discharge plasma on the inactivation of Zygosaccharomyces rouxii and quality of apple juice. (2018). https://doi.org/10.1016/j.foodchem.2018.02.008.
82. Montenegro, J., Ruan, R., Ma, H., Chen, P.: Inactivation of E. coli O157:H7 Using a Pulsed Nonthermal Plasma System. J. Food Sci. 67, 646–648 (2002). https://doi.org/10.1111/j.1365-2621.2002.tb10653.x.
83. Shi, X.M., Zhang, G.J., Wu, X.L., Li, Y.X., Ma, Y., Shao, X.J.: Effect of low-temperature plasma on microorganism inactivation and quality of freshly squeezed orange juice. IEEE Trans. Plasma Sci. 39, 1591–1597 (2011). https://doi.org/10.1109/TPS.2011.2142012.

84. Xu, L., Garner, A.L., Tao, B., Keener, K.M.: Microbial Inactivation and Quality Changes in Orange Juice Treated by High Voltage Atmospheric Cold Plasma. Food Bioprocess Technol. 10, 1778–1791 (2017). https://doi.org/10.1007/s11947-017-1947-7.
85. Pankaj, S.K., Wan, Z., Colonna, W., Keener, K.M.: Effect of high voltage atmospheric cold plasma on white grape juice quality. J. Sci. Food Agric. 97, 4016–4021 (2017). https://doi.org/10.1002/jsfa.8268.
86. Dasan, B.G., Boyaci, I.H.: Effect of Cold Atmospheric Plasma on Inactivation of Escherichia coli and Physicochemical Properties of Apple, Orange, Tomato Juices, and Sour Cherry Nectar. Food Bioprocess Technol. 11, 334–343 (2018). https://doi.org/10.1007/s11947-017-2014-0.
87. Hou, Y., Wang, R., Gan, Z., Shao, T., Zhang, X., He, M., Sun, A.: Effect of cold plasma on blueberry juice quality. (2019). https://doi.org/10.1016/j.foodchem.2019.03.123.
88. Starek, A., Pawłat, J., Chudzik, B., Kwiatkowski, M., Terebun, P., Sagan, A., Andrejko, D.: Evaluation of selected microbial and physicochemical parameters of fresh tomato juice after cold atmospheric pressure plasma treatment during refrigerated storage. Sci. Rep. 9, 1–11 (2019). https://doi.org/10.1038/s41598-019-44946-1.
89. Los, A., Ziuzina, D., Akkermans, S., Boehm, D., Cullen, P.J., Van Impe, J., Bourke, P.: Improving microbiological safety and quality characteristics of wheat and barley by high voltage atmospheric cold plasma closed processing. Food Res. Int. 106, 509–521 (2018). https://doi.org/10.1016/j.foodres.2018.01.009.
90. Devi, Y., Thirumdas, R., Sarangapani, C., Deshmukh, R.R., Annapure, U.S.: Influence of cold plasma on fungal growth and aflatoxins production on groundnuts. (2017). https://doi.org/10.1016/j.foodcont.2017.02.019.
91. Wan, Z., Chen, Y., Pankaj, S.K., Keener, K.M.: High voltage atmospheric cold plasma treatment of refrigerated chicken eggs for control of Salmonella Enteritidis contamination on egg shell. (2016). https://doi.org/10.1016/j.lwt.2016.10.051.
92. Georgescu, N., Apostol, L., Gherendi, F.: Inactivation of Salmonella enterica serovar Typhimurium on egg surface, by direct and indirect treatments with cold atmospheric plasma. (2017). https://doi.org/10.1016/j.foodcont.2017.01.005.
93. Moritz, M., Wiacek, C., Koethe, M., Braun, P.G.: Atmospheric pressure plasma jet treatment of Salmonella Enteritidis inoculated eggshells. (2017). https://doi.org/10.1016/j.ijfoodmicro.2017.01.004.
94. Patange, A., Boehm, D., Bueno-Ferrer, C., Cullen, P.J., Bourke, P.: Controlling Brochothrix thermosphacta as a spoilage risk using in-package atmospheric cold plasma. Food Microbiol. 66, 48–54 (2017). https://doi.org/10.1016/j.fm.2017.04.002.
95. Bauer, A., Ni, Y., Bauer, S., Paulsen, P., Modic, M., Walsh, J.L., Smulders, F.J.M.: The effects of atmospheric pressure cold plasma treatment on

microbiological, physical-chemical and sensory characteristics of vacuum packaged beef loin. (2017). https://doi.org/10.1016/j.meatsci.2017.02.003.

96. Choi, S., Puligundla, P., Mok, C.: Effect of corona discharge plasma on microbial decontamination of dried squid shreds including physico-chemical and sensory evaluation. (2016). https://doi.org/10.1016/j.lwt.2016.08.063.
97. Wan, Z., Misra, N.N., Li, G., Keener, K.M.: High Voltage Atmospheric Cold Plasma Treatment of Listeria innocua and Escherichia coli K-12 on Queso Fresco (Fresh Cheese). LWT. 111406 (2021). https://doi.org/10.1016/j.lwt.2021.111406.
98. Pan, Y., Cheng, J.H., Lv, X., Sun, D.W.: Assessing the inactivation efficiency of Ar/O2 plasma treatment against Listeria monocytogenes cells: Sublethal injury and inactivation kinetics. LWT. 111, 318–327 (2019). https://doi.org/10.1016/j.lwt.2019.05.041.
99. Zhang, Z., Xu, Z., Cheng, C., Wei, J., Lan, Y., Ni, G., Sun, Q., Qian, S., Zhang, H., Xia, W., Shen, J., Meng, Y., Chu, P.K.: Bactericidal Effects of Plasma Induced Reactive Species in Dielectric Barrier Gas–Liquid Discharge, (2017).
100. Huang, H.W., Lung, H.M., Yang, B.B., Wang, C.Y.: Responses of microorganisms to high hydrostatic pressure processing, (2014).
101. Xu, Z., Cheng, C., Shen, J., Lan, Y., Hu, S., Han, W., Chu, P.K.: *In vitro* antimicrobial effects and mechanisms of direct current air-liquid discharge plasma on planktonic Staphylococcus aureus and Escherichia coli in liquids. Bioelectrochemistry. 121, 125–134 (2018). https://doi.org/10.1016/j.bioelechem.2018.01.012.
102. Xu, Z., Shen, J., Cheng, C., Hu, S., Lan, Y., Chu, P.K.: *In vitro* antimicrobial effects and mechanism of atmospheric-pressure He/O2 plasma jet on Staphylococcus aureus biofilm. J. Phys. D. Appl. Phys. 50, 105201 (2017). https://doi.org/10.1088/1361-6463/aa593f.
103. Dolezalova, E., Lukes, P.: Membrane damage and active but nonculturable state in liquid cultures of Escherichia coli treated with an atmospheric pressure plasma jet. Bioelectrochemistry. 103, 7–14 (2015). https://doi.org/10.1016/j.bioelechem.2014.08.018.
104. Patange, A., Boehm, D., Ziuzina, D., Cullen, P.J., Gilmore, B., Bourke, P.: High voltage atmospheric cold air plasma control of bacterial biofilms on fresh produce. Int. J. Food Microbiol. 293, 137–145 (2019). https://doi.org/10.1016/j.ijfoodmicro.2019.01.005.
105. Ziuzina, D., Han, L., Cullen, P.J., Bourke, P.: Cold plasma inactivation of internalised bacteria and biofilms for Salmonella enterica serovar Typhimurium, Listeria monocytogenes and Escherichia coli. Int. J. Food Microbiol. 210, 53–61 (2015). https://doi.org/10.1016/j.ijfoodmicro.2015.05.019.
106. Imlay, J.A.: Transcription Factors That Defend Bacteria Against Reactive Oxygen Species. Annu. Rev. Microbiol. 69, 93–108 (2015). https://doi.org/10.1146/annurev-micro-091014-104322.
107. Thorpe, G.W., Fong, C.S., Alic, N., Higgins, V.J., Dawes, I.W.: Cells have distinct mechanisms to maintain protection against different reactive oxygen

species: Oxidative-stress-response genes. Proc. Natl. Acad. Sci. U. S. A. 101, 6564–6569 (2004). https://doi.org/10.1073/pnas.0305888101.
108. Turnbull, A.L., Surette, M.G.: Cysteine biosynthesis, oxidative stress and antibiotic resistance in Salmonella typhimurium. Res. Microbiol. 161, 643–650 (2010). https://doi.org/10.1016/j.resmic.2010.06.004.
109. Perinban, S., Orsat, V., Raghavan, V.: Nonthermal Plasma–Liquid Interactions in Food Processing: A Review, (2019).
110. Zhang, Q., Liang, Y., Feng, H., Ma, R., Tian, Y., Zhang, J., Fang, J.: A study of oxidative stress induced by non-thermal plasma-activated water for bacterial damage. Appl. Phys. Lett. 102, 203701 (2013). https://doi.org/10.1063/1.4807133.
111. Lymar, S. V., Hurst, J.K.: CO2-Catalyzed One-Electron Oxidations by Peroxynitrite: Properties of the Reactive Intermediate. Inorg. Chem. 37, 294–301 (1998). https://doi.org/10.1021/ic970946i.
112. Joshi, S.G., Cooper, M., Yost, A., Paff, M., Ercan, U.K., Fridman, G., Friedman, G., Fridman, A., Brooks, A.D.: Nonthermal dielectric-barrier discharge plasma-induced inactivation involves oxidative DNA damage and membrane lipid peroxidation in Escherichia coli. Antimicrob. Agents Chemother. 55, 1053–1062 (2011). https://doi.org/10.1128/AAC.01002-10.
113. Marnett, L.J.: Lipid peroxidation - DNA damage by malondialdehyde. Mutat. Res. - Fundam. Mol. Mech. Mutagen. 424, 83–95 (1999). https://doi.org/10.1016/S0027-5107(99)00010-X.
114. Mai-Prochnow, A., Clauson, M., Hong, J., Murphy, A.B.: Gram positive and Gram negative bacteria differ in their sensitivity to cold plasma. Sci. Rep. 6, 1–11 (2016). https://doi.org/10.1038/srep38610.
115. Kostya (ken) Ostrikov, Zhou, R., Zhou, R., Wang, P., Xian, Y., Mai-Prochnow, A., Lu, X., Cullen, P.J., Ostrikov, K. (Ken), Bazaka, K.: Plasma-activated water: Generation, origin of reactive species and biological applications, (2020).
116. Sen, Y., Onal-Ulusoy, B., Mutlu, M.: Aspergillus decontamination in hazelnuts: Evaluation of atmospheric and low-pressure plasma technology. Innov. Food Sci. Emerg. Technol. 54, 235–242 (2019). https://doi.org/10.1016/j.ifset.2019.04.014.
117. Mir, S.A., Dar, B.N., Shah, M.A., Sofi, S.A., Hamdani, A.M., Oliveira, C.A.F., Hashemi Moosavi, M., Mousavi Khaneghah, A., Sant'Ana, A.S.: Application of new technologies in decontamination of mycotoxins in cereal grains: Challenges, and perspectives. Food Chem. Toxicol. 148, 111976 (2021). https://doi.org/10.1016/j.fct.2021.111976.
118. Wu, Y., Cheng, J.H., Sun, D.W.: Blocking and degradation of aflatoxins by cold plasma treatments: Applications and mechanisms. Trends Food Sci. Technol. 109, 647–661 (2021). https://doi.org/10.1016/j.tifs.2021.01.053.
119. Martins, C., Assunção, R., Cunha, S.C., Fernandes, J.O., Jager, A., Petta, T., Oliveira, C.A., Alvito, P.: Assessment of multiple mycotoxins in breakfast cereals available in the Portuguese market. Food Chem. 239, 132–140 (2018). https://doi.org/10.1016/j.foodchem.2017.06.088.

120. Zain, M.E.: Impact of mycotoxins on humans and animals. J. Saudi Chem. Soc. 15, 129–144 (2011). https://doi.org/10.1016/j.jscs.2010.06.006.
121. Shephard, G.S.: Impact of mycotoxins on human health in developing countries. Food Addit. Contam. Part A. 25, 146–151 (2008). https://doi.org/10.1080/02652030701567442.
122. Ostry, V., Malir, F., Toman, J., Grosse, Y.: Mycotoxins as human carcinogens—the IARC Monographs classification. Mycotoxin Res. 33, 65–73 (2017). https://doi.org/10.1007/s12550-016-0265-7.
123. Rushing, B.R., Selim, M.I.: Aflatoxin B1: A review on metabolism, toxicity, occurrence in food, occupational exposure, and detoxification methods, (2019).
124. Misra, N.N., Yadav, B., Roopesh, M.S., Jo, C.: Cold Plasma for Effective Fungal and Mycotoxin Control in Foods: Mechanisms, Inactivation Effects, and Applications. Compr. Rev. Food Sci. Food Saf. 18, 106–120 (2019). https://doi.org/10.1111/1541-4337.12398.
125. Gavahian, M., Sarangapani, C., Misra, N.: Cold plasma technology for mitigating agrochemical and pesticide residue in food and water: Similarities with ozone and ultraviolet processing, considerations, and research needs. Food Res. Int. 141, 110138 (2021). https://doi.org/10.1016/j.foodres.2021.110138.
126. Iqdiam, B.M., Abuagela, M.O., Boz, Z., Marshall, S.M., Goodrich-Schneider, R., Sims, C.A., Marshall, M.R., MacIntosh, A.J., Welt, B.A.: Effects of atmospheric pressure plasma jet treatment on aflatoxin level, physiochemical quality, and sensory attributes of peanuts. J. Food Process. Preserv. 44, e14305 (2020). https://doi.org/10.1111/jfpp.14305.
127. Shi, H., Cooper, B., Stroshine, R.L., Ileleji, K.E., Keener, K.M.: Structures of Degradation Products and Degradation Pathways of Aflatoxin B1 by High-Voltage Atmospheric Cold Plasma (HVACP) Treatment. J. Agric. Food Chem. 65, 6222–6230 (2017). https://doi.org/10.1021/acs.jafc.7b01604.
128. Siciliano, I., Spadaro, D., Prelle, A., Vallauri, D., Cavallero, M., Garibaldi, A., Gullino, M.: Use of Cold Atmospheric Plasma to Detoxify Hazelnuts from Aflatoxins. Toxins (Basel). 8, 125 (2016). https://doi.org/10.3390/toxins8050125.
129. Sen, Y., Onal-Ulusoy, B., Mutlu, M.: Detoxification of hazelnuts by different cold plasmas and gamma irradiation treatments. Innov. Food Sci. Emerg. Technol. 54, 252–259 (2019). https://doi.org/10.1016/j.ifset.2019.05.002.
130. Puligundla, P., Lee, T., Mok, C.: Effect of corona discharge plasma jet treatment on the degradation of aflatoxin B1 on glass slides and in spiked food commodities. LWT. 124, 108333 (2020). https://doi.org/10.1016/j.lwt.2019.108333.
131. Wielogorska, E., Ahmed, Y., Meneely, J., Graham, W.G., Elliott, C.T., Gilmore, B.F.: A holistic study to understand the detoxification of mycotoxins in maize and impact on its molecular integrity using cold atmospheric plasma treatment. Food Chem. 301, 125281 (2019). https://doi.org/10.1016/j.foodchem.2019.125281.

132. FAO-WHO, C.A.: Pesticide Database |.
133. Misra, N.N., Pankaj, S.K., Walsh, T., O'Regan, F., Bourke, P., Cullen, P.J.: In-package nonthermal plasma degradation of pesticides on fresh produce. J. Hazard. Mater. 271, 33–40 (2014). https://doi.org/10.1016/j.jhazmat.2014.02.005.
134. Zhu, W.C., Wang, B.R., Xi, H.L., Pu, Y.K.: Decontamination of VX surrogate malathion by atmospheric pressure radio-frequency plasma jet. Plasma Chem. Plasma Process. 30, 381–389 (2010). https://doi.org/10.1007/s11090-010-9221-z.
135. Cong, L., Huang, M., Zhang, J., Yan, W.: Effect of dielectric barrier discharge plasma on the degradation of malathion and chlorpyrifos on lettuce. J. Sci. Food Agric. 101, 424–432 (2021). https://doi.org/10.1002/jsfa.10651.
136. Moutiq, R., Pankaj, S.K., Wan, Z., Mendonca, A., Keener, K., Misra, N.N.: Atmospheric Pressure Cold Plasma as a Potential Technology to Degrade Carbamate Residues in Water. Plasma Chem. Plasma Process. 40, 1291–1309 (2020). https://doi.org/10.1007/s11090-020-10093-z.
137. Zheng, Y., Wu, S., Dang, J., Wang, S., Liu, Z., Fang, J., Han, P., Zhang, J.: Reduction of phoxim pesticide residues from grapes by atmospheric pressure non-thermal air plasma activated water. J. Hazard. Mater. 377, 98–105 (2019). https://doi.org/10.1016/j.jhazmat.2019.05.058.
138. Wardenier, N., Gorbanev, Y., Van Moer, I., Nikiforov, A., Van Hulle, S.W.H., Surmont, P., Lynen, F., Leys, C., Bogaerts, A., Vanraes, P.: Removal of alachlor in water by non-thermal plasma: Reactive species and pathways in batch and continuous process. Water Res. 161, 549–559 (2019). https://doi.org/10.1016/j.watres.2019.06.022.
139. Zhou, R., Zhou, R., Yu, F., Xi, D., Wang, P., Li, J., Wang, X., Zhang, X., Bazaka, K., Ostrikov, K. (Ken): Removal of organophosphorus pesticide residues from Lycium barbarum by gas phase surface discharge plasma. Chem. Eng. J. 342, 401–409 (2018). https://doi.org/10.1016/j.cej.2018.02.107.
140. Phan, K.T.K., Phan, H.T., Boonyawan, D., Intipunya, P., Brennan, C.S., Regenstein, J.M., Phimolsiripol, Y.: Non-thermal plasma for elimination of pesticide residues in mango. Innov. Food Sci. Emerg. Technol. 48, 164–171 (2018). https://doi.org/10.1016/j.ifset.2018.06.009.
141. Saberi, M., Modarres-Sanavy, S.A.M., Zare, R., Ghomi, H.: Amelioration of Photosynthesis and Quality of Wheat under Non-thermal Radio Frequency Plasma Treatment. Sci. Rep. 8, 1–8 (2018). https://doi.org/10.1038/s41598-018-30200-7.
142. Houée-Lévin, C., Bobrowski, K., Horakova, L., Karademir, B., Schöneich, C., Davies, M.J., Spickett, C.M.: Exploring oxidative modifications of tyrosine: An update on mechanisms of formation, advances in analysis and biological consequences, https://www.tandfonline.com/doi/abs/10.3109/10715762.2015.1007968, (2015).
143. Wang, Y., Ye, Z., Li, J., Zhang, Y., Guo, Y., Cheng, J.H.: Effects of dielectric barrier discharge cold plasma on the activity, structure and conformation

of horseradish peroxidase (HRP) and on the activity of litchi peroxidase (POD). LWT. 141, 111078 (2021). https://doi.org/10.1016/j.lwt.2021.111078.
144. Segat, A., Misra, N.N., Cullen, P.J., Innocente, N.: Effect of atmospheric pressure cold plasma (ACP) on activity and structure of alkaline phosphatase. Food Bioprod. Process. 98, 181–188 (2016). https://doi.org/10.1016/j.fbp.2016.01.010.
145. Han, Y., Cheng, J.H., Sun, D.W.: Activities and conformation changes of food enzymes induced by cold plasma: A review. Crit. Rev. Food Sci. Nutr. 59, 794–811 (2019). https://doi.org/10.1080/10408398.2018.1555131.
146. de Castro, D.R.G., Mar, J.M., da Silva, L.S., da Silva, K.A., Sanches, E.A., de Araújo Bezerra, J., Rodrigues, S., Fernandes, F.A.N., Campelo, P.H.: Dielectric barrier atmospheric cold plasma applied on camu-camu juice processing: Effect of the excitation frequency. Food Res. Int. 131, 109044 (2020). https://doi.org/10.1016/j.foodres.2020.109044.
147. Farias, T.R.B., Rodrigues, S., Fernandes, F.A.N.: Effect of dielectric barrier discharge plasma excitation frequency on the enzymatic activity, antioxidant capacity and phenolic content of apple cubes and apple juice. Food Res. Int. 136, (2020). https://doi.org/10.1016/j.foodres.2020.109617.
148. Khani, M.R., Shokri, B., Khajeh, K.: Studying the performance of dielectric barrier discharge and gliding arc plasma reactors in tomato peroxidase inactivation. J. Food Eng. 197, 107–112 (2017). https://doi.org/10.1016/j.jfoodeng.2016.11.012.
149. Illera, A.E., Chaple, S., Sanz, M.T., Ng, S., Lu, P., Jones, J., Carey, E., Bourke, P.: Effect of cold plasma on polyphenol oxidase inactivation in cloudy apple juice and on the quality parameters of the juice during storage. Food Chem. X. 3, (2019). https://doi.org/10.1016/j.fochx.2019.100049.
150. Malinowski, S., Herbert, P.A.F., Rogalski, J., Jaroszyńska-Wolińska, J.: Laccase enzyme polymerization by soft plasma jet for durable bioactive coatings. Polymers (Basel). 10, 1–19 (2018). https://doi.org/10.3390/polym10050532.
151. Ji, H., Dong, S., Han, F., Li, Y., Chen, G., Li, L., Chen, Y.: Effects of Dielectric Barrier Discharge (DBD) Cold Plasma Treatment on Physicochemical and Functional Properties of Peanut Protein. Food Bioprocess Technol. 11, 344–354 (2018). https://doi.org/10.1007/s11947-017-2015-z.
152. Choi, S., Attri, P., Lee, I., Oh, J., Yun, J.H., Park, J.H., Choi, E.H., Lee, W.: Structural and functional analysis of lysozyme after treatment with dielectric barrier discharge plasma and atmospheric pressure plasma jet /631/45 / 631/57 /101 /82 /145 article. Sci. Rep. 7, (2017). https://doi.org/10.1038/s41598-017-01030-w.
153. Bußler, S., Ehlbeck, J., Schlüter, O.K.: Pre-drying treatment of plant related tissues using plasma processed air: Impact on enzyme activity and quality attributes of cut apple and potato. Innov. Food Sci. Emerg. Technol. 40, 78–86 (2017). https://doi.org/10.1016/j.ifset.2016.05.007.
154. Chutia, H., Kalita, D., Mahanta, C.L., Ojah, N., Choudhury, A.J.: Kinetics of inactivation of peroxidase and polyphenol oxidase in tender coconut water

by dielectric barrier discharge plasma. LWT. 101, 625–629 (2019). https://doi.org/10.1016/j.lwt.2018.11.071.

155. Porto, E., Alves Filho, E.G., Silva, L.M.A., Fonteles, T.V., do Nascimento, R.B.R., Fernandes, F.A.N., de Brito, E.S., Rodrigues, S.: Ozone and plasma processing effect on green coconut water. Food Res. Int. 131, (2020). https://doi.org/10.1016/j.foodres.2020.109000.

156. Gu, Y., Shi, W., Liu, R., Xing, Y., Yu, X., Jiang, H.: Cold plasma enzyme inactivation on dielectric properties and freshness quality in bananas. Innov. Food Sci. Emerg. Technol. 69, 102649 (2021). https://doi.org/10.1016/j.ifset.2021.102649.

157. Tolouie, H., Mohammadifar, M.A., Ghomi, H., Hashemi, M.: Argon and nitrogen cold plasma effects on wheat germ lipolytic enzymes: Comparison to thermal treatment. Food Chem. 346, 128974 (2021). https://doi.org/10.1016/j.foodchem.2020.128974.

158. Tolouie, H., Mohammadifar, M.A., Ghomi, H., Yaghoubi, A.S., Hashemi, M.: The impact of atmospheric cold plasma treatment on inactivation of lipase and lipoxygenase of wheat germs. Innov. Food Sci. Emerg. Technol. 47, 346–352 (2018). https://doi.org/10.1016/j.ifset.2018.03.002.

159. Zhang, H., Xu, Z., Shen, J., Li, X., Ding, L., Ma, J., Lan, Y., Xia, W., Cheng, C., Sun, Q., Zhang, Z., Chu, P.K.: Effects and Mechanism of Atmospheric-Pressure Dielectric Barrier Discharge Cold Plasma on Lactate Dehydrogenase (LDH) Enzyme OPEN. (2015). https://doi.org/10.1038/srep10031.

160. Tappi, S., Gozzi, G., Vannini, L., Berardinelli, A., Romani, S., Ragni, L., Rocculi, P.: Cold plasma treatment for fresh-cut melon stabilization. Innov. Food Sci. Emerg. Technol. 33, 225–233 (2016). https://doi.org/10.1016/j.ifset.2015.12.022.

161. Giannoglou, M., Xanthou, Z.M., Chanioti, S., Stergiou, P., Christopoulos, M., Dimitrakellis, P., Efthimiadou, A., Gogolides, E., Katsaros, G.: Effect of cold atmospheric plasma and pulsed electromagnetic fields on strawberry quality and shelf-life. Innov. Food Sci. Emerg. Technol. 68, (2021). https://doi.org/10.1016/j.ifset.2021.102631.

162. Moosavi, M.H., Khani, M.R., Shokri, B., Hosseini, S.M., Shojaee-Aliabadi, S., Mirmoghtadaie, L.: Modifications of protein-based films using cold plasma. Int. J. Biol. Macromol. 142, 769–777 (2020). https://doi.org/10.1016/j.ijbiomac.2019.10.017.

163. Pankaj, S.K., Bueno-Ferrer, C., Misra, N.N., O'Neill, L., Tiwari, B.K., Bourke, P., Cullen, P.J.: Dielectric barrier discharge atmospheric air plasma treatment of high amylose corn starch films. LWT. 63, 1076–1082 (2015). https://doi.org/10.1016/j.lwt.2015.04.027.

164. Oh, Y.A., Roh, S.H., Min, S.C.: Cold plasma treatments for improvement of the applicability of defatted soybean meal-based edible film in food packaging. Food Hydrocoll. 58, 150–159 (2016). https://doi.org/10.1016/j.foodhyd.2016.02.022.

165. Romani, V.P., Olsen, B., Pinto Collares, M., Meireles Oliveira, J.R., Prentice, C., Martins, V.G.: Cold plasma and carnauba wax as strategies to produce improved bi-layer films for sustainable food packaging. Food Hydrocoll. 108, 106087 (2020). https://doi.org/10.1016/j.foodhyd.2020.106087.
166. Romani, V.P., Olsen, B., Pinto Collares, M., Meireles Oliveira, J.R., Prentice-Hernández, C., Guimarães Martins, V.: Improvement of fish protein films properties for food packaging through glow discharge plasma application. Food Hydrocoll. 87, 970–976 (2019). https://doi.org/10.1016/j.foodhyd.2018.09.022.
167. Cold Plasma in Food and Agriculture - 1st Edition, https://www.elsevier.com/books/cold-plasma-in-food-and-agriculture/misra/978-0-12-801365-6.
168. Misra, N.N., Yepez, X., Xu, L., Keener, K.: In-package cold plasma technologies, https://doi.org/10.1016/j.jfoodeng.2018.09.019, (2019).
169. Han, L., Boehm, D., Amias, E., Milosavljević, V., Cullen, P.J., Bourke, P.: Atmospheric cold plasma interactions with modified atmosphere packaging inducer gases for safe food preservation. Innov. Food Sci. Emerg. Technol. 38, 384–392 (2016). https://doi.org/10.1016/j.ifset.2016.09.026.
170. Xu, L., Garner, A.L., Tao, B., Keener, K.M.: Microbial Inactivation and Quality Changes in Orange Juice Treated by High Voltage Atmospheric Cold Plasma. Food Bioprocess Technol. 10, 1778–1791 (2017). https://doi.org/10.1007/s11947-017-1947-7.
171. Ziuzina, D., Misra, N.N., Han, L., Cullen, P.J., Moiseev, T., Mosnier, J.P., Keener, K., Gaston, E., Vilaró, I., Bourke, P.: Investigation of a large gap cold plasma reactor for continuous in-package decontamination of fresh strawberries and spinach. Innov. Food Sci. Emerg. Technol. 59, (2020). https://doi.org/10.1016/j.ifset.2019.102229.
172. Kim, S.Y., Bang, I.H., Min, S.C.: Effects of packaging parameters on the inactivation of Salmonella contaminating mixed vegetables in plastic packages using atmospheric dielectric barrier discharge cold plasma treatment. J. Food Eng. 242, 55–67 (2019). https://doi.org/10.1016/j.jfoodeng.2018.08.020.
173. Moutiq, R., Misra, N.N., Mendonça, A., Keener, K.: In-package decontamination of chicken breast using cold plasma technology: Microbial, quality and storage studies. Meat Sci. 159, (2020). https://doi.org/10.1016/j.meatsci.2019.107942.
174. Wang, X., Wang, Z., Zhuang, H., Nasiru, M.M., Yuan, Y., Zhang, J., Yan, W.: Changes in color, myoglobin, and lipid oxidation in beef patties treated by dielectric barrier discharge cold plasma during storage. Meat Sci. 176, 108456 (2021). https://doi.org/10.1016/j.meatsci.2021.108456.
175. Huang, M., Wang, J., Zhuang, H., Yan, W., Zhao, J., Zhang, J.: Effect of in-package high voltage dielectric barrier discharge on microbiological, color and oxidation properties of pork in modified atmosphere packaging during storage. Meat Sci. 149, 107–113 (2019). https://doi.org/10.1016/j.meatsci.2018.11.016.

176. Choi, M.S., Jeon, E.B., Kim, J.Y., Choi, E.H., Lim, J.S., Choi, J., Park, S.Y.: Impact of non-thermal dielectric barrier discharge plasma on Staphylococcus aureus and Bacillus cereus and quality of dried blackmouth angler (Lophiomus setigerus). J. Food Eng. 278, 109952 (2020). https://doi.org/10.1016/j.jfoodeng.2020.109952.
177. Ziuzina, D., Misra, N.N., Cullen, P.J., Keener, K., Mosnier, J.P., Vilaró, I., Gaston, E., Bourke, P.: Demonstrating the potential of industrial scale in-package atmospheric cold plasma for decontamination of cherry tomatoes. Plasma Med. 6, 397–412 (2016). https://doi.org/10.1615/PlasmaMed.2017019498.
178. Albertos, I., Martin-Diana, A.B., Cullen, P.J., Tiwari, B.K., Ojha, K.S., Bourke, P., Rico, D.: Shelf-life extension of herring (Clupea harengus) using in-package atmospheric plasma technology. Innov. Food Sci. Emerg. Technol. 53, 85–91 (2019). https://doi.org/10.1016/j.ifset.2017.09.010.
179. Jahid, I.K., Han, N., Ha, S. Do: Inactivation kinetics of cold oxygen plasma depend on incubation conditions of Aeromonas hydrophila biofilm on lettuce. Food Res. Int. 55, 181–189 (2014). https://doi.org/10.1016/j.foodres.2013.11.005.
180. Han, L., Ziuzina, D., Heslin, C., Boehm, D., Patange, A., Sango, D.M., Valdramidis, V.P., Cullen, P.J., Bourke, P.: Controlling Microbial Safety Challenges of Meat Using High Voltage Atmospheric Cold Plasma. Front. Microbiol. 7, 977 (2016). https://doi.org/10.3389/fmicb.2016.00977.
181. El Kadri, H., Costello, K.M., Thomas, P., Wantock, T., Sandison, G., Harle, T., Fabris, A.L., Gutierrez-Merino, J., Velliou, E.G.: The antimicrobial efficacy of remote cold atmospheric plasma effluent against single and mixed bacterial biofilms of varying age. Food Res. Int. 141, (2021). https://doi.org/10.1016/j.foodres.2021.110126.
182. Patenall, B.L., Hathaway, H., Sedgwick, A.C., Thet, N.T., Williams, G.T., Young, A.E., Allinson, S.L., Short, R.D., Jenkins, A.T.A.: Limiting pseudomonas aeruginosa biofilm formation using cold atmospheric pressure plasma. Plasma Med. 8, 269–277 (2018). https://doi.org/10.1615/PLASMAMED.2018028325.
183. Zhu, Y., Li, C., Cui, H., Lin, L.: Feasibility of cold plasma for the control of biofilms in food industry, https://doi.org/10.1016/j.tifs.2020.03.001, (2020).
184. Bourke, P., Ziuzina, D., Han, L., Cullen, P.J., Gilmore, B.F.: Microbiological interactions with cold plasma. J. Appl. Microbiol. 123, 308–324 (2017). https://doi.org/10.1111/jam.13429.
185. Múgica-Vidal, R., Sainz-García, E., Álvarez-Ordóñez, A., Prieto, M., González-Raurich, M., López, M., López, M., Rojo-Bezares, B., Sáenz, Y., Alba-Elías, F.: Production of Antibacterial Coatings Through Atmospheric Pressure Plasma: a Promising Alternative for Combatting Biofilms in the Food Industry, https://doi.org/10.1007/s11947-019-02293-z, (2019).
186. Rowan, N.J., Espie, S., Harrower, J., Anderson, J.G., Marsili, L., MacGregor, S.J.: Pulsed-plasma gas-discharge inactivation of microbial pathogens in

chilled poultry wash water. J. Food Prot. 70, 2805–2810 (2007). https://doi.org/10.4315/0362-028X-70.12.2805.
187. Kazemi, M., Taghvaei, H.: A novel multi array dielectric barrier discharge plasma gas diffuser for wastewater treatment: The role of reactive species. Sep. Purif. Technol. 260, 1383–5866 (2021). https://doi.org/10.1016/j.seppur.2020.118236.
188. Patange, A., Boehm, D., Giltrap, M., Lu, P., Cullen, P.J., Bourke, P.: Assessment of the disinfection capacity and eco-toxicological impact of atmospheric cold plasma for treatment of food industry effluents. Sci. Total Environ. 631–632, 298–307 (2018). https://doi.org/10.1016/j.scitotenv.2018.02.269.

5

Utilization of Magnetic Fields in Food Industry

S. Abinaya[1], Anil Panghal[2], Roopa H.[1], Navnidhi Chhikara[3]*, Anju Kumari[1] and Rakesh Gehlot[1]

[1]Centre of Food Science and Technology, CCS Haryana Agricultural University, Hisar, Haryana, India
[2]Department of Processing and Food Engineering, AICRP-PHET, Chaudhary Charan Singh Haryana Agricultural University, Hisar, Haryana, India
[3]Department of Food Technology, Guru Jambheshwar University of Science and Technology, Hisar, Haryana, India

Abstract

Non-thermal food processing technologies (NTFPT) are appealing to food scientists, processors, and consumers due to their minimal impact on the nutritional and sensory qualities of foods while extending shelf life by suppressing or destroying microbes. NTFPT are more energy efficient and maintain better quality than traditional thermal methods. The food industries are undergoing continuous transformation as part of its ongoing evolution. The evolution to meet the demands and address the major issues posed by an ever-increasing global population and food claimants in terms of both quantity and quality. Magnetic fields (MF) are one of the NTFPT that are widely used in the food industry to protect food quality. It has numerous advantages in terms of ease of use, such as the fact that it is non-toxic, non-polluting, and safe. Magnetic fields have a broad array of applications in the food processing sector due to the favorable biological effects. Weak magnetic fields are more effective at activating bacteria and stimulating their proliferation in food processing than strong magnetic fields, which are commonly used as a physical non-thermal sterilizing method. Depending on how they behave in food systems over time, magnetic fields can be classified as oscillating magnetic fields, static magnetic fields, or pulsed magnetic fields. The ability to apply various types of magnetism requires a thorough understanding of how materials react to magnetic fields. The background information on the generation of magnetic fields, the basic principle of magnetism, and the classification of magnetic fields was presented in this chapter. In addition, this chapter

**Corresponding author*: navnidhi24@gmail.com

Navnidhi Chhikara, Anil Panghal and Gaurav Chaudhary (eds.) Novel Technologies in Food Science, (171–234) © 2023 Scrivener Publishing LLC

focuses on magnetic field research and applications in the food processing and preservation industries. It provides an overview of current experimental results for the compositional and structural analysis of various food constituents using a magnetic field. Following a discussion of compositional and structural analysis, the use of the magnetic field in various food processing methods such as freezing, drying, frying, fermentation, and quality inspection of various commodities is discussed. The effect of the magnetic field on microbes and enzymes was discussed in relation to food preservation. Magnetic fields as an alternative food processing technique have not gained widespread commercial acceptance due to conflicting results on microbial growth and death kinetics. Furthermore, there is a significant knowledge gap in the field of machine development. However, there is a significant research gap in these areas, and future progress is expected to bring solutions to unsolved issues.

Keywords: Food industry, food preservation, magnetism, non-thermal food processing techniques (NTFPT), oscillating magnetic fields, static magnetic fields, pulsed magnetic fields

5.1 Introduction

Traditionally in food industries and households, for enzyme and microbial inactivation, heat processing such as blanching, pasteurization, and sterilization had been employed as functional, cost-effective, dependable, and safe food preservation methods. Heat application, on the other hand, has some drawbacks due to its negative impact on product quality. Heat causes a variety of biochemical reactions, resulting in adverse changes in nutritional and sensory properties of heat sensitive foods [1]. As a result, food processors and customers are interested in new and innovative technologies that can work around thermal processing constraints. Consumer demand for microbially safe products with exclusive look, flavor, high nutritional content and longer storage life has fueled the innovation of novel non-thermal food processing techniques as a replacement to traditional thermal processing techniques [2]. Several methods of mild preservation have been established in the past. One of which is gamma irradiation. Even though the procedure has proven to be effective in decontaminating food, consumer acceptance of the method is poor. Non-thermal processing systems that pasteurize and sterilize foods have emerged as a viable alternative to traditional thermal processes in recent years. The majority of these innovative processes are geared toward energy-saving and environmentally sustainable applications [3]. In addition, non-thermal techniques have been found to be successful at inactivating spoilage and pathogenic microorganisms while preserving nutritional and sensory qualities in a variety of food products [2]. Some of the possible mild preservation methods such as high hydrostatic pressure, high-intensity light pulses, X-ray, ohmic and inductive heating, oscillating magnetic field (OMF), microwave, and radiofrequency. Some of the novel technologies listed above are now in commercial use, while others are only in the research and development stages.

One of these novel non-thermal processing methods is the use of magnetic fields to process food [4]. However, only a few researches on the implementation and commercialization of this technology have been conducted. By holding on unstable nutrients, reducing energy need, inactivating microbes, and in few instances, potentially treating foodstuffs within flexible packages, the technique could generate fresh-like qualities in foodstuffs [5]. Controlling cellular growth and inhibition with magnetic fields in the food fermentation industry could be a huge success. They have been applied on enzymes, microorganisms and food products to deliver safe and quality food to the ultimate consumer [6]. For food preservation using magnetic field, pulses of 1-100, frequency of 5-5000 kHz, temperature of 0-50°C and exposure time of 25μs- 10-milli sec, which is efficient enough to achieve overall acceptability of the food products. Before understanding the mechanism of the magnetic field in food applications, the knowledge of magnetism is important to know.

5.2 Magnetism

Magnetism is a process in which magnetic materials attract or repel other magnetic materials and it is caused by electron orbital and spin motions, as well as electron interactions. Describing how materials react to magnetic fields is the best way to introduce the various forms of magnetism. Hence, materials are classified as diamagnetic, paramagnetic, or ferromagnetic, based on how they react to magnetic field. The properties of different types of magnetic materials were compared in Table 5.1. All material has diamagnetic property as a fundamental feature, albeit a weak one. It is caused by orbiting electron's non-cooperative activity when they are exposed to a magnetic field. The materials with diamagnetic properties are generally made of atoms that do not have any net magnetic moments (i.e., there are no lone electrons because the shells of the orbits are occupied). Water is an excellent example of a material that has diamagnetic properties. However, when subjected to a magnetic field, a negative magnetization is developed, resulting in a negative susceptibility. Because of lone electrons in partly occupied orbitals, some atoms or ions in the material with paramagnetic property exhibit a net magnetic moment. The best example of an unpaired electron is oxygen. Specific magnetic moments, on the other hand, do not interfere

Table 5.1 Comparison of different types of magnetic materials.

Properties	Diamagnetic	Paramagnetic	Ferromagnetic
Magnetic behaviour	Repulsion	Attraction	Attraction
Magnetic moment	Very low	Very low	Very high
Magnetic field intensity	Very low	Very low	Very high
Magnetic field(B)	$B<B_0$	$B>B_0$	$B>>>>B_0$
Permeability (μ_r)	$\mu_r<1$	$\mu_r>1$	$\mu_r>>>1$
Susceptibility (X)	Negative & less	Positive & less	Positive & maximum
Examples	Cu, Pb, H_2O, Bi, Nacl	Na, $Cucl_2$, O_2	Fe, Ni, Co
	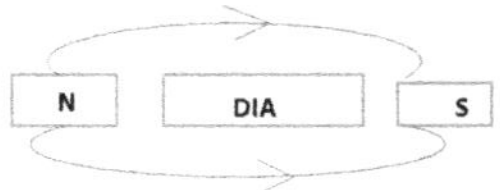	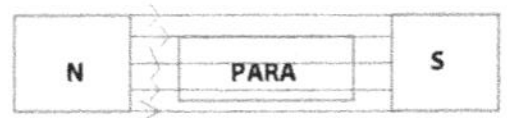	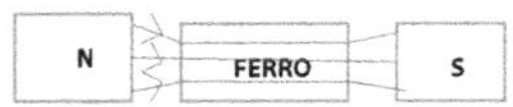

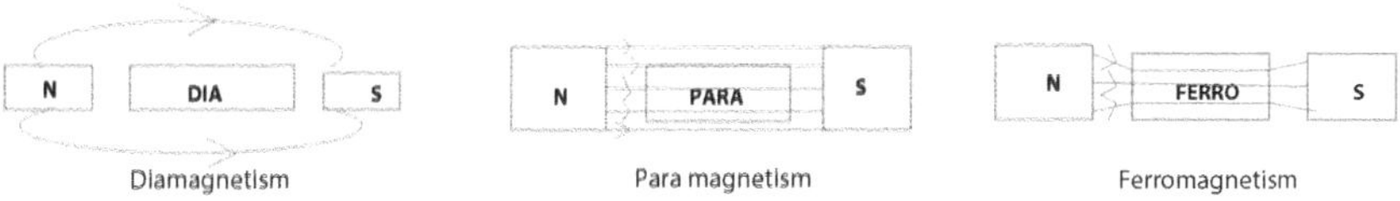

Figure 5.1 Comparison of different types of magnetic materials.

magnetically, and like diamagnetism, revert to zero when the field is eliminated. When a magnetic field is applied to the materials exhibiting ferromagnetic property, the atoms in the material are now partly ordinates in the orientation of the applied magnetic field and thus results in net positive magnetization and susceptibility. In contrast to paramagnetic materials, ferromagnetic materials have extremely strong interactions between their atomic moments. Electronic exchange forces create these interactions, which result in atomic moments aligning parallel or anti-parallel. The exchange forces are enormous, corresponding to a field of 1000 Tesla or nearly 100 a million times the strength of the earth's magnetic field. The exchange force is said to be a quantum mechanical event because of the parallel orientation of the rotation of two electrons. The materials with ferromagnetic property exhibit a parallel orientation of moments, even when there is no magnetic field, resulting in considerable net magnetization. The most common ferromagnetic materials include iron, nickel, cobalt, and various steels [7]. A force that surrounds electric current circuit is known as magnetic field (MF). The ancient Greeks as well as the Chinese were the first to record magnetic phenomena, describing irons affinity for lodestone [8]. Hans Christian, a Danish scientist, proved that the movement of electric and magnetic fields are linked. Magnetic forces are caused by the passage of electrical charges, according to a study conducted by ampere. According to a study conducted by ampere, magnetic forces are caused by the passage of electrical charges [8]. The action of a magnetic field on particles resting within it determines its existence. The most important consequences are force exerted on an electrostatic charged particle in motion, as the magnetic field changes, it exerts force on a stationary charged particle and the mutual force that exists between two objects [8]. The force imposed by a magnetic field on a charged particle in a motion like an electron or iron can be used to describe the magnetic field generated around the magnet. The comparison of different types of magnetic materials were shown in Figure 5.1.

5.2.1 Classification of Magnetic Fields

Magnetic fields can be classified as Oscillating Magnetic Field (OMF), Static Magnetic Field (SMF) and Pulsed Magnetic Field (PMF) depending on how they behave on food system over time. SMF has constant field strength over time, whereas in OMF field intensity is given in the pattern of constant and decaying amplitude sinusoidal waves. Further, Static magnetic fields are classified as super weak (range between 100nT-0.5mT), weak (less than 1mT), moderate (range between 1mT-1T), strong (range between 1-5T), and an ultra strong magnetic field (more than 5T) depending on its magnetic flux density [6]. Additionally, magnetic fields can be classified as homogenous or heterogeneous. Homogenous magnetic field provides uniform field strength all over the region surrounded by the magnetic field coil. On the other hand, the intensity of field in case of heterogeneous magnetic field is non-uniform; also, if the distance of the coil from centre increases there is a decrease in intensities [7].

5.2.2 Generation of Magnetic Field

All magnetic fields are generated by moving charged particles. In the early nineteenth century, it was found that moving electric charges produce magnetic fields [8]. During the experiment, it was discovered that when an electric current passed over a wire, the direction of a closest compass changed.

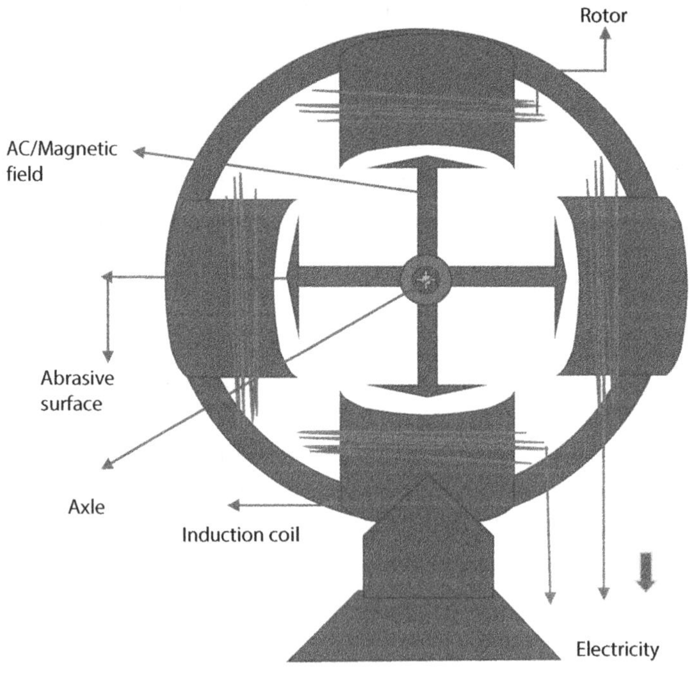

Figure 5.2 Magnetic field generator.

However, the compass would return to its natural north or South Pole with the magnetic field of earth once the current was switched off. As a result, it was found that the electrons flowing over the wire created a magnetic field that did not exist after the current was switched off. It is worth noting that charged particles only create magnetic field when they are in motion [9]. The model of a magnetic field generator was shown in Figure 5.2.

5.2.3 Magnetic Field Around a Current Carrying Conductor

An experiment in the nineteenth century demonstrated that a wire carrying an electric current is a magnet. Because the electrons all moves in the uniform direction along the wire, a well-defined magnetic field intensity surround in it. When a magnetic compass near a current carrying conductor, then the needle of the magnetic compass experiences a torque and deflects to align in the direction of the magnetic field at that point. For a current carrying conductor, the nature of the magnetic field is like concentric circles having their common centre on the axis of the conductor as shown in Figure 5.3. The intensity of magnetic field is proportional to the current flowing through the wire. If the strength of the current increased then the intensity of magnetic field will also increase. On the other hand, the intensity of the magnetic field decreases as the distance from the conductor increases [7]. It is given as;

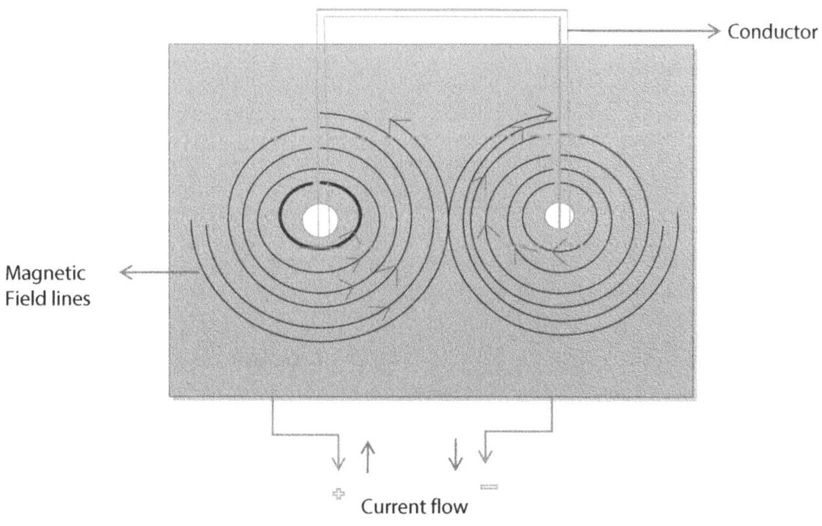

Figure 5.3 Magnetic field around a current carrying conductor.

$$B = \mu_0 I / 2 \pi r,$$

where, μ_0 is a constant and its value is $4\pi \times 10^{-7}$ Tm/A.

An electromagnet is a magnet, which generates magnetic field using an electric current. An electromagnet which is just a wire, but the magnetic field is too unstable to perform anything useful. If the wire is bend around to form a coil, on the other hand, the magnetic field of the loops will focus on the centre. The strength of the magnetic field is greater, when there is a greater number of loops on wire and the area of each loop and the current passes between the wire is higher [10]. When the centre of ferromagnetic material

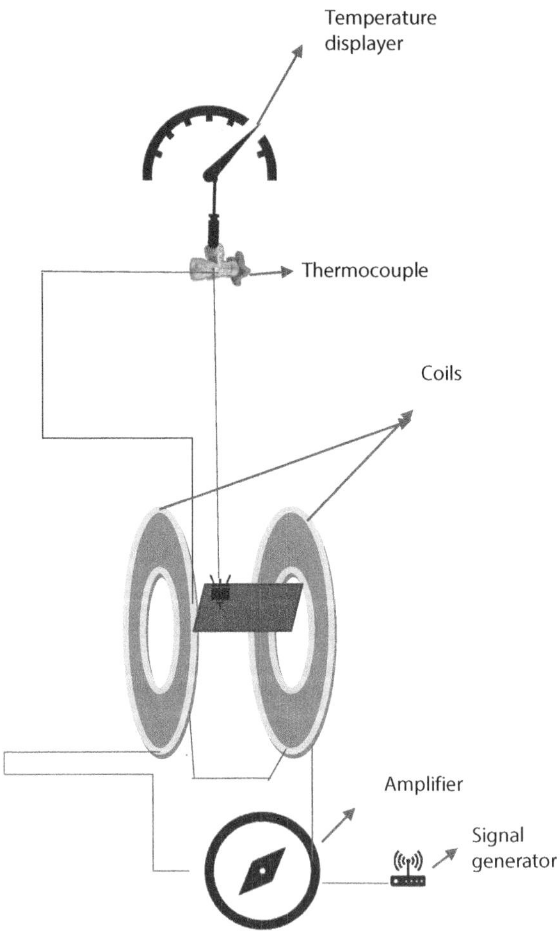

Figure 5.4 Magnetic field system.

such as iron is set inner part of the coil, the magnetic field is generated with higher intensity that is up to 3Tesla. Here the magnetic field strengthened close to the core that makes more intense than in the coil. When the current flows between the coil, magnetic domain (magnetic region) is changed along with applied field leads to an increased magnetic field intensity. Whenever increasing the current, magnetic domain finally turn out to be aligned, which is known as saturation. If the core gets saturated, an additional rise in current promoted only small rise in magnetic field. Accordingly, when iron is placed in the core it produces saturation level of above 3Tesla compared to air-core coils. Hence, air-core coils are only used to generate high intensities. There is a need for superconducting coils when more than 20Tesla of magnetic strength is required. In general, these coils are constructed with superconducting filaments such as niobium and titanium, because these materials have the critical temperature of 10K and withstand up to 15 T. The critical temperature is an important property of superconducting coil, if the temperature or intensity outreaches the critical point, leads to the loss of superconductivity [11]. To maintain the critical temperature of superconducting coils, coolants such as liquid helium and liquid nitrogen were used. The advantage of using superconducting coils than resistive electromagnets is that consume less power, smaller in size, more stable and in the steady state the consumption of power is nil [11]. On the other hand, there is a need for hybrid magnets to generate magnetic fields of above 30 Tesla. They are constructed by combining superconductive magnetic coil and water-cooled resistive magnets [12]. The model of a magnetic field system was presented in Figure 5.4.

5.2.4 Effect of Magnetic Fields in Biological Systems

Factors affecting magnetic field efficiency

Furthermore, the structure and composition of food system has a major impact on its susceptibility to magnetic field. When a magnetic field is applied on food, it behaves differently based on isotropy and magnetic susceptibility [8]. For example organic molecules present in food shows anisotropic susceptibility, which inferred distinct sensitivity to magnetization. Whereas carbon atoms present in food shows isotropic susceptibility which implies level of magnetization is same through the three rectangular axes. Another intriguing property is how some compounds are influenced by applying magnetic field under vacuum conditions. Since the majority of organic and inorganic molecules are diamagnetic, so they are more sensitive to magnetization in vacuum [13]. On the other hand, free radicals and compounds containing elements with a partially filled d sub-shell on its atom, exhibit paramagnetic property, which indicates they are less susceptible to magnetization under vacuum conditions. A number of factors including resistivity, electric conductivity, thickness, and magnetic susceptibility [8] influences magnetic field interaction with food systems.

5.2.4.1 *Effect on Microorganisms*

The aim of applying magnetic fields on microorganisms is to bring down the pathogenic or spoilage-causing microbes while also encouraging the growth of beneficial microbes that produce various fermented foods [6]. The impact of magnetic field on the inactivation of microbes can be divided into three; i) stimulatory ii) inhibitory and iii) no effects [14]. The experimental results on the effect of magnetic field on microorganisms were presented in the Table 5.2. There are two important theories, which explain the mechanisms of microbial inactivation by magnetic field. The first one is that when applying weak oscillating magnetic field might weaken the interactions between ions and protein present in microbial cells. The ions and proteins are essential for the active metabolism of microbial cells. On the other hand, the effects of OMF on biological systems are strong over specific frequencies such as the cyclotron resonance frequency of ions or in the presence of stable magnetic field. In case of cyclotron frequency, the energy is particularly shifted from the field of force to the ions present in cells, which then penetrate, into the tissues and other organs causing cell death. The second one is related to the impact of magnetic field on the calcium binding protein (calmodulin).in the calcium-binding site of calmodulin, calcium ions are constantly vibrating at an equilibrium point. The degree of vibration of calcium ions rotates in the magnetic field direction at a frequency that is half of cyclotron frequency. The addition of vibrating magnetic field at the cyclotron frequency causes the calmodulin-calcium ion bond to weaken [15]. Barbosa-Canovas *et al.* [16] demonstrated the study using the pulsed magnetic field to inactivate spoilage causing microbes in sealed foods. They used PMF conditions such as pulses (1-10pulses), frequency (5-500Hz), temperature (0-50°C), intensity (5-50Tesla), and exposure time of 0-50 and 25-100μs and observed that there is a reduction of microbes by 2 log CFU. Lins *et al.* [17] demonstrated the use of a moderate pulsed magnetic field to evaluate the microbiological quality of aerobic refrigerated stored ground beef. The study divided the samples into three categories control (without PMF treatment), PMF- 2 hours,

Table 5.2 Effect of magnetic fields on Microorganisms.

Food/microbes	Equipment	Research aim	Result	Reference
Sealed foods	PMF (Frequency-5 to 500Hz, Pulses-1 to 10, Magnetic field intensity-5 to 50Tesla)	Inactivation of Pathogenic microbes	Reduction of microbes by 2log cycle	Barbose *et al.* [16]
Aerobic refrigeration stored ground beef	PMF (Frequency-1Hz, Magnetic field intensity-10mT)	Analysis of Total aerobic count	Showed slow increase in TAC in PMF treated samples than untreated samples	Lins *et al.* [17]
Heterotrophic bacteria	Magnetic field (Frequency- 30,50,100 and 300Hz)	Sterilization effect on Heterotrophic bacteria	Low frequency square waves proven to showed the bacterial inhibitory effect	Liu *et al.* [18]
Escherichia coli	SMF (Exposure time-0 to 60 minutes, Temperature 25-45%, Magnetic field induction B- 45 to 3500mT)	Testing of viability of *E.coli* cells	Maximum exposure time of 60minutes and higher temperature of 45 and magnetic field induction of 350mT	Ji *et al.* [19]

(*Continued*)

Table 5.2 Effect of magnetic fields on Microorganisms. (*Continued*)

Food/microbes	Equipment	Research aim	Result	Reference
Listeria grayi	PMF (Magnetic field intensity-2.5T and 25 pulses)	Inactivation mechanism of *Listeria grayi*	Structure of bacterial cells was altered by damaging the cell wall membrane and PMF induced imbalance in nitrogen, energy and metabolism of L.grayi results in cell death	Wu *et al.* [20]
Yeast (*Saccharomyces cerevisiae*)	OMF	Effect of OMF on yeast cells	Alter the enzyme function of yeast cells and some cases even cell death may cause	Bayraktar [21]
Yeast (*Saccharomyces cerevisiae*)	SMF	Effect of OMF on ethanol production of yeast cells	Significant change in the conversion of ethanol to sucrose	Da Motta *et al.* [22]
Cheese whey permeate	MF	*Lactococcus lactis* (Nisin production)	Inhibition of antibiotic production (Nisin)	Alvarez *et al.* [23]

and PMF - 12 days. In these PMF-12 days, packages were stored at 4°C with the continuous exposure of PMF whereas the control and 2 hours PMF treated packages were stored at 4°C in a darker place inside the same refrigerator. The total aerobic count (TAC) was analyzed in three samples and observed that sample at the initial day (before treatment) the average TAC was found to be 3.3log 10 CFU/g and remains the same until the 6th day under refrigerated condition. On the 12th day, there is an increase in TAC in control samples of about 2.0log10 cycles, and the average TAC increased by 5.3log10 CFU/g whereas for PMF (2hour) and PMF (12day) samples showed a lower increase in TAC of about 0.7 and 1.5 log10 cycles respectively. The study concluded that a lower increase in TAC in PMF-treated samples indicates a reduction in microbial count along with refrigerated storage. Liu *et al.* [18] investigated the effect of solenoid magnetic field in the sterilization of heterotrophic bacteria in circulating cooling water. Square waves were used in this study because it exhibits strong bacterial activity among all other waveforms. The sterilization process was divided into three categories to compare the results as; parallel sterilizing groups, anti-parallel sterilizing groups, and control groups. The operating conditions used: water temperature of 30°C - 40°C), rate of water flow (100,180 and 250 litre/hour), pulse voltage (10,15 and 20 volt) and pulse frequency of 30,50, 100 and 300Hz). The study observed that the low-intensity square wave produced magnetic field causes powerful shock waves causing an alteration in bacterial proteins results in shutting down of bacterial cell wall metabolism leads to the death of the microbial cell. It is also observed that when the water moves parallel to the produced field of force, the cells of bacteria tend to migrate gradually and remain more time in the magnetic field. On the other hand, when the water flows in the opposite direction to that of the field of force, the cells of bacteria tend to migrate rapidly and remain temporarily in the magnetic field. As a result, they concluded that overall bacterial inhibitory action was found when the water moves parallel to the produced field of force than in the opposite direction of flow.

Ji *et al.* [19] investigated the effects of static magnetic fields (SMF) on cells of Escherichia *coli* (E. coli). To further explain the findings, transmission electron microscopy (TEM) and scanning electron microscopy (SEM) were employed to examine the structure of the cells. This study aimed to assess the viability of E. coli cells and to determine the dependency of colony-forming unit (CFU) towards the time of exposure, temperature, and the strength of applied magnetic field. The study used a maximum exposure time of 60 min, treatment temperature of 25-45°C, intensities of magnetic field of 45 mT, 450 mT, 1200 mT, 1800 mT, and 3500 mT. The study observed that when bacteria were exposed to the static magnetic field at an exposure time of 30 minutes, their viability was reduced. Also observed that count of CFU was reduced by exposing the bacterial cells to a higher temperature for prolonged exposure time. This indicates that at a higher temperature, E. coli cells were more susceptible to SMF. These findings could be explained by Rosen's membrane theory. Based on this theory, the impact of SMF on living cells depends on the diamagnetism of phospholipids present in the cell membrane. During the SMF treatment, these molecules reoriented and lead to the distortion of enclosed ion channels, thus results in a change in activation kinetics. Consequently, by rising the temperature, the heat produced rotational stimulation of the hydrocarbon chain makes shifting of these molecules simpler [24]. As a result, the impact of SMF on living cells becomes strengthened. Further, the study observed by changing the magnetic field induction B, different maximum values of the relative number of CFU existed. These findings could be explained by the ion interference mechanism. This mechanism explained that the separation chances of ion-protein complexes are linked to the properties of the magnetic field [25]. The structure of the bacterial cells was determined using SEM and TEM and observed that there is an occurrence of cell surface damage when bacterial cells are exposed to SMF.

Wu *et al.* [20] demonstrated the study on PMF to determine the inactivation mechanism of *Listeria grayi* by examining biological effects and monitoring the structure, membrane permeability, and intracellular protein of bacterial cells. The study analyzed the structure of *L. grayi* cells by using TEM and Laser Scanning Confocal Microscopy (LSCM) and observed that the structure of bacterial cells was altered by damaging the cell wall membrane. It is also increased the cell membrane permeability, resulting in an increase in intracellular Ca_2+ concentration under optimum conditions of PMF treatment (2.5 Tesla of intensity and 25 pulses). It was also reported that the concentration of intracellular protein also increased after being exposed to PMF. Further, the proteomic studies explained that the function of the cell membrane, intracellular protein stability, and ATP formation and transportation get affected after PMF treatment. Hence, the study revealed that PMF activated imbalance in nitrogen, energy, and metabolism of *L. grayi* might be correlated with the death of cells. According to Bayraktar [21] the enzymes present in yeast cells is altered by the application of oscillating magnetic field (OMF), also restricting their cell division and in some cases even leads to cell disintegration. Da Motta *et al.* [22] reported that by applying a static magnetic field (SMF) during the ethanol production of yeast (*Saccharomyces cerevisiae*) there is a change in the conversion of the ethanol-glucose process. Alvarez *et al.* [23] revealed that the application of a magnetic field on *Lactococcus lactis* inhibits the production of antibiotic nisin. Iwasaka *et al.* [26] stated that during fermentation in presence of a magnetic field, it is essential to recognize the paramagnetic property of O_2 molecules and the diamagnetic property of CO_2 molecules, because it restricted the mass transfer mechanism.

5.2.4.2 *Operating Conditions*

The crucial operating parameters that influence the efficacy of magnetic field application on microbial community have yet to be established. The important parameters evaluated were properties of applied magnetic field, electrical resistivity, microbiological morphology, growth phase of microbes, time- temperature combinations during processing.

5.2.4.3 *Characteristics of Magnetic Field*

Magnetic fields can either promote or hinder microbe growth and reproduction. Pulsed magnetic field intensities of 5–50 T with frequencies ranging from 5–500 kHz minimizes the count of microbes as a minimum of two log cycles. Magnetic fields of high strength can affect the fluidity of cell membranes and other cell properties of microbes. The heterogeneous type of magnetic field heterogeneity may also have a major impact on the destruction of microbial cells [8].

5.2.4.4 *Temperature*

Microbial destruction is a poorly understood as a function of temperature. Berk *et al.* [27] studied the impact of applying different intensities of between static magnetic fields on three harmful amoebae and found that after 72 hours of treatment to magnetic fields at temperatures of 20°C or higher, the count was drastically reduced. Magnetic-field-exposed amoebae had a final count of 9 to 72%, which was lower than the control sample. Exposure to the magnetic field had no effect when the temperature was held constant at 20°C for 48 hours.

5.2.4.5 *Microbial Growth Stage*

For *E. coli* bacterial cultures, a growth phase or treatment time-dependent reaction was identified. In order to test if the applied magnetic fields may operate as a stress factor, cells were subjected to mild heat treatments following incubation time of 30 minutes under magnetic field exposure. However, there were no distinctions between the treated and control samples [8].

5.2.4.6 *Electrical Resistivity*

The electrical properties of a food product, as with all electromagnetic non-thermal technologies, play a big role in its potential to be processed. A well-known fact that every electric field is accompanied by a magnetic field and vice versa. In the case of OMF, an electrical current is generated in the exposed product as the magnetic flux varies. Electrical resistivity, also known as actual electrical resistance, is a measurement of a material's resistance to electrically charged particles passing through it. As a result, the intensity of magnetic field strength is determined by the electrical resistivity and depth of the food product exposed to the magnetic field. To allow OMF microbial inactivation, a food's electrical resistivity must be greater than 10–25 Ohms-cm. The electrical resistivity of most foods falls within this range (30 Ω.cm) [8].

5.2.4.7 *Effect on Enzymes*

Enzymes are naturally occurring proteins that perform a specific function that can be useful or harmful to the quality of food. Thus, by choosing magnetic fields (MFs) action levels would therefore be determined by whether to activate or inactivate the enzyme activity. The investigations on the application of magnetic field and the effects on enzymes were presented in the Table 5.3. Ma *et al.* [28] looked into the impact of pulsed magnetic field on the action and structure of the peroxidase enzyme found in Horseradish (HPR). For the study, they used pulsed magnetic field intensity of 1-4Tesla and 5-40 pulses at 30°C. The study found that the peroxidase activity was remarkably reduced by increasing the number of pulses and intensity from 2T to 4T at 30°C. The reduction rate of peroxidase activity (67.67%) was found to be higher when by exposing the enzyme at 4T for pulses. On the other hand, the intensity of 2T and 3T for 40pulses showed a reduction rate of about 29.30% and 36.23% respectively at 30°C. It should be noted that the PMF treatment time was higher than the lower limit for the inactivation of HPR. On the other hand, activity of PMF-treated HPR was analyzed after a storage period of 48 hours at 4°C. It was discovered that there is no significant change in behavior between PMF exposed samples immediately after exposure and those that have been stored at 4°C for 48 hours. Hence, the findings show that the

Table 5.3 Effect of magnetic fields on enzymes.

Source	Enzymes	Equipment	Operating conditions	Research aim	Results	Reference
Horse radish	Peroxidase (POD)	PMF	Magnetic field intensity-1 to 4T, 5-40 pulses, Temperature-30°C	Study the structure and activity of peroxidase	Activity: At 2-4T, reduction in activity of POD Maximum inactivation (67.67%) at 4T of 40 pulses At 2 and 3T of 40 pulses inactivation percentage was about 29.30% & 36.23% respectively Structure: conformational change in tertiary structure	Ma *et al.* [28]
Algae	cellulase	AMF	Frequency-50Hz Magnetic field intensity-2.2 to 4.2mT Time-20 minutes	Study the activity and enzymatic parameter (Km & Vm of cellulase	Activity: At 50Hz frequency the maximum enzyme activity was about 25% and the minimum activity of 8% Enzymatic parameters: Km value was increased at 4.2mT thus inhibits the enzyme-substrate activity resulting in inactivation of enzyme	Zhang *et al.* [29]

(Continued)

Table 5.3 Effect of magnetic fields on enzymes. (*Continued*)

Source	Enzymes	Equipment	Operating conditions	Research aim	Results	Reference
Bacillus subtilis	α-amylase	SMF	Magnetic field intensity of 0.15, 0.30 and 0.45T	Study the activity, enzymatic parameter and structural changes in enzyme	Activity: activity ofα-amylase was increased about 27%, 34.1% and 37.8% at the magnetic field intensity of 0.15, 0.30 and 0.45T Enzymatic parameters: Both Km &Vm get reduced from 2.20×10-2 to 0.87×10-2 and 2.0×10-3 to 1.1×10-3 g/min respectively Structure: conformational change in secondary structure	Jia *et al.* [30]

PMF trated samples resulted in the enzyme's irreversible inactivation. In the same study, various models were utilized to assess the experimental outcomes to explain the inactivation mechanism of PMF-treated HPR and the models were Bigelow, Weibull, and Hulsheger. Among the three models, the Weibull model was found to be the fitter experimental data with the regression value closer to 1 and also provide a better accuracy factor with experimental results. Besides, the conformational change in the molecular structure of PMF treated and untreated HPR samples were analyzed using Ultra visible (UV) and intrinsic fluorescence spectra. The HPR treated at 3T for 40 pulses at 30°C was taken for the analysis under UV-spectroscopy. It was found that HPR showed two peaks of absorbance at 203nm and 402nm. The absorbance peaks of untreated HPR at 203 nm changed to 202nm but this change was not clear. On the other hand, the absorbance peak at 402nm does not show any change after PMF treatment, but it is obvious that the intensity changed. The intensity dropped from 0.149 to 0.081 after PMF treatment. Another technique to detect the conformational change in molecular structure was found to be fluorescence spectroscopy. Hence HPR samples analyzed under fluorescence spectroscopy showed two peaks of absorbance at 410 nm and 540nm and observed that the absorbance peak of untreated HPR at 410 nm doesn't show any change whereas the absorbance peak at 540nm showed change that is increased from 543.423 to 613.89. The results obtained from these two spectroscopic techniques indicate that there is a conformational alteration in its molecular arrangement of HPR after PMF exposure and these structural changes lead to alterations in enzyme activity.

Zhang *et al.* [29] investigated the application of alternating magnetic field (AMF) on the enzyme cellulase. The study observed that the AMF at the frequency of 50Hz was greatly influenced the activity of cellulase and also found that the magnetic field of 2.2 and 4.2mT applied at the time of 20minutes showed maximum and minimum activities of cellulase. The maximum activity of cellulase was about 25% and the minimum activity was found to be 8%. On the other hand, the enzymatic parameters such as Km and Vm were analyzed using the Michaelis-Menten model. As a result, the values of Km acquired at the field intensity of 4.2mT were found to be higher than that of the field intensity of 2.2mT. They found that the Km values were found to be significantly higher with a magnetic field of 4.2mT than the control samples. It was concluded that the magnetic field intensity of 4.2mT can reduce enzyme-substrate activity thus inhibits the activity of an enzyme. Jia *et al.* [30] demonstrated the impact of static magnetic field on the action of α-amylase enzyme. The study observed that both the magnetic field intensity and exposure period had a significant effect on α-amylase behavior, enzymatic parameters (Km, Vm), and conformational change in molecular structure. The activity of the enzyme was increased (27%, 34.1%, and 37.8%) by applying magnetic field intensity of 0.15, 0.30 & 0.45T respectively than the control sample. Also, found that by increasing the magnetic field, there is a decrease in enzymatic parameters (Km, Vm). The value of Km reduced from 2.20×10^{-2} to 0.87×10^{-2} whereas the value of Vm reduced from 2.0×10^{-3} g/min to 1.1×10^{-3} g/min. Furthermore, they observed that by increasing the magnetic field intensity, there was a change in secondary conformation structure.

5.3 Potential Applications of Magnetic Fields in Food Industry

5.3.1 Compositional Analysis

5.3.1.1 Water

Nuclear magnetic resonance spectroscopy has been studied extensively to investigate the amount of water present in food samples [31]. In reality, the structural and molecular diffusion of water in food system can be visualized using magnetic resonance imaging (MRI) technique. For instance, MRI offers a two-dimensional and limited image of the interior structure of specific foods (especially with high water content). For example, MRI provides a bi-dimensional and partial picture of the internal arrangement of specific foods (especially foods with high water content). The magnitude of the signals in various zones of the food samples may be related to the number of water molecules present. The presence of less water in an image might be indicated by darkened areas. The reason for darkened areas is due to the presence of a huge number of ions in that area [32]. Water content and distribution can also be quantified using magnetic resonance parameters (i.e., relaxation time), which offer information regarding physical and chemical properties of water. Using repeated measurements of the same samples, MRI can identify alterations in concentration of water as well as variations in water interaction with the cellular tissues of these foods. This variance could be due to fruit and vegetable ripeness, damage, decay, or other variables that affect quality [32].

Ramos-cabrer *et al.* [33] studied moisture movement in two separate multi-component food complex that were successfully monitored using 2D MRI: one had a low water activity difference, which resulted in delayed moisture movement (hours), whereas the other had a high-water activity difference, which resulted in rapid moisture movement (minutes). Comparative studies have been conducted to assess the efficiency and fineness of NMR in the application water content determination. NMR analysis and AOAC approved analytical procedures, which included forced air oven drying, were used by Keeton *et al.* [34] to analyze the moisture levels of beef, deboned chicken with skin, pork, and all-beef hot dogs. The study observed that results obtained by NMR analysis of moisture content of beef, chicken, and pork were found to be 40.42 percent, 74.37 percent, and 73.75 percent, respectively, and was in good accordance with the AOAC approved Methods (i.e., 40.39 percent, 74.57 percent, and 73.94 percent, respectively).

5.3.1.2 Fat

Nuclear magnetic imaging (NMR) technique can be used to determine total fat content as well as the fatty acid composition of various foods. This technique has been used to measure fat and oil in several agri-food items. The fat content of their milk sample was estimated to be 3.60.1 percent using 1 H NMR and a pre-saturation process to remove the signal of H_2O [35]. The results acquired by NMR technique were very similar to that calculated by the Rose–Gottlieb process [35]. For fatty acid analysis, gas chromatography (GC) has long been acknowledged as a trustworthy scientific technique, and it is frequently used to confirm NMR results. Keeton *et al.* [34] observed that the NMR analysis of fat content of beef, chicken, and pork was found to be 46.00 percent, 7.15 percent, and 3.88 percent, respectively, which were consistent with the values obtained by Soxhlet extraction method which were 45.84 percent, 7.24 percent, and 3.74 percent, respectively. This study revealed that NMR fat content analysis was quite comparable to the AOAC methods. Leffler *et al.* [36] reported that as no single method is perfect; combining NMR with other analytical methods can yield more accurate results. Adam-Berret *et al.* [37] reported that combining ^{1}H NMR and ^{13}C NMR was able to shed light on the polymorphs of triacylglycerol (α, β1 & β forms) which affect the final product quality. Also, reported that ^{1}H NMR was able to distinguish between α and β forms by assessing proton spin-interlace, which was verified by high-field ^{13}C NMR, which provides information on each triacylglycerol carbon site. All of the research that used NMR to characterize and quantify fat provides substantial evidence of NMR's adaptability, performance, and specificity in such applications.

5.3.1.3 Protein

The composition of amino acid is thought to be a key factor in determining the characteristics of protein-based foods. Only some of these factors include nutrient quality, thermal stability, and functional features. The probability of determining the total amount of protein in foods is also possible with ^{1}H NMR [38]. Using low-field ^{1}H nuclear magnetic resonance (LF-NMR), Pedersen *et al.* [39] discovered that the average protein content in mustard seeds and 20 rapeseeds was found to be 28.02% and 20.80%, respectively. NMR can also be utilized to learn more about the amino acid makeup in food. Consonni *et al.* [40] discovered that while analyzing Amarone wine, various amino acids exhibit distinct ^{1}H NMR spectral peaks, including leucine at 0.88, 1.44, and 3.37 ppm, threonine at 1.42 and 4.38 ppm, alanine at 1.45 and 3.65 ppm, proline at 2.01, 2.08, 2.35, 3.38, and 4.11 ppm, and choline at 3.20 ppm. These data again prove that NMR was found to provide accurate results while dealing with a large number of samples.

5.3.2 Structure Analysis

The use of NMR allows not only to study composition of foods but also helpful for analyzing the structure of foods. The NMR relaxation time is used in Magnetic Resonance Imaging (MRI) to quantify image contrast, allowing for visual observation of alterations in food structure that occurs throughout processing and storage. These findings are frequently utilized to identify the required structures, which might have an impact on the food's ultimate physical and mechanical attributes [38]. Oztop *et al.* [41] employed Magnetic Resonance Imaging (MRI), a noninvasive real-time imaging technology, to assess the spatial diffusion of water molecules in whey protein-based gels in the course of swelling at varied pH 2.5, 7, and 10 solutions. They observed that when water diffuses into gels at a pH of 2.5, it produces more substantial micro-structural changes in the gel than when the pH is higher. These findings suggest that MRI/NMR could be used to monitor and provide a better understanding of whey protein gel during swelling, as well as other properties of the gel. Another use for NMR/MRI is the analysis of starch crystalline structures and discovered that the crystalline structure of starch makes it possible to determine the botanical origins of the starch [42]. Therien-Aubin *et al.* [43] reported that ^{13}C CP/MAS NMR spectroscopy with a frequency of 75.46 MHz can discern these starch crystalline structures and also determining the capabilities of this technique to recognize the sources of starches and providing information regarding the alterations of these crystalline structures when subjected to various processing environment.

5.4 Food Processing

5.4.1 Freezing

Tang *et al.* [44] looked into how different magnetic fields affected the freezing properties of cherries. The intensity of the Permanent Magnetic Field (PMF) and Alternating Magnetic Field (AMF) used was in the range between 0-20mT and 0-2mT respectively. The conditions of frozen surroundings were: Temperature (-30°C) and rate of cooling (4°C/min). The study observed that there is no change in freezing temperature by PMF whereas there is an increase in freezing temperature up to -3.2 by application of 1.26mT intensity of AMF. On the other hand, the applied magnetic field has a significant impact on time taken for phase change; hence applying PMF reduces phase change time, whereas by applying AMF there is an increase in phase change time. Furthermore, the drip loss was found to be less with the intensities of 10mT (4.79%) and 1.26mT (4.64%) of PMF and AMF respectively. Whereas PMF consumes lesser energy compared to AMF. Magnetic use while freezing cherries results in reduced loss of nutrients, and smaller and uniform ice crystals as compared to untreated samples. The mean size reduction in ice crystals was found to be 67% by PMF and 78% by AMF. Overall, the study concluded that to evaluate different freezing parameters, the optimum intensities required are 10mT and 1.26mT of PMF and AMF respectively.

Choi *et al.* [45] studied the alteration in structure and sensory properties of frozen stored beef (lion and round portion) by the use of ElectroMagnetic Resonance (EMR) and air-blast freezing. The study observed that, in the early stage of frozen storage (0 months), the size of ice crystals of the beef lion by EMR was found to be 42.08μm, which is lesser compared to air blast freezing (46.61μm). The storage after 8 months of EMR freezing showed an increase in ice crystal size of the lion (42.08-90.20μm), but these values seem to be lower compared to air-blast freezing (46.61-107.20 μm). Further, the amount of ice crystal size by EMR freezing was found to be smaller (214.3%) compared to air blast freezing (232.5%) after 8 months of frozen storage. In the case of EMR freezing of beef round during the start of frozen storage (0month), the ice crystal size was found to be 46.79μm, which was smaller compared to air blast freezing (51.08μm). In case of beef lion, the size of ice crystal by EMR freezing increased from 46.79-51.08μm, but it was lower compared to air-blast freezing. It indicates upon prolonged frozen storage the ice crystal size of both beef round and lion showed a minor increase. The smaller and uniform ice crystals produced by EMR freezing, the quality of meat is not much affected during thawing. The water holding capacity (WHC) has a greater impact on meat quality. It has been estimated that fibrous protein (actin and myosin) accounts for about 65% of the water holding capacity of meat. The initial frozen storage of beef lion by EMR freezing was found to be 62.24% and these values were reduced to 58.42% after 8 months. Overall, the water holding capacity of EMR freezing and air blast freezing of beef lion throughout frozen storage were 58.42-62.24% and 55.49-59.59% respectively. This indicates the Water Holding Capacity (WHC) of the beef lion by EMR freezing was higher compared to air-blast freezing. Whereas the WHC of beef round by EMR freezing was found to be 63.32% in the initial period of storage and reduced to 58.03% after 8 months. The WHC values of beef round by EMR freezing and air blast freezing during frozen storage ranges from 58.03-63.32% and 59.72-61.32%. As a result, the WHC of both beef lion and round by EMR freezing was found to be higher than air-blast freezing. Farouk *et al.* [46] reported that the time for storage has passed and the tiny ice crystals had formed bonds with one another. Because of the recrystallization that resulted in the formation of large crystals, the water holding capacity was influenced. The thawing loss of beef lion and round by EMR freezing ranges between 0.66-2.43% to 1.25-3.10% respectively, whereas the thawing loss for air-blast freezing ranges between 0.92-3.10 to 2.13-4.39% respectively. The thawing loss of beef lion and round by EMR freezing was found to be lesser than air-blast freezing. Regarding organoleptic characteristics, the flavor and taste of the beef lion by EMR freezing seems to be higher than the beef lion by air-blast freezing. Overall, the study concluded that by the use of electromagnetic resonance freezing there is a reduction in the organoleptic characteristic's destruction and ultrastructure of beef during frozen storage for a long time. As a result, electromagnetic resonance freezing had a greater overall acceptance than air-blast freezing.

Tang *et al.* [47] demonstrated the application of the magnetic field on the process of freezing pork. The study observed that various magnetic field types have different effects on the freezing of pork samples. The nucleation temperature lowered by the use of Permanent Magnetic Field (PMF). By applying intensities in the range of 15.98 mT and 1.74 mT PMF and AMF respectively, the maximum difference in nucleation temperature was 4.12 °C. When comparing the freezing temperatures of AMF and PMF, it was discovered that AMF reduced freezing temperatures more effectively than PMF. Various interactions with water molecules in the presence of various magnetic field types can influence the generation and dispersion of ice crystals. Finer and uniform ice crystals can form when the supercooling temperature is higher, meaning that superior-quality foods can be produced with PMF. It was discovered that PMF and AMF have different phase change times. The lowering degree of AMF was lower because of the difference in nucleation rates. Moreover, in the presence of AMF, the variance pattern of the sub-cooling period was close to that of PMF. The total freezing duration in the presence of AMF was substantially longer than in the presence of PMF. It is worth noting that the alternating current coil's high heat output was detrimental to food freezing. In addition, when compared to the PMF, the AMF can consume more energy. As a result, it appears that the persistent magnetic field is better for freezing food.

Tang *et al.* [48] demonstrated the application of permanent magnetic field and an alternating magnetic field on the freezing process of blueberries. In the absence and presence of PMF and Alternating Magnetic Field (AMF), the structure of the cells, as well as freezing parameters was compared. When the intensity of PMF (10 mT) and AMF (0.05 mT, 50 Hz) were applied to the blueberry, the average size of ice crystal was decreased by 33.6% and 53.8%, respectively. Furthermore, as the PMF intensity increased, the size of ice crystal reduced first and then increased, while as the intensity of AMF increased, the size of ice crystal reduced first and then increased. With varying PMF intensities, all of the freezing parameters were substantially different. When the intensity of PMF was 10 mT, the temperature of freezing and nucleation both dropped, accompanied by a maximum supercooling temperature of 7.51°C. The time it took to go from 1-5°C and the time it took to shift phases were also decreased, which was good for quality of frozen foods (with regard to size of ice crystal). With the exception of the phase change time with variable AMF intensities, none of the freezing characteristics differed significantly. Both the nucleation and freezing temperature had a consistent variance pattern, decreasing in the beginning, then rising as the intensity of AMF increased. As the strength of AMF increased, the phase transition time increased, providing a difficulty for food freezing. The heat generated by the alternating current coils may cause the AMF aided freezing to require more electricity. The study concluded that the PMF intensity of 10 mT was shown to be the best for increasing food quality of frozen foods when all of the assessment parameters were taken into account.

5.4.2 Drying

Drying is the method of removing volatile substances including moisture from products to get solid products [49]. Various drying technologies have been used commercially for a variety of food products but each one has its own set of benefits and drawbacks. Moisture content and water activity are the two important parameters of dried product's quality and safety [50]. To manufacture superior quality food products, it is important to comprehend the water dynamics and their alterations throughout the drying process. LF-NMR (low frequency nuclear magnetic resonance imaging) and MRI (magnetic resonance imaging) are the most effective non-destructive methods for identifying and quantifying water levels and distribution in food system. The use of online NMR in the drying process seems to be having the desired impact, and it was recently deployed to reduce the time of drying, provide consistent drying throughout the product, and enhances the efficiency of the drying process. Song *et al.* [51] reported that reduced values of hardness and chewiness in abalone during rehydration were due to the rehydration period. Song *et al.* [51] also discovered that the results of (LFNMR) and (MRI) technology reduced the rehydration period of abalone from 144 hours to 120 hours, which had a significant effect on the final product's consistency. Shi *et al.* [52] reported that the moisture status of wild mushrooms in hot air oven drying was effectively monitored in real-time using a combination of LFNMR and MRI techniques. The study found that the moisture present in wild mushrooms was not uniformly distributed throughout the drying process. It was observed that the moisture dissolution flowed primarily from the center to the surface. Another study conducted by Lv *et al.* [53] used an online NMR technique with eight components. To track and regulate the Microwave Vacuum Drying (MVD) of corn kernels, they used a vacuum gauge with a diameter of 10 mm, a microwave oven with a volume of 25 liters, a cylindrical drying chamber with a height and diameter of 200 & 38mm, and a heater of hot air flow with a power of 1000 W. A vacuum pump with a power of 150W was used. The drying procedures were completed in the MVD compartment and then moved to the NMR system at a pre-determined time. The temperature was automatically monitored throughout the drying and detection process. The moisture and water content of the corn kernels were calculated every 10 minutes using an NMR technique with a detection period of 2 minutes to assure a precise result. The study reported that NMR analysis showed an unequal drying and microwave power absorption in the initial stage because of the non-uniform moisture diffusion and structure of the corn kernel. The moisture distribution appears to be more uniform as the process advances, and it was found to match the remaining moisture content of the corn kernel. In addition, the integral magnitude of NMR showed an error of less than 0.01 g moisture/gram of solid. Also, the study observed that as the drying advanced, the pace of drying, in addition the moisture content of the corn kernel, become reduced. As a result, the method was found to provide a high level of precision and can be used to estimate the residual moisture content in corn kernels. Another study conducted by Lv *et al.* [54] used Low frequency-nuclear magnetic resonance technique to evaluate the moisture present in six different vegetables such as potato, carrot, mushroom, corn, edamame, and lotus. It was discovered that the obtained results were unable to detect the moisture level of all six different vegetable species. Further, the signal amplitude A2 and the accompanying moisture content M1 of each vegetable in the distinct microwave vacuum drying (MVD) stage exhibits a significant correlation.

5.4.3 Frying

Food frying usually involves frying food in hot oil, which transfers heat to the food and melts fat while also converting water in the food to steam [55]. The sensory attributes such as food texture, color, taste, and nutritional value may be affected by under-frying at a lower temperature for a shorter period as well as over-frying at a higher temperature for a longer period [56]. Chen *et al.* [57] reported that the use of LFNMR and MRI techniques helps evaluate and monitor the impact of the process of frying and frying time on food products. Fan *et al.* [58] used the LFNMR technique to evaluate the impact of various pretreatments on vacuum fried carrots and observed that pretreatments had a substantial impact on the proximate composition of vacuum fried carrot chips. Zhang *et al.* [59] used LFNMR in conjunction with other techniques such as mercury porosimeter, confocal laser scanning microscopy, and digital picture analysis to determine the consumption of oil in the course of frying of potato chips. The study mainly aimed to determine whether the rate of oil absorption is affected by frying time, microstructure, or pore formation. The study observed that the oil chunks, oil contents, and its distribution were all influenced by frying time during the frying process and also observed that by increasing the frying time for 2 minutes there is an increase in structural oil content (40.44%) as well as total oil content (6.44%) in fried potato chips. Wang *et al.* [60] used the LFNMR technique to detect the polar components in fried camellia seed oil and observed that the number of polar compounds after the deep-frying process seems to be higher and attained the threshold limit after 24 hours. Also reported that the LFNMR was chosen as an efficient method for the quick detection of polar compounds in fried oils than column chromatography.

5.4.4 Fermentation

Fermentation is a food processing system that transforms and stabilizes food materials by the use of microorganisms. Food fermentation improves the shelf life, flavor, and texture profiles of food products also it is proven to provide possible health benefits. The study conducted by Zhang *et al.* [61] investigated the online assessment of water content in sausage during fermentation using the LFNMR approach. The relaxation time T1- weighted MR photographs revealed that water distribution in sausages was not uniform during fermentation because of the evaporation of some water and the relaxation time T2 could describe the movement of three separate water phases such as immobilized water (T22), bound water (T21) and free water (T23). Also observed that T22 and T23 were switched to the lowest relaxation time during the fermentation of sausages. Ramanjooloo *et al.* [62] used a proton-nuclear magnetic resonance technique to examine the direct conversion of sugarcane or syrup into lactic acid. The study aimed to figure out how much lactic acid was released during the fermentation process. As a result, the relaxation delay and the lactic acid-mandelic acid ratio were discovered to be essential NMR characteristics in precisely estimating the amount of lactic acid. Wu *et al.* [63] used LHNMR T2 relaxation values to show that the four protons related to different states of water molecules during the fermentation of natto. On the other hand, the MRI technique was helped to imagine the water molecules distribution inside the fermented natto and observed a complete hydration mechanism within the natto. It was also discovered that water molecules inside the natto infiltrate slowly from the periphery to the centre. Based on the finding, the fermentation process of natto might be classified into three phases, thus rendering the process more intense. Another study conducted by Goetz *et al.* [64] used the MRI technique and volumetric measurements for the online detection of bread and cake dough during fermentation and observed that during fermentation, the interior portion of both bread and cake dough was found to be less dense than the outer surface. MRI, as opposed to volumetric measures, offered the average dough volume as well as uniformity in structure by signal propagation. This knowledge will aid in the improvement of dough preparations as well as the observation of porous structure during the proofing stage.

PMF- Pulsed magnetic field, AMF-Alternating magnetic field, EMR-Electro magnetic resonance LFNMR-Low field nuclear magnetic resonance, MRI- Magnetic resonance imaging, GC-Gas chromatography, HR NMR-High resolution nuclear magnetic resonance, PLS-NMR- Partial least square-nuclear resonance imaging, EGCG-Epigallocatechin gallate, ECG- Epigallocatechin-3-gallate.

5.4.5 Extraction

The goal to achieve valuable data as the basis for evaluating the quality of trace levels of analytes has fuelled the development of current analytical procedures, such as solid-phase extraction. Wierucka & Biziuk [65] investigated the use of magnetic nanoparticles (MNPs) in the preparation of biological, environmental, and dietary samples for magnetic solid-phase extraction. When utilised in small quantities, these materials have a high selectivity and can give good analyte yield even in higher quantities. Using an external magnetic field, MNPs allow for simple and fast isolation of analytes. MNPs are now commonly used in a variety of fields, due to their ease of use, ease of surface alteration, and flexibility. These materials are used in magnetic SPE to isolate and/or enrich analytes from mixture of compounds. The key problem in research is to eliminate naturally existing substances found in several samples so as to lower the limit of detection of analytes. Interferences in the food system includes lipids, amino acids, carbohydrates were found to be a hindering factor since they can able to cause ion suppression or enhancement, making accurate spotting of desired analytes difficult. Hence pre-treatment is an important step before analysis of samples. Analyzing food samples without treatment is rare, even in sophisticated separation procedures. Even the distillates obtained using accelerated solvent extraction (ASE) or microwave-assisted extraction (MAE) procedures need to be refined using a different analytical procedure (e.g., SPE, SPME, or MSPD), as these are time-consuming, increases the analytical cost, produce more debris and inaccuracy, and can add contamination to the distillates obtained. The co-adsorption of sol will alter the properties of sorbents used in traditional techniques. The binding of different biomolecules on the surface of MNPs allows for more selective, easier, and simpler analyte separation while reducing the use of organic solvents. MSPE I a single step process, which identifies, separates and enhances analytes. To achieve more efficient extraction, the number of MNPs, pH, ionic pressure, time of extraction and form of eluent need to be monitored & optimised. Much research has been done on using magnetite MNPs in SPE for the extraction of pollutants in various samples. Though, several systematic & technological hurdles must be overcome before they can be used in food analysis. MNPs' possible health threats should be thoroughly investigated.

In the liquid–liquid extraction process, the effect of an oscillating magnetic field on mass transfer from magnetite nanofluid single drops was investigated by Saien *et al.* [66] studied the influence of oscillating magnetic field (OMF) on the transfer of mass from MNPs tiny drops in the liquid-to-liquid extraction method. The chemical system was toluene–acetic acid–water including hydrophobic MNPs present in the chemical free diffused form. The intensity of magnetic field ranging from 0.36 to 1.45 T, with varying concentrations of nanoparticles ranging from 0.001 to 0.005 wt percent. The study observed that the intensity of magnetic field of 1.45 T and a concentration of 0.005 wt % of nanoparticles, the rate of mass transfer was increased by about 65 percent on average and by 121 percent for the smallest drops.

5.4.6 Packaging

The characteristics of the packaging are directly correlated to the packaged product's quality with good choice of packaging material and optimization resulting in desirable maintenance, transportation, and handling of food products. NMR has the potential to explore the properties of packaging materials and optimize packaging materials suitable for foods. Pentimalli *et al.* [67] used NMR spectroscopy to determine the impact of gamma-irradiation with a dose range of 1-100kGy on food packaging materials such as polystyrene, poly-butadiene, high impact polystyrene, acrylonitrile–butadiene–styrene, and styrene-acrylonitrile and observed that Polystyrene was found to be an ideal polymer for packaging of irradiated foods. Lamanna *et al.* [68] used the 1H NMR technique to assess the deteriorative changes that happened in Robiola cheese when exposed to air and in a composite paper foil package stored at a room temperature of seven days. The results obtained by NMR spectrum show that compared to cheese stored in packages, the aqueous extract of cheese exposed to air revealed more differences in the amount of some chemical constituents during storage for 7days at room temperature. All these studies revealed that the NMR has the potential to assess the efficiency of various types of packages in terms of their ability to preserve food in various storage environments, as well as its potential as a tool for determining desired food package materials. The summary of the application of magnetic field on different areas of food proceesing was presented in Table 5.4.

5.5 Quality Inspection

5.5.1 Fruits

The phenomenon of Nuclear Magnetic Resonance (NMR) and its applications in evaluating the characteristics of food system depends on the intrinsic magnetic characteristics of individual atomic nuclei within a

Table 5.4 Applications of magnetic fields in different food processing methods.

Applications	Products	Research objective	Technique	Observation	Reference
i) Processing methods					
Freezing	Cherry	To study the effect of different freezing parameters in cherry	PMF Magnetic field intensity- 0 to 20mT Frozen Environment: Temperature (-30, cooling rate-4/min	No change in freezing temperature Phase change reduced Drip loss was less (4.79%) at 10mT Average size reduction of ice crystal by about 67%	Tang *et al.* [44]
	Cherry	To study the effect of different freezing parameters in cherry	AMF Magnetic field intensity- 0 to 2mT Frozen Environment: Temperature (-30, cooling rate-4/min	Increase in freezing temperature by -3.2 at 1.26mT Phase change increased Drip loss was less (4.64%) at 1.26mT Average size reduction of ice crystal by about 78%	Tang *et al.* [44]

(Continued)

Table 5.4 Applications of magnetic fields in different food processing methods. (*Continued*)

Applications	Products	Research objective	Technique	Observation	Reference
	Beef (lion, round)	To study the alterations in the ultra-structure and organoleptic characteristic of frozen stored beef	EMR	Reduces the ice crystal size (42.08μm) Prolonged frozen storage results in minor increase in crystal size Water holding capacity was increased (58.42%-62.24%) Thawing loss was reduced –lion (0.66-2.43%), round (1.25-3.10%) Organoleptic characteristic was increased	Choi *et al.* [45]
Drying	Corn kernels	To determine the moisture status	LF-NMR combined microwave vacuum drying Frequency-23.137MHz, Magnetic field intensity-0.55T	Initial stage of drying-unequal drying and microwave power absorption Later stage of drying-more uniform moisture distribution and the rate of drying and moisture content reduced	Lv *et al.* [53]

(*Continued*)

Table 5.4 Applications of magnetic fields in different food processing methods. (*Continued*)

Applications	Products	Research objective	Technique	Observation	Reference
	Abalone	To evaluate the rehydration period	LF-NMR, MRI Frequency-23.1MHz and magnetic field intensity-0.5T	Reduced rehydration period from 144 hours to 120 hours Improves final product consistency	Song *et al.* [51]
	Six different varieties of vegetables (Mushroom, lotus, potato, edamame, carrot, vegetable corn)	To determine the moisture content	LF-NMR Frequency-23.137MHz and magnetic field intensity-0.55T	The moisture content was not able to detect directly	Lv *et al.* [54]
Fermentation	Sausage	To detect the moisture content and distribution during fermentation	LF-NMR Frequency-22.4MHz	T2 area per sample weight exhibited good linearity with water content MRI maps shows that water distribution in sausage was non-uniform due to vaporizatiom	Zhang *et al.* [61]
	Natto	To determine the different status of water molecules during fermentation	LF-NMR Frequency-18.16MHz	MRI images shows complete hydration mechanism inside natto and also water molecules permeation	Wu *et al.* [63]

(*Continued*)

Table 5.4 Applications of magnetic fields in different food processing methods. (*Continued*)

Applications	Products	Research objective	Technique	Observation	Reference
Frying	Vacuum fried Carrot chips	To evaluate the effect of different pre-treatments on the property of carrots during frying	LF-NMR Frequency-23.137MHz and magnetic field intensity-0.55T	Pre-treatments shows significant effect on water content, oil content and water activity in vacuum fried carrot chips	Fan *et al.* [58]
	Potato chips	To measure the oil absorption of potatoes during frying process	LF-NMR Frequency-23.1MHz and magnetic field intensity-0.5T	Oil chunks, oil content is affected by frying process and time. By increasing frying time for 2 minutes results in increase in structural as well as total oil content. Surface oil absorption and pore formation in potato chips were closely related	Fan and Zhang [69]
	Camellia seed oil	To determine the concentration of polar compounds during frying process	LF-NMR Frequency-14.8MHz	Amount of polar compounds in oil is increased after frying	Wang *et al.* [60]

(*Continued*)

Table 5.4 Applications of magnetic fields in different food processing methods. (*Continued*)

Applications	Products	Research objective	Technique	Observation	Reference
ii) Compositional Analysis					
Fatty acid	palm, olive, safflower, and corn oils	Comparative study to determine the composition of n-6 polyunsaturated fatty acids(PUFA) in the palm, olive, safflower, and corn oils	^{13}C-NMR (300 MHz) and GC	NMR analysis-different oils contain11.2%, 14.3%, 43.7%, and 60.2 %of n-6 PUFA. GC analysis-13.5%, 15.5%, 44.9%, and 60.6% of n-6 PUFA. The results obtained by NMR analysis seems to be quite similar to GC analysis	Miyake *et al.* [70]
Fatty acid	Unrefined bonito and tuna oils from Japan, salmon oil from Norway	Comparative study to determine the composition of n-3fatty acid content in unrefined bonito and tuna oils as well as salmon oil	^{1}H-NMR (500 MHz) and GC analysis	Strong correlation between ^{1}H-NMR (500 MHz) data and those obtained by GC.	Igarashi [71]
Fat content	Beef, chicken, and pork	Comparative study to determine the fat content cf beef, chicken and pork samples	NMR and Soxhlet ether extraction method	Strong correlation between NMR data and those obtained by soxhlet ether extraction method	Keeton *et al.* [34]

(*Continued*)

Table 5.4 Applications of magnetic fields in different food processing methods. (*Continued*)

Applications	Products	Research objective	Technique	Observation	Reference
Protein	17 winter mustard seeds and 20 spring rapeseeds	To determine the mean protein content	LF-NMR	The LF-NMR results show mean protein content in 17 winter mustard seeds and 20 spring rapeseeds was 28.02% and 20.08%	Pedersen *et al.* [39]
Amino acid	Amarone wine	To analyze the amino acid composition of Amarone wine	^{1}H NMR	Some amino acids have different ^{1}H NMR spectral peaks, including leucine at 0.88, 1.44 & 3.37ppm, threonine at 1.42 & 4.38ppm, alanine at 1.45 & 3.65 ppm, proline at 2.01, 2.08, 2.35, 3.38 & 4.11ppm and choline at 3.20ppm	Consonni *et al.* [40]
Moisture content	Beef, deboned chicken with skin, pork, and Beef hot dogs	To determine the moisture content and distribution	NMR	The NMR moisture results for beef, chicken, and pork were in good accordance with the AOAC Official Methods	Keeton *et al.* [34]

(*Continued*)

Table 5.4 Applications of magnetic fields in different food processing methods. (*Continued*)

Applications	Products	Research objective	Technique	Observation	Reference
iii) Structural Analysis					
Moisture content	whey protein-based gels	To quantify the spatial distribution of water and micro-structural changes in whey protein-based gels during swelling at different pH 2.5, 7, and 10 solutions.	MRI	The MRI results show when water diffuses into gels at a pH of 2.5, it produces more substantial micro-structural changes in the gel than when the pH is higher.	Oztop *et al.* [41]
Starch	Different types of starch from various food sources	To analyze the crystalline structure of different types of starch to determine the botanical origin	MRI	MRI results revealed A-type crystalline structure from cereal starches, B-type from fruits and tubers, C-type from legumes and V-type from modified starches.	Barsby *et al.* [42]

(*Continued*)

Table 5.4 Applications of magnetic fields in different food processing methods. (*Continued*)

Applications	Products	Research objective	Technique	Observation	Reference
iv) Packaging					
Packaging of foods	Different packaging materials (polystyrene, poly-butadiene, high impact polystyrene, acrylonitrile–butadiene–styrene,&styrene-acrylonitrile)	To determine the effects of ϒ-irradiation at a dose range of 1-100kGy on food packaging polymers	NMR	NMR results observed that Polystyrene was found to be an ideal polymer for packaging of irradiated foods	Pentimalli *et al.* [67]
	Robiola cheese	To assess the deteriorative changes in Robiola cheese when exposed to air and in a composite paper foil package of 7-day storage at room temperature	^{1}H NMR	NMR spectrum show compared to cheese stored in packages,the aqueous extract of cheese exposed to air revealed more differences in the concentration of some chemical compounds.	Lamanna *et al.* [68]

(*Continued*)

Table 5.4 Applications of magnetic fields in different food processing methods. (*Continued*)

Applications	Products	Research objective	Technique	Observation	Reference
Other food applications					
	Tomato juice and pulp from two cultivars(Red Setter and Ciliegino)	To determine the Chemical composition which also includes several organic acids and aliphatic groups	HR NMR–to analyze the fluid portion, and magic angle spinning (MAS) NMR was used to analyze the tomato pulps	-Resonances in high-field region are primarily due to aliphatic groups of amino acids and organic acids from the tricarboxylic acid metabolic cycle. -Resonances in the middle area(from 3.2 to 5.3 ppm) are mainly due to saccharides and major components (D-glucose and D-fructose). Resonances in the lower region are mainly due to minor components of tomato juice, such as aromatic amino acids, phenolic compounds, and formate. -All water-soluble substances present in both juice and pulp	Sobolev *et al.* [72]

(*Continued*)

Table 5.4 Applications of magnetic fields in different food processing methods. (*Continued*)

Applications	Products	Research objective	Technique	Observation	Reference
	Coffee	To quantify the major constituents in coffee (Caffiene, formic acid, trigonelline, and 5-hydroxymethylfurfural)	^{1}H NMR Frequency-500 MHz	The coefficients of difference for ten measurements by ^{1}HNMR are found to be 4.2%, 2.6%, 2.4%, and 7.3% for caffeine, formic acid, trigonelline, and 5-hydroxymethylfurfural respectively.	Del Campo *et al.* [73]
	Mozzarella cheese	To determine the water mobility	NMR	Two phases of water were observed: The high mobile bulk water entrapped inside the casein matrix and less mobile water-bound directly to the casein	Kuo *et al.* [74]
	Cheese	To estimate the dissolution of water as well as fat globules in cheese	Pulsed gradient NMR	Observation: The size of serum pockets was found to be about 10 mm and the size of fat globules to be about 2–3 mm using this method, which is consistent with microscopic observations.	Callaghan *et al.* [75]

(*Continued*)

Table 5.4 Applications of magnetic fields in different food processing methods. (*Continued*)

Applications	Products	Research objective	Technique	Observation	Reference
	Japanese Green Tea	To determine the levels of chemical constituents in Japanese green tea in various frequency ranges	^{1}H NMR	-In the low-frequency range of 0.5-0.3ppm- Amino acid theanine (1.10, 2.12, 2.39, 3.19 ppm) and quinic acid (1.88, 1.97, 2.05 ppm) signals were observed. -In the middle frequency range of 3.0–4.5 ppm- Caffeine (3.27, 3.43, 3.89 ppm), arginine (3.22, 3.47 ppm), Myo-inositol (3.30 ppm), chlorogenic acid (3.89, 4.22 ppm), and quinic acid (3.56 and 4.01 ppm) signals were observed -In the high-frequency range of 5.0–8.0 ppm-2-O-β-L arabinopyranosyl-myoinositol (5.19), p-coumarylquinic acid and/or cinnamic acid (7.51, 7.75 ppm), EGCG (6.61, 7.02, 7.14 ppm), and ECG (6.61, 7.02 ppm) signals were observed	Tarachiwin *et al.* [76]

(*Continued*)

Table 5.4 Applications of magnetic fields in different food processing methods. (*Continued*)

Applications	Products	Research objective	Technique	Observation	Reference
	Beer	To quantify the major organic acids in beer, acetic, citric, lactic, malic, pyruvic and succinic acids	PLS-NMR	The PLS1-NMR method proved to be a quick and efficient way to analyse the major organic compounds in beer.	Duarte *et al.* [77]

field of force [78]. NMR imaging technology can produce superior quality images of the interior assembly of horticultural crops, making it ideal for distinguishing interior damages and the quality attributes including warm injury, surface damage, bruises, maturity and ripening stage, senescence, and the presence of seeds, voids, and pits.

5.5.1.1 Apples

Wang *et al.* [79] investigated the normal and water cored tissues of apples using NMR imaging technology and observed that the water cored tissue come about 20mm out from the center of the fruit and also detected the most damaged area in the range between 5-10mm out from the center. McCarthy *et al.* [80] investigated the bruised apples using the NMR imaging technique and observed that the bruised part of the apple looks brighter than the unaffected region. They also studied the time-related difference between a bruised and unaffected region of apples by pulse sequence technique and observed that time-related difference in both regions found to be increased. Finally, by differentiating the images obtained, they found that the variations in time-related changes within these two regions may contribute to a small shift in diamagnetic susceptibility. Specifically, the fluid released from the bruised region replaces air in the intracellular region, thus the entire area of the network over which magnetic field occurs reduced. Chen *et al.* [81] reported that the rate at which the process begins depends on the degree of damage caused and the character of cell structure present in the tissue. Generally bruising is characterized by a water-soaked area that shows a greater signal difference, whereas void space is identified in the absence of signal. On the other hand, maturation takes place as a result of a rise in water as well as other constituents such as carbohydrates, fats, etc. In the case of over-ripened pineapple, translucency and free water content can be easily detected by NMR imaging technology. Mealiness is characterized by the impact of desegregated tissue and lack of juiciness in fruits, which considered to adversely affecting the sensory textural characteristics of the fruit [82]. Barreiro *et al.* [83] assessed the mealiness in apples using NMR imaging technology. They observed that T2 values for mealy apples were found to be minimum than that of fresh apples, this is due to desegregated tissue and lack of juiciness in mealy apples.

5.5.1.2 Citrus Fruits

The existence of seeds in mandarins and oranges are the key worries for citrus growers. Seeds are particularly unappealing in mandarins due to market demand [84]. Obtaining MR images that differentiate seeds from the juice segment is the most auspicious technique to apply when it comes to MRI identification of seeds in citrus. Hernandez *et al.* [85] used MRI to examine mandarins for the existence of seeds found that design produced under static conditions can be easily exported to online conditions. Under MRI static conditions, they discovered that 88.9% of mandarins were without seeds and 86.7% of mandarins were with seeds. On the other hand, under driving conditions, they found that 92.5% were without seeds and 79.5% with seeds. In the same study, by imaging the fruits when moving at 54mm/s, the output under online conditions was assessed. As a result, the diagonal relaxation time-weighted quick low-angle shot images (703ms of acquisition time) were used efficiently to improve the difference between seeds and pulp. Finally, they concluded that under static conditions, the obtained transverse MR quick images at the acquisition time of 703 and 390ms is proven to provide a good difference between seeds present in citrus fruits, and also there is a drop in image quality when fruits moved at a rate of 54mm/s compared to static images. Wang *et al.* [86] used LFNMR imaging technique to examine the alterations in moisture content and its mobility in different methods of treating naval orange in storage and found that the water migration shows a strong correlation in conjunction with naval orange fraudulence. As a result, MRI is a viable method for obtaining the high-resolution difference between seeds and pulp even so it takes time and thus limited to offline conditions at the juicy state. Barreiro *et al.* [87] examined two forms of quick MRI sequences to obtain better results such as a gradient echo and spiral-radial sequences at the acquisition time of 484ms and 240ms respectively to obtain MR images of the inside anatomy of mandarins to detect seeds without damaging the fruits.

5.5.1.3 *Kiwifruit*

Since the epicarp of the kiwifruit is extremely permeable to water mass transfer, thus leads to water loss during prolonged storage or transportation. At the rate of 0.5% loss of water, enzymes present in the cell wall get increased so far, and an additional increase of water loss speed up respiration and ethylene synthesis as well as volatile loss [78]. Aside from water loss, very minor damages which are caused by bruising lead to the activation of ethylene production which enhances the ripening process. As a result, kiwifruit needs special attention throughout the harvest and post-harvest operations. Thus, MRI can be an efficient method for non-destructively tracking minor changes in structures in kiwifruit during the ripening process. Burdon and Clark [88] used the MRI technique to analyze the water level of kiwifruit and observed that the loss of water and alterations in moisture boundaries were dependent on the original moisture content of the kiwifruit and by determining relaxation time (T1, T2) revealed major differences were seen in the fruit's core, inner and outer regions. On the other hand, there are no such differences that arise on MRI parameters during analysis. Newman *et al.* [89] used a nuclear spin relaxation technique to found that the softening of cell wall pectin of kiwifruit during the early stages of ripening before solubilization in the later period of ripening. Another study conducted by Taglienti *et al.* [90] used the MRI technique to examine the consequence of storage conditions on least changes in the texture of kiwifruit tissue and observed that MRI was found to be an efficient method for detecting minor textural changes associated with water movement. This proceeding changes in the kiwifruit softening even stored at the temperature of 0°C may be due to these least textural changes, which are likely caused by a wound ethylene reaction. Vaientini *et al.* [91] used MRI technology to differentiate kiwifruits exposed with plant growth substances and control sample (without treatment) and also used to differentiate the plant growth regulators such as auxin and cytokinin.

5.5.2 Vegetables

5.5.2.1 Tomato

A bruising is a condition of failure of subcutaneous tissue in tomatoes that do not result in skin rupture. Various techniques have been developed to investigate and measure bruise damage in the case of tomatoes such as thermal imaging, blue laser light scatter imaging and 1 H-NMR imaging [78]. Researchers claimed that 1 H-NMR imaging technology was the most effective tool to detect bruise damage in tomatoes (immature as well as red ripe fruits). For red ripe as well as fragile tomatoes, bruises appeared as highly water-soaked areas under the skin that could spread to the central part of the tomatoes. On the other hand, the effect of bruises above the locule also damaged the whole pericarp wall [92]. Another study conducted by [93] used non-invasive nuclear magnetic resonance imaging technology to assess the maturity of tomatoes. They took the ripened tomatoes for the assessment and observed that variations in water content have a direct correlation in the developmental changes in the pericarp, seed, and other tissues. Ishida *et al.* [94] used the NMR technique to measure the ripening stage, amount of carbon dioxide and ethylene synthesis, the thickness of the pericarp wall, firmness of locular tissue as well as pigments such as lycopene and chlorophyll content. It was observed that higher signal was concurrent with the dissolution of locular tissue and also increase in signal was associated with the pericarp of ripened red tomatoes than immature green ones. They finally concluded that the increased signal strength was linked to placental tissue dissolution during the stages of ripening whereas the wall of the pericarp grittiness was linked to the reduced wall thickness. As a result, NMR images of mature-immature tomatoes help to distinguish between mature red and immature green tomatoes.

5.5.2.2 *Potatoes*

The quality of cooked potatoes is an important concern among food manufactures. Both the structural and chemical composition of potatoes largely influenced by a variety of factors such as variety, soil type, maturity, growing season, processing, and storage conditions [78]. Specifically, textural characteristics are one of the most significant factors in determining both consumer's and food manufacturer's acceptance [95]. MRI technology was used to examine changes in the interior portion of fruits during developmental changes, as well as report internal damages and their spread using high-quality images [96] but it also used to investigate the organoleptic characteristics of cooked potatoes [97]. As the texture of cooked potatoes is an important organoleptic character, it can be determined using pulsed low field 1 H-NMR of raw potatoes. The texture consistency of cooked potatoes can be analyzed from the relaxation curves of raw potato samples since these curves also provide the knowledge of moisture content which is contrarily related to the starch content of the potatoes [98]. Thybo *et al.* [98] also reported that low field NMR appears to be marginally efficient at characterizing the adhesion and springiness attributes than chemical variants. Because of its susceptibility to the moisture content present in the sample, low-field NMR can explore those mouths feel attributes that no other instrumental techniques can calculate. On the other hand, Thybo *et al.* [99] used LFNMR and NMR imaging technique to identify different potato samples and analyzed dry matter content in potatoes, and observed that images obtained by NMR showed exactly wide contrast between varieties, cellular tissues and maturation. In their study, image analyses obtained depends on histogram study and were not related to the dry matter content of the whole potatoes. As a result, the relation between NMR details and dry matter content in potato samples was not complemented by MR imaging. This implies that the method used to determine the spatial diffusion of water in the potato samples did not account for all of the details about the dry matter. As a result, MR-image processing would take further attention in the upcoming trends. Because of the success story of MRI technology in various food applications, Straadt *et al.* [100] used a combination of techniques of (LFNMR) and Confocal laser Scanning Microscopy (CLSM) to study the effects of the brining on the dry matter content of potato samples. They demonstrated that the capability of LFNMR & CLSM can be used to identify and explain structural as well as water migration changes in the potato tissue and found that combining NMR and CLSM yielded valuable knowledge about the origins of NMR water populations. As salting resulted in the loss of weight and turgor pressure in raw potato cells and the case of high dry matter containing potatoes after the salting process, the total water loss was found to be the least. Hence, the LFNMR methodology clearly shows that this is due to a faster T2 relaxation period of both cytoplasmic and extracellular water content in potatoes with high dry matter content, indicating that water was more limited. Furthermore, in the same study, they noticed a trend towards a quick T2 relaxation period with the increasing brining time in case of low and high dry matter content potatoes, indicating a salt-developed reduction of the cytoplasmic and extracellular water content.

5.5.3 Cereal and Cereal Products

Food companies around the globe are constantly introducing novel products in all areas of food industry. Cereals and cereal products are one of them and they play a significant role in food innovation. Concerning consumer acceptance, cereal products such as rusk, bread, cookies, and snacks have essential features such as crispiness, crunchiness, etc [78]. The reason behind the loss of such essential features in cereal products is due to the water absorption, migration and distribution during storage. Due to this, the understanding of microstructure in cereals and water migration is important to minimize the losses that occurred during processing and storage. Hence, there is a need for non-destructive techniques to visualize the microstructures, structural modifications, water transport, and diffusion. Magnetic resonance imaging technology is proven to provide real-time visualization of cereal grains and their product's microstructures and modifications caused as a result of processing and storage. Horigane *et al.* [101] investigated the water diffusion and alteration in the structure of cooked rice grains using proton-NMR imaging technology. They found a hollow in the internal structure of all the analyzed cooked rice grains and the hollow form is due to the closing of laceration caused by the gelatinization of starch granules and enlargement of grains during the cooking process. The hollows of cooked grains are considered cracks or fissures. As a result, they concluded, that NMR micro imaging has a lot of promise in whole grain research. For better outcomes, Mohoric *et al.* [102] to obtain good results used MRI and X-Ray micro tomography in conjunction to analyze the microstructure and hydration behavior of cooked rice kernel and observed that MRI technology provides three-dimensional, real-time visualization of water passage into rice kernels during cooking process. Hwang *et al.* [103] analyzed the moisture migration and distribution during the tempering process of a single rice kernel by using the MRI technique and the study revealed that MRI technology was found to help examine the dynamic moisture migration within individual rice kernels. The study also reported that the MRI technology can be used as a guide and validation for the moisture migration simulation models as well as an effective methodology to investigate moisture migration mechanisms within the cereal grains. Ishida *et al.* [104] investigated the moisture loss in harvested seeds of rice by MRI technique at different temperatures and observed that the embryo and endosperm region of rice seeds occupies most of the water content. The reduction in water content at 50°C was found to be higher in the outside region than the central part of the rice seeds. They also found that the reduction in water content from rice kernels was raised greatly when the temperature reaches close to 60°C and then rise slightly when the temperature reaches above 60°C. All these research findings revealed that the MRI technique may be an effective method for the non-destructively investigating mechanism of moisture migration and distribution in the cereal grains.

5.5.4 Seafood

The use of MRI technology to evaluate sea-foods is not a novel concept. Since MRI is a technique that allows for the creation of high-resolution transverse section images of intact species, it has been used to gain a basic understanding of a variety of topic including anatomical, compositional, structural and thermal characteristics of seafoods [105]. A device to extract pin bones will benefit pre-rigor processing of Atlantic salmon fillets. MRI images of salmon fillets were obtained using the 1H MRI mode to examine the position of the tendon connections in pre-rigor Atlantic salmon [106]. Bock *et al.* [107] reported that using MRI to image live unanesthetized fish in a flow tube through an MRI scanner is possible. The effect of temperature on the circulatory output of Baltic Sea eelpout was studied using flow-weighted 1H-MRI technology. Flow-weighted MRI can also be used in marine animals for angiography and to observe blood flow changes in larger vessels [107]. Foucat *et al.* [108] used 1H-MRI to determine the water distribution in white rainbow trout muscle with soft meat, a known situation that generally happened in farmed fish. Veliyulin *et al.* [109] used 1H-MRI technology to study the belly-bursting phenomenon in real-time and observed that neither the stomach nor the zooplankton in the stomach is the source of enzymes that induce belly bursting in herring. Finally, he concluded that belly bursting in herring is caused by two anatomical structures such as the ventral muscle and the part of the intestine. 1H-MRI can provide details on alterations in texture of fish after subjected to freezing and thawing. Veliyulin *et al.* [110] reported that the relaxation times (T1 & T2); also, the portion of water which is not frozen in the fish sample was measured as a constituent of temperature in an MRI analysis of frozen cod fillet segments. With the temperature (0-4°C), relaxation time and portions of water which is not frozen decreased rapidly below the freezing point. Foucat *et al.* [111] studied a variety of fresh and frozen-thawed samples of trout (frozen storage time- 1 to 44 days) at the same time. In all samples, different variables such as transfer rate of magnetic field, T2 relaxation period, and diffusion constant, which is parallel and perpendicular to muscle fiber direction, were analyzed. As a result, only the average value of T2 and diffusion perpendicular was remarkably influenced by the freezing-thawing process. On the other hand, only the diffusion values in perpendicular direction were influenced by a prolonged frozen storage of 41 days compared to fresh samples. In addition, the loss of diffusion anisotropy after frozen period of 41 days resulted in the increased disorder of the muscle fiber arrangement that was cleared by partial denaturation of protein.

Since the addition of salt inhibits the activity of microorganisms, it has long been one of the most essential food preservation techniques. It is well known that excessive salt consumption leads to negative health consequences and food manufacturers are considering reducing salt content in their products as a key strategy. Chemical methods for measuring salt content have significant flaws and there is a clear understanding of rapid and non-destructive methods. 23Na MR imaging is an effective method for optimizing industrial salting techniques as well as monitoring entire concentration of salt and its diffusion in processed meat products [112]. The use of MRI to determine the total concentration of salt in biological substances is a difficult experiment. Traditional MRI procedures based on the observation of spin-echo total volume of sodium content are frequently underrated, a phenomenon known as incomplete MRI detection of sodium ions. The existence of 23Na signal components with exceedingly shorter lifetimes and the impotence to detect these signals with standard

Table 5.5 Applications of magnetic field in quality inspection.

Products	Applications	Reference
FRUITS		
Apple	Detection of bruised and non-bruised regions in apple Assessment of mealiness Examination of water core loss	McCarthy *et al.* [80] Barreiro *et al.* [83] Wang *et al.* [79]
Citrus	Identification of seeds	Hernandez-Sanchez *et al.* [85], Barreiro *et al.* [87]
Kiwifruits	Assessment of water loss and minimal textural changes during storage conditions	Burdon & Clark [88], Taglienti *et al.* [90]
Mango	Observation of bruising, chilling injury and damages by insects	Mazucco *et al.* [113]
Pear	Assessment of internal browning and breakdown of core	Lammertyn *et al.* [114], Hernandez-Sanchez *et al.* [115]
VEGETABLES		
Potato	Analysis of sensory characteristics Distinguish between potato samples Differentiation of dry matter content in potato samples Analysis of texture characteristics in cooked potatoes	Thybo *et al.* [97] Thybo *et al.* [99] Thybo *et al.* [99] Thybo *et al.* [98]
Tomato	Detection of bruises Differentiate mature and immature fruits	Almeida & Huber [92] Saltveit [93]
Carrot	Analysis of change in pH	Evans & Hall [116]
Cereals		

(*Continued*)

Table 5.5 Applications of magnetic field in quality inspection. (*Continued*)

Products	Applications	Reference
Wheat	Estimation of moisture distribution during storage	Song *et al.* [117]
Corn	Detection of water movement and distribution Detection of stress cracking	Ruan and Litchfield [118] Song & Litchfield [119]
Rice Kernel	Detection of moisture loss during storage Determination of microstructure and hydration behavior Determination of water diffusion & alterations in structure during cooking Assessment of Moisture migration & distribution during tempering	Ishida *et al.* [104] Mohoric *et al.* [102] Horigane *et al.* [101] Hwang *et al.* [103]
Fish products	Impact of freeze-thawing of frozen trouts Study of anatomy Observation of belly bursting in herring Effects of salt content on cod and salmon fillets	Veliyulin *et al.* [110] Bock *et al.* [107] Veliyulin *et al.* [109] Veliyulin *et al.* [109]
Meat products	Diffusion of sodium ions Water activity (aw)	Bertram *et al.* [112] Venturi *et al.* [120]

Data from Patel *et al.* [78], Melvina *et al.* [121].

MRI procedures causes this effect. The level of sodium underestimation was found to be a measure of salt concentration and muscle structure ranging from 4-79% in homogenously salted cod and salmon fillet parts at brine strength of 5-25% [122]. The application of magnetic field in quality inspection of various commodities were presented in Table 5.5.

5.5.5 Other Food Applications

Sobolev *et al.* [72] used 1 H HR-MAS NMR (High resolution-magic angle spinning nuclear magnetic resonance) to determine the chemical composition (which also includes several organic acids and aliphatic groups) of juices and pulp of tomato from two cultivars such as Red Setter and Ciliegino. They used the HR NMR technique to analyze the fluid portion, and magic angle spinning (MAS) NMR was used to examine the tomato pulps. Especially Valine, isoleucine, threonine, alanine, glutamate, glutamine, aminobutyrate (4-aminobutanoate), asparagine, aspartate, citrate (2-hydroxypropane-1,2,3-tricarboxylate), acetate, and malate were discovered. In addition, the study observed that the resonances in the proton spectrum's high-field region are primarily because of amino acids and organic acids, whereas the resonances in the middle area (from 3.2 to 5.3 ppm) are mainly because of sugars. On the other hand, the resonances in the lower region are mainly because of minor constituents of tomato juice. 1H HR-MAS NMR has many benefits, including the ability to analyze semi-liquid & solid samples without going through any extraction procedure and achieve a good resolution images comparable to that of a liquid sample [38].

Coffee is a common beverage that contains caffeine, the most commonly consumed psychoactive compound in the world. Caffeine, formic acid, trigonelline, and 5-hydroxymethylfurfural are the major constituents in coffee that have piqued scientists' interest. Caffeine and trigonelline are responsible for bitter taste, formic acid for acidity, and 5-hydroxymethylfurfural is a consistency indicator of coffee during roasting. Using 1H NMR techniques, several major constituents contained in coffee (such as caffeine, formic acid, and trigonelline) were analyzed. Del campo *et al.* [73] used 1H NMR to estimate the content of these major constituents in commercialized soluble coffees and observed that the results of 1H NMR spectra showed 4.2% caffeine, 2.6% formic acid, 2.4% trigonelline, and 7.3% 5-HMF. While another study conducted by Bosco *et al.* [123] only documented the qualitative analysis of caffeine, formic acid, and trigonelline using the 1H NMR technique. This was the foremost research to describe the empirical measurements of the NMR method for the chemical evaluation of coffee [73].

Budiman *et al.* [124] reported that the HR-MS-NMR has the potential to determine the amount of moisture and fat in cheese and in addition, it is also used to assess the ripening process by assessing several chemical compounds in cheese. Kuo *et al.* [74] used NMR (Nuclear magnetic resonance) imaging technique to determine the moisture movement in Mozzarella cheese and observed that a minimum of two boundaries of water was present: one is the high bulk mobile water entrapped inside the casein matrix whereas the less mobile water-bonded directly to the casein.

Weekley *et al.* [125] discovered the principal proton-containing constituents in a complete bottle of spoiled wine are water, ethanol, and acetic acid, by assessing the intensity of their peaks in the ^{1}H NMR spectrum might be possible to evaluate the level of wine spoilage. Also reported that the spoilage of bottled wine is also determined by calculating acetic acid, complex sugars, phenols, and trace elements present in wine. The combination of ^{1}H NMR and FTIR attenuated total reflectance (ATR) can be used to distinguish between various beer samples depending on their alcohol content. This offered rapid information on different stages of beer fermentation, which is an essential for the understanding of beer production [77]. Duarte *et al.* [77] observed various organic acids such as acetic, citric, lactic, malic, pyruvic, and succinic acids in beer samples. Of which organic acids was found to be essential in beer quality because they contribute to flavour, colour, and aroma while also serving as good indicators of fermentation progress. Rodrigues *et al.* [126] reported that the well-established PLS-NMR (Partial least square-nuclear resonance imaging) technique for determining number of organic acids present in beer samples which offers useful details about the quality and history of the product.

Well-trained panelists using sensory evaluation methods frequently do green tea quality analysis. Since it is a subjective operation, it may be inconsistent. Tarachiwin *et al.* [76] employed 1H NMR to develop new methods such as metabolic profiling and fingerprinting. In this study, green tea leaves were dried, ground, added with Deuterium Oxide (Heavy Water), and then subjected to further treatment to obtain a supernatant. Following that, the supernatant was subjected to1H NMR technique to determine the level of chemical constituents present in green tea at various frequency ranges. The study observed that the amino acids theanine (1.10, 2.12, 2.39, 3.19 ppm), which is found in green tea (20), and quinic acid (1.88, 1.97, 2.05 ppm) were mainly found in the low-frequency range of 0.5–3.0 ppm. Caffeine (3.27, 3.43, 3.89 ppm), arginine (3.22, 3.47 ppm), Myo-inositol (3.30 ppm), chlorogenic acid (3.89, 4.22 ppm), and quinic acid (3.56 and 4.01 ppm) signals were present in the middle-frequency area at 3.0–4.5 ppm after removing sugar signals. The 2-O-β-L arabinopyranosyl-myoinositol (5.19), p-coumaryl quinic acid and/or cinnamic acid (7.51, 7.75 ppm), EGCG (6.61, 7.02, 7.14 ppm), and ECG (6.61, 7.02 ppm) signals were observed in the high-frequency region at 5.0–8.0 ppm. Because of least soluble in water, signals of EGC (6.05, 6.08, 6.61 ppm) and EC (6.05, 6.08, 6.91, 7.02 ppm) were negligible in this area [76].

5.6 Conclusion

Consumer demand for minimally processed foods is increasing, so the food processing industry is on the lookout for novel developments of mild processing technology. Because of contradictory results on microbial growth and death kinetics, magnetic fields as an alternative food processing technique have not received widespread commercial acceptability. Furthermore, there is a considerable shortage of knowledge on the subject and machine development. The effectiveness of magnetic fields in deactivating microbes must first be proven by food experts. Using superconducting technology, the application of magnetic fields could be practical and advantageous for biomass production at a controlled growth rate. The fermentation and pharmaceutical sectors would benefit from this. The structural alterations of proteins and lipids could be accomplished by subjecting them to magnetic fields. Though, there is a considerable paucity of research in these areas, thus future progress is hoped to bring solutions to criticizing issues.

References

1. Zhang, Z., Thermal food processing: New technologies and quality issues. *Trends in Food Science & Technology,* ***19***(1), 48, 2008.
2. Zhao, Y.M., De Alba, M., Sun, D.W., & Tiwari, B., Principles and recent applications of novel non-thermal processing technologies for the fish industry—a review. *Critical Reviews in Food Science and Nutrition,* 1–15, 2018.
3. Awsi Jan., & Sajad Ahmad Sofi., Non-thermal processing in food applications: A review. *International Journal of Food Science and Nutrition,* ***2,*** 171-180, 2017.
4. Morris, C., Aaron, L., & Brody Louise., Non-thermal food processing/preservation technologies. *Packaging Technology and Science,* **20,** 275-286, 2007.
5. Martin, F. S., Harte, F., Barbosa-Canovas, G. V., & Swanson, B. G., Magnetic field applications to foods. *Encyclopedia of Agricultural, Food and Biological Engineering,* 580–584, 2003.

6. Minano, H. L. A., Silva, A. C., de. S., Souto, S. P. A., & Costa, E. J. X., Magnetic fields in food processing perspectives. *Applications and Action Models. Processes,* ***8***(7), 814, 2020.
7. Jasim Ahmed., & Hosahalli, S., Ramaswamy., Application of magnetic field in food preservation, CRC press, 855-865, 2007.
8. Grigelmo-Miguel, N., Soliva-fortuny, R., Barbosa-Canovas, G. V., & Martin-Belloso, O., Use of Oscillating magnetic fields in food preservation. *Nonthermal Processing Technologies for Food,* 222–235, 2011.
9. How magnetic fields are creates. Retrieved from https://study.com/academy/lesson/how-magnetic fields-are-created.html, 2018.
10. Gersdorf, R.., DeBoer, F.R., Wolfrat, J.C., Muller, F.A., & Roeland, L.W., The high magnetic facility of the University of Amsterdam, high field magnetism. In: Proceedings International symposium on High Field Magnetism, 277–287, 1983.
11. Asner, F.M., High Field Superconducting Magnets. Oxford University Press, 264, 1999.
12. Iwasa, Y., Hybrid magnets: a magnet engineer's experience and a proposal for the next generation of hybrids. *Physica B,* 216(3), 186-192, 1996.
13. Mulay, L.N., Basic concepts related to magnetic fields and magnetic susceptibility. In: Biological effects of magnetic fields, Plenum Press, 33–55, 1964.
14. Yoshimura, N., Application of magnetic action for sterilization of food. *Shokukin Kihatsu* ***24***(3), 46–48, 1989.
15. Hofmann, G.A., Deactivation of microorganisms by an oscillating magnetic field. U.S. Patent 4, 524, 1985.
16. Barbosa-Canovas, G. V., Schaffner, D. W., Pierson, M. D., & Zhang, Q. H., Oscillating magnetic fields. *Journal of Food Science,* **65**, 86–89, 2000.
17. Goldschmidt Lins, P., Aparecida Silva, A., Marina Piccoli Pugine, S., Ivan Cespedes Arce, A., Jose Xavier Costa, E., & Pires De Melo, M., Effect of exposure to pulsed magnetic field on microbiological quality, color and oxidative stability of fresh ground beef. *Journal of Food Process Engineering,* ***40***(2), 12405, 2016.
18. Liu, Z., Gao, X., Zhao, J., & Xiang, Y., The sterilization effect of solenoid magnetic field direction on heterotrophic bacteria in circulating cooling water. *Procedia Engineering,* ***174***, 1296–1302, 2017.
19. Ji, W., Huang, H., Deng, A., & Pan, C., Effects of static magnetic fields on *Escherichia coli. Micron,* ***40***(8), 894–898, 2009.
20. Wu, P., Qu, W., Abdualrahman, M. A. Y., Guo, Y., Xu, K., & Ma, H., Study on inactivation mechanisms of *Listeria grayi* affected by pulse magnetic field via morphological structure, Ca2+ trans-membrane transport and proteomic analysis. *International Journal of Food Science & Technology,* ***52***(9), 2049–2057, 2017.
21. Bayraktar, V.N., Magnetic field on yeast *Saccharomyces cerevisiae* activity at grape must fermentation. *Biotechnologia Acta*, **6**(1), 125-135, 2013.

22. Da Motta, M. A., Muniz, J. B. F., & Schuler, A., Static magnetic fields enhancement of *Saccharomyces cerevisae* ethanolic fermentation. *Biotechnology Progress,* ***20***(1), 393–396, 2008.
23. Alvarez, D. C., Perez, V. H., Justo, O. R., & Alegre, R. M., Effect of the extremely low frequency magnetic field on nisin production by *Lactococcus lactis* subsp. *lactis* using cheese whey permeate. *Process Biochemistry,* ***41***(9), 1967–1973, 2006.
24. Rosen, A.D., Membrane response to static magnetic fields: effect of exposure duration. *Biochimica et Biophysica Acta (BBA)—Biomembranes,* ***1148***, 317–320, 1993.
25. Binhi, V. N., Alipov, Y. D., & Belyaev, I. Y., Effect of static magnetic field on E. coli cells and individual rotations of ion-protein complexes. *Bioelectromagnetics,* ***22***(2), 79–86, 2001.
26. Iwasaka, M., Ikehata, M., Miyakoshi, J., & Ueno, S., Strong static magnetic field effects on yeast proliferation and distribution. *Bioelectrochemistry,* ***65***(1), 59–68, 2004.
27. Berk, S. G., Srikanth, S., Mahajan, S. M., & Ventrice, C. A., Static uniform magnetic fields and amoebae. *Bioelectromagnetics,* ***18***(1), 81–84, 1997.
28. Ma, H., Huang, L., & Zhu, C., The effect of pulsed magnetic field on horseradish peroxidase. *Journal Of Food Process Engineering,* ***34***(5), 1609–1622, 2009.
29. Jialan Zhang., Shajin Wang., & Baoguo Xu., Effect of alternating magnetic field treatment on enzymatic parameters of cellulase. *Journal on Food and Agricultural Science,* ***92***, 1384-1388, 2012.
30. Jia, S., Liu, Y., Wu, S., & Wang, Z., Effect of static magnetic field on α-amylase activity and enzymatic reaction. *Transactions of Tianjin University,* ***15***(4), 272–275, 2009.
31. Pearce, K. L., Rosenvold, K., Andersen, H. J., & Hopkins, D. L., Water distribution and mobility in meat during the conversion of muscle to meat and ageing and the impacts on fresh meat quality attributes- A review. *Meat Science,* ***89***(2), 111-124, 2011.
32. Butz, P., Hofmann, C., & Tauscher, B., Recent developments in noninvasive techniques for fresh fruit and vegetable internal quality analysis. *Journal of Food Science,* ***70***(9), 131–141, 2005.
33. Ramos-Cabrer, P., Van Duynhoven, J. P., Timmer, H., & Nicolay, K., Monitoring of moisture redistribution in multicomponent food systems by use of magnetic resonance imaging. *Journal of Agricultural and Food Chemistry,* ***54***(3), 672–677, 2006.
34. Keeton, J. T., Hafley, B. S., Eddy, S. M., Moser, C. R.., McManus, B. J., & Leffler, T. P., Rapid determination of moisture and fat in meats by microwave and nuclear magnetic resonance analysis. *Journal of AOAC International,* ***86***(6), 1193–1202, 2003.
35. Hu, F., Furihata, K., Kato, Y., & Tanokura, M., Nondestructive quantification of organic compounds in whole milk without pretreatment by 2-D NMR

Spectroscopy. *Journal of Agricultural and Food Chemistry,* ***55***(11), 4307–4311, 2007.

36. Leffler, T. P., Moser, C. R., McManus, B. J., Urh, J. J., Keeton, J. T., & Claflin, A., Determination of moisture and fat in meats by microwave and nuclear magnetic resonance analysis: Collaborative study. *Journal of AOAC International,* ***91***(4), 802–810, 2008.
37. Adam-Berret, M., Rondeau-Mouro, C., Riaublanc, A., & Mariette, F., Study of triacylglycerol polymorphs by nuclear magnetic resonance: Effects of temperature and chain length on relaxation parameters. *Magnetic Resonance in Chemistry,* ***46***(6), 550–557, 2008.
38. Marcone, M. F., Wang, S., Albabish, W., Nie, S., Somnarain, D., & Hill, A., Diverse food-based applications of nuclear magnetic resonance (NMR) technology. *Food Research International,* ***51***(2), 729–747, 2013.
39. Pedersen, H. T., Munck, L., & Engelsen, S. B., Low-field 1H nuclear magnetic resonance and chemometrics combined for simultaneous determination of water, oil, and protein contents in oilseeds. *Journal of the American Oil Chemists' Society,* ***77***(10), 1069–1077,2000.
40. Consonni, R., Cagliani, L. R., Guantieri, V., & Simonato, B., Identification of metabolic content of selected Amarone wine. *Food Chemistry,* ***129***(2), 693–699, 2011.
41. Oztop, M. H., Rosenberg, M., Rosenberg, Y., McCarthy, K. L., & McCarthy, M. J., Magnetic resonance imaging (MRI) and relaxation spectrum analysis as methods to investigate swelling in whey protein gels. Journal of Food Science, 75(8), 508–515, 2010.
42. Barsby, T. L., Donald, A. M., & Frazier, P. J., Starch: Advances in structure and function. Cambridge: Royal Society of Chemistry, 233, 2001.
43. Therien-Aubin, H., & Zhu, X. X., NMR spectroscopy and imaging studies of pharmaceutical tablets made of starch. *Carbohydrate Polymers,* ***75***(3), 369–379, 2009.
44. Junyan Tang., Hainan Zhang., Changqing Tian., & Shuangquan Shao., Effects of different magnetic fields on the freezing parameters of cherry. *Journal of Food Engineering,* ***273***, 109949, 2020.
45. Choi, Y.S., Ku, S.K., Jeong, J.Y., Jeon, K.H., & Kim, Y.B., Changes in ultrastructure and sensory characteristics on electro-magnetic and air blast freezing of beef during frozen storage. *Korean Journal for Food Science of Animal Resources,* ***35***(1), 27–34, 2015.
46. Farouk, M. M., and Swan, J. E., Effect of rigor temperature and frozen storage on functional properties of hotboned manufacturing beef. *Meat Science,* ***49***, 223-247, 1998.
47. Tang, J., Shao, S., & Tian, C., Effects of the magnetic field on the freezing parameters of the pork. *International Journal of Refrigeration,****107***,31-38, 2019a.
48. Tang, J., Shao, S., & Tian, C., Effects of the magnetic field on the freezing process of blueberry. *International Journal of Refrigeration,****113***, 288-295, 2019b.

49. Ran, X.L., Zhang, M., Wang, Y.,& Liu, Y., A comparative study of three drying methods on drying time and physicochemical properties of chicken powder. *Drying Technology,* 1–14, 2018.
50. Jia, C., Wang, L., Li, R., & Liu, C., Experimental study on drying characteristics of wheat by low-field nuclear magnetic resonance. *Drying Technology,* **35**(10), 1258-126, 2016.
51. Song, Y., Zang, X., Kamal, T., Bi, J., Cong, S., Zhu, B., & Tan, M., Real-time detection of water dynamics in abalone (*Haliotis discus hannai Ino*) during drying and rehydration processes assessed by LF-NMR and MRI. *Drying Technology,* **36**(1), 72–83, 2017.
52. Shi, F., Li, Y., Wang, L., Yang, Y., Lu, K., Wu, S., & Ming, J., Measurement of moisture transformation and distribution in tricholoma matsutake by low field nuclear magnetic resonance during the hot-air drying process. *Journal of Food Processing and Preservation,* **42**(3), 13565, 2017.
53. Lv, W., Zhang, M., Wang, Y., & Adhikari, B., Online measurement of moisture content, moisture distribution, and state of water in corn kernels during microwave vacuum drying using novel smart NMR/ MRI detection system. *Drying Technology,* 1–11, 2018.
54. Lv, W., Zhang, M., Bhandari, B., Li, L., & Wang, Y., Smart NMR method of measurement of moisture content of vegetables during microwave vacuum drying. *Food and Bioprocess Technology,* **10**(12), 2251–2260, 2017.
55. Fan, L. P., Zhang, M., Xiao, G. N., Sun, J. C., & Tao, Q., The optimization of vacuum frying to dehydrate carrot chips. *International journal of food science & technology,* **40**(9), 911–919,2005.
56. Choe, E., & Min, D. B., Chemistry of deep-fat frying oils. *Journal of Food Science,* **72**(5), 77–86, 2007.
57. Chen, Q., Zhang, C., Zhao, J., & Ouyang, Q., Recent advances in emerging imaging techniques for non-destructive detection of food quality and safety. *TrAC Trends in Analytical Chemistry,* **52**, 261–274, 2013.
58. Fan, L., Zhang, M., & Mujumdar, A. S., Effect of various pretreatments on the quality of vacuum-fried carrot chips. *Drying Technology,* **24**(11), 1481–1486, 2006.
59. Zhang, Y., Zhang, T., Fan, D., Li, J., & Fan, L., The description of oil absorption behavior of potato chips during the frying. *LWT-Food science and Technology,* **96**, 119–126, 2018.
60. Wang, J., Liu, C., & Sun, D., Study on relationship between polar compounds and LF-NMR properties in fried camellia seed oil. *Journal of Food and Nutrition Research,* **6**(7), 433-438, 2018.
61. Zhang, Q. Q., Li, W., Li, H. K., Chen, X. H., Jiang, M., & Dong, M. S., Low-field nuclear magnetic resonance for online determination of water content during sausage fermentation. *Journal of Food Engineering,* **212**, 291–297, 2017.

62. Ramanjooloo, A., Bhaw-Luximon, A., Jhurry, D., & Cadet, F., 1H NMR quantitative assessment of lactic acid produced by biofermentation of cane sugar juice. *Spectroscopy Letters,* ***42***(6-7), 296–304, 2009.
63. Wu, J., Li, Y., & Gao, X., Monitoring a typical fermentation process of natto by low-field nuclear magnetic resonance (LF-NMR) and magnetic resonance imaging (MRI) techniques. *Analytical Methods,* ***8***(39), 7135–7140, 2016.
64. Goetz, J., Gro, D., & Koehler, P., On-line observation of dough fermentation by magnetic resonance imaging and volumetric measurements. *European Food Research and Technology,* ***217***(6), 504–51, 2003.
65. Wierucka, M., & Biziuk, M., Application of magnetic nanoparticles for magnetic solid-phase extraction in preparing biological, environmental and food samples. *TrAC Trends in Analytical Chemistry,* ***59***, 50–58,2014.
66. Saien, J., Bamdadi, H., & Daliri, S., Liquid–liquid extraction intensification with magnetite nanofluid single drops under oscillating magnetic field. *Journal of Industrial and Engineering Chemistry,* ***21***, 1152–1159, 2015.
67. Pentimalli, M., Capitani, D., Ferrando, A., Ferri, D., Ragni, P., & Segre, A., Gamma irradiation of food packaging materials: an NMR study. *Polymer,* ***41***(8), 2871–2881, 2000.
68. Lamanna, R., Piscioneri, I., Romanelli, V., & Sharma, N., A preliminary study of soft cheese degradation in different packaging conditions by1H-NMR. *Magnetic Resonance in Chemistry,* ***46***(9), 828–831, 2008.
69. Fan, K., & Zhang, M., Recent developments in the food quality detected by non-invasive nuclear magnetic resonance technology. *Critical Reviews in Food Science and Nutrition,* 1–12, 2018.
70. Miyake, Y., Yokomizo, K., & Matsuzaki, N., Determination of unsaturated fatty acid composition by high-resolution nuclear magnetic resonance spectroscopy. *Journal of the American Oil Chemists' Society,* ***75***(12), 1091–1094, 1998.
71. Igarashi, M., Spin–spin relaxation process of 23Na-NMR in Na-loaded NaY. *Molecular Crystals and Liquid Crystals,* ***341***, 1241–1244, 2000.
72. Sobolev, A. P., Segre, A., & Lamanna, R., Proton high-field NMR study of tomato juice. *Magnetic Resonance in Chemistry,* ***41***(4), 237–245, 2003.
73. Del Campo, G., Berregi, I., Caracena, R., & Zuriarrain, J., Quantitative determination of caffeine, formic acid, trigonelline and 5-(hydroxymethyl) furfural in soluble coffees by 1H NMR spectrometry. *Talanta,* ***81***(1-2), 367–371, 2010.
74. Kuo, M.I., Gunasekaran, S., Johnson, M., & Chen, C., Nuclear magnetic resonance study of water mobility in pasta filata and non-pasta filata mozzarella. *Journal of Dairy Science,* ***84***(9), 1950–1958, 2001.
75. Callaghan, P. T., Jolley, K. W., & Humphrey, R. S., Diffusion of fat and water in cheese as studied by pulsed field gradient nuclear magnetic resonance. *Journal of Colloid and Interface Science,* ***93***, 521–529, 1983.

76. Tarachiwin, L., Ute, K., Kobayashi, A., & Fukusaki, E.,1H NMR based metabolic profiling in the evaluation of Japanese green tea quality. *Journal of Agricultural and Food Chemistry,* ***55***(23), 9330–9336, 2007.
77. Duarte, I., Barros, A., Almeida, C., Spraul, M., & Gil, A. M., Multivariate analysis of NMR and FTIR data as a potential tool for the quality control of beer. *Journal of Agricultural and Food Chemistry,* 52(5), 1031–1038, 2006.
78. Patel, K. K., Khan, M. A., & Kar, A., Recent developments in applications of MRI techniques for foods and agricultural produce—an overview. *Journal of Food Science and Technology,* ***52***(1), 1–26, 2013.
79. Wang, S. Y., Wang, P. C., & Faust, M., Non-destructive detection of watercore in apple with nuclear magnetic resonance imaging. *Scientia Horticulturae,* ***35***(3-4), 227–234, 1988.
80. McCarthy, M. J., Zion, B., Chen, P., Ablett, S., Darke, A. H., & Lillford, P. J., Diamagnetic susceptibility changes in apple tissue after bruising. *Journal of the Science of Food and Agriculture,* ***67***(1), 13–20, 1995.
81. Chen, P., McCarthy, M. J., & Kauten, R., NMR for internal quality evaluation of fruits and vegetables. *Trans. ASAE,* ***32***(5), 1747-1753, 1989.
82. Smith, R.C., & Lange, R.C., Understanding magnetic resonance imaging, CRC press, 1998.
83. Barreiro, P., Ortiz, C., Ruiz-Altisent, M., Ruiz-Cabello, J., Fernandez-Valle, M., Recasens, I., & Asensio, M., Mealiness assessment in apples and peaches using MRI techniques. *Magnetic Resonance Imaging,* ***18***(9), 1175–1181, 2000.
84. Gambhir, P.N., Choi, Y.J., & McCarthy, M.J., Development of rapid and non-invasive nuclear magnetic resonance method for identifying freeze damaged citrus fruits. IFT Annual Meeting, Las Vegas, USA, 2004.
85. Hernandez-Sanchez, N., Barreiro, P., & Ruiz-Cabello, J., On-line identification of seeds in mandarins with magnetic resonance imaging. *Biosystems Engineering,* ***95***(4), 529–536, 2006.
86. Wang, N., Lin, X. Y., & Yuan, R.S., Study on navel orange storage processing with NMR. *Food Sci* ***28***(7), 83–87, 2007.
87. Barreiro, P., Zheng, C., Da-Wen, S., Hernandez-Sanchez, N., Perez-Sanchez, J.M.,& Ruiz-Cabello, J., Non-destructive seed detection in mandarins: comparison of automatic threshold methods in FLASH and COMSPIRA MRIs. *Postharvest Biol Technol,* ***47***, 189–198, 2008.
88. Burdon, J., & Clark, C., Effect of postharvest water loss on "Hayward" kiwifruit water status. *Postharvest Biology and Technology,* ***22***(3), 215–225, 2001.
89. Newman, R. H., & Redgwell, R. J., Cell wall changes in ripening kiwifruit: 13C solid state NMR characterisation of relatively rigid cell wall polymers. *Carbohydrate Polymers,* ***49***(2), 121–129, 2002.
90. Taglienti, A., Massantini, R., Botondi, R., Mencarelli, F., & Valentini, M., Postharvest structural changes of hayward kiwifruit by means of magnetic resonance imaging spectroscopy. *Food Chemistry,* ***114***(4), 1583–1589, 2009.

91. Vaientini, M., Sequi, P., Ciampa, A., Ritota, M., Taglienti, A., Cozzolino, S., Conte, L., &Terlizzi, M., Magnetic resonance imaging for kiwifruit quality evaluation: shelf-life and PGR detection. *Italus Hortus* **16**(5), 324–328, 2009.
92. Almeida, D. P. F., & Huber, D. J., Transient increase in locular pressure and occlusion of endocarpic apertures in ripening tomato fruit. *Journal of Plant Physiology, 158(2),* 199–203, 2001.
93. Saltveit, M. E., Determining tomato fruit maturity with nondestructive in vivo nuclear magnetic resonance imaging. *Postharvest Biology and Technology,* **1**(2), 153–159, 1991.
94. Ishida, N., Kobayashi, T., Koizumi, M., & Kano, H., 1H-NMR imaging of tomato fruits. *Agricultural and Biological Chemistry,* **53**(9), 2363–2367, 1989.
95. Povlsen, V. T., Rinnan, A., van den Berg, F., Andersen, H. J., & Thybo, A. K., Direct decomposition of NMR relaxation profiles and prediction of sensory attributes of potato samples. *LWT - Food Science and Technology,* **36**(4), 423–432, 2003.
96. Thybo, A. K., Jespersen, S. N., Lærke, P. E., & Stødkilde-Jørgensen, H. J., Nondestructive detection of internal bruise and spraing disease symptoms in potatoes using magnetic resonance imaging. *Magnetic Resonance Imaging,* **22**(9), 1311–1317, 2004a.
97. Thybo, A., Bechmann, I., Martens, M., & Engelsen, S., Prediction of sensory texture of cooked potatoes using uniaxial compression, near infrared spectroscopy and low field1H NMR spectroscopy. *LWT - Food Science and Technology,* **33**(2), 103–111, 2000.
98. Thybo, A. K., Szczypinski, P. M., Karlsson, A. H., Donstrup, S., Stodkilde-Jorgensen, H. S., & Andersen, H. J., Prediction of sensory texture quality attributes of cooked potatoes by NMR-imaging (MRI) of raw potatoes in combination with different image analysis methods. *Journal of Food Engineering,* **61**(1), 91–100, 2004b.
99. Thybo, A., Andersen, H., Karlsson, A., Donstrup, S., & Stodkilde-Jorgensen, H., Low-field NMR relaxation and NMR-imaging as tools in differentiation between potato sample and determination of dry matter content in potatoes. LWT - Food Science and Technology, 36(3), 315–322, 2003.
100. Straadt, I. K., Thybo, A. K., & Bertram, H. C., NaCl-induced changes in structure and water mobility in potato tissue as determined by CLSM and LF-NMR. *LWT - Food Science and Technology,* **41**(8), 1493–1500, 2008.
101. Horigane, A. K., Toyoshima, H., Hemmi, H., Engelaar, W. M. H. G., Okubo, A., & Nagata, T., Internal hollows in cooked rice grains (*Oryza sativa* cv. Koshihikari) observed by NMR micro imaging. *Journal of Food Science,* **64**(1), 1–5, 2011.
102. Mohoric, A., Vergeldt, F., Gerkema, E., Dalen, G. van., Doel, L. R. van den., Vliet, L. J. van.,& Duynhoven, J. van., The effect of rice kernel microstructure on cooking behaviour: A combined μ-CT and MRI study. *Food Chemistry,* **115**(4), 1491–1499, 2009.

103. Hwang, S.S., Cheng, Y.C., Chang, C., Lur, H.S., & Lin, T.T., Magnetic resonance imaging and analyses of tempering processes in rice kernels. *Journal of Cereal Science,* ***50***(1), 36–42, 2009.
104. Ishida, N., Naito, S., & Kano, H., Loss of moisture from harvested rice seeds on MRI. *Magnetic Resonance Imaging,* ***22***(6), 871–875,2004.
105. Hills, B., Food processing: An MRI perspective. *Trends in Food Science & Technology,* ***6***(4), 111–117, 1998.
106. Thielemann, J., Kirkhus, T., Kavli, T., Schumann-Olsen, H., Haugland, O., & Westavik, H., System for estimation of pin bone positions in pre-rigor salmon. *Lect Notes Comput Sci,* ***4678***, 888–896, 2007.
107. Bock, C., Sartoris, F.J., & Portner, H.O., In vivo MR spectroscopy and MR imaging on non-anaesthetized marine fish: techniques and first results. *Magnetic Resonance Imaging,* ***20***(2), 165–172, 2002.
108. Foucat, L., Taylor, R. G., Labas, R., & Renou, J. P., Soft flesh problem in freshwater rainbow trout investigated by magnetic resonance imaging and histology. *Journal of Food Science,* ***69***(4), 320–327, 2004.
109. Veliyulin, E., Felberg, H., Digre, H., & Martinez, I., Non-destructive nuclear magnetic resonance image study of belly bursting in herring (*Clupea harengus*). *Food Chemistry,* ***101***(4), 1545–1551, 2007.
110. Veliyulin, E., Borge, A., Singstad, T., Gribbestad, I., & Erikson, U., Postmortem studies of fish using magnetic resonance imaging. *Modern Magn Reson,* 949–956, 2006.
111. Foucat, L., Taylor, R.G., Labas, R., & Renou, J.P., Characterization of frozen fish by NMR imaging and histology. *Am Lab* ***33***(16), 38–43, 2001.
112. Bertram, H.C., Holdsworth, S.J., Whittaker, A.K., & Andersen, H.J., Salt diffusion and distribution in meat studied by Na-23 nuclear magnetic resonance imaging and relaxometry. *J Agric Food Chem* **53** (20), 7814–7818, 2005.
113. Mazucco, R.A., Joyce, D.C., & Hockings, P.D., Magnetic resonance imaging applied to harvested mango fruit. Proc. Australasian Postharvest Conf., Gatton, Australia, 355–358, 1993.
114. Lammertyn, J., Dresselaers, T., Van Hecke, P., Jancsok, P., Wevers, M., & Nicolai, B., MRI and x-ray CT study of spatial distribution of core breakdown in "Conference" pears. *Magnetic Resonance Imaging,* ***21***(7), 805–815, 2003.
115. Hernandez-Sanchez, N., Hills, B. P., Barreiro, P., & Marigheto, N., An NMR study on internal browning in pears. *Postharvest Biology and Technology,* ***44***(3), 260–270, 2007.
116. Evans, S., & Hall, L., Evaluation of a range of MRI-active pH indicators using a multiple-sample method. *AIChE Journal,* ***51***(5), 1541–1547, 2005.
117. Song, H. P., Delwiche, S. R., & Line, M. J., Moisture distribution in a mature soft wheat grain by three-dimensional magnetic resonance imaging. Journal of Cereal Science, ***27***(2), 191–197, 1998.

118. Ruan, R.,& Litchfield, J.B., Determination of water distribution and mobility inside maize kernels during steeping using magnetic resonance imaging. *Cereal Chem,* ***69***(1), 13-17, 1992.
119. Song, H. P., & Litchfield, J. B., Measurement of stress cracking in maize kernels by magnetic resonance imaging. Journal of Agricultural Engineering Research, ***57***(2), 109–118, 1994.
120. Venturi, L., Rocculi, P., Cavani, C., Placucci, G., Rosa, M. D., & Cremonini, M. A., Water absorption of freeze-dried meat at different water activities: a multi-analytical approach using sorption isotherm, differential scanning calorimetry, and nuclear magnetic resonance. *Journal of Agricultural and Food Chemistry,* **55**(26), 10572–10578, 2007.
121. Melvina, C., Ezeanaka ., John Nsor-Atindana., & Min Zhang., Online lowfield nuclear magnetic resonance (LF-NMR) and magnetic resonance imaging (MRI) for food quality optimization in food processing. *Food and Bioprocess Technology,* 1-17, 2019.
122. Veliyulin, E., & Aursand, I. G.,1H and23Na MRI studies of Atlantic salmon (*Salmo salar*) and Atlantic cod(*Gadus morhua*) fillet pieces salted in different brine concentrations. *Journal of the Science of Food and Agriculture,* ***87***(14), 2676–2683, 2007.
123. Bosco, M., Toffanin, R., Palo, D. D., Zatti, L., & Segre, A., High-resolution 1H NMR investigation of coffee. *Journal of the Science of Food and Agriculture,* 79(6), 869-878, 1999.
124. Budiman, M., Stroshine, R. L., & Cornillon, P., Moisture measurement in cheese analogue using stretched and multi-exponential models of the magnetic resonance T2 relaxation curve. Journal *of Dairy Research,* ***69***, 619–632, 2002.
125. Weekley, A. J., Bruins, P., Sisto, M., & Augustine, M. P., Using NMR to study full intact wine bottles. *Journal of Magnetic Resonance,* ***161***(1), 91–98, 2003.
126. Rodrigues, J. E., Erny, G. L., Barros, A. S., Esteves, V. I., Brandao, T., & Ferreira, A. A., Quantification of organic acids in beer by nuclear magnetic resonance (NMR)-based methods. *Analytica Chimica Acta,* ***674***(2), 166–175, 2010.

6

Microwaves Application to Food and Food Waste Processing

Cristina Barrera, Pedro J. Fito, Marta Castro-Giráldez, Noelia Betoret and Lucía Seguí*

Institute of Food Engineering for Development (IIAD), Universitat Politècnica de València, Valencia, Spain

Abstract
The use of radiation in the microwave spectrum has been developed since the beginning of the 20th century with radars and communication systems. The expansion of this technology in other fields has allowed more and more possibilities for food industry applications. Applications of microwave technology to food processing include common unit operations such as drying, cooking, tempering, baking or frying; but it has also shown great potential for food process monitoring, food waste valorization processes through components extraction or biorefinery approaches, as well as for the development of functional foods and ingredients. Volumetric heating, faster reaction rates or MW promoted structural changes, are some of the characteristics of this technology which makes it interesting to be used in food and food waste processing. This chapter reviews the possibilities of the use of radiation in the microwave range in the food industry, considering present applications and prospects.

Keywords: Microwaves, food processing, food process monitoring, food waste, functional foods, biowaste pretreatments

**Corresponding author*: lusegil@upvnet.upv.es

Navnidhi Chhikara, Anil Panghal and Gaurav Chaudhary (eds.) Novel Technologies in Food Science, (235–270) © 2023 Scrivener Publishing LLC

6.1 Introduction to Microwave Technology. Basis of Photon-Matter Interaction in the Microwave Range

Microwave (MW) technology became famous with the development of fast heating systems in the 70s of the 20th century. The application of MW represents the use of photon fluxes at an electromagnetic range of $1.24 \cdot 10^{-6}$ to $1.24 \cdot 10^{-3}$ eV or 300 MHz to 300 GHz. This range comprises the radiofrequency at lower frequencies and the infrared at higher frequencies. Radiofrequency, microwaves and infrared make up the low energy spectrum of photons. The radiofrequency range has traditionally been used for communication (antennas or radars), while the infrared spectrum has been used for heating. In the radiofrequency range, the interactions of low-energy photons with the chemical species predominate. These interactions are based on the orientation of molecular spins, which facilitate transmission over long distances. While in the case of the infrared range the photons cause an excitation of the molecular bonding orbitals that causes a fast increase in the internal energy of the organic chemical species, generating a fast heating; in the MW range, photons generate interactions that can work for communication and for heating. In this frequency range, the interactions or relaxations between photons and the chemical species that are in their track are γ-dispersion and ionic conductivity. γ-dispersion is caused by the induction and orientation of dipolar molecules, water being the most important in biological systems. These phenomena generate electric energy accumulation caused mainly by water spin reorientation. On the other hand, a part of the electrical energy of photons is transformed into other energies (mechanical or thermal) due to the collisions or frictions associated to the increase of molecular mobility. Moreover, the vibration of chemical species with very high ionic strength causes a part of the losses of electrical energy; this is called ionic conductivity (Figure 6.1).

The interactions of photons with matter can be modeled by Schrodinger's equation [1] according to the quantum theory. However, at the macroscopic level, it is possible to apply the Maxwell's equations [2], where the physical property that describes the electric effect is the complex permittivity whereas the complex permeability describes the magnetic effect [3]. These physical properties are vectorial since they represent the displacement of the polarity or photon flux path. In the case of permittivity, expressed as a complex number, the electric energy accumulation is represented by the real part of the permittivity (ε) or dielectric constant, whereas the

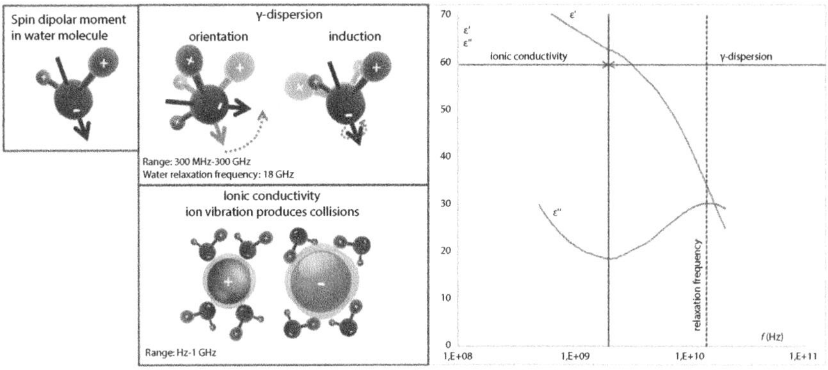

Figure 6.1 Photon matter interactions in the microwave range (left) and permittivity spectra in microwave range of poultry breast tissue (right).

imaginary part of the permittivity represents the photon energy dissipated in mechanical or thermal energy due to the molecular collisions or frictions, i.e. the loss factor (ε). Dielectric constant is important for energy storage and communication applications, and loss factor is important for monitoring and processing applications. Nowadays, photonics in the MW range is used for communication systems (4G, 5G, satellites, Wi-Fi systems), for energy storage (batteries), or for heating and monitoring biological processes (food processing, medicine, sports, materials science).

Complex permittivity changes with frequency (Figure 6.1), reaching the maximum interaction photon/dipolar molecule at relaxation frequency of γ-dispersion. Moreover, below the frequency range of GHz, the

Microwave's parameters	Microwave's models to obtain relaxation parameters
Loss tangent $tg\,\delta = \frac{\varepsilon'}{\varepsilon''} = \frac{\lvert\varepsilon\rvert\, sen\,\delta}{\lvert\varepsilon\rvert\, cos\,\delta}$ Conductivity (S/m) $\sigma = \varepsilon_0 \varepsilon'' \omega$ Penetration depth $\beta = \frac{\lvert\varepsilon\rvert}{2\pi f}\frac{\sqrt{\varepsilon'}}{\varepsilon''}$	Debye's model $\varepsilon' - j\varepsilon''(\omega) = \varepsilon_\infty - \frac{\sigma}{\varepsilon_0 \omega} + \sum_{n=1}^{3} \frac{\Delta\varepsilon_n}{1 + j\omega\tau_n}$ Traffano-Schiffo's model $log\,\varepsilon'(\omega) = log\,\varepsilon'_\infty + \sum_{n=1}^{3} \frac{\Delta log\,\varepsilon'_n}{1 + e^{(log\,\omega^2 - log\,\tau_n^2)\cdot\alpha_n}}$

Figure 6.2 Microwave's parameters obtained from permittivity measurements and models proposed in the literature to obtain relaxation parameters [4, 5].

conductivity effect appears in the loss factor. This is related with the ionic strength of the analyzed media, and it can be estimated as indicated in Figure 6.2, where some models to fit the permittivity and to obtain the relaxation frequency are given. In addition to permittivity, other dielectric parameters could be useful, such as the loss tangent or the microwave penetration depth.

6.2 Microwaves Applications to Food Process Monitoring

The possibilities of the use of MW in food process monitoring can be understood by means of the γ-dispersion. In the γ-dispersion the interactions of photons with water are higher the higher mobility and the higher the quantity of water molecules, while below 1 GHz it is possible to analyze the mobility of the ions present in the medium. As an example, in Figure 6.3a a loss factor spectrum of pork loin during a drying process is shown. As drying progresses, the value of the loss factor is lower as well as its relaxation frequency, since drying reduces the number of water molecules. However, on the side of ionic conductivity it also decreases although the concentration of ions increases. This occurs because conductivity depends on ionic strength, which in turn depends on molecular

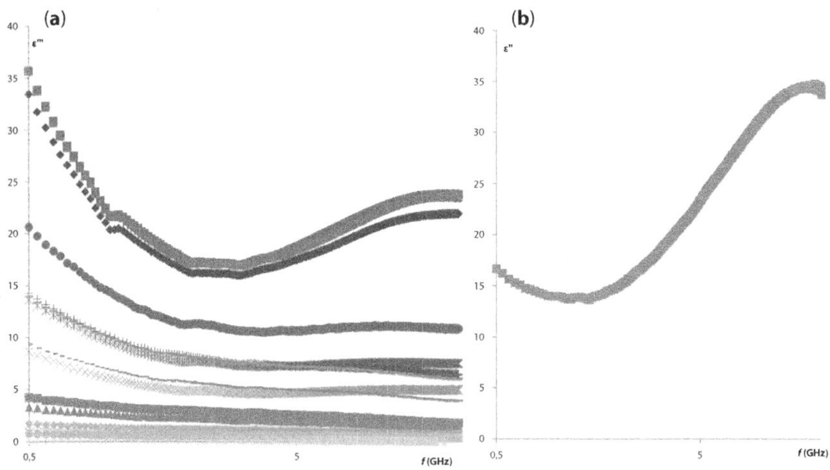

Figure 6.3 Loss factor spectra in microwave range, (a) Pork loin during the hot air-drying process and (b) Mandarin juice.

mobility, which is reduced with water removal. Figure 6.3b shows mandarin juice, with a low ion content and high water activity. This composition causes a high loss factor in γ-dispersion but a low conductivity value.

If the food is subjected to a low-power photon flux and this signal is analyzed, some food products and processes can be monitored. This technique has been used in different food industry sectors, such as the meat, post-harvest or the dairy sectors, for different purposes. It was used to non-destructively determinement quality [6–8] and ageing [9], to determine added water in pork [10] or to predict water activity of meat during drying process [11]. It also demonstrated to be useful to determine the maturity state of fruits [12, 13]. This technique has also been widely used to determine water content in agricultural products [14–18] or animal feed [19]. In addition, it has been used to monitor compositional changes during beer production [20] (Figure 6.4). In fact, the alcohol content of alcoholic beverages was clearly identified at MW frequencies [21].

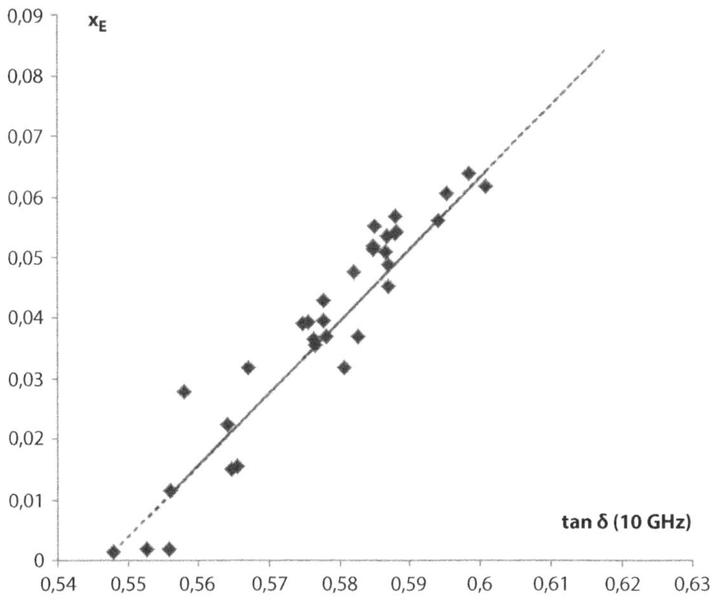

Figure 6.4 Measured (◆) and predicted (line) ethanol content with regard to loss tangent at 10 GHz (adapted from Veláquez-Varela *et al.*, [20]).

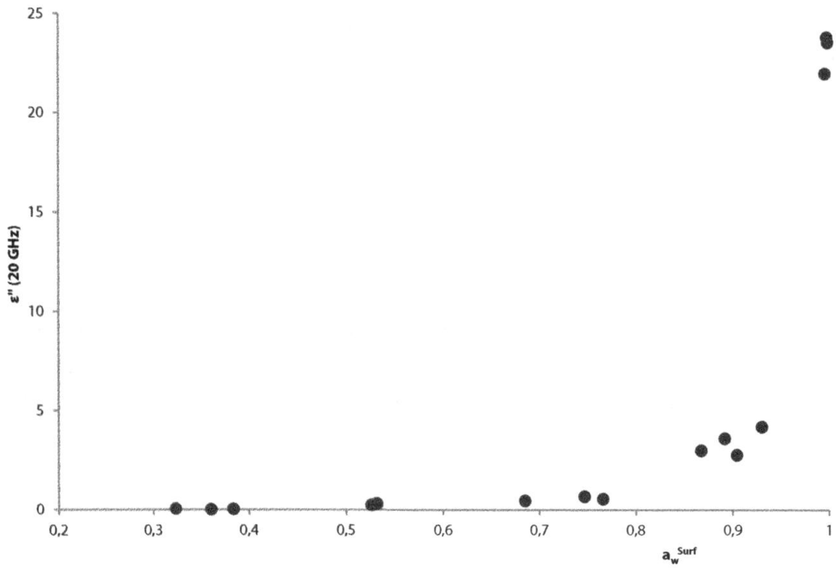

Figure 6.5 Dielectric isotherm of pork meat (adapted from Traffano-Schiffo *et al.*, [11]).

In Traffano-Schiffo *et al.* [11] it was demonstrated that moisture content in dry basis is related with loss factor at 20 GHz, making it possible to develop a dielectric isotherm to be used as a food design tool (Figure 6.5). This technique has also been used to determine nitrate and phosphate concentration in water [22], to predict the water/sucrose states during apple candying [23], to monitor the salting process of foods [24, 25], or to detect food contamination and quality [26]. It has also been applied to seeds detection mandarin fruits [27].

There are numerous applications of MW to food process monitoring. However, this technique has limitations when measuring without contact since measurements with head space between probe and food induces loses in accuracy. Hence, this technology has experienced enormous growth in other areas such as communications or batteries and shows a promising future, as long as current problems are solved.

6.3 Microwaves in Food Processing

MW processing techniques have been extensively used in the food industry [28]. Applications of MW technology in food processing include drying, concentration, pasteurization, sterilization, thawing, tempering, baking, roasting, cooking, precooking, among others [29]. Main advantage of MW heating as compared to conventional one is the faster and volumetric heating capability of MW electromagnetic field, which significantly reduces the cooking time, together with a relatively safe, ease and low maintenance handling [30]. MW heating has also been reported to substantially reduce thermal degradation of essential nutrients, while increasing the retention of food quality factors such as flavour, colour and texture. Major disadvantage recognized in MW heating is the non-uniform temperature distribution resulting in over- and under- heated spots in the product [31], which is especially prevalent when processing solid foods in batches and within a cavity oven [32]. In Table 6.1, application of MW technology to several food processes and products is summarized.

Dehydration is one of the oldest methods for food preservation. Indeed, reducing the water content of foods slows spoilage reactions while substantially reducing storage and transport costs. As a pre-treatment for further solvent extraction, dehydration increases the yield of the process and significantly reduces solvent consumption. Convective hot air drying is the most commonly used in drying foodstuffs despite presenting some disadvantages such as high energy consumption and long drying time, which may lead to a decline in both nutritional and organoleptic properties of the product [33]. To overcome these problems, the coupling with both intermittent and continuous MW energy has been proposed as a solution since it allows moisture to faster diffuse towards the surface while shortening the exposure time of the product to the air stream [34]. Moreover, the high internal pressure due to MW heating often involves the expansion and puffing of structured materials, thus reducing their density while increasing their porosity and their capacity to absorb water [34]. High porous-dried products are more likely to occur when MW are used in the initial stage of the drying process, while reduction in both the total drying time and shrinkage is more evident when MW are applied at the final stage [35]. All these benefits attributed to MW-assisted drying might get enhanced under vacuum conditions due to the lower boiling temperature of water, and the benefits of a low-oxygen atmosphere. MW can also be combined with other drying techniques so that applications of MW-far infrared drying or MW-freeze drying are also reported in the literature. Expansion of the food structure when MW are used in the drying process have also been reported for MW vacuum-dried samples [37] and MW-freeze-dried ones [38]. In all cases, MW drying is a complex

Table 6.1 Applications of microwave technology to food processes and products.

Process	Food	References
Dehydration	Apple, restructured fish product, banana, jamun pulp, pumpkin, blueberries, sour cherries, persimmon, bilberry press cake	[36–44]
Blanching	Red cabbage Fruit peels (mango, apple, orange, and banana) Kiwifruit puree Sweet corn	[45–48]
Pasteurization and sterilization	Pomelo juice, apple juice, orange juice-milk beverages	[49–51]
Baking	Bread, cakes, legume-cakes, rice cake	[52–54]
Thawing and tempering	Meat, fish	[55–58]
Roasting	Sunflower seeds, peanuts, hazelnuts, coffee beans	[59–63]
Frying	Potatoes (French fries, slices, potato chips)	[64–67]
Alkalinization	Cocoa	[68]

process involving heat and mass transfer, which is based on volumetric heating [28].

Regarding preservation of bioactive constituents, coupling air drying with MW technology has been reported to benefit the retention of bioactive constituents of foods. Paul & Das [39], obtained better retention of bioactive compounds (total phenols, monomeric anthocyanins, total antioxidants) by applying MW-convective hot air drying of jamun (*Syzygiumcumini* L.) pulp-skin paste, as comparted to vacuum drying, dehumidified air drying and freeze drying, especially when combining air at 70 °C and 1 W/g of MW power. According to Aghilinategh *et al.* [34], MW drying in both continuous, but mainly intermittent regime, improved total phenol content of dried apple slices. Zhang *et al.* [40] reported reduced carotenoid losses when applying intermittent MW irradiation as compared to air drying at 70 °C. Applied under vacuum conditions, MW drying implies less prolonged exposure to oxygen. Zielinska & Michalska [41] proved this method to be effective in providing high anthocyanins content and related antioxidant activity of blueberries (*Vacciniumcorymbosum* L.). Decreasing MW power beyond the critical moisture content has been reported to significantly reduce overheating of sour cherries (*Prunus cerasus* L.) subjected to vacuum-microwave drying and to ensure the highest content of phenolic compounds, quercetin derivatives, anthocyanins, and antioxidant capacity [42]. In contrast, initial hot-air drying at 70 °C until 10% of moisture content followed by intermittent MW drying at 10.7 W/g until 5% of moisture content reduced the β-carotene content of persimmon chips to a greater extent than air drying at 70 °C [43]. Other authors also reported negative effects of MW-assisted drying, this is the case of Oliveira *et al.* [44], regarding anthocyanins degradation during drying of bilberry press cake into powder, and Kroehnke *et al.* [36], who reported fewer carotenoids and polyphenol content in MW-dried apple slices, together with a reduced antioxidant active Maillard products.

Blanching is an essential thermal pre-treatment to many preservation processes (drying, canning, freezing…) that is intended to inactivate enzymes and reduce microbial load, while eliminating the air in the intracellular space and softening the tissues [69]. As an alternative to traditional blanching by immersion in hot water (70-100 °C) for some minutes (HWB), MW blanching (MB) implies lesser consumption of water and energy, and higher retention of water-soluble nutrients. MB reduces the amount of water required as well as the duration of the treatment, which implies benefits regarding both energy saving and bioactive compounds retention. Wang *et al.* [69] reported MB (900W/100 s) to be as efficient as

HWB (90 °C 1 min) when applied to red bell pepper slabs to reduce the enzyme activity, while allowing a higher retention of both red pigments (capsanthin, capsorubin) and ascorbic acid. MB samples showed increased antioxidant activity when MW intensity increased, together with cell wall disruption. On broccoli florets, duration of the blanching process to inactivate 90% of peroxidase enzyme was reduced from 90 s to 50 s when MB (900 W) was compared to HWB [70]. MB also induced a considerable increase in ascorbic acid and total phenols, evidencing an increased extractability of such compounds from plant tissue. Extractability of bioactive compounds after MW blanching has been reported to increase for several products such as red cabbage [45], fruit peels [46], or kiwifruit puree [47]. Kachhadiya *et al.* [48] also reported higher retention of total sugars and ascorbic acid in MB sweet corn as compared to HWB. On the one hand, an increased extractability may favor bioavailability of these compounds; however, this must especially be considered if further thermal treatment is to be applied since free compounds are usually more thermolabile than bound ones.

MW can also contribute to **pasteurization and sterilization** processes. These preservation techniques are mainly applied to kill key pathogens, but also to inactivate vegetative bacteria and enzymes to make food safe for consumption and to increase its shelf-life [51, 71]. As an alternative to conventional pasteurization and sterilization, MW heating provides greater energy efficiency and reduces heating times, which results in products with better sensory and nutritional quality. This has been confirmed in several studies carried out on milk, fruit juices and ready-to-eat foods. In contrast, uneven heating is a main drawback, since uniformity during heating results essential for pasteurization and sterilization processes to ensure microbiological safety. Regarding food quality and preservation of natural constituents, Kumar *et al.* [49] reported better physicochemical properties of MW pasteurized pomelo juice as compared to conventional pasteurization. Indeed, the naringin content and the total phenolic content were observed to be less affected by MW heating pasteurization than by conventional one. Also, MW-assisted pasteurization of cloudy apple juice minimized the flavour changes related to conventional thermal processing due to the shorter time required to reach the target temperature [50]. MW heating was also proved to be an alternative to conventional pasteurization in orange juice-milk beverages [51] since it resulted in a lower browning index and higher levels of ascorbic acid, total phenolics, and carotenoids, higher antioxidant activity, and greater α-amylase, α-glucosidase, and antihypertensive activity than the conventional pasteurized product, while no significant differences were observed in the volatile organic compounds and fatty acids levels. MW heating is particularly suitable for pasteurization of pre-packaged heat-sensitive, high-viscous, semisolid, solid, multi-component ready-to-eat meals [72] due to processing time reduction.

MW can also be applied to **thawing or tempering** stages. Tempering is generally used in product conditioning (frozen meat, fish, vegetables, fruit, butter, juice concentrates) before further processing such as cutting or grinding. The advantages of MW when applied to thawing or tempering treatments rely mainly in the reduction of the time needed to complete the process; in addition, the use of MW generally implies lower temperature of the air surrounding the product, thus reducing the risk for microbial spoilage and chemical deterioration [73]. MW thawing or tempering has been successfully applied to frozen meat [55, 56] or fish [57, 58]. Heating uniformity and end temperature are critical points of this process.

MW can also be used to assist in **baking**. It may provide several advantages such as saving time, energy, and space, as well as preserving nutrients. However, some quality problems such as fast staling, tough texture, low volume, dryness or lack of color or crust formation are common, mainly when MW technology is used alone. It has been applied to bread, cakes, legume-cakes or rice-cake, as well as combined with other techniques such as infrared baking, which contributes to browning and crust formation [52–74]. Combination of infrared heating with MW has also been proposed for hazelnut roasting [62]. Other applications of MW technology to the roasting process are peanuts [60], sunflower [59], pumpkin [61], or coffee beans [63]. According to Goszkiewicz *et al.* [59], MW roasting significantly increased the quality of the fat contained in sunflower seeds and, as no open pores were formed, the product was less susceptible to rancidity related changes.

Frying can also benefit from the faster rates provided by MW. Again, uneven heating causing oil deterioration is the main challenge to face [75]. As compared to conventional fried French fries, MW frying has been reported to produce a lighter and yellower product [66]; in contrast, Sansano *et al.* [67] reported moisture values and texture characteristics more similar to potato chips than to French fries. In addition, the latter also evaluated fat content and acrylamide formation during MW frying, reporting lower values of both parameters as compared to deep fat frying. MW technology may have limited applications to frying when used alone, but it has been effectively combined with other technologies such as vacuum frying. The combination of MW with vacuum frying has been proposed to produce high quality products with reduced oil uptake [64, 65]. A more recent application of MW in the food industry is their use in the **alkalinization** process of cocoa. This has been recently proposed by Valverde [68], who have developed a fast MW heating cocoa alkalinization method, obtaining products with similar physicochemical, functional and sensory characteristics than traditionally alkalized ones.

6.4 Microwaves Contribution to Food Waste Valorization Processes

Current food transformation processes must focus on the valorization of by-products to reintroduce them in the economic cycle, thus contributing to circularity and resource efficiency of Food Systems and the development of a more sustainable and resilient food industry [76]. Present developments focus on new processes that contribute to the valorization of food wastes to generate social and economic value [77]. On the one hand, food waste is classified as lignocellulosic biomass, thus it can be investigated for biorefinery approaches; on the other, food wastes are rich in fiber and other bioactive compounds, for which theyare of interest for the recovery of bioactive compounds.

6.4.1 Microwaves as A Pretreatment for Food Waste Transformation Into Biofuels and Other Value-Added Products

MW have been proposed for biowaste processing in several ways: in assisting thermochemical processes of biomass conversion into biofuels, for enhancing the production of biodiesel, and as a pretreatment for biological conversion of bio-wastes into biofuels [78]. Present trends in biofuels production are centered on the use of renewable resources. European directive EU/2015/1513 particularly encourages the use of non-food cellulosic materials as a source for biofuels. In addition to biofuels, sugars can also be transformed into other valuable compounds such as enzymes, bioplastics, acids, or lipids via biorefinery pathways [149].

The effective bioconversion of lignocellulosic biomass into profitable energy or other valuable products highly depends on pretreatment results.

Lignocellulose recalcitrance is the main barrier for the efficient and economical conversion of biomass, the main goals of pretreatment being reducing cellulose crystallinity, decreasing lignin content, and increasing materials porosity for an efficient subsequent enzyme action [79]. Subsequent processes (saccharification, fermentation, downstream processes) are constrained to pretreatment results [80]. In lignocellulosic materials, cellulose is the main polymer susceptible to be hydrolyzed to fermentable sugars. A pretreatment process should ideally improve the enzymatic hydrolysis reaction, produce minimal or no inhibitory compounds, require acceptable costs as well as low energy input and minimal waste generation [79].

MW have been suggested among the alternative methods for biomass pretreatment [81]. Different studies have investigated MW positive effects on biomass to improve further bioconversion [82–86]. MW pretreatment enables a volumetric, targeted, and faster heating than conventional one. Nevertheless, the non-uniform temperature distribution at specific moments of the treatment results in both hot and cold spots in the heated product [31]. MW interacts with polar molecules and ions present in the waste material, resulting in both thermal and non-thermal effects that modify material properties and reaction rates. Thermal effect is responsible for temperature and subsequent pressure increases, which implies disruption of lignin and the cell walls, facilitating further enzymatic hydrolysis; non-thermal effects are the results of MW forcing dipoles to align with the generated field, which implies hydrogen bonds disruption [78].

Structural analysis and microscopic observations of MW pre-treated biomass have shown structural changes which facilitate further enzyme action such as reduced particle size, reduced density and increased tissue porosity and surface area, or evidence lignin degradation [81–83, 86, 87]. Temperature and pressure increases have been said to break cell walls due to MW thermal effects. Some of these structural changes have been attributed to explosion in the hot spots leading to an increased porosity. These results have been obtained in several lignocellulosic materials, such as sugarcane bagasse, industrial pineapple wastes, cassava rhizome, corn stover, or herbal lignocellulose [83, 85–90].

In Conesa *et al.* [83], MW were applied as a pretreatment for further saccharification of pineapple cannery and juice residues. Results proved that MW pretreatments intermediate powers and exposure times enhances further saccharification. Low powers and short exposure times provided insufficient energy to disrupt the lignocellulosic structure and therefore promote subsequent enzyme action, whereas increasing the power applied or lengthening the treatment resulted in lower sugar yields. The latter was observed to be a consequence of sugar hydrothermal degradation and increased tissue compactness due to biowaste dehydration. In contrast, mild MW powers and intermediate times implied increased sugar yields. Similar results were obtained by Binod *et al.* [82], who reported that short duration MW pretreatments of sugarcane bagasse enhanced fermentable sugars yield. Aguilar-Reynosa *et al.* [87] studied MW heating as a pretreatment for the autohydrolysis of corn stover, which were treated with low enzyme loads and high substrate loads. They found that the highest conversion (95.1% for MW pretreated solids). MW-pretreated samples also offered better ethanol yield when being subjected to simultaneous saccharification and fermentation process.

The use of MW as a pretreatment of biomass has been applied either alone, or in combination with chemical or physical pretreatments in order improve the results. Among chemicals, acid or basic assisted MW treatments remain the most applied ones, whereas physical pretreatments include sonication or steam explosion [78]. Reaction rates are increased when combining MW with a chemical reagent. MW-alkali pretreatments have reported shorter times to optimize biomass bioconversion [82, 87, 91–93]. Therefore, the combination of MW with alkali have been proposed as promising techniques for the bioconversion of biomass into sugars for biofuel and other biorefinery approaches [94]. As compared to water, presence of ionic compounds such as sodium hydroxide may increase the loss tangent factor enhancing MW heating ability [91]. Among basic reagents, NaOH characterizes by providing high reaction rates and a greater degree of hydrolysis than other alkaline pretreatments. Alkali alter the degree of polymerization of lignocellulose, modifying the physical properties of biomass (increased surface area, decreased crystallinity). NaOH effectively cleaves the linkage between lignin and hemicellulose in lignin-carbohydrate complexes [80].

Thermal effects of MW heating have been analyzed by the used of infrared thermography [83, 87]. The study of temperature evolution during MW heating evidences different heating periods: a first heterogenous heating period during which temperatures remain below 100 °C; a second period, in which temperatures are maintained close to 100 °C; and a third heterogeneous period characterized by non-centered hot spots and temperatures higher than 100 °C. In Figure 6.6, thermographs for MW and MW-alkali pretreated fruit residues are shown. It is evidenced that exposure to the MW electromagnetic field implies heating from the inside to the outside

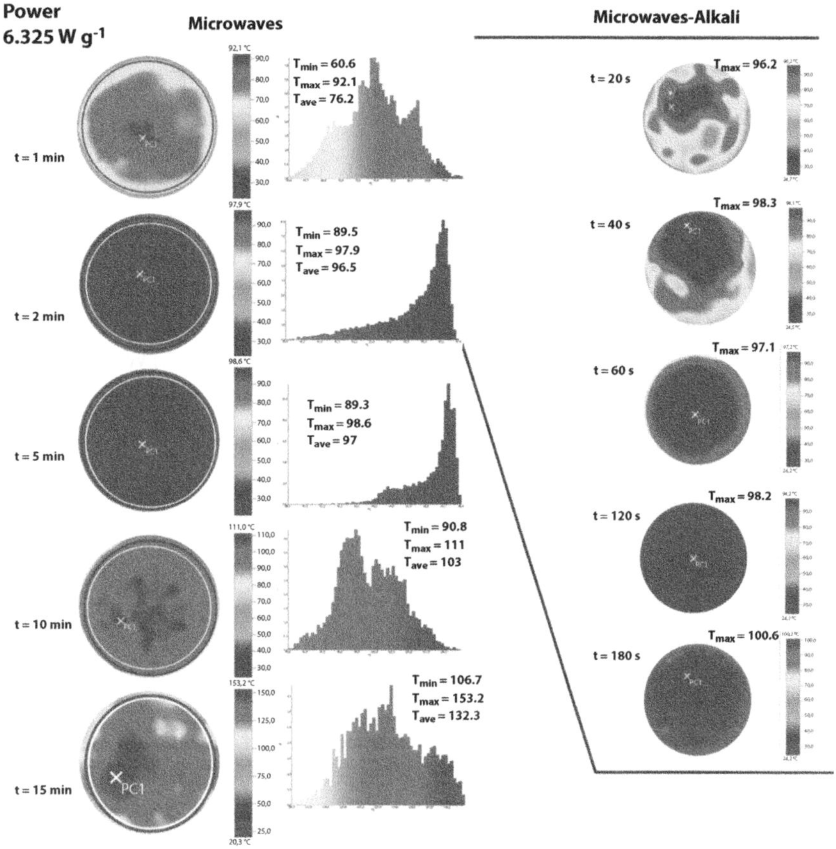

Figure 6.6 Thermographs and temperature histogram of industrial fruit biowaste along MW and MW-alkali pretreatments. The white cross indicates hot spot locations during MW heating.

of the sample. Homogeneity/heterogeneity of colors evidence the heating stages defined. The higher power applied, the earlier each stage is reached. MW-alkali pretreatments significantly reduce the time needed to heat up the waste, thus providing faster reaction rates. Once the third heterogenous phase begins, hot spot displacement occurs due to water evaporation and regions of different moisture content being distinguished [95]. This stage precedes sample calcination phenomena. Difference among both treatments could be explained by the different dielectric properties of the samples when NaOH is added [91], thus increasing MW absorption and fastening heating [82, 87, 91–93].

In general, release of simple sugars after saccharification is increased thanks to MW pretreatments but lengthening the treatment or reaching high temperatures leads to a further decrease in sugars release [81, 83, 96, 97]. MW have a different impact on sugar release after saccharification depending on the sugar being analyzed. While it promotes fructose and glucose increase, it reduces xylose content and other sugars such as sucrose and arabinose are not such affected by MW pretreatments. MW pretreatment have been said to cause hemicellulose solubilization and xylose degradation. Conesa *et al.* [83] found a markedly reduction of this pentose in the hydrolyzed samples when being MW-pretreated. Hemicellulose depletion during MW treatment have been generally reported [98–100]. On the other hand, MW pretreatments reduces the lignin content of the lignocellulosic biomass. Delignification is particularly enhanced when MW are assisted with alkali, as reported by Hu and Wen [81] or Liu *et al.* [101]. In fact, alkaline pretreatment of lignocellulose is considered one of the most effective methods to increase digestibility having several advantages such as effective delignification, low interaction with hemicellulose, and mild reaction conditions [80].

Release of inhibitory compounds during biomass pretreatment is another important factor to consider. Substances that may act as microbial inhibitors include phenolic compounds, furan aldehydes and weak acids [84, 102]. In this regard, both degradation of lignin into phenolic constituents and sugar degradation into furfural and 5-hydroxymethilfurfural (5-HMF) after MW pretreatments have been reported. Some authors suggest that their concentration may remain below values considered to be inhibitory for microorganisms under controlled conditions [87, 103], whereas others [96], have reported them to condition further fermentation yields.

When combining MW with alkali, it is found that total sugars increase is produced earlier because of the faster reaction rates; nevertheless, this is followed by a remarkable sugar depletion short after some minutes of treatment, suggesting that hydrothermal degradation of sugars could be enhanced when combining MW with alkali [87]. When used alone, 5-HMF appeared only at long exposure times, the more significant the higher power applied, thus suggesting partial solubilization of hemicellulose and cellulose during MW heating and further sugar degradation during the same MW stage. In contrast, 5-HMF is present at very short times when MW are combined with alkali, indicating that polymer solubilization to monomeric sugars due to alkali exposure also makes sugars more susceptible to further degradation. Conesa *et al.* [87] and Lü and Zhou [104] found furfural to be untraceable after MW-alkali pretreatments. Regarding phenolic constituents, the combination with alkali has a more significant impact than MW power or exposure time so that samples subjected to MW-alkali pretreatment contained a marked phenol content.

6.4.2 Microwaves Applied to the Recovery of Bio-Compounds from Food Wastes

Consumers are becoming more and more aware of the relationship between food and health and the prevention of nutrition-related diseases. In addition, the recent outbreak of the COVID-19 pandemic has boosted the interest of consumers in immunity promotion through the diet [105]. Good nutrition is considered essential for the success of all the SDGs, but particularly contributes to the SDG "Ensure healthy lives and promote well-being for all at all ages". Increasing the nutritional value of foods may be achieved by the use of natural ingredients with biological activity for food formulation. The wastes generated by the agri-food industry are especially rich in bioactive compounds, for which optimization of the extraction and recovery techniques becomes an essential part of the process. In addition, integral valorization of biowastes for the obtention of functional food ingredients through drying is another opportunity for MW to participate from food waste valorization processes.

Main factors to be considered for improving extraction efficiency are the type of solvent, processing time, temperature and pressure, the solvent to solid ratio and the sample matrix [106]. The environment protection recognition [107] together with the increasing legislative restrictions on the use of organic solvents [108], have boosted the search for alternative extraction technologies for more effective and sustainable recovery processes, among which MW-assisted extraction is included, together with supercritical fluid extraction and high-hydrostatic pressure, ultrasounds, pulsed electric fields, or high-voltage electric discharges assisted extraction [109].

The use of MW-assisted extraction (MWAE) started in the late 1980s and has become one of the most popular and cost-effective extraction methods available [110]. It basically involves the use of MW energy to heat the solvent, thus improving the diffusion of the solvent into the sample while accelerating the transfer of components from the sample to the solvent [111]. Mechanisms enhancing extraction include ionic conduction and dipole rotation but also cell disruption due to the internal pressure increase and increased diffusion rates due to tissue softening. Extraction efficiency of MWAE increases with both the nominal power and the treatment duration. This was evidenced by Plazzotta *et al.* [112] when recovering antioxidant compounds from frozen or air-dried peach juice waste. They also found that extracts from air-dried waste resulted richer in polyphenols but poorer in flavonoids, anthocyanins, and vitamin C than those from frozen waste. In comparison with control extracts, MWAE significantly improved total phenol (1.5 times), total flavonoids (3 times), total anthocyanin content (26 times) and ascorbic acid (2 times). Compared to ultrasound assisted extraction MWAE required half extraction time and implied a reduced impact in terms of greenhouse gas emission and energy requirements. Promising results of this technique have led to MW being widely applied to extract phenolics from different plant materials including citrus peels [113, 114], grape pomace [115], apple skins [116], avocado seeds [117], tomato peel waste [118], soluble dietary fibre from coffee silverskin [119], pectin from mango peel [120] and lipids from microalgae [121], among others. However, extraction of thermolabile compounds during MWAE may be reduced if exposure to the MW field is extended or the power applied is too high, as evidenced by Bachtler & Bart [122] in the extraction of oleanolic and ursolic acid from elder bark, which increased with MW power rising from 80 W to 400 W, but decreased when increasing to 800 W. Vasu *et al.* [123] also reported a decrease in the yield of bixin extracted from annatto seeds in ethyl acetate when the power was risen from 80 W to 800 W. Total flavonoid recovery and the antioxidant capacities of Vernonia amygdalina leaf extracts in distilled water did not further increase over 500 W [124].

Solvent polarity also determines MWAE extraction yield, ethanol having higher ability than methanol to absorb MW [113]. The addition of water increases the dielectric constant of the solvent and the absorption of MW energy, thus leading to a higher temperature, the rupture of cells and the easier release of antioxidant compounds. This was evidenced by Rodsamran & Sothornvit [114] in dried lime peel powder extracts, since the use of pure ethanol resulted in a lower total phenol content and antioxidant activity than 50-60% ethanol:water. Energy consumption and processing costs during MWAE can be reduced by using repeated extraction steps [114], stirring the solid and solvent mixture inside the microwave oven [113], or working in closed pressure vessels which allow heating above the atmospheric pressure boiling point of the solvent [125]. Combining MWAE with other eco-friendly techniques, such as ultrasounds [119] or pre-heating-irradiation [124] is another way to reach a more effective process. Another green feature of MWAE is the possibility of using no additional solvents by using moisture content as the solvent [126]. This was applied by Périno-Issartier *et al.* [128] to extract essential oils from plant material. MWAE of dried apple skins also allowed to obtain a less pollutant solid residue with higher calorific value, thus simplifying waste handling and increasing the efficiency of its potential use in waste-to-energy plants [116].

MW can also be applied to food waste valorization by assisting in the drying process to obtain ingredients in powdered form. Fruit and vegetable waste powders are rich in fibre and other bioactive constituents, which makes them interesting as functional food ingredients [128].

6.5 Microwaves for Functional Food Development and Increased Bioaccessibility

As stated by the European Commission Concerted Action on Functional Food Science in Europe, a food can be rated as functional, 'if it is satisfactorily demonstrated to affect beneficially one or more target functions in the body, beyond adequate nutritional effects, in a way that is relevant to either an improved state of health and well-being and/or reduction of risk of disease' [129]. Accordingly, the term functional foods encompasses from fruits and vegetables, with a naturally high content of phytochemical compounds, to processed foods in which a specific ingredient has been added, eliminated, increased or decreased. Functional food market continues to grow since its appearance in the early 1990s, together with consumers awareness that food is not only necessary for the development, growth, and maintenance of the body, but also to guarantee certain quality of life, improving physical and mental conditions, as well as reducing the risk of suffering from certain nutrition-related diseases. As a result of this change in mentality, much research is focused on the identification of biologically active components, as well as on the effect of food processing on such beneficial components in order to increase their content and their bio-accessibility.

Application of MW technology to improve food functionality derives from either its effect on the chemical structure of macronutrients (proteins and polysaccharides) and/or on the kinetics of their hydrolysis reactions, or their effect on the kinetics of heat and mass transfer phenomena together with structural changes in food matrices. MW treatment can influence chemical and enzymatic hydrolysis of polysaccharides and proteins, determining the chemical nature of the final products and consequently their physico-chemical (solubility, reactivity, viscosity, water retention capacity ...) and functional properties (antiradical capacity, cholesterol binding capacity, bio-accessibility ...). The search for alternative sources of protein and the obtention of bioactive peptides from sources other than animals such as insects has gained interest in the last years. However, peptides from insect sources may cause reactions of the immune system. The combination of enzymatic proteolysis with a thermal treatment may induce chemical and conformational changes in proteins and/or peptides reducing their immunoreactivity [130]. Hall & Liceaga [131] studied the effect of MW-assisted enzymatic hydrolysis of cricket (*Gryllodes sigillatus*) protein on the peptides' immune reactivity. They concluded that MW-assisted enzymatic hydrolysis can be a useful method for generating bioactive peptides from insect proteins and lowering their immunoreactivity. Most by-products from the vegetable industry are important sources of fiber, and many are rich in protein of acceptable nutritional quality and other bioactive compounds such as phytochemicals. However, in many cases the extracted fiber is mostly insoluble, with poorer functional properties than the soluble one. Leão *et al.* [132] evidenced that MW used as a pretreatment shortens the extraction times, reduces energy consumption and solvent volume in pectin extraction from plant residues. MW can also improve the properties of the extracted fiber. Lin *et al.* [133] documented the effect of mixed fermentation by lactic acid bacteria and MW treatment on the composition, structure, and functional properties of okra dietary fiber. Although it has already been demonstrated that lactic acid bacteria produce a range of enzymes that generate an acidic environment which allows to degrade some insoluble dietary fiber during fermentation [134], Lin *et al.* [133] showed that combined fermentation and MW treatment significantly increases the content and composition of soluble dietary fiber in okra samples. This result arises as a good opportunity to improve the functionality of the fiber extracted from biowastes and expand their use.

The functional effect of carbohydrates like starch is conditioned by its hydrolysis rate, generating simple sugars during the digestive process.

One accepted hypothesis is that the bulk density of starch structure governs their digestion rate by affecting the diffusion of enzymes. In complex foods, the higher slowly digestible starch/resistant starch level and lower digestion rate could slow glucose release, decreasing food glycemic index [135]. Li *et al.* [136] affirmed that MW treatment could endow the starch with enhanced slowly digestible characteristics, relative to conventional heating and storage. In the case of oligosaccharides with prebiotic effect it is reported that linear structures with a lower degree of polymerization, such as fructooligosaccharides and galactooligosaccharides are faster fermented by colon microbiota causing intestinal discomfort or even diarrhea [137]. However, highly branched, randomly linked polydextrose fermentation by colonic bacteria has been demonstrated to degrade slowly and result in a better gut health compared to other prebiotics [138]. Wang *et al.* [139] applied MW treatment to synthesize branched, randomly linked and highly polymerized oligosaccharides (polydextrose, poly-galactose and poly-mannose) from simple sugars (glucose, galactose, and mannose), with favorable effects on the gut microbiota such as a higher production of short chain fatty acids. MW can also act in reducing antinutritional factors, which is a disadvantage of some legumes and cereals traditionally used as a source of dietary protein in many developing countries. These include phytic acid, polyphenols, saponins, protease inhibitors, phyto-lectins and oligosaccharides, which impede protein digestibility. Compared to other physical treatments such as soaking, grinding and ultrasounds, MW treatment resulted more effective despite its effect in decreasing the protein water solubility [140]. They attributed the effect to changes in the starch structures or decreased β-sheet structure and increased random coil conformation of the proteins.

Another mechanism by which MW may contribute to functional foods development is their effect on the kinetics of the heat and mass transfer phenomena coupled with structural changes (porosity, texture ...) in food matrices. As being faster, MW heating reduces thermal damage of micronutrients such as vitamins, antioxidants, and other phytochemicals, compared to conventional heating, as discussed in section 3. Regarding changes in the food structure, MW treatment contributes to food digestibility and favors the release of bound compounds which become more amenable to extraction and assimilation [40]. MW treatment may significantly affect the functional properties of high dietary fiber flour and increase the acceptability of food products with high bran content [141]. Patrignani *et al.* [142] evidenced that MW may disrupt the cell wall material in brewers' spent grain leading to improved water absorption, processing quality, and product texture, in addition to increasing the proportion of bioaccesible antioxidant compounds. This treatment made it possible to use the brewers' spent grain as an ingredient of wheat-based products. Pratap-Singh *et al.* [143] applied vacuum MW-drying to the same by-product, evidencing denaturation and conformational changes in protein structure which positively affected both water and oil holding capacities. Patrigiani *et al.* [142] also reported an improved water absorption and better texture after applying MW to brewery spent grain. On orange peel, Talens *et al.* [144] combined MW with air drying to produce orange peel fiber powder. Results evidenced a reduction in the time and energy needed for the process, together with an improvement of the fiber physicochemical properties, especially its swelling capacity.

MW vacuum drying has been proposed as an alternative to deep-fat frying to obtain potato chips with reduced fat content. The application of vacuum together with the increased internal heating caused by MW promotes water removal reducing the oil gain. MW frying of potato strips has also been reported to reduced acrylamide formation from 37 to 83% compared to conventional deep-fat frying [67]. This is explained in terms of the enhanced outflow of water in vapor state occurring during microwave frying, which might drag part of the generated acrylamide and its precursors. Erdoĝdu *et al.* [145] obtained similar results when performing frying after a MW pre-cooking step. Duarte-Correa *et al.* [146] showed that it was possible to produce potato chips fortified with calcium, vitamin C, and vitamin E, by combining vacuum impregnation and MW vacuum drying. Barreto *et al.* [147] studied the production of oil-free potato snacks using successive cycles of MW heating vacuum pulses. During the vacuum pulse, water quickly evaporates from the inside of the food matrix, creating a porous microstructure that favors product crispness and increases the drying rate.

By altering the product microstructure, MW may also impact nutrients bio-accessibility after gastrointestinal digestion [148]. Total phenolics, flavonoids and anthocyanins of *Rhodomyrtus tomentosa* berries dried first at 136 W for 2.5 h (and later with air at 70 °C for 1 h) were reported to be, respectively, 1.57, 1.82 and 1.86 times higher than in berries dried at 70 °C for 12 h, and 1.15, 1.16 and 1.3 times higher than in samples dried at 136 W for 4 h; conversely, the phenolic bio-accessibility after gastrointestinal digestion was found to be maximum (59.2 ± 1.4%) for hot air dried berries and minimum (45.2 ± 1.7%) for combined MW-hot air dried berries [148]. In a similar study on pumpkin slices [40], carotenoid losses after drying with air at 70 °C resulted significantly higher than after intermittent MW irradiation; however, cell wall degradation and complete chromoplast organelle disruption contributed to the improvement to the liberation and micellization of carotenoids.

6.6 Conclusions and Prospects

In this chapter, applications of MW technology to food and food waste processing have been reviewed. MW have shown great potential for food processing applications, from food process monitoring and their application to assist or replace conventional treatments used in the food industry, to their contribution to food waste valorization processes and the development of new functional foods, showing its capacity to address contemporary problems and participate from the Sustainable Development Goals Development.

As a non-conventional technology, MW has shown applications in assisting many unit operations commonly applied in the food industry. As reviewed in this chapter, most of the advantages of MW to pasteurization, cooking, drying, frying, baking, tempering and others, derive from increased reaction rates, volumetric heating, and structural modifications due to local pressure increases in the food matrix. This allows using shorter treatments and lower temperatures, which imply energy savings and, in many cases, have beneficial effects with respect to preservation of nutritional properties. However, non-uniform heating continues to be the main drawback of MW heating, for which developments in this sense are needed to achieve a more uniform heating, avoiding over and under-heated spots which generate uneven results. As for radiation monitoring systems, the future in this sector relies on developing small magnetrons allowing the generation of photons in all the MW spectrum, fast and economically. In addition, there is a need for developing measurement probes for the analysis of food without contact but maintaining the accuracy of measurements. If achieved, the use of this technology will allow monitoring all food transformation processes based on the state of water.

MW irradiation have been extensively studied for the pretreatment of biowastes to be subjected to biological transformation into biofuels or other products via biorefinery pathways. MW may improve biomass digestibility as a result of thermal and non-thermal effects which promote structural changes of biowaste as well as changes in conformation and solubilization of lignocellulose constituents. It is essential to select the proper MW power and treatment duration to avoid sugar degradation and the production of inhibitory constituents, in addition to the waste of energy this may represent. MW assisted extraction of bioactive components from biowastes has also been discussed as a promising technology. Mechanisms enhancing extraction include ionic conduction and dipole rotation but also cell disruption due to the internal pressure increase and higher diffusion rates due to tissue softening. MW extraction significantly reduces solvent volume, and it is considered among greener extraction technologies, for which its applications in this field are expected to grow. MW may also contribute to improve nutritional properties of foods. In addition to contribute to the preservation of nutritional characteristics, new foods and ingredients with improved functionality can be obtained by means of MW treatments. MW may benefit food functionality and bioactive compounds bio-accessibility by altering the chemical structure of macronutrients and the kinetics of hydrolysis reactions, but also because of their impact on food matrix structure and heat and mass transfer phenomena. Hence, MW technology is a potential tool to be used in the food industry and related areas, with a great potential of expansion in the near future.

References

1. Roychoudhuri, C., Kracklauer, A.F. and Creath, K. The Nature of Light: what Is a Photon. CRC Press, New York. 2008.
2. Horie, K., Ushiki, H. andWinnik, F.M. Molecular Photonics. Wiley & Sons, New York (2000).
3. Baker-Jarvis, J. and Kim, S. The interaction of radio-frequency fields with dielectric materials at macroscopic to mesoscopic scales. *J. Res. Natl.*, 117, 1 2012.
4. Debye, P. ,Bewilogua, L. and Ehrhardt F. , Physik. Z., 30 (84); Sächs. Akad. Ber., 81 (1929) p29.
5. Traffano-Schiffo, M.V., Castro-Giraldez, M., Colom, R.J., and Fito, P.J., Development of a Spectrophotometric System to Detect White Striping Physiopathy in Whole Chicken Carcasses. *Sensors*, 17, 5, 1024, 1-14, 2017.
6. Castro-Giráldez M., Aristoy M.C., Toldrá F., and Fito P. Microwave dielectric spectroscopy for the determination of pork meat quality. *Food Res. Int.*, 1, 2369-2377, 2010a.
7. Samuel, D., Trabelsi, S., Karnuah, A.B., Anthony, N.B., and Aggrey, S.E. (2012). The use of dielectric spectroscopy as a tool for predicting meat quality in poultry. *Int. J. Poult. Sci.*, 11, 551, 2012.

8. Traffano-Schiffo, M.V., Castro-Giraldez, M., Colom, R.J., and Fito, P.J. Innovative photonic system in radiofrequency and microwave range to determine chicken meat quality. *J. Food Eng.*, 239, 1-7, 2018a.
9. Traffano-Schiffo, M.V., Castro-Giraldez, M., Colom, R.J., Talens, P. and Fito, P.J. New methodology to analyze the dielectric properties in radiofrequency and microwave ranges in chicken meat during postmortem time. *J. Food Eng.*, 292, 110350, 2021.
10. Kent M., Peymann A., Gabriel C., and Knight A. Determination of added water in pork products using microwave dielectric spectroscopy. *Food Control*, 13, 143,149, 2002.
11. Traffano-Schiffo, M.V., Castro-Giraldez, M., Colom, R.J., and Fito, P.J. Study of the application of dielectric spectroscopy to predict the water activity of meat during drying process. *J. Food Eng.*, 166, 285-290, 2015.
12. Castro-Giraldez, M., Fito, P.J., Chenoll, C., and Fito, P. Development of a dielectric spectroscopy technique for the determination of apple (Granny Smith) maturity. *Innovative Food Sci. Emerg. Technol.*, 11, 4,749-754, 2010c.
13. Redzwan, S., Perez, M.D., Velander, J., and Augustine, R. Study of Maturity Fruit Assessment using Permittivity and Microwave Reflectivity Measurements for Quality Classification. *IEEE CAMA*, 1-3, 2018.
14. Kon, S., Kato, Y., and Horibe, M. Dynamic measurement of moisture content using microwaves for moisture evaluation of agricultural products. *IEEJ Trans. Electr. Electron. Eng.*, 15, 166-171, 2020.
15. Nelson, S.O., Kraszewski, A.W., Trabelsi, S., and Lawrence, K.C. Using cereal grain permittivity for quality sensing by moisture determination, *IEEE IMTC P*, 49, 237-42, 2003.
16. Nelson, S.O., Trabelsi, S., and Kraszewski, A.W. Advances in Sensing Grain Moisture Content By Microwave Measurements Trans. *ASAE*, 41, 483-7, 2013.
17. Pelletier, M.G., Wanjura, J.D., and Holt, G.A. Microwave moisture sensing of seedcotton: Part 1: Seedcotton microwave material properties. *Sensors*, 16(11), 1843, 2016.
18. Ramli, N.A.M., Rahiman, M.H.F., Kamarudin, L.M., Zakaria, A., and Mohamed, L. A Review on Frequency Selection in Grain Moisture Content Detection, *IOP Conf. Ser.: Mater. Sci. Eng.*, 705, 012002, 2019.
19. Bekal., A., GouriBalsubramaniam, B., Awaghade, V., and Ghute, S. Application of Microwave Moisture Sensor for DOC and Animal Feed. *IEEE Sens. J.*, 20(24), 14809-14816, 2020.
20. Velazquez-Varela J., Castro-Giraldez M., and Fito P.J. Control of the brewing process by using microwaves dielectric spectroscopy. *J. Food Eng.*, 119, 633-639, 2013.
21. Kataria, T.K., Sosa-Morales, M.E., Olvera-Cervantes, J.L., and Corona-Chavez, A. Dielectric properties of tequila in the microwave frequency range (0.5-20 GHz) using coaxial probe. *Int. J. Food Prop.*, 20(1), S377-S384, 2017.

22. Harnsoongnoen, S., Wanthong, A., Charoen, U., and Siritaratiwat, A. Microwave Sensor for Nitrate and Phosphate Concentration Sensing, *IEEE Sens. J.*, 19, 8, 2950-2955, 2019.
23. Tomas-Egea, J.A., Fito, P.J., and Castro-Giraldez, M. Analysis of Apple Candying by Microwave Spectroscopy. *Foods*, 4, 8(8), 316, 2019.
24. Castro-Giraldez M., Fito P.J., and Fito P. Application of microwaves dielectric spectroscopy for controlling pork meat (*Longissimus dorsi*) salting process. *J. Food Eng.* 4, 484-490, 2010b.
25. Velazquez-Varela J., Castro-Giraldez M., Cuibus L., Tomas-Egea J.A., Socaciu C., and Fito P.J. Study of the cheese salting process by dielectric properties at microwave frequencies. *J. Food Eng.*, 224, 121-128, 2018.
26. Krraoui, H., Mejri, F. and Aguili, T. Nondestructive measurement of complex permittivity by a microwave technique: detection of contamination and food quality, *J. Electromag. Waves Appl.*, 31, 16, 1638-1657, 2017.
27. Traffano-Schiffo, M.V., Castro-Giraldez, M., Colom, R.J., and Fito, P.J. New Spectrophotometric System to Segregate Tissues in Mandarin Fruit. *Food Bioprocess. Technol.*, 11, 399-406, 2018b.
28. Guo, D.W.S., Cheng, J.H., and Han, Z. Microwave processing techniques and their recent applications in the food industry. *Trends Food Sci. Technol.*, 37, 236-247, 2017.
29. Chandrasekaran, S., Ramanathan, S., and Basak, T. Microwave food processing-A review. *Food Res. Int.*, 52(1), 243-261, 2013.
30. Chizoba-Ekezie, F-G., Sun, D-W., Han, Z., and Cheng, J-H. Microwave-assisted food processing technologies for enhancing product quality and process efficiency: A review of recent developments. *Trends Food Sci. Technol.*, 67, 58-69, 2017.
31. Kumar, C., Joardder, M., Karim, A., Millar, G.J., and Amin, Z. Temperature redistribution modelling during intermittent microwave convective heating. *Procedia Eng.*, 90, 544-549, 2014.
32. Salazar-González, C., San Martín-González, M.F., López-Malo, A. and Sosa-Morales, M.E. Recent Studies Related to Microwave Processing of Fluid Foods. *Food Bioprocess Tech.*, 5, 31-46, 2012.
33. Dehghannya, J., Hosseinlar, S.H., and Heshmati, M.K. Multi-stage continuous and intermittent microwave drying of quince fruit coupled with osmotic dehydration and low temperature hot air drying. *Innovative Food Sci. Emerg. Technol.*, 45, 132-151, 2018.
34. Aghilinategh, N., Rafiee, S., Gholikhani, A., Hosseinpur, S., Omid, M., Mohtasebi, S.S., and Maleki, N. A comparative study of dried apple using hot air, intermittent and continuous microwave: evaluation of kinetic parameters and physicochemical quality attributes. *Food Sci. Nut.*, 3, 6, 519-526, 2015.
35. Feng, H., Yin, Y., and Tang, J. Microwave Drying of Food and Agricultural Materials: Basics and Heat and Mass Transfer Modeling. *Food Eng. Rev.*, 4, 89-106, 2012.

36. Kroehnke, J., Szadzińska, J., Stasiak, M., Radziejewska-Kubzdela, E., Biegańska-Marecik, R., and Musielak, G. Ultrasound- and microwave-assisted convective drying of carrots - Process kinetics and product's quality analysis. *Ultrason. Sonochem.*, 48, 249-258, 2018.
37. Wang, Y., Zhang, M., Mujumdar, A.S., and Mothibe, K.J. Quality Changes of Dehydrated Restructured Fish Product from Silver Carp (*Hypophthalmichthys molitrix*) as Affected by Drying Methods. *Food Bioprocess. Technol.*, 6, 1664-1680, 2013.
38. Jiang, H., Zhang, M., and Mujumdar, A. S. Physico-chemical changes during different stages of MFD/FD banana chips. *J. Food Eng.*, 101(2), 140-145, 2010.
39. Paul, I.D. and Das, M. Effect of freeze, microwave-convective hot air, vacuum and dehumidified air drying on total phenolics content, anthocyanin content and antioxidant activity of jamun (*Syzygiumcumini* L.) pulp. *J. Food Sci. Tech.*, 55(7), 2410-2419, 2018.
40. Zhang, Z., Wang, X., Li, Y., Wei, Q., Liu, C., Nie, M., Li, D., Xiao, Y., Liu, C., Xu, L., Zhang, M. and Jiang, N. Evaluation of the impact of food matrix change on the in vitro bioaccessibility of carotenoids in pumpkin (*Cucurbita moschata*) slices during two drying processes. *Food Funct.*, 8, 4693-4702, 2017.
41. Zielinska, M. and Michalska, A. Microwave-assisted drying of blueberry (*Vacciniumcorymbosum* L.) fruits: Drying kinetics, polyphenols, anthocyanins, antioxidant capacity, colour and texture. *Food Chem.*, 212, 671-680, 2016.
42. Wojdyło, A., Figiel, A., Lech, K., Nowicka, P., and Oszmiański, J. Effect of Convective and Vacuum–Microwave Drying on the Bioactive Compounds, Color, and Antioxidant Capacity of Sour Cherries. *Food Bioproc. Tech.*, 7, 829-841, 2014.
43. Jia, Y., Khalifa, I., Hu, L., Zhu, W., Li, J., Li, K., and Li, C. Influence of three different drying techniques on persimmon chips' characteristics: A comparison study among hot-air, combined hot-air-microwave, and vacuum freeze drying techniques. *Food Bioprod. Process.*, 118, 67-76, 2019.
44. Oliveira, G., Eliasson, L., Ehrnell, M., Höglund, E., Andlid, T., and Alminger, M. Tailoring bilberry powder functionality through processing: Effects of drying and fractionation on the stability of total polyphenols and anthocyanins. *Food Sci. Nut.*, 7, 1017-1026, 2018.
45. Oerlemans, K., Barrett, D.M., Bosch-Suades, C., Verkerk, R., and Dekker, M. Thermal degradation of glucosinolates in red cabbage. *Food Chem.*, 95, 19-29, 2006.
46. Dibanda, R.F., Akdowa, E.P., Ashwini, R.P., Tongwa, Q.M., and Mbofung, C.M. Effect of microwave blanching on antioxidant activity, phenolic compounds, and browning behaviour of some fruit peelings. *Food Chem.*, 302, 125308, 2020.

47. Benlloch-Tinoco, M., Igual, M., Rodrigo, D., & Martínez-Navarrete, N. Comparison of microwaves and conventional thermal treatment on enzymes activity and antioxidant capacity of kiwifruit puree. *Innovative Food Sci. Emerg. Technol.*, 19, 166-172, 2013.
48. Kachhadiya, S., Kumar, N., and Seth, N. Process kinetics on physico-chemical and peroxidase activity for different blanching methods of sweet corn. *J. Food Sci. Technol.*, 55(12), 4823-4832, 2018.
49. Kumar, S., Khadka, M., Mishra, R., Kohli, D., and Upadhaya, S. Effects of Conventional and Microwave Heating Pasteurization on Physiochemical Properties of Pomelo (*Citrus maxima*) Juice. *J. Food Process. Technol.*, 8(7), 2017.
50. Siguemoto, E.S., Purgatto, E., Hassimotto, N.M.A., and Gu, J.A.W. Comparative evaluation of flavour and nutritional quality after conventional and microwave-assisted pasteurization of cloudy apple juice. *LWT-Food Sci. Tech.*, 111, 853-860, 2019.
51. Martins, C.P.C., Cavalcanti, R.N., Cardozo, T.S.F., Couto, S.M., Guimaraes, J.T., Balthazar, C.F., Rocha, R.S., Pimentel, T.C., Freitas, M.Q., Raíces, R.S.L., Silva, M.C., Esmerino, E.A., Granato, D., and Cruz, A.G. Effects of microwave heating on the chemical composition and bioactivity of orange juice-milk beverages. *Food Chem.*, 345, 128746, 2021.
52. Sumnu, G., Sahin, S., and Sevimli, M. Microwave, infrared and infrared-microwave combination baking of cakes. *J. Food Eng.*, 71(2), 150-155, 2005.
53. Keskin, S.O., Sumnu, G. and Sahin, S. Bread baking in halogen lamp–microwave combination oven. *Food Res. Int.*, 37(5), 489-495, 2004.
54. Turabi, E., Sumnu, G., and Sahin, S. Optimization of baking of rice cakes in infrared microwave combination oven by response surface methodology. *Food Bioprocess. Tech.*, 1, 64-73, 2008.
55. Taher, B.J. and Farid, M.M. Cyclic microwave thawing of frozen meat: experimental and theoretical investigation. *Chem. Eng. Process.*, 40(4), 379-389, 2001.
56. Wang, B., Du, X., Kong, B., Liu, W., Li, F., Pan, N., Xia, X and Zhang, D. Effect of ultrasound thawing, vacuum thawing, and microwave thawing on gelling properties of protein from porcine longissimus dorsi. *Ultrason. Sonochem.*, 64, 104860, 2020.
57. Cai, L., Cao, M., Cao, A. Regenstein, J., Li, J. and Guan, R. Ultrasound or microwave vacuum thawing of red seabream (*Pagrus major*) fillets. *Ultrason. Sonochem.*, 47, 122-132, 2018.
58. Cai, L., Zhang, W., Cao, A., Cao, M. and Li, J. Effects of ultrasonics combined with far infrared or microwave thawing on protein denaturation and moisture migration of *Sciaenops ocellatus* (red drum), *Ultrason. Sonochem.*, 55, 96-104, 2019.
59. Goszkiewicz, A., Kołodziejczyk, E., and Ratajczyk, F. Comparison of microwave and convection method of roasting sunflower seeds and its effect on

sensory quality, texture and physicochemical characteristics. *Food Structure*, 25, 100144, 2020.
60. Yoshida, H., Hirakawa, Y., Tomiyama, Y., and Mizushina, Y. Effects of microwave treatment on the oxidative stability of peanut (*Arachishypogeae*) oils and the molecular species of their triacylglycerols. *Eur. J. Lipid Sci. Technol.*, 105, 351-358, 2003.
61. Yoshida, H., Tomiyama, Y., Hirakawa, Y., and Mizushina, Y. Microwave roasting effects on the oxidative stability of oils and molecular species of triacylglycerols in the kernels of pumpkin (Cucurbita spp.) seeds. *J. Food Compos. Anal.*, 19, 330-339, 2006.
62. Uysal, N., Summu, G., and Sahin, S. Optimization of microwave–infrared roasting of hazelnut. *J. Food Eng.*, 90(2), 255-261, 2009.
63. Nebesny, E. and Budryn, G. Antioxidative activity of green and roasted coffee beans as influenced by convection and microwave roasting methods and content of certain compounds. *Eur. Food Res. Technol.*, 217, 157-163, 2003.
64. Su, Y., Zhang, M., Zhang, W., Adhikari, B., and Yang, Z. Application of novel microwave-assisted vacuum frying to reduce the oil uptake and improve the quality of potato chips. *LWT-Food Sci. Technol.*, 73, 490-497, 2016.
65. Quan, X., Zhang, M., Zhang, W., and Adhikari, B. Effect of microwave-assisted vacuum frying on the quality of potato chips. *Dry. Technol.*, 32, 1812-1819, 2014.
66. Parikh, A. and Takhar, P.S. Comparison of microwave and conventional frying on quality attributes and fat content of potatoes. *J. Food Sci.*, 81, E2743-E2755, 2016.
67. Sansano, M., De los Reyes, R., Andrés, A., and Heredia, A. Effect of Microwave Frying on Acrylamide Generation, Mass Transfer, Color, and Texture in French Fries. *Food Bioprocess Tech.*, 11, 1934-1939, 2018.
68. Valverde D. Application of new technologies to cocoa alkalization. UniversitatPolitècnica de València, 2020, available in: https://doi.org/10.4995/Thesis/10251/140311
69. Wang, J., Yang, X.-H., Mujumdar, A.S., Wang, D., Zhao, J.-H., Fang, X.-F., Zhang, Q., Xie, L., Gao, Z.-J., and Xiao, H.-W. Effects of various blanching methods on weight loss, enzymes inactivation, phytochemical contents, antioxidant capacity, ultrastructure and drying kinetics of red bell pepper (*Capsicum annuum* L.). *LWT-Food Sci. Technol.*, 77, 337-347, 2017.
70. Severini, C., Giuliani, R., De Filippis, A., Derossi, A., and De Pilli, T. Influence of different blanching methods on colour, ascorbic acid and phenolics content of broccoli. *J. Food Sci. Tech.*, 53(1), 501-510, 2016.
71. Ahmed, J. and Ramaswamy, H.S. Microwave Pasteurization and Sterilization of Foods, in: M. Shafiur Rahman ed., *Handbook of Food Preservation* (2nd edition), CRC Press, pp. 691-711, 2007.
72. Tang, J., Hong, Y.-K., Inanoglu, S., and Liu, F. Microwave pasteurization for ready-to-eat meals. Curr. Opin. *Food Sci.*, 23, 133-141, 2018.

73. Li, B. and Sun, D.W. Novel methods for rapid freezing and thawing of foods - a review. *J. Food Eng.*, 54(3), 175-182, 2002.
74. Ekezie, F.G.C., Sun, D.W., Han, A. and Cheng, J.H. Microwave-assisted food processing technologies for enhancing product quality and process efficiency: A review of recent developments. *Trends Food Sci. Technol.*, 67, 58-69, 2017.
75. Pankaj, S.K. and Keener, K.M. A review and research trends in alternate frying technologies, *Curr. Opin. Food Sci.*, 16, 74-79, 2017.
76. The 2030 Agenda for Sustainable Development. Retrieved from: https://en.unesco.org/
77. Scheel, C. Beyond sustainability. Transforming industrial zero-valued residues into increasing economic returns. *J. Clean Prod.*, 131(10), 376-386, 2016.
78. Bundhoo, Z.M.A. Microwave-assisted conversion of biomass and waste materials to biofuels. *Renewable Sustainable Energy Rev.*, 82, 1149-1177, 2018.
79. Kim T.H. Pretreatment of lignocellulosic biomass, in S.T. Yang, H.A. El-Enshasy, N.Thongchul, Y. Martin, eds., *Bioprocessing Technologies in Integrated Biorefinery for Production of Biofuels, Biochemicals, and Biopolymers from Biomass*, Wiley, pp. 91-109, 2013.
80. Kim, J.S., Lee, Y.Y., and Kim, H. A review on alkaline pretreatment technology for bioconversion of lignocellulosic biomass. *Bioresour. Technol.*, 199, 42-48, 2016.
81. Hu, Z. and Wen, Z. Enhancing enzymatic digestibility of switchgrass by microwaveassisted alkali pretreatment. *Biochem. Eng. J.*, 38, 369-378, 2008.
82. Binod, P., Satyanagalakshmi, K., Sindhu, R., Janu, K.U., Sukumaran, R., and Pandey, A. Short duration microwave assisted pretreatment enhances the enzymatic saccharification and fermentable sugar yield from sugarcane bagasse. *Renew. Energ.*, 37, 109-116, 2012.
83. Conesa, C., Seguí, L., Laguarda-Miró, N., and Fito, P. Microwaves as a pretreatment for enhancing enzymatic hydrolysis of pineapple industrial waste for bioethanol production. *Food Bioprod. Process.*, 100, 203-213, 2016a.
84. Ethaib, S., Omar, R., Kamal, S.M.M., and Biaki, D.R.A. Microwave-assisted pretreatment of lignocellulosic biomass: a review. *J. Eng. Sci. Tech.*, 97-109, 2015.
85. Peng, H., Chen, H., Qu, Y., Li, H., and Xu, J. Bioconversion of different sizes of microcrystalline cellulose pretreated by microwave irradiation with/without NaOH. *Appl. Energ.*, 117, 142-8, 2014.
86. Sombatpraiwan S., Junyusen T., Treeamnuk T. and Junyusen P. Optimization of microwave-assisted alkali pretreatment of cassava rhizome for enhanced enzymatic hydrolysis glucose yield. *Food Energy Sec.*, 8:e00174. 1-12, 2019.
87. Conesa, C., Seguí, L., Laguarda-Miró, N., and Fito, P. Microwave-Assisted Alkali Pretreatment for Enhancing Pineapple Waste Saccharification. *Bioresour.*, 11(3), 6518-6531, 2016b.

88. Aguilar-Reynosa, A., Romaní, A., Rodríguez-Jasso, R.M., Aguilar, C.M. and Garrote, G., Ruiz, H.A. Comparison of microwave and conduction-convection heating autohydrolysis pretreatment for bioethanol production. *Bioresource Technology*. 2017, 243, 273-283.
89. Cheng, X.Y and Liu, C.Z. Enhanced biogas production from herbal-extraction process residues by microwave-assisted alkaline pretreatment. *J. Chem. Technol. Biotechnol.*, 85, 127-31, 2010.
90. Chen, W.H., Tu, Y.J., and Sheen, H.K. Disruption of sugarcane bagasse lignocellulosic structure by means of dilute sulfuric acid pretreatment with microwave-assisted heating. *Appl. Energy*, 88, 2726-2734, 2011.
91. Jin, S., Zhang, G., Zhang, P., Li, F., Wang, S., Fan, S., and Zhou, S. Microwave assisted alkaline pretreatment to enhance enzymatic saccharification of catalpa sawdust. *Bioresour. Technol.*, 221, 26-30, 2016.
92. Singh, R.D., Bhuyan, K., Banerjee, J., Muir, J., and Arora, A. Hydrothermal and microwave assisted alkali pretreatment for fractionation of arecanut husk. *Ind. Crops Prod.*, 102, 65-74, 2017.
93. Tiwari, G., Sharma, A., Kumar, A., and Sharma, S. Assessment of microwave-assisted alkali pretreatment for the production of sugars from banana fruit peel waste. *Biofuels*, 10(1), 3-10, 2018.
94. Agu, O.S., Tabil, L. G., Meda, V., Dumonceaux, T., and Mupondwa, E. Pretreatment of Crop Residues by Application of Microwave Heating and Alkaline Solution for Biofuel Processing: A Review, in E. Jacob-Lopes and L. Queiroz-Zepka, eds., *Renewable Resources and Biorefineries*, IntechOpen, pp. 47-67, 2019.
95. Liu, S.T, Chen, X.G., Zhang, A.B., Yan, K.K., and Ye, Y. Electromagnetic Performance of Rice Husk Ash. *Bioresources*, 9(2), 2328-2340, 2007.
96. Jung, Y.H., Kim, I.J., Kim, H.K., and Kim, KH. Dilute acid pretreatment of lignocellulose for whole slurry ethanol fermentation. *Bioresour. Technol.*, 132,109-14, 2016.
97. Lin, R., Cheng, J., Song, W., Ding, L., Xie, B., Zhou, J. and Cen, K. Characterisation of water hyacinth with microwave-heated alkali pretreatment for enhanced enzymatic digestibility and hydrogen/methane fermentation. *Bioresource. Technol.*, 182, 1-7, 2015.
98. Fan, Y.T., Zhang, Y.H., Zhang, S.F., Hou, H.W., and Ren, B.Z. Efficient conversion of wheat straw wastes into biohydrogen gas by cow dung compost. *Bioresour. Technol.* 97, 500-505, 2006.
99. Hendriks, A.T.W.M. and Zeeman, G. Pretreatments to enhance the digestibility of lignocellulosic biomass. *Bioresour. Technol.*, 100 (1), 10-18, 2009.
100. Li, H., Chen, X., Ren, J., Deng. H., Peng, F. and Sun, R. Functional relationship of furfural yields and the hemicellulose-derived sugars in the hydrolysates from corncob by microwave-assisted hydrothermal pretreatment. *Biotechnol. Biofuels*, 8, 127, 2015.

101. Liu, J., Takada, R., Karita, S., Watanabe, T., Honda, Y., and Watanabe, T. Microwave-assisted pretreatment of recalcitrant softwood in aqueous glycerol. *Bioresour. Technol.*, 101, 9355-60, 2010.
102. Jonsson, L.J., Alriksson, B., and Nilvebrant, N.O. Bioconversion of lignocellulose: inhibitors and detoxification. *Biotechnol. Biofuels*, 6, 16, 2013.
103. Mikulski, D., Kłosowski, G., Menka, A., and Koim-Puchowska, B. Microwave-assisted pretreatment of maize distillery stillage with the use of dilute sulfuric acid in the production of cellulosic ethanol. *Bioresour. Technol.*, 278, 318-328, 2019.
104. Lü, J. and Zhou, P. Optimisation of microwave-assisted $FeCl_3$ pretreatment conditions of rice straw and utilisation of *Trichodermaviride* and *Bacillus pumilus* for production of reducing sugars. *Bioresour. Technol.*, 102(13), 6966-6971, 2011.
105. Galanakis, C.M. Functionality of Food Components and Emerging Technologies. *Foods*, 10, 128, 2021.
106. Moreira, M.M., Morais, S., and Delerue-Matos, C. Environment-Friendly Techniques for Extraction of Bioactive Compounds from Fruits, in Alexandru Mihai Grumezescu and Alina Maria Holban, eds., *Soft Chemistry and Food Fermentation*, Academic Press, pp. 21-47, 2017.
107. Chemat, F., Vian, M.A., and Cravotto, G. Green Extraction of Natural Products: Concept and Principles. *Int. J. Mol. Sci.*,13(7), 8615-8627, 2012.
108. Papadaki, S., Kyriakopoulou, K., Tzovenis, I., and Krokida, M. Environmental impact of phycocyanin recovery from Spirulina platensis cyanobacterium. *Innov. Food Sci. Emerg. Technol.*, 44, 217-223, 2017.
109. Barba, F.J., Brianceau, S., Turk, M., Boussetta, N., and Vorobiev, E. Effect of Alternative Physical Treatments (Ultrasounds, Pulsed Electric Fields, and High-Voltage Electrical Discharges) on Selective Recovery of Bio-compounds from Fermented Grape Pomace. *Food Bioprocess Technol.*, 8, 1139-1148, 2015.
110. Delazar A., Nahar L., Hamedeyazdan S., and Sarker S.D. Microwave-Assisted Extraction in Natural Products Isolation, *Methods Mol. Biol.*, 864, 89-115, 2012.
111. Pacheco-Fernández, I. and Pino, V. Extraction with Ionic Liquids-Organic Compounds, in Colin F. Poole, ed., *Liquid-Phase Extraction*, Elsevier, pp. 499-537, 2020.
112. Plazzotta, S., Ibarz, R., Manzocco, L., Martín-Belloso, O. Optimizing the antioxidant biocompound recovery from peach waste extraction assisted by ultrasounds or microwaves. *Ultrason. Sonochem.*, 63, 104954, 2020.
113. Kaderides, K., Papaoikonomou, L., Serafim, M., and Goula, A.M. Microwave-assisted extraction of phenolics from pomegranate peels: Optimization, kinetics, and comparison with ultrasounds extraction. *Chem. Eng. Process. Process Intensif.*, 137, 1-11, 2019.

114. Rodsamrana, P. and Sothornvit, R. Extraction of phenolic compounds from lime peel waste using ultrasonic-assisted and microwave-assisted extractions. *Food Biosci.*, 28, 66-73, 2019.
115. Garrido, T., Gizdavic-Nikolaidis, M., Leceta, I., Urdanpilleta, M., Guerrero, P., De la Caba, K., and Kilmartin, P.A. Optimizing the extraction process of natural antioxidants from chardonnay grape marc using microwave-assisted extraction. *Waste Manage.*, 88, 110-117, 2019.
116. Casazza, A.A., Pettinato, M., and Perego, P. Polyphenols from apple skins: A study on microwave-assisted extraction optimization and exhausted solid characterization. *Sep. Purif. Technol.*, 240, 116640, 2020.
117. Araújo, R.G., Rodriguez-Jasso, R.M., Ruiz, H.A., Govea-Salas, M., Pintado, M.E., and Aguilar, C.N. Process optimization of microwave-assisted extraction of bioactive molecules from avocado seeds. *Ind. Crop. Prod.*, 154, 112623, 2020.
118. Tranfić-Bakić, M., Pedisić, S., Zorić, Z., Dragović-Uzelac, V., and Ninčević-Grassino, A. Effect of Microwave-Assisted Extraction on Polyphenols Recovery from Tomato Peel Waste. *ActaChim. Slov.*, 66, 367-377, 2019.
119. Wen, L., Zhang, Z., Zhao, M., Senthamaraikannan, R., Padamati, R.B., Sun, D.W., and Tiwari, B.K. Green extraction of soluble dietary fibre from coffee silverskin: impact of ultrasound/microwave-assisted extraction. *Int. J. Food Sci. Tech.*, 55, 2242-2250, 2020.
120. Sommano, S.R., Ounamornmas, P., Nisoa, M., Sriwattana, S., and Colelli, G. Characterisation and physiochemical properties of mango peel pectin extracted by conventional and phase control microwave-assisted extractions. *Int. Food Res. J.*, 25(6), 2657-2665, 2018.
121. Kapoore, R.V., Butler, T.O., Pandhal, J., and Vaidyanathan, S. Microwave-Assisted Extraction for Microalgae: From Biofuels to Biorefinery. *Biology*, 7(1), 18, 2018.
122. Bachtler, S. and Bart, H.J. Increase the yield of bioactive compounds from elder bark and annatto seeds using ultrasound and microwave assisted extraction technologies. *Food Bioprod. Process.*, 125, 1-13, 2021.
123. Vasu, S., Palaniyappan, V, Kothandam, H.P., and S. Badami, S. Microwave facilitated extraction of Bixin from Bixaorellana and it's in-vitro antioxidant activity. *Der Pharm. Lettre*, 2(2), 479-485, 2010.
124. Alara, O.R., Abdurahman, N.H., and Olalere, O.A. Optimization of microwave-assisted extraction of flavonoids and antioxidants from *Vernonia amygdalina* leaf using response surface methodology. *Food Bioprod. Process.*, 107, 36-48, 2018.
125. Vandenburg, H.J. Polymers: Supercritical Fluid Extraction, in Ian D. Wilson, ed., *Encyclopedia of Separation Science*, Academic Press, pp. 3915-3921, 2000.

126. Kusuma, H., Putri, D., Dewi, I., and Mahfud, M. Solvent-free microwave extraction as the useful tool for extraction of edible essential oils. *Chem. Chem. Technol.*, 10(2), 213-218, 2016.
127. Périno-Issartier, S., Huma, Z., Abert-Vian, M., and Chemat, F. Solvent Free Microwave-Assisted Extraction of Antioxidants from Sea Buckthorn (*Hippophaerhamnoides*) Food By-Products. *Food Bioprocess. Tech.*, 4, 1020-1028, 2011.
128. Bas-Bellver, C., Barrera, C., Betoret, N., and Seguí, L. Turning agri-food cooperative vegetable residues into functional powdered ingredients for the food industry. *Sustainability*, 12(4), 1-15, 2020.
129. Henry, C.J. Functional foods. *Eur. J. Clin. Nutr.*, 4, 657-659, 2010.
130. Izquierdo, F.J., Peñas, E., Baeza, M.L.and Gomez, R. Effects of combined microwave and enzymatic treatments on the hydrolysis and immunoreactivity of dairy whey proteins. *Int. Dairy J.*, 18(9), 918-922, 2008.
131. Hall, F. and Liceaga, A. Effect of microwave-assisted enzymatic hydrolysis of cricket (*Gryllodessigillatus*) protein on ACE and DPP-IV inhibition and tropomyosin-IgG binding. *J Funct. Foods*, 64, 103634, 2020.
132. Leão, D., Botelho, B.G., Oliverira, L.S., and Franca, A.S. Potential of pequi (*Caryocarbrasiliense*Camb.) peels as sources of highly esterified pectins obtained by microwave assisted extraction. *LWT*, 87, 575-580, 2018.
133. Lin, D., Long, X., Huang, Y., Yang, Y., Wu, Z., Chen, H., Zhang, Q., Wu, D., Wen Qin, W. and Tu, Z. Effects of microbial fermentation and microwave treatment on the composition, structural characteristics, and functional properties of modified okara dietary fiber. *LWT*, 123, 109059, 2020.
134. Tu, Z., Chen, L., Wang, H., Ruan, C., Zhang, L. and Kou, Y. Effect of fermentation and dynamic high pressure microfluidization on dietary fibre of soybean residue. *J. Food Sci. Tech.*, 51(11), 3285-3292, 2014.
135. Lehmann, U. and Robin, F. Slowly digestible starch - its structure and health implications: a review. *Trends Food Sci. Tech.*, 18(7), 346-355, 2007.
136. Li, N., Wang, L., Zhao, S., Qiao, D., Jia, C., Niu, M., Lin, Q. and Zhang, B. An insight into starch slowly digestible features enhanced by microwave treatment. *Food Hydrocolloid.*, 103, 105690, 2020.
137. Hernot. D.C.. Boileau, T.W., Bauer, L.L., Middelbos, I.S., Murphy, M.R., Swanson, K.S., and Fahey, G.C. In Vitro Fermentation Profiles, Gas Production Rates, and Microbiota Modulation as Affected by Certain Fructans, Galactooligosaccharides, and Polydextrose. *J. Agric. Food Chem.*, 57(4), 1354-1361, 2009.
138. Costabile, A., Fava, F., Röytiö, H., Forssten, S.D., Olli, K., Klievink, J., Rowland, I.R., Ouwehand, A.C., Rastall, R.A., Gibson, G.R., and Walton, G.E. Impact of polydextrose on the faecal microbiota: a double-blind, crossover, placebo-controlled feeding study in healthy human subjects. *Br. J. Nutr.*, 108(3), 471-81, 2012.

139. Wang, H., Ren, P., Mang, L., Shen, N., Chen, J., and Yiwen Zhang, Y. *In vitro* fermentation of novel microwave-synthesized non-digestible oligosaccharides and their impact on the composition and metabolites of human gut microbiota. *J. Funct. Foods*, 55, 156-166, 2019.
140. Sun, X., Ohanenye, I.C., Ahmed, T., Udenigwe, C.C. Microwave treatment increased protein digestibility of pigeon pea (*Cajanuscajan*) flour: Elucidation of underlying mechanisms. *Food Chem.*, 329, 127196, 2020.
141. Zhang, H., Wang, H., Cao, X., and Wang, J. Preparation and modification of high dietary fiber flour: A review. *Food Res. Int.*, 113, 24-35, 2018.
142. Patrignani, M., Brantsen, J.F., Awika, J.M., and Conforti, P.A. Application of a novel microwave energy treatment on brewers' spent grain (BSG): Effect on its functionality and chemical characteristics. *Food Chem.*, 346, 128935, 2021.
143. Pratap-Singh, A., Mandal, R., Shojaei, M., Singh, A., Kowalczewski, P.Ł., Ligaj, M., Pawlicz, J., and Jarzębski, M. Novel Drying Methods for Sustainable Upcycling of Brewers' Spent Grains as a Plant Protein Source. *Sustainability*, 12(9), 3660, 2020.
144. Talens, C., Arboleya, J.C. Castro-Jiráldez, M., and Fito, P.J. Effect of microwave power coupled with hot air drying on process efficiency and physicochemical properties of a new dietary fibre ingredient obtained from orange peel. *LWT*, 77, 110-118, 2017.
145. Erdoğdu, S.B., Palazoğlu, T.K., Gökmen, V., Şenyuva, H.Z., and Ekiz, H.I. Reduction of acrylamide formation in French fries by microwave precooking of potato strips. *J. Sci. Food Agric.*, 87(1), 133-137, 2007.
146. Duarte-Correa, Y., Díaz-Osorio, A., Osorio-Arias, J., Sobral, P.J.A., Vega-Castro, O. Development of fortified low-fat potato chips through Vacuum Impregnation and Microwave Vacuum Drying. *Innovative Food Sci. Emerg. Technol.*, 64, 102437, 2020.
147. Barreto, I.M.A., Tribuzi, G., Marsaioli Junior, A., Carciofi, B.A.M., and Laurindo, J.B. Oil-free potato chips produced by microwave multiflash drying. *J. Food Eng.*, 261, 133-139, 2019.
148. Zhao, G., Zhang, R., Liu, L., Deng, Y., Wei, Z., Zhang, Y., Ma, Y., and Zhang, M. Different thermal drying methods affect the phenolic profiles, their bioaccessibility and antioxidant activity in *Rhodomyrtustomentosa* (Ait.) Hassk berries. *LWT - Food Sci. Technol.*, 79, 260-266, 2017.
149. EU/2015/1513 Directive (EU) 2015/1513 of the European parliament and of the council amending Directive 98/70/EC relating to the quality of petrol and diesel fuels and amending Directive 2009/28/EC on the promotion of the use of energy from renewable sources.

7

Radio-Frequency Technology in Food Processing

Aastha Dewan[1], Anil Panghal[2], Bahareh Dabaghiannejad[3], Vivek Ranga[2], Naveen Kumar[4] and Navnidhi Chhikara[1]*[1*]

[1]Department of Food Technology, Guru Jhambheshwar University of Science & Technology, Hisar, India
[2]Department of Processing & Food Engineering, AICRP-PHET, Chaudhary Charan Singh Haryana Agricultural University, Hisar, India
[3]Department of Electronics and Communication Engineering, Guru Jhambheshwar University of Science & Technology, Hisar, India
[4]Airports Authority of India, New Delhi, India

Abstract

To prevent food spoilage, increase the shelf life and inactivate harmful biological hazards, various techniques are used in food industry involving preservatives and heat treatments. Conventional methods of heat treatment commonly expose the food products to hot air and moisture; however in case of solid and semisolid foods the transmission of heat energy to the core is almost prevented. The exposure of the surface to the heat mostly results in cracking before the core is heat treated effectively which leads to the degradation of the food quality. Radio Frequency treatment as a nondestructive technique uses the advantage of electromagnetic waves and their properties of penetrating and interacting with the basis of the materials which are electrons and nucleus. By changing the direction of the waves during time and causing friction inside the material a huge amount of heat can be created in a short time. Since the electromagnetic waves are generated in different directions the food is heat treated at all portions at the same time. Many studies are done to explore the advantage of RF technique in the food industry and extend the applications from drying and baking to disinfestation and more. This chapter explains the physics behind the performance of Radio Frequency technique, its different applications and their effects on the food quality, challenges and future scope of this technology in the food industry.

**Corresponding author*: navnidhi24@gmail.com

Navnidhi Chhikara, Anil Panghal and Gaurav Chaudhary (eds.) Novel Technologies in Food Science, (271–306) © 2023 Scrivener Publishing LLC

Keywords: Radio frequency, microwave, electromagnetic field, food industry

7.1 Introduction

Fresh produce especially vegetables and fruits are very perishable, having high content of moisture at the time of harvesting, because of which the output becomes a low quality, low shelf-life product. Ample of preservation and processing techniques are used for the products in food industry to produce a safe food and ingredients with high quality and extended life. Mostly thermal energy is used to complete all the actions like inactivate various enzymes, bacteria and killing of insects. Commonly these techniques are known as thermal processing techniques in the food industry. Furthermore, the conventional practices in the food processing use steam of high temperature, water and air for transmission of heat to the products. Low diffusivity foods are solid and semi solids and hinder the transmission of heat to the center of the product. When the surface of a food product is exposed to a high temperature source of heat the quality degradation occurs in the food material. Thus, for making the products that stand on the parameters of the food safety the firms of food have to compromise with their food products. In addition, to meet up the further demand of the consumers they have to increase the output as well as ensure the food product with high quality. Radio Frequency (RF) heating is the use of electromagnetic waves to generate thermal energy inside a food sample. The term Radio Frequency describes the oscillation of electromagnetic waves over time and varies from 3 Hz to 300 GHz. The application of RF technology in food industry goes back to 1940s [1] where 150 MHz waves were used to cook processed food [2], blanch vegetables or heat loaves of bread. Later on, other potentials of RF technique were discovered such as applications in post-baking, tempering and defrosting, pasteurization, disinfestations. Electromagnetic waves have a very wide frequency range, however for the RF heating treatment in the food industry, only waves in range 3 kHz to 300 MHz are used, and to prevent possible interference to the telecommunication system the Federal Communication Commission (FCC) has allocated only 13.56 ± 0.00678, 27.12 ± 0.16272 and 40.68 ± 0.02034 MHz frequencies to Industrial, Scientific, Medical (ISM) [3].

7.2 RF Technology and Principle

The bases of all matter in nature (gas, liquid, solid) are molecules and atoms. Molecules are made up of one or more atoms, and atoms include electrons with negative charges and nucleus with positive charges. These constituent particles of the material can have a symmetrical or asymmetrical structure. These particles also have different movements, including electronic, ionic, molecular or atomic. An asymmetric molecule inherently acts like a very small magnet since the positive and negative charge are placed at opposite ends e.g. water molecule. These molecules are known as polar molecules. When a polar molecule is exposed to an external electric field, it tries to align itself with that field. If the direction of the electric field changes with time, the molecule constantly changes its direction. If a substance containing such molecules is placed in an alternating external electric field, all its molecules simultaneously tend to align with the external field. Since the change speed of the external field is very high with time (millions of times per second), the molecules are forced to move rapidly which causes internal friction. The friction created between the molecules eventually leads to heat production of thermal [4].

Electromagnetic waves can be reflected, propagated or absorbed depending on the material being irradiated. For example, metallic materials reflect these waves (due to presence of free mobile charges), non-metallic materials such as glass and plastic pass through them, and substances made up of liquids, such as foods, and even humans themselves absorb it. Because these waves are so absorbed in substances that contain liquids, they raise the temperature very high, so they are also used for heating and cooking in the food industry. The extent to which electromagnetic waves are absorbed into matter depends on the dielectric properties of the material, including: permittivity, permeability and conductivity [5]. Hence the RF heating is referred to as dielectric heating.

Permittivity

One of the important properties of dielectric materials is permittivity, which indicates the ability of that material to store electrical energy when exposed to an external electric field. This property is generally referred to as relative complex permittivity and is presented with the following formula:

$$\varepsilon = \varepsilon' - i\varepsilon'' \qquad (7.1)$$

Where ε' is the dielectric constant and represents the lossless interaction and storage of energy, and ε'' represents the dielectric loss which results in heat production.

Permeability

Simply put the ability of a substance to become a magnet when exposed to an external electromagnetic field is called permeability and like permittivity it is the form of relative complex permeability presented as follows:

$$\mu = \mu' - i\mu'' \tag{7.2}$$

Where μ' is the magnetic storage part and tendency of matter to become a magnet and μ'' holds the glossy interaction if any. However, permeability plays a very small role in the process of RF heating thus it could be neglected.

Conductivity

Electrical conductivity is another important property of a material which indicates the ease with which electrical current passes through a material and it is shown by the following equation:

$$\sigma = 2\pi f\varepsilon'' \tag{7.3}$$

Where f is the frequency of the electromagnetic field.

Penetration depth

The measurement of how much electromagnetic wave can penetrate inside a material until its energy falls to 1/e (about 37%) of its original value at the surface is known as penetration depth which is one of the important factors while using RF technology and it is presented as:

$$\delta = \sqrt{\frac{1}{\pi f \mu \sigma}}$$

Where f is the frequency of the electromagnetic field, μ and σ are the relative complex permeability and the conductivity of the material.

7.2.1 Types and Equipment

There are various uses for the conversion of electrical energy into heat in industry. One of its applications is in the food industry for heating and drying food. In this way, the target food is placed inside an alternating electromagnetic field. The reaction between this field and the polar molecules and charged ions of the food produces heat. The RF heating treatment has 2 main types: the open-circuit or free-running oscillator system and the 50Ω system. The open-circuit RF system is generally divided into three main parts, which include the power supply, the control circuit and a pair of parallel plates. The distance between parallel plates can be changed

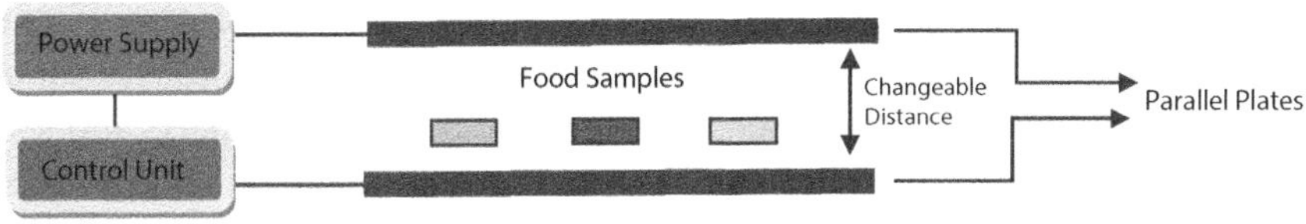

Figure 7.1 Simple illustration of a RF heating system.

according to the use and the food is placed between them (Figure 7.1). Two parallel plates act as a capacitor, and the food plays the role of a dielectric material that stores the electric energy and its resistance converts the electric energy to heat.

In this method, the food is placed on a movable belt between two the parallel plates. According to the requirements of the application (drying, baking, etc.) the control unit adjusts the frequency of the work, and a potential difference is applied between the two plates, which causes the production of electromagnetic waves.

The 50Ω consists of a RF generator (RFG), standard 50Ω cables, a matching network and a pair of electrodes. Radiofrequency energy is generated by the RFG and then transmitted to the matching network through standard 50Ω cables. The matching network transmits this energy to the electrodes, where RF waves are generated. The role of the matching network is to keep the input impedance of the load to 50 Ω. Since the output impedance of the RFG is 50 Ω the maximum power is transferred from the RFG to the load. This method is new compared to the conventional RF heating system. The 50Ω system is costly with more equipment needed, however it provides precise control over frequency and power since the matching network ensures the load impedance to stay stable throughout the process. Therefore it is quite possible to achieve accurate characteristics of the final product such as moisture content by controlling the power and heating rate [6].

The open-circuit system is cheaper and simpler in case of assembly and maintenance. It is also possible to set a system for a high volume of samples to be heat treated. Whether the open-circuit or the 50Ω is used the voltage difference between the electrodes generates electromagnetic waves that penetrate the food target and the interaction between the RF waves and the molecular moments generates internal heat. This is unlike old methods that conduct the heat from surface to the core which takes long time. Therefore, one of the advantages of RF heating is its high speed. However, as every technology comes with its drawbacks this technology is no exception. One of the major problems of RF heating is the non-uniformly distribution of heat inside the food sample. Since samples come in different size and shape the distribution of heat inside them is not uniform and this causes some hot spots inside the target which leads to more heat at some parts and less at some others. This non-uniform spread of heat affects the final quality of the product. Many studies have been done to solve the problem of non-uniform distribution and changes in the process or suggested, but most of the proposed solutions are very expensive and time consuming. As a result, it is generally believed that RF heating is only suitable for some products and many times changes are needed to achieve a uniform heating [3].

7.2.2 RF vs. Microwave (MW) Heating

Even though RF and Microwave are both subdivisions of Radio Frequency range, RF heating and MW heating are different. The electromagnetic waves generated in RF system are propagated in one direction only between the two parallel plates however in MW the waves are generated using special oscillator tubes and transmitted in all directions. In a microwave system, food is placed in a metal cabinet. Electromagnetic waves are generated by an electron tube called the magnetron. The generated waves are transmitted to the cabinet through a waveguide. These waves propagate in all directions when they strike the metal wall of the chamber. A metal fan simultaneously evacuates the heat and moisture produced during the heating process. The frequency at which most of microwaves work is 2.45 GHz. Since these waves are absorbable by human body and have harmful affects on eyes, a good shielding and closed chamber is required to prevent the leakage of waves.

7.3 Application of RF in Processing

Extensive research has been conducting in the application of RF heating technology (Table 7.1) in food processing industry for drying [7–9], sterilization [10], pasteurization [11, 12], cooking [13, 14], blanching [15, 16], roasting [17], disinfestations [18, 19], post-baking [20], enzymes inactivation [21], inactivating anti-nutritional factors [22] and its application is continuously evolving to solve the drawbacks of traditional thermal methods (Figure 7.2).

7.3.1 Drying

Conventional ways of drying are based on heat convection in which heat is continuously generated from a heat source and transferred to the desired

Table 7.1 A brief summarization of RF processing in food processing.

Method	Material	RF system processing parameters (frequency, power)	Bacterial species/log reduction and other results	Operating conditions and time-temperature combination	Reference
Drying	Fresh Sichuan pepper kernels	27.12-MHz, 8-kW RF system with plate applicators	TPC of 5.46 log CFU/g in fresh to 2.45 log CFU/g in RF treated product	50 °C for 180 s (electrode gap 115mm, sample thickness 60mm)	[9]
	Kiwi fruit	27.12 MHz, 3 kW free-running oscillator RF-vacuum drying system	Better quality retention in terms of color, vitamin C content, and rehydration capacity	60°C for 245 minutes with electrode gap of 60 mm, vacuum pressure of 0.02 MPa, and kiwifruit slice thickness of 8 mm	[7]
	Apple slices	Air impingement combined with hot air assisted radio-frequency 6 kW, 27.12 MHz pilot scale RF unit	Better quality attributes of apple slices in terms of color, Vitamin-C and total phenolic content	RF drying 65-60 °C for total duration of 200 min reduced moisture content from 88 g/100 g to < 15 g/100 g in two stages	[8]

(Continued)

Table 7.1 A brief summarization of RF processing in food processing. (*Continued*)

Method	Material	RF system processing parameters (frequency, power)	Bacterial species/log reduction and other results	Operating conditions and time-temperature combination	Reference
Pasteurization	Cocoa powder	27.12 MHz, 6 kW RF heating unit together with hot air system	*Enterococcus faecium* with 5-log reduction	90°C for 7.6 min RF heating plus 48 min insulated holding	[12]
	Wheat flour	27.12-MHz, 8-kW RF system with electrode gap 150 mm	6.6-log reduction of *S. Enteritidis* by targeting 5D of *E. faecium*	85°C for 25 min (*E. faecium*) or 18 min (*S. Enteritidis*)	[11]
Sterilization	Caixin (*Brassica campestris* L.)	6 kW, 27.12 MHz coupled with hot air system	Reduction of total colony number of Caixin by 3–4 log	60°C with hot air speed of $3 ms^{-1}$ for 20 min with electrode gap 20 mm	[10]
	Vacuum packed fish (Raw Pacific sauries)	10 kW 27MHz RF system	5-log reduction of *B. subtilis* spores	120 °C for 18 min	[40]

(*Continued*)

Table 7.1 A brief summarization of RF processing in food processing. (*Continued*)

Method	Material	RF system processing parameters (frequency, power)	Bacterial species/log reduction and other results	Operating conditions and time-temperature combination	Reference
Heating	Egg white powder	6-kW 27.12-MHz RF system heating system with a free running oscillator	Mesh approach was more applicable with higher efficiency than discrete approach for RF heating	550 g of Egg white powder moving with a speed of 2.1 cm per min ensure uniform heating to make it microbiologically safe	[13]
	Pizza	6-kW, 27.12-MHz RF system with electrode gap of 110mm	Fast RF heating was observed resulting in low moisture loss due to reheating	Heating temperature- 25°C to 85°C for 7.33 min with 110mm electrode gap.	[14]
Baking	Cookies	2 kW 27.12 MHz RF tunnel oven 2 kW 27.12 MHz RF tunnel oven 2 kW 27.12 MHz RF tunnel oven 2kW 27.12 MHz RF tunnel oven	RF post baking yielded cookies with lower level of acrylamide	After conventional baking, 8 and 9 min cookies were post dried	[20]

(*Continued*)

Table 7.1 A brief summarization of RF processing in food processing. (*Continued*)

Method	Material	RF system processing parameters (frequency, power)	Bacterial species/log reduction and other results	Operating conditions and time-temperature combination	Reference
Roasting	Cashew kernels	12 kW, 27.12 MHz pilot scale free-running RF system	Moisture content of cashew nut kernels reduced from 6.2 g/100 g to 1.5 g/100 g with well-acceptable quality of roasted kernels	RF roasting at 120–130 °C for 30 min with 5 cm sample thickness	[17]
Thawing	Frozen Pork	27.12 MHz 800 W RF system	At -20°C half-cylindrically shaped meat was thawed without surface burning	Temperature for thawing chamber varied by -5, -10, and -20°C	[56]
Tempering	Frozen block of shrimp	27.12 MHz with a maximum power of 2 kW RF system	RF was more suitable than microwave tempering for large sample due to deep penetration	RF tempering was done between -5 and -3 °C for 11 and 7 min with electrode gap of 160 and 150 mm respectively	[57]

(*Continued*)

Table 7.1 A brief summarization of RF processing in food processing. (*Continued*)

Method	Material	RF system processing parameters (frequency, power)	Bacterial species/log reduction and other results	Operating conditions and time-temperature combination	Reference
	Grass crap mince and Tuna	600W, 27.12 MHz	On experimental comparison of microwave frequency with RF, RF was better in terms of temperature uniformity with deeper penetration	Tempering time was set for 7 min to reach sample center temperature -4°C with electrode gap of 7cm	[58]
Deinfestation	Dates	3.5 kW, 27.12 MHz RF system with adjustable range of electrode gap from 60-120mm	Date disinfestations (infested by *C. hemipterus*)was achieved in 6 min	RF chamber temperature was kept at 60°C with conveyor belt speed ranging from 3.3-6.7 mm/s	[18]
	Dried apricots	6 kW, 27.12 MHz free running oscillator RF system	Indian meal moth larvae were killed by the optimized conditions	50°C for 5 min with electrode gap of 10cm	[19]

(*Continued*)

Table 7.1 A brief summarization of RF processing in food processing. (*Continued*)

Method	Material	RF system processing parameters (frequency, power)	Bacterial species/log reduction and other results	Operating conditions and time-temperature combination	Reference
Blanching	Carrot cubes	A free running 12 kW 27.12 MHz RF system	Improved rehydration capability, enhance color, retain texture and vitamin-C	95% reduction of peroxidase activity for blanching 3.0-7.0 min at different electrode gaps (8.0-8.6 cm).	[15]
	Apple slices	A free running 12 kW 27.12 MHz RF system with 120mm electrode gap	Inactivated polyphenolase enzyme with no color change	85°C for 111s	[16]

(*Continued*)

Table 7.1 A brief summarization of RF processing in food processing. (*Continued*)

Method	Material	RF system processing parameters (frequency, power)	Bacterial species/log reduction and other results	Operating conditions and time-temperature combination	Reference
Inactivating anti-nutritional factors	Soybean	2.1 kW, 27.12 MHz RF system with electrode gap of 45mm	Inactivation rate of lipoxygenase, urease and trypsin inhibitor was found to be more than 90%	RF heating at 113°C for 150s with holding time of 120s	[22]
	Rice bran	RF system (12 kW, 27.12 MHz) with a built-in hot air system with fixed electrode gap 9.2 cm	Inactivated enzymes while maintaining good nutritional quality, antioxidant ability and microstructure of RB	Two different combination of RF heating at 100°C and 115°C with holding time 15min and 6min respectively	[21]

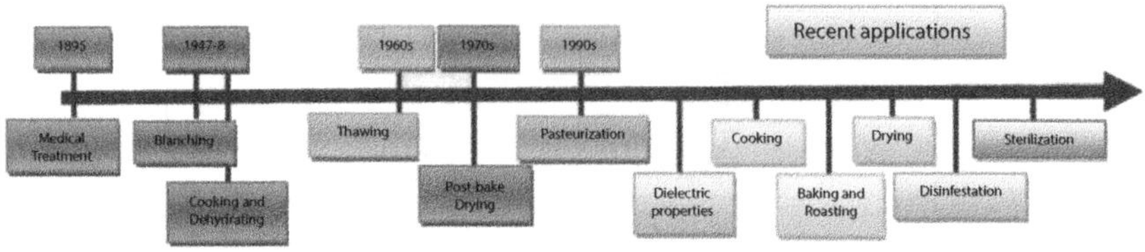

Figure 7.2 Evolution of RF technology in food industry.

material through the environment contacting the surface first and then penetrating slowly inside. However, in RF heating the heat starts inside the food itself with the help of electromagnetic waves. The electromagnetic waves penetrate inside the food, start interacting with the polar molecules of water and generate heat. In conventional methods, the surface of the food is constantly exposed to the heat which results in more dryness, browning and cracks at the surface while the inside might stay high in moisture content. Comparatively the RF technique has a volumetric heating property that leads to more uniform heating in much shorter time and the moisture in all portions of the material evaporates almost at the same time. The energy of the RF is selective according to the required level of dryness and this allows for a far better quality products.

RF heating of food product is governed by food properties like dielectric properties, geometry, and homogeneity of the constituents in the food material which in turn interact with electromagnetic waves defining the charge density under the electric field. These properties help in determining the processing parameters thereby the quality of foods. There is exponentially inverse relationship between the absorption of energy with the moisture content of the material during RF drying. Due to this, region with higher portion of moisture content with absorb more energy and heat up at increasing rate, leads to uneven temperature distribution during drying. Factors influencing the dielectric properties of food like composition and distribution of food components must be understood to apply RF heating effectively.

From 1940s to 1980s, RF drying was used for dehydrating vegetables then for post baking of cookies and snack foods [23]. Currently, food and non-food industries is using RF heating technology for wood, paper, textiles, spools, pharmaceuticals products and many others [24, 25]. Technology of RF drying is being continuously improving by combining with other accessories or technology like hot air oven, water bath, vacuum assisted and many more to improve its version and make it more compatible and efficient for the designated task.

In order for RF heating to be cost-effective, it is better to be combined with conventional techniques because it alone is expensive. Computer Assisted Design (CAD) techniques can help for simulation and developing applicators for power distribution. Some commercial numerical software such as COMSOL, FEMLAB and QW3D are beneficial for electromagnetic fields study. Having the size, shape and dielectric and thermal properties of the food target a simulation will give a good understanding of the final product. However as one of the most important concerns in RF heating is the non-uniform distribution of heat, which is related to the dielectric properties of the food the exact properties of the material under process are needed. Due to the great variation and enhancement in food product varieties, the database of dielectric information of materials is not complete. Hence computer processing remains incomplete. Therefore, more studies and measuring of dielectric properties of materials are necessary.

7.3.2 Baking

On exploring the product quality parameters, product surface color, appearance, texture, aroma and taste are all important for consumer perception and acceptance. Kocadağli *et al.* [20] claimed that RF post-drying of partially baked cookies minimizes the surface browning without affecting the hardness and fracturability. Even though RF installation is generally more expensive than conventional ovens, the advantages over the long-term usage compensate the initial investment. One of the benefits of using post-baking in bakery products is the prevention of cracking at the surface. Biscuits and cookies are prone to get cracks while baked in conventional ovens due to the moisture content difference between the crust and the middle part. The crust which is the outer layer of the product is dried first and by time the heat reaches the inner portion. The dryness of the crust mostly acts as an insulator preventing the heat to dry the core resulting in 90% of checking (cracking) in some cases. Using the RF post-baking has shown to reduce this effect to 1% and give a high-quality product. Another advantage of RF post-baking is that the biscuit production line is easily scalable.

To increase the throughput line in the conventional oven heating adding a length of ovens a long route of ovens must be added to the existing production line however RF system can be situated and fitted in the existing line for the equal increase of production.

In conventional baking the color and moisture content of the product are linked and often-high temperatures cause the surface to turn dark brown. Using RF technique helps the baker to control the amount of moisture at the desirable level and prevent extra browning. Eliminating the moisture from the center-bone of the product has the effect of longer shelf life.

In conventional heating, a large part of the heat produced is spent on heating the body of the oven rather than the food product and when a product is at the last stage of baking, the hot air produced cannot be returned to the oven which causes a lot of energy loss. However, this high energy loss does not occur in RF heating giving the advantage of energy saving. The baking process is a complex physical and chemical phenomenon at the same time. The bakery product undergoes different changes including water evaporation, volume expansion, porosity formation, gelatinization of starch content, crust formation, denaturation of protein [26], sugars caramelization and volume shrinkage. Maillard type reaction and sugar caramelization at low moisture and high-baking temperatures develops browning in bakery products [27]. To simulate and optimize the color and moisture of the final product it is important to have a good knowledge reactions involved in browning process. For instance, the bioavailability of amines in proteins is significantly affected by the Maillard reactions during baking process. Methionine and tryptophan (essential amino acids) can undergo oxidation and destruction also while crust formation cross-linking of proteins occurs that impairs the digestibility of protein and ultimately lowers the nutritional value of the final product [27–29].

Acrylamide (a carcinogenic substance) is highly formable in bread, cookies, crackers and other starch-based bakery products upon heating above 120°C [30, 31]. RF heating has shown significant reduction of acrylamide in short dough biscuits and leavened cakes [20, 32]. Using RF heating can increase the product throughput up to 40%, while having significant reduction in acrylamide level. The formation of acrylamide is strongly related to the temperature, sugar content and amino acid asparagines. Studies have been done to reduce the formation of acrylamide and suggestions have been made in this regard. Changing the sugar type to sucrose, reducing the sugar content, adding amino acids or protein hydrolysates, using sodium bicarbonate as the leavening agent instead of ammonium bicarbonate, reducing temperature, and etc. however most of these changes especially in the acidity and temperature bring low quality in the end product [29]. Therefore, using the RF post-baking technique at the last stage of baking while the product is still high in moisture can help to improve the quality by reducing the acrylamide formation and preventing the extra browning.

7.3.3 Sterilization & Pasteurization

In the sterilization process, time and temperature are two important variables. Depending on the type of product sterilization is done with high temperature in a short time or with low temperature in long time. For pasteurization, the temperature is comparatively lower than sterilization process however in both treatments the heat is conducted through surface using moist or dry heat. The rise in the temperature of the canned/packaged product is non-uniform and slow. The non-uniform distribution of heat inside the food results in overheating and under-heating of different layers and leads to degrading the quality of the product. The use of RF heating has the advantage of fast heat generation in bulk, which saves lots of energy and time as well as the product end quality. One of the problems while killing microorganisms is in dry products were the microbes are in dormant state, however the RF is a very effective way for treating dry material such as nuts, flour, protein supplements while it keeps the quality and shelf-life.

Raw, semi-processed and processed food products can be contaminated and re-contaminated while harvesting, handling/processing, storage and post processing depending upon the type of food. The food products like raw meat, poultry, seafood and their products (raw egg) are often contaminated by pathogens such as *Escherichia coli* O157:H7, *Campylobacter*, *Clostridium perfringens*, *Salmonella*, *Vibrio parahaemolyticus*, *Listeria monocytogenes*, and *Staphylococcus aureus*. Unhygienic handling during processing, packaging, storage and transportation leads to recontamination even after the providing the adequate thermal treatment to the food commodity [33].

The illustration of microbial inactivation on the application of thermal treatment have been evolved in 1900s by [34–37].

a) The vital enzymes and proteins are degraded due to protein denaturation which takes place at higher temperature (above 95°C) [38] melting the DNA (deoxyribonucleic acid), RNA (ribonucleic acid) and deteriorating cytoplasmic membrane resulting in the cell death.
b) Toxins are produced by microorganism due to acceleration of metabolism on receiving thermal treatment causing the cell death due to cellular intoxication.
c) Melting of essential lipids due to high temperature leading to the destruction of life supporting system to the microbes.

RF pasteurization experiments have been initiated since 1947, but commercial application is still at its budding stage due to challenges in RF system designing for uniform heating to packaged foods, lack of precise accessories and application of pressure during RF heating. Currently many researchers are developing proper protocol with optimized process parameters to achieve successful safe log reduction of pathogens in various food products like cocoa powder [12], wheat flour [39], vaccum packed fish [40], corn [41], cumin seeds [42] and many other.

RF heating system design plays a key role in the successful sterilization and pasteurization. While designing of the RF heating system and process parameters following points are need to be considered.

a) Uniform heating to avoid cold spot, temperature loss and electrical break down resulting in arching [43] is the foremost priority for the designing of RF heating equipment.
b) To maintain a uniform heat load inside the container.
c) Avoidance of fringe effect between packed food and air results in overheating of edges of food containers. It was solved by Wang *et al.*, [43] by immersing the food container in the water.
d) Microbial safety of RF treated food product and reliable temperature monitoring accessories for validation and recording is FDA's requirement
e) Dielectric properties of food products governed by its composition, geometry and ingredients distribution is required for efficient designing of RF system and operating parameters.

Apart from RF system designing, there are important processing parameters, which are, need to be controlled to achieve effective destruction of pathogens and spoilage microorganisms. The control parameters to be considered RF heating system are product dielectric properties, product mass and thickness, initial temperature of product, electrode gap, system pressure and processing time including heating and holding duration to achieve targeted sterility. The process parameters to control when combining water circulatory system with RF heating system are water flow rate, water temperature [44] and water conductivity to reduce the edge heating and minimize the loss factors [43]. Thus, more focus is required for uniform heating with optimized parameters to achieve target sterility which can go with combinations as in hurdle technology for spore destructions even to increase its storage life with enhanced food safety.

7.3.4 Roasting

Roasting is a dry heat cooking process that requires a higher temperature to achieve Maillard browning and caramelization leading to develop desirable flavor, color, texture, and appearance of food products. As the process requires a higher temperature which results in the volatilization of many polyphenolic compounds and destruction of heat sensitive nutrients. This leads to explore those techniques which achieve the result of roasting with minimum destruction impact. Various modern roasting techniques including superheated steam, microwave, Revtech roaster, hot-air assisted RF (HA-RF), infrared hot-air, and Forced Convection Continuous Tumble (FCCT) roasting have been used for efficacious roasting diverse food products without missing the nutritional values [45]. Hot air assisted radio frequency is a new method explored for roasting tree nuts and seeds such as almonds, cashew nuts [17, 46], peanuts [47], sunflower seeds [48]. Different types of nuts have been roasted using HA-RF with much lower temperature yet keeping the desired flavor compounds. In consideration of high roasting efficiency with low pollution, desirable product quality with easy process control, HA-RF technology has great potential for snack and nut roasting. However, more to improve roasting uniformity and shelf-life more studies are needed.

7.3.5 Blanching

Blanching is a mild heat treatment for inactivating enzymes like polyphenol oxidases, catalase, phenolase and peroxidase present mainly in fruits and vegetables. It is an essential step to be performed prior to food processing and storage to prevent browning and off-flavor development [49]. This traditional method of dipping produce in hot water or exposing to steam results in loss of nutrients due to leaching and exposure to high temperature. Many researchers have exploited RF for enzyme inactivation without affecting the quality of the product [15, 16]. Apart from fruits and vegetables, grain products are also prone to quality deterioration as the substrate comes in contact with enzymes after milling, crushing or grinding process. [50]. RF heating has been actively employed for inactivation of enzymes in grain based products like potato [51], rice bran [52], soybean [53] and wheat germ [54]. Although it has many benefits, such as shorter processing times and reduced water usage, the capital cost of the equipment and the lack of systematic research in experiment development are still the largest obstacles for commercial uses of RF heating in blanching.

7.3.6 Thawing and Defrosting

Conventional methods of defrosting food are often time consuming as heat slowly enters the frozen material from the surface to the core and the time increases as the size of material increases. In this condition, some bacteria grow and degrade the quality of the product. Using rapid volumetric RF heating system brings the advantage of shortening the time. Another advantage of RF defrosting is its capability to defrost the product without removing the packaging. Cathcart *et al.* [55] pioneered in RF thawing on food products like fruits, vegetables, frozen eggs and fish wrapped in cardboard containers, cellophane coverings and in their original form at -17°C to 10°C for 2 to 15 min. The reduction in thawing time, drip losses, color and loss in flavors were observed by the research. The gaps left by the researcher were the consideration of cold spot (i.e. slowest heating point) dependent on the shape of the commodity leading to under or over thawed regions in the samples.

Advantages of RF thawing are a) uniform and rapid temperature distribution for regular geometrical shape; b) reduce thawing time, c) bulk thawing, d) efficient continuous process with less water consumption, e) reduced drip losses and f) minimum space requirement and processing cost. The challenges to be encountered during RF thawing are a) heat runaway due to heterogeneity in the dielectric property of the various food product which results in non-uniform heating, b) arching due to surface irregularity leading to burning of the product [55], c) selection of suitable packaging material, and d) high installation and maintenance cost [4]. Recent researches on thawing of meat [56] and seafood [57, 58] using RF technology have focused on solving the challenges of non-uniform thawing with minimum drip loss and maximum nutrient, color and flavor retention in the thawed meat and seafood products.

RF thawing and defrosting are energy-efficient techniques with quality strong points of maintaining cellular structures, thus preventing nutrient loss through dripping, along with retention of color and flavor compounds. RF thawing has been very well adopted at commercial level due to its various benefits. There is scope of investigation on dielectric properties of foods and food components, eliminating and revealing solutions for the challenges faced during RF thawing.

7.3.7 Inhibition of Anti-Nutritional Factors

Thermal treatments such as hot air drying and roasting are employed to reduce anti nutritional factors in legumes like soybean [59] resulting in loss of overall nutritional quality parameters. As the traditional thermal method require temperature ranging from 100°C to 200°C due to the thermal stability of anti-nutritional factors and low thermal conductivity of heat treatment [60]. To minimize the undesirable effects of traditional heating techniques, novel techniques such as microwave treatment [61], infrared technique [62], radio frequency [22], have been applied to improve the product quality. Among these techniques, RF has proven to be better in terms of improving the overall product quality along with overcoming the deleterious effects resulted from other heat treatment due to its greater deep penetration with bulk heating. Jiang *et al.* [22] effectively inactivated more than 90% anti-nutritional factors including lipoxygenase, urease and trypsin inhibitor using RF technology within 300s at 113°C while maintaining the functional properties of soybean protein isolate extracted from RF treated soybean along with the improvement in the sensory characteristics. The application of RF technology for anti-nutritional factor is less explored so this provides researchers with a scope to develop a standard protocol to be followed at commercial level.

7.3.8 Disinfestation

Insect infestation on agricultural and food products is one of the major problems, especially in the tropics. One traditional method of overcoming this problem is to use chemical fumigants however they have harmful effects on the environment. Heat treatment as another way of battling insects can result in deterioration of the product. Exploration of electromagnetic energy in pest disinfestations at selected frequencies shortens the treatment time and thermal adverse effect as it heat the insect faster than the product itself [63]. As an environmental friendly method RF technology can be used to kill the pests and insects selectively without causing any harm to the agricultural products. Many studies have been conducted for disinfestations utilizing RF technology in food commodities like food grains [64], legumes [65], tree nuts [66] and fruits [18, 19]. RF disinfestations treatment operated at the temperature between 50-65°C which resulted in no significant changes on the overall nutritional as well as sensory quality of the food commodities [67]. Dry food products were more successfully treated for the RF disinfestation treatment due to uniformity in heating as the fresh food like fruits were more sensitive towards temperature fluctuations [68].

RF applications for disinfestations have taken a deep research direction in the food grains, nuts, perishable commodities and their products processing. Many researchers are considering the relevant point, such as thermal death kinetics of insect pests, radio frequency heating processing parameters of stored-grain insects, the optimization and confirmation of insect control treatments for small quantity in batch mode with their effects on quality parameters of food commodity, have been conducted. There is scope of optimized protocol to be set up for industrial application with the parameters to ensure uniform radio frequency heating, product quality, energy efficiency, process throughput and cost.

7.4 Effect on Food Quality

The food industry is always looking for new, cost-effective and high-quality technologies to process food. These methods often have to do the least damage to the final quality of the product. They should extend the shelf-life of the product as much as possible while maintaining the nutritional properties. The conventional heating method has long been used in the food industry to extend and stabilize the life of the product, but today the food market, especially processed foods, is expanding rapidly and increasing demand has led the food industry to use several novel methods since the conventional method is no longer able to meet the quality, time and scale of the market. As a result, science and technology, in collaboration with each other, continuously examine various thermal and non-thermal methods with the ability to improve and maintain the quality of food.

7.4.1 Microbiological Quality

Radiofrequency heating is a novel innovative technique based on electromagnetic energy which kills the microorganism. According to the "dielectric rupture theory" an electric potential is induced in a membrane of a cell while exposed to an external electric field. The additional potential is larger than the cell's natural potential. When the membrane potential is sufficient, the membrane ruptures thus several pores are formed, permeability is increased and the cell is no longer integrated [69] therefore the function of the microorganisms are disrupted. Another method for cell destruction was presented by Kinosita and Tsong [70] using electric pulses to generate pores. Oscillating magnetic field was studied by [71] investigating the patent of [72] which demonstrates that an oscillating magnetic field with multiple oscillations energizes the active parts of large biological molecules. With the presence of a large number of magnetic dipoles in a molecule sufficient energy is transferred to the molecule to break the covalent bond. DNA and proteins are those molecules that can be destroyed or inactivated using this treatment. Many studies have shown the susceptibility of cells and membranes to the electromagnetic field. A change in the permittivity of erythrocytes [73] and phospho-lipid vesicles [74] was reported. In some studies, the destructive effect of the electromagnetic field on the membrane is not attributed to its thermal effect, for example liposome vesicles were exposed to the radio frequency resulting in rupture of cells and discharge of 6-carboxyfluorescein reported by Ponne *et al.* [75].

As one of the food-borne pathogens Escherichia coli O157:H7 is responsible for different illnesses such as hemolytic uremic syndrome, hemorrhagic colitis [76]. Several studies have investigated the effect of radio frequency in different parts of food processing on inactivating *E. Coli.* Guo *et al.* [77] studied the cooking effect of RF on ground beef in compare with the hot water bath cooking. The pasteurization effect of RF for inactivating *E. Coli* in apple juice and apple cider was investigated by Geveke and Brunkhorst [78, 79]. Table 7.2 shows a list of studies over the effect of RF technology on inactivating the pathogens. Though many researchers still try to develop applications of RF in the food industry the usage of this novel technology is not as fast as expected.

7.4.2 Nutritional Quality

Organoleptic and nutritional characteristics of the processed food are the important keys determining their value, acceptability and popularity in market. Food companies as well as consumers are inclined towards manufacturing and demanding respectively for the food with pleasant appearance, aesthetic appeal, mouth feel, high nutritional value and overall acceptability. Novel technology like RF has been explored by many researchers to produce such processed food products.

Studies so far show that RF heating in comparison with conventional heating methods is able to provide high-quality products while keeping nutritional values. Affordability by industry, time saving some other benefits of RF heating compared with conventional systems under identical operating conditions are other advantages of this technology. However, the technique is not adopted by the food industry on a grand scale. Cardello [80] has talked about some of the reasons for not accepting this system to some food processors producing a safe product is the most important concern in adopting any new technique or technology. Peer-reviewed researches indicate that negative effects from RF heating in comparison with conventional heating are very few. Color, texture, nutritional, and sensory quality maintain with RF heating resulting in better quality products. Table 7.3 summarizes some of the studies on the effect of RF over food products quality.

Table 7.2 Summarization of research work conducted to inactivate pathogens using radio frequency heating.

Food commodity	Target microorganism	RF system and process parameters (electric field strength, frequency, time-temperature combination)	Reduction log	Reference
Ground beef	*Escherichia coli* O157:H7, *E. coli* K12	RF cooking at 72°C (4.25±0.25 min)	4.0 log10 cfu/g and 1.93 log10 cfu/g	[77]
Apple Juice	*E. coli*	RF system (26kV/cm, 15-70kHz). RF treatment at 60°C for 3s	2.5 log cfu/ml	[78]
Orange Juice	*E. coli*	RF system (15-kW and 20-kW at 21, 30 and 40 kHz) 60°C for 3s	2.1 log	[91]
Liquid Suspension	*Saccharomyces cerevisiae*	RF system (30kV/cm, 20kHz) 35 to 55°C	30-kV/cm electric field at 35°C reduced by 0.8 log units	[92]
Milk	*E. coli.* and *Listeria monocytogenes*	2kW, 27.12 MHz RF heater heated milk at a temperature of 65°C	With residence time of 55.5 and 26 s at 65°C, upto 5 and 7 log reduction for *Listeria* and *E. coli.*, respectively at 1200 W	[93]
Corn flour	*Salmonella Enteritidis* PT 30 (*S. Enteritidis* PT30) and *Enterococcus faecium* NRRL B-2354 (*E. faecium*)	27.12-MHz, 6-kW. RF heating at 85°C for 10 mins	*S. Enteritidis* PT30 by 6.59 ± 0.21 log and *E. faecium* by 4.79 ± 0.17 log.	[94]

(*Continued*)

Table 7.2 Summarization of research work conducted to inactivate pathogens using radio frequency heating. (*Continued*)

Food commodity	Target microorganism	RF system and process parameters (electric field strength, frequency, time-temperature combination)	Reduction log	Reference
Mashed Potatoes	*Clostridium sporogenes*	6-kW, 27.12-MHz pilot-scale RF sterilization system. 121°C for 2.8 min	4.56 log reduction	[44]
Corn	*Aspergillus parasiticus*	27.12 MHz, 6 kW RF heating system/70°C for 12 min	5 log reduction	[41]
Wheat flour	*Enterococcus faecium* NRRL B-2354	27.12 MHz, 6 kW RF heating system. RF treatment at 75 °C, 80 °C, and 85 °C for 10 min 1.0-, 2.5-, and 4.9-log reduction of FDE, respectively.	1.0-, 2.5-, and 4.9-log reduction at 75 °C, 80 °C, and 85 °C, respectively	[39]
Cumin seeds	*Salmonella enterica*	6 kW, 27.12 MH free oscillating RF system/108, 120 and 127 s around 100 °C	More than 5-log reduction	[42]

Table 7.3 A summarization of changes in food due to RF processing.

RF processing	Food product	Purpose of RF processing with nutritional and physico-chemical changes	References
RF Heating	Carrot sticks	RF heating was done at 60°C for 2 min to reduce total microbial while maintaining good color, taste	[81]
	Dry milk Powder	RF heating was done for 1-5 log reduction of *Cronobacter sakazakii and Salmonella* spp at 75-90°C for 5-125 min with better retention of whey protein nitrogen index and color while maintaining desired nutritional value	[82]
	Corn	RF heating was done at 50°C, 55°C and 60°C which resulted in enchanced nutritional value by increasing the availability of Ca, Fe and decreasing the anti-nutritional factors	[83]
	Barley grass	RF pasteurization for microbial inactivation at 80°C for 10 min. Negative impact on the color value, flavonoid and chlorophyll content, while reduced the sourness, astringency and bitterness	[84]
Thawing/ Defrosting/ Tempering	Fish meal	No protein loss, lipid oxidation and moisture losses observed	[85]
	Seafood	Reduced drip loss, Minimum impact on texture and other quality parameters	[86]

(*Continued*)

Table 7.3 A summarization of changes in food due to RF processing. (*Continued*)

RF processing	Food product	Purpose of RF processing with nutritional and physico-chemical changes	References
Pasteurization/ Sterilization	Tomato puree	RF pasteurization resulted in better nutritional quality retention in terms of L-ascorbic acid with less formation of hydroxymethylfurfural (HMF)	[87]
Enzyme inactivation/ Blanching	Aqueous food model system	RF treatment was provided to inactivate Polyphenoloxidase activity and lipoxygenase.	[53]
	Mustard seeds	RF heating was applied to eliminate pungency by inactivating myrosinase enzyme. No significant effect on the protein content, amino acid fractions, and antioxidant activity	[88]
Baking/ Postbaking	Cookies	RF postbaking lowered the acrylamide content while maintaining the nutritional value, color and texture	[20]
Disinfestaion	Mangoes	Mexican fruit fly was disinfested at 48°C for 4-8 min while maintaining the firmer texture with no changes in color, TSS and moisture content	[89]

(*Continued*)

Table 7.3 A summarization of changes in food due to RF processing. (*Continued*)

RF processing	Food product	Purpose of RF processing with nutritional and physico-chemical changes	References
	Oranges	Mediterranean fly was disinfested at 48°C, 50°C, and 52°C with varying time with no significant loss of weight, firmness, color, TSS, pH and volatile compounds	[90]
Inactivating anti-nutritional factors	Soybean	Reduced anti-nutritional factors including trypsin inhibitor, tannins, saponins, and phytic acid, while no significant losses in basic nutrition including amino acid, and fatty acid composition	[61]

7.5 Future Scope/Prospectus

The RF dielectric heating has found a special place in the food industry due to its high various capabilities. However improvement and optimization is always a part of each technology. Apart from all the advantages that RF brings to the food industry there are still aspects of it that can be improved such as uniformity in heat distribution which is one of the major concerns. An all time issue in each industry is the installation/operation and maintenance costs that can be reduced through further studies and the use of cost effective device production line. Eliminating extra steps would be beneficial in terms of finance and time. Therefore if researchers could develop RF techniques combining two or more steps of processing that would bring a revolution in the food industry.

7.6 Conclusion

Despite the fact that RF technology has a wide range of applications in the food industry and still more are investigated as this technology has not been able to completely take over the conventional methods. Applications such as thawing, defrosting, cooking and etc. can be done in a fraction of time compared to the conventional methods, however there are uncertain parameters that have not been studied and lack of data and uncertainty of the final effect on the human health which is the most important concern in the food industry hinders the growth of this remarkable technology. This technology has a shining horizon ahead with the help of advanced studies.

References

1. McCormick, "Dielectric heat seeks low moisture applications," *Prep. Foods*, vol. 162, no. 9, pp. 139–140, 1988.
2. J. Moyer and E. Stotz, "The blanching of vegetables by electronics," *Food Technol.*, vol. 1, pp. 252–257, 1947.
3. Z. Huang, F. Marra, J. Subbiah, and S. Wang, "Computer simulation for improving radio frequency (RF) heating uniformity of food products: A review," *Crit. Rev. Food Sci. Nutr.*, vol. 58, no. 6, pp. 1033–1057, Apr. 2018, doi: 10.1080/10408398.2016.1253000.
4. G. B. Awuah, H. S. Ramaswamy, and J. Tang, *Radio-Frequency Heating in Food Processing*. 2015.
5. S. Ryynänen, "The electromagnetic properties of food materials: A review of the basic principles," *J. Food Eng.*, vol. 26, no. 4, pp. 409–429, 1995, doi: https://doi.org/10.1016/0260-8774(94)00063-F.
6. T. T. Moirangthem, R. Macana, and O.-D. Baik, "Characterization of 50-ohm radio frequency heating of bulk canola seeds (Brassica napus. L) in a tubular applicator with parallel electrodes and post-treatment quality," *Innov. Food Sci. Emerg. Technol.*, vol. 64, p. 102409, 2020, doi: https://doi.org/10.1016/j.ifset.2020.102409.
7. X. Zhou *et al.*, "Radio Frequency-Vacuum Drying of Kiwifruits: Kinetics, Uniformity, and Product Quality," *Food Bioprocess Technol.*, vol. 11, no. 11, pp. 2094–2109, 2018, doi: 10.1007/s11947-018-2169-3.
8. J. Peng, X. Yin, S. Jiao, K. Wei, K. Tu, and L. Pan, "Air jet impingement and hot air-assisted radio frequency hybrid drying of apple slices," *LWT*, vol. 116, p. 108517, 2019, doi: https://doi.org/10.1016/j.lwt.2019.108517.
9. S. Liu, H. Wang, S. Ma, J. Dai, Q. Zhang, and W. Qin, "Radiofrequency-assisted hot-air drying of Sichuan pepper (Huajiao)," *LWT*, vol. 135, p. 110158, 2021, doi: https://doi.org/10.1016/j.lwt.2020.110158.
10. Q. Liu, M. Zhang, B. Xu, Z. Fang, and D. Zheng, "Effect of radio frequency heating on the sterilization and product quality of vacuum packaged Caixin," *Food Bioprod. Process.*, vol. 95, pp. 47–54, 2015, doi: https://doi.org/10.1016/j.fbp.2015.03.007.

11. S. Liu *et al.*, "Microbial validation of radio frequency pasteurization of wheat flour by inoculated pack studies," *J. Food Eng.*, vol. 217, pp. 68–74, 2018, doi: https://doi.org/10.1016/j.jfoodeng.2017.08.013.
12. K. Ballom, N. Dhowlaghar, H.-C. Tsai, R. Yang, J. Tang, and M.-J. Zhu, "Radiofrequency pasteurization against Salmonella and Listeria monocytogenes in cocoa powder," *LWT*, vol. 145, p. 111490, 2021, doi: https://doi.org/10.1016/j.lwt.2021.111490.
13. J. Chen, S. K. Lau, S. R. Boreddy, and J. Subbiah, "Modeling of radio frequency heating of egg white powder continuously moving on a conveyor belt," *J. Food Eng.*, vol. 262, pp. 109–120, 2019, doi: https://doi.org/10.1016/j.jfoodeng.2019.05.029.
14. R. Lan, Y. Qu, H. S. Ramaswamy, and S. Wang, "Radio frequency reheating behavior in a heterogeneous food: A case study of pizza," *Innov. Food Sci. Emerg. Technol.*, vol. 65, p. 102478, 2020, doi: https://doi.org/10.1016/j.ifset.2020.102478.
15. C. Gong, Y. Zhao, H. Zhang, J. Yue, Y. Miao, and S. Jiao, "Investigation of radio frequency heating as a dry-blanching method for carrot cubes," *J. Food Eng.*, vol. 245, pp. 53–56, 2019, doi: https://doi.org/10.1016/j.jfoodeng.2018.10.004.
16. X. Zhang *et al.*, "Developing radio frequency blanching process of apple slice," *J. Food Eng.*, vol. 273, p. 109832, 2020, doi: https://doi.org/10.1016/j.jfoodeng.2019.109832.
17. M. Liao, Y. Zhao, C. Gong, H. Zhang, and S. Jiao, "Effects of hot air-assisted radio frequency roasting on quality and antioxidant activity of cashew nut kernels," *LWT*, vol. 93, pp. 274–280, 2018, doi: https://doi.org/10.1016/j.lwt.2018.03.047.
18. F. Garbati Pegna *et al.*, "Radio frequency irradiation treatment of dates in a single layer to control Carpophilus hemipterus," *Biosyst. Eng.*, vol. 155, pp. 1–11, 2017, doi: https://doi.org/10.1016/j.biosystemseng.2016.11.011.
19. Z. Huang, Y. Chen, and S. Wang, "Numerical studies on the electromagnetic and thermal performances of radio frequency disinfestation treatments for dried apricots," *Postharvest Biol. Technol.*, vol. v. 163, p. 2020 v.163, 2020, doi: 10.1016/j.postharvbio.2019.111116.
20. T. Koray Palazoğlu, Y. Coşkun, T. Kocadağlı, and V. Gökmen, "Effect of radio frequency postdrying of partially baked cookies on acrylamide content, texture, and color of the final product.," *J. Food Sci.*, vol. 77, no. 5, pp. E113-7, May 2012, doi: 10.1111/j.1750-3841.2012.02664.x.
21. M. Liao *et al.*, "Hot air-assisted radio frequency heating for stabilization of rice bran: Enzyme activity, phenolic content, antioxidant activity and microstructure," *LWT*, vol. 131, p. 109754, 2020, doi: https://doi.org/10.1016/j.lwt.2020.109754.
22. Y. Jiang *et al.*, "Highly effective inactivation of anti-nutritional factors (lipoxygenase, urease and trypsin inhibitor) in soybean by radio frequency

treatment," *Int. J. Food Sci. Technol.*, vol. 56, no. 1, pp. 93–102, 2021, doi: https://doi.org/10.1111/ijfs.14605.
23 J. Rice, "RF technology sharpens Bakery's competitive edge," *Food Process*, vol. 6, pp. 18–24, 1993.
24. C. Nindo, T. Sun, S. Wang, J. Tang, and J. Powers, "Evaluation of drying technologies for retention of physical quality and antioxidants in asparagus (Asparagus officinalis, L.)," *LWT - Food Sci. Technol.*, vol. 36, pp. 507–516, Aug. 2003, doi: 10.1016/S0023-6438(03)00046-X.
25. W. Yan, M. Zhang, L. Huang, J. Tang, A. S. Mujumdar, and J. Sun, "Study of the optimisation of puffing characteristics of potato cubes by spouted bed drying enhanced with microwave," *J. Sci. Food Agric.*, vol. 90, no. 8, pp. 1300–1307, Jun. 2010, doi: https://doi.org/10.1002/jsfa.3940.
26. S. S. Sablani, M. Marcotte, O. D. Baik, and F. Castaigne, "Modeling of Simultaneous Heat and Water Transport in the Baking Process," *LWT - Food Sci. Technol.*, vol. 31, no. 3, pp. 201–209, 1998, doi: https://doi.org/10.1006/fstl.1997.0360.
27. O. R. Fennema, *Food Chemistry*, 3rd ed. New York: Marcel Dekker, 1996.
28. F. J. Morales, Ö. Ç. Açar, A. Serpen, G. Arribas-Lorenzo, and V. Gökmen, "Degradation of Free Tryptophan in a Cookie Model System and Its Application in Commercial Samples," *J. Agric. Food Chem.*, vol. 55, no. 16, pp. 6793–6797, Aug. 2007, doi: 10.1021/jf071036e.
29. E. Purlis, "Browning development in bakery products – A review," *J. Food Eng.*, vol. 99, no. 3, pp. 239–249, 2010, doi: https://doi.org/10.1016/j.jfoodeng.2010.03.008.
30. M. Friedman, "Chemistry, biochemistry, and safety of acrylamide. A review.," *J. Agric. Food Chem.*, vol. 51, no. 16, pp. 4504–4526, Jul. 2003, doi: 10.1021/jf030204+.
31. T. Kocadbli, T. K. Palazolu, and V. Gökmen, "Mitigation of acrylamide formation in cookies by using Maillard reaction products as recipe modifier in a combined partial conventional baking and radio frequency post-baking process," *Eur. Food Res. Technol.*, vol. 235, pp. 711–717, 2012.
32. M. Anese, S. Sovrano, and R. Bortolomeazzi, "Effect of radiofrequency heating on acrylamide formation in bakery products," *Eur. food Res. Technol. = Zeitschrift fur Leb. und -Forschung. A*, vol. 226, no. 5, pp. 1197–1203, 2008, doi: 10.1007/s00217-007-0693-x.
33. ICMSF, *Microorganisms in Foods 6: Microbial Ecology of Food Commodities*. Glasqow: Blackie Academic & Professional, 1998.
34. A. P. Dunnigan, "Practical Food Microbiology and Technology," *J. Assoc. Off. Agric. Chem.*, vol. 46, no. 1, p. 67, Feb. 1963, doi: 10.1093/jaoac/46.1.67.
35. J. Seidemann, "Weiser, H. H.: Practical Food Microbiology and Technology. (Praktische Lebensmittel-Mikrobiologie und -Technologie.) Avi Publishing Comp., Inc. Westport, Connecticut (USA). 1962, 345 S., 37 Abb., 36 Tabellen, Ganzleinen $ 12.–," *Starch - Stärke*, vol. 15, no. 8, p. 309, Jan. 1963, doi: https://doi.org/10.1002/star.19630150809.

36. G. J. Mountney and W. A. Gould, *Practical food microbiology and technology.*, no. Ed. 3. New York: Van Nostrand Reinhold Co. Inc. (AVI), 1988.
37. B. B. Banwart, *Basic Food Microbiology*. 1989.
38. B. M. Mackey, C. A. Miles, S. E. Parsons, and D. A. Seymour, "Thermal denaturation of whole cells and cell components of Escherichia coli examined by differential scanning calorimetry.," *J. Gen. Microbiol.*, vol. 137, no. 10, pp. 2361–2374, Oct. 1991, doi: 10.1099/00221287-137-10-2361.
39. J. Xu, S. Liu, J. Tang, S. Ozturk, F. Kong, and D. H. Shah, "Application of freeze-dried Enterococcus faecium NRRL B-2354 in radio-frequency pasteurization of wheat flour," *LWT*, vol. 90, pp. 124–131, 2018, doi: https://doi.org/10.1016/j.lwt.2017.12.014.
40. K. Uemura, S. Kanafusa, C. Takahashi, and I. Kobayashi, "Development of a radio frequency heating system for sterilization of vacuum-packed fish in water," *Biosci. Biotechnol. Biochem.*, vol. 81, no. 4, pp. 762–767, Apr. 2017, doi: 10.1080/09168451.2017.1280660.
41. A. Zheng, L. Zhang, and S. Wang, "Verification of radio frequency pasteurization treatment for controlling Aspergillus parasiticus on corn grains.," *Int. J. Food Microbiol.*, vol. 249, pp. 27–34, May 2017, doi: 10.1016/j.ijfoodmicro.2017.02.017.
42. L. Chen, S. Irmak, B. D. Chaves, and J. Subbiah, "Microbial challenge study and quality evaluation of cumin seeds pasteurized by continuous radio frequency processing," *Food Control*, vol. 111, p. 107052, 2020, doi: https://doi.org/10.1016/j.foodcont.2019.107052.
43. Y. Wang, T. D. Wig, J. Tang, and L. M. Hallberg, "Sterilization of Foodstuffs Using Radio Frequency Heating," *J. Food Sci.*, vol. 68, no. 2, pp. 539–544, Mar. 2003, doi: https://doi.org/10.1111/j.1365-2621.2003.tb05708.x.
44. K. Luechapattanaporn *et al.*, "Microbial Safety in Radio-frequency Processing of Packaged Foods," *J. Food Sci.*, vol. 69, no. 7, pp. 201–206, Sep. 2004, doi: https://doi.org/10.1111/j.1365-2621.2004.tb13621.x.
45. N. U. Sruthi, Y. Premjit, R. Pandiselvam, A. Kothakota, and S. V Ramesh, "An overview of conventional and emerging techniques of roasting: Effect on food bioactive signatures.," *Food Chem.*, vol. 348, p. 129088, Jun. 2021, doi: 10.1016/j.foodchem.2021.129088.
46. M. Liao, Y. Zhao, Y. Xu, C. Gong, and S. Jiao, "Effects of hot air-assisted radio frequency roasting on nutritional quality and aroma composition of cashew nut kernels," *LWT*, vol. 116, p. 108551, Aug. 2019, doi: 10.1016/j.lwt.2019.108551.
47. S. Jiao, D. Zhu, Y. Deng, and Y. Zhao, "Effects of Hot Air-assisted Radio Frequency Heating on Quality and Shelf-life of Roasted Peanuts," *Food Bioprocess Technol.*, vol. 9, no. 2, pp. 308–319, 2016, doi: 10.1007/s11947-015-1624-7.
48. D. WANG, C. JIANG, M. LIAO, Y. XU, and S. JIAO, "Effects of Hot Air-assisted Radio Frequency Roasting on Quality and Flavor of Sunflower Seeds," *Farm Prod. Process.*, vol. 22, 2017.

49. H. K. Vishwanathan, G. K. Giwari, and H. U. Hebbar, "Infrared assisted dry-blanching and hybrid drying of carrot," *Food Bioprod. Process.*, vol. 91, no. 2, pp. 89–94, 2013, doi: https://doi.org/10.1016/j.fbp.2012.11.004.
50. A. F. Doblado-Maldonado, O. A. Pike, J. C. Sweley, and D. J. Rose, "Key issues and challenges in whole wheat flour milling and storage," *J. Cereal Sci.*, vol. 56, no. 2, pp. 119–126, 2012, doi: https://doi.org/10.1016/j.jcs.2012.02.015.
51. Z. Zhang *et al.*, "Effects of radio frequency assisted blanching on polyphenol oxidase, weight loss, texture, color and microstructure of potato.," *Food Chem.*, vol. 248, pp. 173–182, May 2018, doi: 10.1016/j.foodchem.2017.12.065.
52. Y. Jiang *et al.*, "Inactivation of lipoxygenase in soybean by radio frequency treatment," *Int. J. Food Sci. Technol.*, vol. 53, no. 12, pp. 2738–2747, Dec. 2018, doi: https://doi.org/10.1111/ijfs.13885.
53. L. Manzocco, M. Anese, and M. C. Nicoli, "Radiofrequency inactivation of oxidative food enzymes in model systems and apple derivatives," *Food Res. Int.*, vol. 41, no. 10, pp. 1044–1049, 2008, doi: https://doi.org/10.1016/j.foodres.2008.07.020.
54. B. Ling, J. G. Lyng, and S. Wang, "Radio-frequency treatment for stabilization of wheat germ: Dielectric properties and heating uniformity," *Innov. Food Sci. Emerg. Technol.*, vol. 48, pp. 66–74, 2018, doi: https://doi.org/10.1016/j.ifset.2018.05.012.
55. W. H. Cathcart, J. J. Parker, and H. G. Beattie, "The treatment of packaged bread with high frequency heat," *Food Technol.*, vol. 1, no. 2, pp. 174–177, 1947.
56. J. Kim *et al.*, "Study of Radio Frequency Thawing for Cylindrical Pork Sirloin," *J. Biosyst. Eng.*, vol. 41, no. 2, pp. 108–115, Jun. 2016, doi: 10.5307/JBE.2016.41.2.108.
57. T. Koray Palazoğlu and W. Miran, "Experimental comparison of microwave and radio frequency tempering of frozen block of shrimp," *Innov. Food Sci. Emerg. Technol.*, vol. 41, pp. 292–300, 2017, doi: https://doi.org/10.1016/j.ifset.2017.04.005.
58. Y. Chen, J. He, F. Li, J. Tang, and Y. Jiao, "Model food development for tuna (Thunnus Obesus) in radio frequency and microwave tempering using grass carp mince," *J. Food Eng.*, vol. 292, p. 110267, 2021, doi: https://doi.org/10.1016/j.jfoodeng.2020.110267.
59. A. W. D. Carvalho *et al.*, "Heat-treatment reduces anti-nutritional phytochemicals and maintains protein quality in genetically improved hulled soybean flour," *Food Sci. Technol.*, vol. 33, no. 2, pp. 310–315, 2013.
60. J. C. Andrade, J. M. G. Mandarino, L. E. Kurozawa, and E. I. Ida, "The effect of thermal treatment of whole soybean flour on the conversion of isoflavones and inactivation of trypsin inhibitors.," *Food Chem.*, vol. 194, pp. 1095–1101, Mar. 2016, doi: 10.1016/j.foodchem.2015.08.115.
61. Y. Zhong, Z. Wang, and Y. Zhao, "Impact of Radio Frequency, Microwaving, and High Hydrostatic Pressure at Elevated Temperature on the Nutritional

and Antinutritional Components in Black Soybeans.," *J. Food Sci.*, vol. 80, no. 12, pp. C2732-9, Dec. 2015, doi: 10.1111/1750-3841.13131.
62. S. Yalcin and A. Basman, "Effects of infrared treatment on urease, trypsin inhibitor and lipoxygenase activities of soybean samples," *Food Chem.*, vol. 169, pp. 203–210, 2015, doi: https://doi.org/10.1016/j.foodchem.2014.07.114.
63. H. Schubert and M. Regier, *The microwave processing of foods.* Cambridge: Woodhead Publishing Ltd, 2005.
64. C. Yang, Y. Zhao, Y. Tang, R. Yang, W. Yan, and W. Zhao, "Radio frequency heating as a disinfestation method against Corcyra cephalonica and its effect on properties of milled rice," *J. Stored Prod. Res.*, vol. 77, pp. 112–121, 2018, doi: https://doi.org/10.1016/j.jspr.2018.04.004.
65. S. Jiao, J. A. Johnson, J. Tang, G. Tiwari, and S. Wang, "Dielectric properties of cowpea weevil, black-eyed peas and mung beans with respect to the development of radio frequency heat treatments," *Biosyst. Eng.*, vol. 108, no. 3, pp. 280–291, 2011, doi: https://doi.org/10.1016/j.biosystemseng.2010.12.010.
66. Y. Mao, P. Wang, Y. Wu, L. Hou, and S. Wang, "Effects of various radio frequencies on combined drying and disinfestation treatments for in-shell walnuts," *LWT*, vol. 144, p. 111246, 2021, doi: https://doi.org/10.1016/j.lwt.2021.111246.
67. B. Ling, T. Cheng, and S. Wang, "Recent developments in applications of radio frequency heating for improving safety and quality of food grains and their products: A review," *Crit. Rev. Food Sci. Nutr.*, vol. 60, no. 15, pp. 2622–2642, Aug. 2020, doi: 10.1080/10408398.2019.1651690.
68. C. Guo, A. S. Mujumdar, and M. Zhang, "New Development in Radio Frequency Heating for Fresh Food Processing: a Review," *Food Eng. Rev.*, vol. 11, no. 1, pp. 29–43, 2019, doi: 10.1007/s12393-018-9184-z.
69. U. Zimmermann, G. Pilwat, and F. Riemann, "Dielectric breakdown of cell membranes," *Biophys. J.*, vol. 14, no. 11, pp. 881–899, Nov. 1974, doi: 10.1016/S0006-3495(74)85956-4.
70. K. Kinosita and T. Y. O. W. Tsong, "Formation and resealing of pores of controlled sizes in human erythrocyte membrane," *Nature*, vol. 268, no. 5619, pp. 438–441, 1977, doi: 10.1038/268438a0.
71. B. Mertens and D. Knorr, "Developments of nonthermal processes for food preservation," *Food Technol.*, vol. 46, no. 5, pp. 124–133, 1992.
72. Anon, "Deactivation of microorganisms by an oscillating magnetic field," World patent 85/02094, 1985.
73. R. P. Liburdy and Vanek Paul F., Jr., "Microwaves and the Cell Membrane: II. Temperature, Plasma, and Oxygen Mediate Microwave-Induced Membrane Permeability in the Erythrocyte," *Radiat. Res.*, vol. 102, no. 2, pp. 190–205, May 1985, doi: 10.2307/3576466.
74. R. P. Liburdy and R. L. Magin, "Microwave-Stimulated Drug Release from Liposomes," *Radiat. Res.*, vol. 103, no. 2, pp. 266–275, Aug. 1985, doi: 10.2307/3576581.

75. C. T. Ponne, M. Balk, Ö. Hancioglu, and L. G. M. Gorris, "Effect of Radio Frequency Energy on Biological Membranes and Microorganisms," *LWT - Food Sci. Technol.*, vol. 29, no. 1, pp. 41–48, 1996, doi: https://doi.org/10.1006/fstl.1996.0006.
76. M. Doyle, "*E. coli* O157:H7, The Nemesis of Cattle Industry," vol. 3, no. 4, pp. 42–45, 1997.
77. Q. Guo, P. Piyasena, G. S. Mittal, W. Si, and J. Gong, "Efficacy of radio frequency cooking in the reduction of Escherichia coli and shelf stability of ground beef," *Food Microbiol.*, vol. 23, no. 2, pp. 112–118, 2006, doi: https://doi.org/10.1016/j.fm.2005.02.004.
78. D. Geveke and C. Brunkhorst, "RFEF pilot plant for inactivation of Escherichia coli in apple juice," *Fruit Process.*, vol. 14, pp. 166–170, Jan. 2004.
79. D. J. Geveke and C. Brunkhorst, "Radio frequency electric fields inactivation of Escherichia coli in apple cider," *J. Food Eng.*, vol. 85, no. 2, pp. 215–221, 2008, doi: https://doi.org/10.1016/j.jfoodeng.2007.06.029.
80. A. V Cardello, "Consumer concerns and expectations about novel food processing technologies: effects on product liking.," *Appetite*, vol. 40, no. 3, pp. 217–233, Jun. 2003, doi: 10.1016/s0195-6663(03)00008-4.
81. V. Orsat, Y. Gariépy, G. S. V Raghavan, and D. Lyew, "Radio-frequency treatment for ready-to-eat fresh carrots," *Food Res. Int.*, vol. 34, no. 6, pp. 527–536, 2001, doi: https://doi.org/10.1016/S0963-9969(01)00068-0.
82. C. Chen *et al.*, "Short communication: Radio frequency dielectric heating of nonfat dry milk affects solubility and whey protein nitrogen index," *J. Dairy Sci.*, vol. 96, no. 3, pp. 1471–1476, 2013, doi: https://doi.org/10.3168/jds.2012-6158.
83. A. B. Hassan, E. Pawelzik, and D. von Hoersten, "Effect of radio frequency heating on nutritional quality and protein solubility of corn," *Food Sci. Nutr.*, vol. 4, no. 5, pp. 686–689, Jan. 2016, doi: 10.1002/fsn3.332.
84. Y. Zhang, X. Xu, P. Zhang, Ling Zhao, H. Qiu, and X. Cao, "Pyrolysis-temperature depended quinone and carbonyl groups as the electron accepting sites in barley grass derived biochar," *Chemosphere*, vol. 232, pp. 273–280, 2019, doi: https://doi.org/10.1016/j.chemosphere.2019.05.225.
85. M. C. Lagunas-Solar *et al.*, "Disinfection of fishmeal with radiofrequency heating for improved quality and energy efficiency," *J. Sci. Food Agric.*, vol. 85, no. 13, pp. 2273–2280, Oct. 2005, doi: https://doi.org/10.1002/jsfa.2250.
86. M. Archer, M. Edmonds, and G. Martin, "Seafood thawing," *Res. Dev. Dep. UK SR598*, 2008.
87. K. Felke, T. Pfeiffer, P. Eisner, and U. Schweiggert-Weisz, "Radio-frequency heating A new method for improved nutritional quality of tomato puree," *Agro Food Ind. Hi. Tech.*, vol. 22, pp. 29–32, May 2011.
88. S.-G. Ildikó, K. A. Klára, T.-M. Marianna, B. Ágnes, M.-B. Zsuzsanna, and C. Bálint, "The effect of radio frequency heat treatment on nutritional and colloid-chemical properties of different white mustard (Sinapis alba L.)

varieties," *Innov. Food Sci. Emerg. Technol.*, vol. 7, no. 1, pp. 74–79, 2006, doi: https://doi.org/10.1016/j.ifset.2005.06.001.

89. M. E. Sosa-Morales, G. Tiwari, S. Wang, J. Tang, H. S. Garcia, and A. Lopez-Malo, "Dielectric heating as a potential post-harvest treatment of disinfesting mangoes, Part II: Development of RF-based protocols and quality evaluation of treated fruits," *Biosyst. Eng.*, vol. 103, no. 3, pp. 287–296, 2009, doi: https://doi.org/10.1016/j.biosystemseng.2009.02.014.
90. S. L. Birla, S. Wang, J. Tang, J. K. Fellman, D. S. Mattinson, and S. Lurie, "Quality of oranges as influenced by potential radio frequency heat treatments against Mediterranean fruit flies," *Postharvest Biol. Technol.*, vol. 38, no. 1, pp. 66–79, 2005, doi: https://doi.org/10.1016/j.postharvbio.2005.06.001.
91. D. J. Geveke, C. Brunkhorst, and X. Fan, "Radio frequency electric fields processing of orange juice," *Innov. Food Sci. Emerg. Technol.*, vol. 8, no. 4, pp. 549–554, 2007, doi: https://doi.org/10.1016/j.ifset.2007.04.012.
92. D. J. GEVEKE and C. BRUNKHORST, "Inactivation of Saccharomyces cerevisiae with Radio Frequency Electric Fields†," *J. Food Prot.*, vol. 66, no. 9, pp. 1712–1715, Sep. 2003, doi: 10.4315/0362-028X-66.9.1712.
93. G. B. Awuah, H. S. Ramaswamy, A. Economides, and K. Mallikarjunan, "Inactivation of Escherichia coli K-12 and Listeria innocua in milk using radio frequency (RF) heating," *Innov. Food Sci. Emerg. Technol.*, vol. 6, no. 4, pp. 396–402, 2005, doi: https://doi.org/10.1016/j.ifset.2005.06.002.
94. S. Ozturk *et al.*, "Inactivation of Salmonella Enteritidis and Enterococcus faecium NRRL B-2354 in corn flour by radio frequency heating with subsequent freezing," *LWT*, vol. 111, pp. 782–789, 2019, doi: https://doi.org/10.1016/j.lwt.2019.04.090.

8

Ultrasound Technology in Food Processing: Technology, Mechanisms and Applications

Kaidi Peng, Olivier Bals, Eugène Vorobiev and Mohamed Koubaa*

Sorbonne University, Université de Technologie de Compiègne, ESCOM, EA TIMR, Centre de Recherche Royallieu, Compiègne Cedex, France

Abstract

Increasing the product's quality along with reducing the energy consumption are among the challenges sought in the food industry. To reach these challenges, numerous emerging technologies are being implemented at industrial scale as alternative to conventional ones. Among these technologies, ultrasound has taken attention due to its several advantages, that include energy saving, its label of "clean" technology, its ability to keep high product quality, and its combination with the use of "green" solvents, among others. Ultrasound technology has been successfully tested in several food processes such as for the extraction of valuable compounds from different matrices, for fermentation, freezing and crystallization, drying, filtration, emulsification, etc. With regard to the diversity of applications in the food industry, this chapter is mainly dealing with discussing the mechanisms of action of ultrasound as well as its application in different food industrial sectors.

Keywords: Ultrasound, cavitation, extraction, freezing, filtration, fermentation, emulsification

8.1 Introduction

Due to the specificity of interactions with the material, ultrasound is widely used in different fields, including the food industry. It has been described in the literature as a key-technology to achieve the objectives

**Corresponding author*: m.koubaa@escom.fr

Navnidhi Chhikara, Anil Panghal and Gaurav Chaudhary (eds.) Novel Technologies in Food Science, (307–332) © 2023 Scrivener Publishing LLC

of a sustainable "green" process [1]. The use of ultrasound in the food industry as alternative to conventional processes has been proven through several applications demonstrating its ability to significantly save time, solvent, and energy, along with keeping or providing a higher final product quality [2–6]. In addition to these benefits, eliminating the post-treatment of waste water, at least partially, and its affordable investment cost, make it a technology of choice to replace many conventional processes. With regards to the above-mentioned benefits, ultrasound technology has been successfully tested in numerous applications such as for the extraction of valuable compounds from different matrices, for fermentation, filtration, drying, emulsification, freezing and crystallization, among others [7–12]. This chapter discusses in detail the mechanisms of action of ultrasound, some equipment currently used for ultrasound treatment, as well as some selected applications presenting high potential to be further developed in the food industry.

8.2 Mechanisms of Action of Ultrasound Technology

The acoustic energy absorbed by the material is converted into heat; therefore, one of the mechanisms of operation of ultrasound of potential biological significance is the thermal mechanism. This thermal effect depends on the absorption and dissipation of ultrasonic energy. Besides the thermal impact of ultrasound, ultrasonic waves are associated with cavitation and stresses [3, 7, 13–15].

Both inertial and non-inertial cavitation (previously called transient and permanent, respectively) may occur simultaneously during the dissipation of ultrasound. The cavitation phenomenon consists in the formation of pulsating vacuum bubbles, which appear as a result of local ruptures of the medium under the influence of high tensile forces (Figure 8.1).

Generated cavitation bubbles grow and pulse in a forced manner during the non-inertial cavitation and may collapse violently during the wave condensation (or compression) phase, which is usually associated with local high-pressure changes and shock waves, also called inertial cavitation. In some cases, short-term local sonoluminescent flashes could be formed. Bubble formation and collapse were recorded using ultrafast photography, where a wide variety of forms of movement as well as active interactions in the liquid and on the boundary surfaces were observed [16]. During their collapse, bubbles showed deformation into funnel-shaped vortices directed with their tip to the liquid/solid

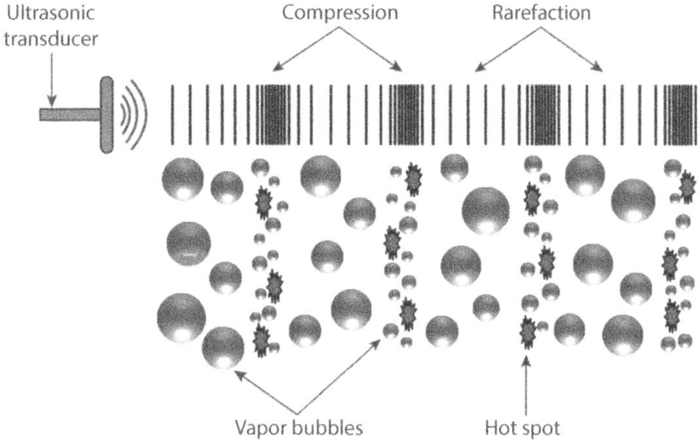

Figure 8.1 The phenomenon of cavitation in a liquid. Hot spots correspond to places with locally increased temperature and pressure, generating shock waves.

interface (Figure 8.2), accelerating the corrosive and erosive processes of solids [13].

Cavitation occurs in a solution when reaching and exceeding a certain cavitation threshold, which depends on the liquid nature, the intensity and frequency of ultrasounds, and the presence of impurities and gas bubbles [16–18].

Different types of acoustically-induced gas bubble activity appear during the non-inertial cavitation, which include bubble growth due to diffusion or aggregation, transient movements of free bubbles, surface distortions (formation of liquid streams and gaseous microbubbles), heat release, and acoustic gas flows inside and out of the liquid that called microflows [13, 19]. When reaching a certain size, cavitation bubbles resonate with the ultrasonic vibrations, wave, and oscillate. These

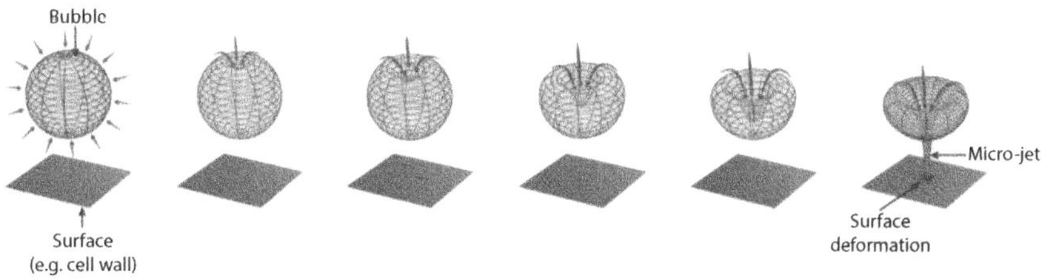

Figure 8.2 Creation of a micro-jet through collapsing acoustic cavitation bubbles. Reproduced with permission from [2].

oscillations occurring on the air-liquid boundary layer cause a swirling (also called a microfluidic) motion in the adjacent liquid, which is characterized by a high velocity gradient. During ultrasound treatment, the cell suspended near the pulsating bubble is pulled to the high velocity gradient area, the vibration causes stream movements in the cell and its exposure to shear stress [20]. Changes in the relative positions of the intracellular organelles, breaks in the cytoplasmic structures, and microfluidic flows in tissues may occur.

Inertial cavitation, in contrast to the non-inertial one, is occurring by applying higher ultrasonic intensities and is more turbulent in nature. In an ultrasonic field, the radius of the cavitation bubbles increases, which leads to their rapid collapse and the local release of a high amount of energy that is accompanied by the increase of temperature and pressure. Increasing the pressure inside the bubble leads to its rupture and therefore to the formation of hydrodynamic shock waves that destroy the biological surface within its range. Under these conditions, water molecules are subjected to a thermal breakdown leading to the release of hydrogen atoms and hydroxyl radicals, as presented in equation (8.1) [21, 22].

$$H_2O \rightarrow \dot{H} + \dot{O}H \tag{8.1}$$

Chain reactions could be initiated by the presence of the radicals formed, where more toxic compounds could be generated. By reacting with each other, the molecules that could be formed include H_2, H_2O, H_2O_2 or attack dissolved molecules, which are reduced or oxidized. Hydrogen and oxygen atoms react together to form the radical HOO•, the acidic form of the superoxide radical anion. As a consequence, the properties of organic compounds present in the ultrasound-treated solution may change by interacting with the radicals formed [17, 22].

The reactions induced by cavitation are generally occurring in three different areas. The first one is inside the collapsing cavitation bubble, where the temperature and the pressure reach several thousand degrees K and several hundred atmospheres, respectively. The hydroxyl radicals and hydrogen atoms formed by the thermal dissociation of water can react with the substances present in the gas phase. The second area is the gas-liquid interface consisting of the area between the collapsing cavitation bubble and the solvent. At this interface, temperature and pressure are still high and may even allow the pyrolyze of low-volatile substances. The thermal decomposition efficiency of these substances in this area depends on their hydrophobicity and therefore on their accumulation at the gas-liquid interface and the activation energy needed to break the bonds. In the third area, consisting of the solution outside the collapsing cavitation bubble, the different free radicals formed as above-mentioned and which did not react in the first or second areas may react with the substances dissolved in the solution. This generates similar products to those present in the water radiolysis [22].

Besides the thermodegradation, the shear forces generated around the collapsing cavitation bubbles induce a mechanochemical degradation and damage of the macrostructures and macromolecules such as polymers [23].

Acoustic cavitation in the liquid medium depends on a large number of sonochemical parameters (e.g., frequency, power, the geometry of the reactor, etc.) and on experimental conditions (e.g., pressure, temperature, solvents, dissolved gas, etc.), which directly affect the sonochemical effects obtained, and therefore the experimental results observed under ultrasound. The main factors affecting the cavitation phenomenon include the ultrasonic frequency, the acoustic power, the hydrostatic pressure, solvents and reaction media, the medium temperature, the gas dissolved in the medium, and the geometry of the sonochemical reactor (also called sonoreactor).

Only the ultrasonic frequency determines the life of the bubbles during transient cavitation. The maximum radius of these bubbles decreases proportionally with the frequency of the ultrasonic wave. The maximum radius at which the bubbles lose their stability corresponds to the resonance radius, reached when their free oscillation frequency (called Minnaert frequency) is higher than that imposed by the acoustic field [24].

Acoustic power is the energy dissipated in the medium in a given ultrasonic irradiation time. It is directly linked to the electrical power imposed by the generator (fixed or adjustable depending on equipment used). Minimum acoustic power is necessary to observe the phenomenon of cavitation, known as the Blake threshold: it is determined from the hydrostatic pressure of the medium and the radius of the cavitation core [25]. The acoustic power also depends on the acoustic impedance of the medium (resistance of the medium to the passage of the ultrasonic wave), which increases when the medium enters cavitation.

In term of hydrostatic pressure, it should be noted that the more the pressure increases in the sonoreactor, the less favorable the cavitation is, but the greater the energy released by the implosion of the bubble and the associated sonochemical effects [26].

Water is usually used as preferred solvent for ultrasound treatment. However, organic solvents could also be used but present the main limitation of a high vapor pressure that is responsible for reducing the intensity of cavitation. The greater the cohesive forces acting in a liquid (e.g., viscosity and surface tension), the more difficult the cavitation phenomenon is to achieve [27]. Besides, the viscosity of the medium is also closely related to the attenuation of the ultrasonic wave.

The temperature of the irradiated medium can play a dual role when it is subjected to ultrasound. On the one hand, increasing the temperature leads to increase all the interactions (e.g., van der Waals forces, hydrogen bonds, dipole attractions, etc.) and improves diffusion phenomena, but on the other hand, cavitation is more easily reached at low temperatures at constant sound power [28]. Generally, the temperature of the medium should not exceed 80 °C to keep the highest efficiency of ultrasound treatment.

Gas bubbles dissolved in a fluid can promote cavitation because they improve the nucleation phase. For this reason, gas could be introduced in solution to saturation, by continuous bubbling, to significantly increase the effects of cavitation [29].

The geometry of the reactor and the height of the solvent can also directly influence the ultrasonic process, depending on the attenuation of ultrasonic waves in the liquid and their reflections on the walls of the reactor [30, 31].

8.3 Equipment Used for Ultrasonic Applications

Ultrasound is produced by a device called a transducer, based on the properties of piezoelectric materials that convert electrical energy into mechanical energy. This mechanical vibration is then transmitted into the liquid medium in the form of an acoustic wave.

Piezoelectric transducers use the inverse piezoelectric effect of natural or synthetic single crystals (e.g., quartz) or ceramics made of barium titanate ($BaTiO_3$), lead metaniobate ($PbNb_2O_6$) or lead titanozirconate (PZT, formula type $Pb(Zr_xTi_{1-x})O_3$). These materials are easily machinable in the form of discs, plates or rings on the faces of which are fixed two metallic electrodes (Figure 8.3). Thus, when an electrical voltage is applied to these electrodes, the material expands or compresses depending on the orientation of

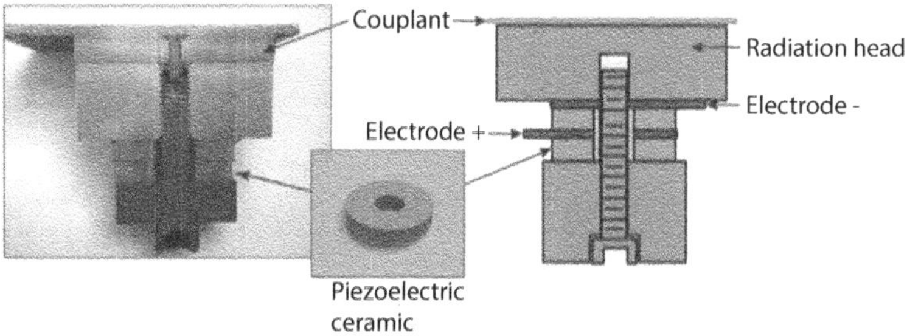

Figure 8.3 A cross section and a schematic representation of a typical piezoelectric ultrasonic transducer. Modified from [32].

the voltage with respect to the polarization of the ceramic, with maximum vibrations depending on the frequency (resonant frequency).

To emit an ultrasonic wave, an electric voltage is therefore applied to the transducer with a frequency equal to its resonant frequency. An electric generator is used to transform the main voltage (220 V-50/60 Hz) into an alternating voltage at the resonant frequency of the system (e.g., 1000 V-20 kHz). The conversion efficiencies of the electrical energy supplied to the generator into acoustic energy transmitted to the medium are currently in the order of 20 to 40% for low frequency ultrasound and 60 to 80% for high frequency ultrasound.

At the laboratory scale, several equipment can be used such as ultrasonic tanks, probes, "cup-horn" reactors, "whistle" reactors, or even transducers suitable for continuous or semi-continuous processes. The most frequently used for ultrasonic applications are ultrasonic bath and probe-type ultrasound equipment (Figure 8.4). In both devices, one or multiple transducers, usually piezoelectric, are used for ultrasound power generation.

Ultrasonic bath consisting of a stainless-steel tank equipped with multiple transducers is the most commonly used device. This equipment is the least expensive and the most widely used in research laboratories to emit ultrasound, mainly to create or break emulsions, dissolve compounds, degas eluents, clean glassware, etc. This type of device produces ultrasound at frequencies between 20 and 60 kHz (low frequencies) with low acoustic intensities (generally 1 to 5 $W.cm^{-2}$) so as not to damage the tank via cavitation. The sources of vibration are generally placed at the bottom of the tank or more rarely on the sides (or both). A liquid, generally water, is used to transmit the acoustic energy to the container used, immersed in the tank (indirect irradiation). The main drawbacks of this equipment are the

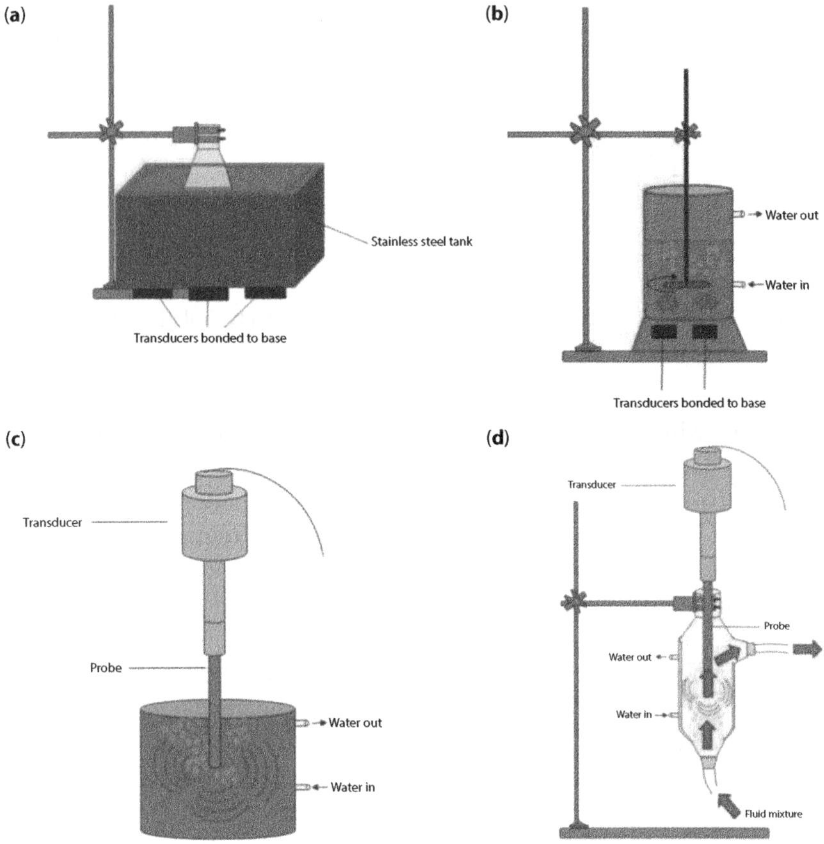

Figure 8.4 Commonly used ultrasonic systems (a: Ultrasound bath, b: Ultrasound reactor with stirring, c: Ultrasound probe, d: Continuous sonication with ultrasound probe). Reproduced with permission from [1].

non-homogeneous nature of the irradiation in the tank and the difficulty in controlling the temperature of the liquid. The ultrasound tank therefore has limited use in terms of reproducibility of experiments [28].

High power ultrasonic probes are preferred to ultrasonic tanks as they deliver higher ultrasound intensity to the sample due to the reduced surface of contact, and the lower water volume in contact of the probe. These systems can produce ultrasound with a power in hundred times greater than that delivered by the ultrasound tanks, directly into the solution studied (direct irradiation). Depending on the application and the volume being sonicated, different ultrasonic probes (e.g., length, diameter, etc.) could be used. Their shape and the geometry of the reactor can have considerable effects on the results obtained during sonication [33]. Usually, a double jacket is used to cool the medium as the temperature increases due to cavitation. This cooling is required to preserve the thermo-sensitive molecules from degradation, and to make the ultrasonic process more efficient [1].

Many parameters can modify the effects of ultrasound, and sometimes it is difficult to replicate experiments described in the literature, mainly because the equipment is different from one research group to another. In addition, some publications do not specifically describe the characteristics of the sonoreactor studied.

The ultrasound frequency is usually indicated by the manufacturer, it is easy to find the frequency of the equipment used (in kHz), in particular to indicate which effects of low or high frequencies are preferentially used. The total electric power could be determined using a wattmeter connected upstream of the generator. It allows measuring the power consumed by the system (in kWh). The acoustic power (P_a, expressed in W) can be estimated in the irradiated medium by calorimetry, by measuring the initial rise in the temperature of the liquid as a function of irradiation time (dT/dt). The acoustic power can then be deduced from the mass (m) and the specific heat capacity (Cp) of the liquid used, according to equation (8.2):

$$P_a = m \times C_p \times (dT/dt) \tag{8.2}$$

The ultrasonic intensity (I_{US}, expressed in $W.cm^{-2}$) is estimated by calculating the ratio of the acoustic power (P_a, in W) by the surface of the emitting probe (S_p, in cm^2), as shown in equation (8.3):

$$I_{US} = \frac{P_a}{S_p} \tag{8.3}$$

Other parameters may also affect the ultrasonic process, depending on the application, as presented in the section below.

8.4 Selected Applications of Ultrasounds in Food Processing

Ultrasound technology can be used in food processing for multiple applications that include extraction, fermentation, drying, filtration, emulsification, and homogenization, among others.

8.4.1 Ultrasound-Assisted Extraction

Ultrasound technology has been widely applied for the extraction of valuable compounds from plant matrices, and reviewed in several papers [34–39]. Usually, this technology is associated with the use of "green" solvents, which fits perfectly well with the current trend of bioeconomy context and the "green" food processing concept. This concept is mainly based on the design of novel extraction processes that allow significantly reducing the water and energy consumption, high quality and safe product, and by-products recycling [40]. In fact, research interest has been widely focused on the recovery of value-added compounds from agri-food by-products, usually using ultrasound-assisted extraction [3]. The other benefits of ultrasound-assisted extraction compared to the conventional extraction, usually called maceration, consists of its ability to intensify the mass transfer, to reduce the extraction time, and to consume less energy when heat is required for the extraction. The molecules extracted from industrial food by-products and using ultrasound include, among others, polyphenols, lipids, essential oils, sugars, carbohydrate polymers, colorants, pigments, etc. It should be noted that many parameters influence the ultrasound-assisted extraction efficiency, which could be physical (e.g. ultrasound power, intensity, frequency, and the shape and size of the ultrasound reactor) as described above, related to the medium (e.g. extraction solvent, temperature, etc.), or related to the matrix itself (e.g. particle size, solid/liquid ratio, etc.) [1].

Ultrasound-assisted extraction systems developed and described in the literature could either be used alone in combination with other technologies (e.g. Soxhlet, Clevenger, microwaves, extrusion, supercritical fluid extraction, etc.) [1, 4, 41, 42].

8.4.2 Ultrasound-Assisted Fermentation

Intensification of several biotechnological processes could occur by applying low-intensity ultrasound (< 10 $W.cm^{-2}$), without damaging the cells that is usually observed for the application of high-intensity ultrasound (10-1000 $W.cm^{-2}$) [43, 44]. The mechanisms explaining this intensification is not yet fully understood, although an important role was associated to the increase of solid-liquid mass transfer (dissolution of a solid in the liquid phase) and gas-liquid mass transfer (liquid phase gas absorption) [45]. Cellular stress occurs by pressure fluctuation during ultrasound treatment, which promotes cell growth and the stimulation of the metabolic processes [45, 46]. Low-intensity ultrasound has been mainly applied in food processing for the improvement of the microorganisms' performances as well as the end product's quality [47–49]. For instance, dairy fermentation assisted by ultrasound showed modifications induced in some macromolecules (e.g. enzymes) [50, 51], shortened time for yogurt fermentation [52, 53], and improved rheological properties of yogurt [54]. Other applications include the production of probiotics (e.g. *Bifidobacterium* sp.), where the application of ultrasound was associated with increased lactose hydrolysis, and transgalactosylation in milk [55, 56].

Production of alcoholic beverages assisted by ultrasound has been also widely described in the literature. In a recent work, the impact of low intensity ultrasound has been studied on *Hanseniaspora* sp. yeast fermentation for cider production [7]. The authors applied ultrasound during the adaptation and exponential fermentation phases, first in a synthetic medium and then in apple juice. An ultrasound treatment chamber was designed and used for cell stimulation (Figure 8.5). The system consisted of pumping the solution from the fermenter, its treatment by ultrasound, and its injection back to the fermenter. The optimal ultrasound treatment parameters in a synthetic medium consisted of a pulse duration of 0.5 s followed by a rest period of 6 s. When applying these parameters

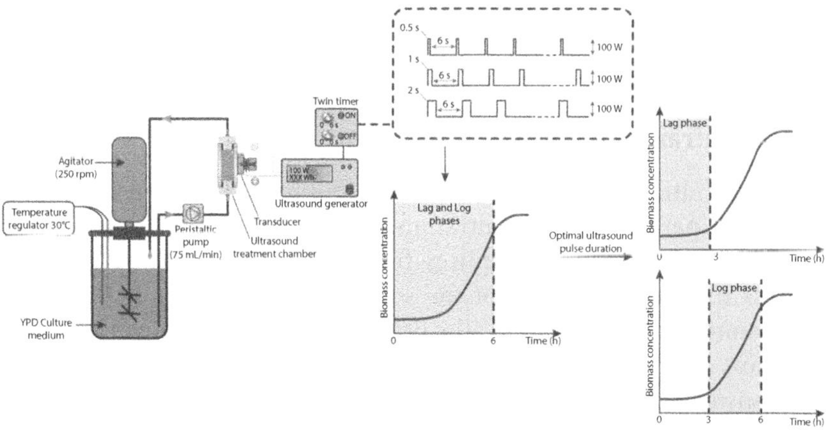

Figure 8.5 Experimental set-up for the optimization of ultrasound-assisted fermentation of *Hanseniaspora* sp. on YPD medium. Reproduced with permission from [7].

during apple juice fermentation, the authors observed increased rates of cellular growth and glucose consumption, and decreased rates of ethanol production. An ultrasound treatment during the first 12 h of fermentation allowed shortening the fermentation time, increasing the maximum biomass growth rate by ten folds, and reducing the ethanol content by 0.55% (v, v).

In another recent study, the application of direct (23 to 32 $W.L^{-1}$) or indirect (1.4 $W.L^{-1}$) low intensity ultrasound showed similar results consisting of reductions in the rates of glucose consumption and ethanol production by the yeasts [57].

It should be noted that the results depend on many factors that include the ultrasound parameters, the strain nature, the culture medium, the fermentation conditions, the growth phase, etc. Therefore, several observations related to alcoholic fermentation assisted by ultrasound have been reported in the literature. For instance, the application of low-intensity ultrasound of 30 $mW.cm^{-2}$ with a frequency of 43 kHz induced a significant shortening (up to 50%) in wine and beer fermentation time, along with an acceleration of ethanol production and a decrease in the dissolved CO_2 [58]. Similar results were observed by Lanchun *et al.* [59] when applying during the exponential growth phase a treatment of low intensity ultrasound (24 kHz, 2 W, pulses of 1 s each 15 s and during 30 min) on *Saccharomyces cerevisiae* cells.

This variability in the results reported for ultrasound-assisted fermentation make understanding the mechanisms behind microbial stimulation by ultrasound difficult, and requiring deeper research studies. Further reading can be found in [10].

8.4.3 Ultrasound-Assisted Filtration

During filtration operations, membrane fouling is a major problem that limits the performance of filtration units, increases the energy consumption of pumps and requires frequent washing, which generates process interruptions. One of the solutions to this issue is the use of ultrasound-assisted filtration. The principle of this technology lies in the emission of ultrasound near the filtration membrane, which increases the transmembrane flow rate and reduces the fouling phenomenon [60]. Ultrasound-assisted filtration has been successfully applied to improve industrial wastewater filtration [61]. When coupling ultrasound with filtration, the life of filters could be extended mainly due to the prevention of membrane caking and fouling [62]. In their work, Camara *et al.* [63]

studied the *in situ* ultrasound-assisted control of polymeric membrane fouling. They investigated the impact of ultrasound frequency, power, and feed solid concentration on permeate flux during ultrafiltration. The authors reported that the maximum increase in permeate flux when applying ultrasound (20 kHz and 0.29 $W.cm^{-2}$) during filtration was equal to 19.7%.

It has been suggested that irradiating the membrane with ultrasound don't influence its intrinsic permeability. Increasing the flux through the membrane by the application of ultrasound is mainly due to the presence of liquid jets that serve as the basis for cleaning. Other cavitational mechanisms are also involved in the increase of this flux, which leads to the release of particles from the fouled membrane. As reported above, several parameters influence the efficiency of ultrasound-assisted filtration, which include the frequency, the power, the feed rate, the properties of the membrane, the cross-flow velocity, as well as the transmembrane pressure and the temperature. The impact of ultrasound on the membrane is controversial, some research groups have reported the membrane degradation by ultrasound irradiation, whereas some others demonstrated the absence of membrane damage even after several uses.

Industrial application of ultrasound-assisted filtration has not yet been widely commercialized. The main reasons behind consist mainly on the development stagnation of transducer technology for membrane filtration and the control of membrane erosion [61].

8.4.4 Ultrasound-Assisted Emulsification

Power ultrasound is a proven technique for preparing stable macro- and nano-emulsions, double emulsions and pickering emulsions. Nanoemulsions are emulsions having a dispersed phase of nanometric size and their preparation requires an intense energy input to disperse the droplets at nanometric size.

High intensity ultrasound has been used by several research groups to stabilize protein emulsions. For instance, this technology has been successfully applied to improve the properties of chicken myofibrillar protein emulsion and to enhance its rheological properties and storage ability [64]. The parameters applied for ultrasound were a frequency of 20 kHz, a power of 450 W, and a treatment time of 6 min. The authors demonstrated that ultrasound-assisted emulsification allowed enhancing the emulsion stability, contributed to form smaller and more uniform emulsion droplets, and increased the interfacial proteins surrounding oil droplets.

In another work, the impact of ultrasound on milk homogenization and fermentation was investigated [65]. The authors demonstrated that power ultrasound provided enhanced homogenization compared with conventional one, and this impact was more pronounced for longer exposure times. After sonication, the total fermentation time was reduced by 30 min. Besides, improved water holding capacity and viscosity as well as reduced syneresis, were observed when increasing the ultrasound power before inoculation.

In another study, ultrasound-assisted emulsification micro-extraction of various preservatives from cosmetics, beverages, and water samples was investigated [66]. The technology was successfully used to determine the content of different compounds consisting of sodium benzoate (SB), p-hydroxybenzoic acid (PHBA), methyl-paraben (MP), ethyl-paraben (EP), and propyl-paraben (PP) in the different samples. Some other related papers have also been published to demonstrate the efficiency of ultrasound-assisted emulsification micro-extraction technique to determine the amount of molecules in different samples [67–70].

To summarize, it could be concluded that ultrasound represents a great potential to be used to improve the emulsification processes as well as the mechanical properties of the emulsified products.

8.4.5 Ultrasound-Assisted Drying

Drying is one of the most energy-consuming unit operations in the food industry. Applying ultrasound to drying allows reducing the drying periods, improving the warmth and mass transfer rate, and maintaining the characteristics of the product [71]. It was applied to meat, fish, fruits and vegetables, among others, and presented a promising technology to substitute traditional drying. The mechanism behind enhancing drying process by ultrasound is mainly related to the evaporation of water by the formation of new micro channels by cavitation. Ultrasound causes air vibration, which enhances the water removal from the sample surface.

Many research works were focused on the application of ultrasound to improve the drying processes. For instance, it was demonstrated that high power ultrasound was effective for drying the brown seaweed *Ascophyllum nodosum* [72]. Drying time was reduced after ultrasound application with the shortest drying times obtained for the samples pre-treated at 75.78 $W.cm^{-2}$. Besides, the authors showed that the color of ultrasound-pretreated dried seaweed samples was lighter compared to the control samples. Similar results were reported for the ultrasound-assisted drying of nectarine, where improved drying rates were observed, along with an improved retention of the polyphenols in the dried samples [73]. The authors reported as well a better preservation of the fruit color, compared to conventional drying.

In another research work, the ultrasound-assisted drying process by microwave was applied to apple slices [74]. The authors demonstrated similarly to the results reported above, an acceleration in the drying kinetics, and a better preservation of the sample color. These observations were also reported for the ultrasound-assisted drying of red peppers, where in addition to reducing the drying time and keeping a fresh-like color of the product, the total yeast and mold counts were significantly reduced in the ultrasound-treated samples [75].

As above-mentioned, several factors may affect the effectiveness of ultrasound-assisted drying. These include the ultrasound power, frequency, amplitude, exposure time, among others [8, 60]. Further reading could be found in [76].

8.4.6 Ultrasound-Assisted Freezing and Crystallization

One of the most applied methods for food preservation is freezing. In the freezer, the temperature of the sample decreases gradually from its initial temperature ($T_{initial}$) to the nucleation one ($T_{nucleation}$). Under certain conditions, the nucleation occurs at a temperature lower than the theoretical freezing temperature of the sample ($T_{freezing}$) (Figure 8.6). This time interval between $T_{freezing}$ and $T_{nucleation}$ is called the super cooled stage, and their temperature difference is called the super cooling degree. The formation of ice crystals starts from the time t_s corresponding to the nucleation temperature ($T_{nucleation}$) to the time t_e reaching back the freezing temperature ($T_{freezing}$). This phenomenon is accompanied with a release of latent heat. The phase change to ice is then occurring until the time t_p, beyond it the system speeds-up cooling until reaching the final sample temperature (T_{final}).

Crystallization phase during which water is converted to ice is the most important compared to the other phases described above, as it influences the efficiency of the freezing process and therefore the quality of the frozen product [77–81]. Crystallization is a spontaneous and stochastic process occurring in two steps, the formation of nuclei (namely, nucleation) and its subsequent growth [82, 83].

One of the efficient methods applied during the phase change stage to induce nucleation and promote the subsequent ice crystal growth is

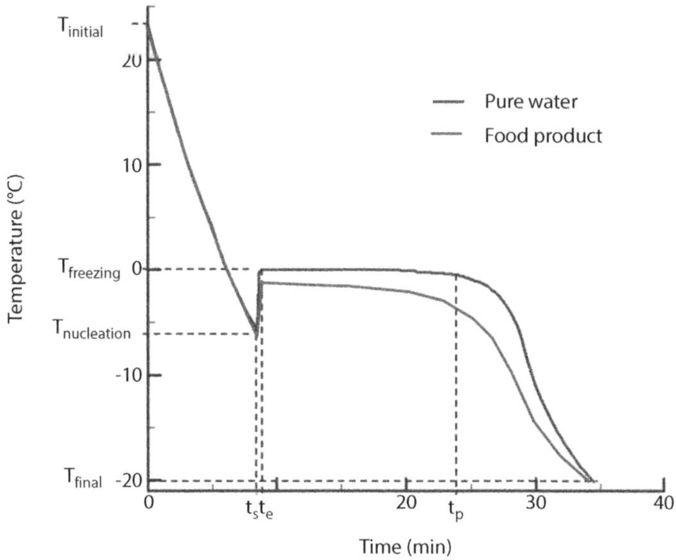

Figure 8.6 Typical freezing curves for pure water and a food product ($T_{initial}$, initial temperature of sample; T_{final}, final temperature of sample; $T_{freezing}$, initial freezing temperature; $T_{nucleation}$, nucleation temperature; t_s, time for beginning of nucleation; t_e, time for end of nucleation; t_p, end of phase change to ice).

ultrasound [84]. It can influence the crystallization process by controlling and predicting its spontaneous and stochastic nature [85], as well as enhancing the freezing efficiency [36].

The review of the literature shows that power ultrasound may enhance the freezing process. This occurs through the initiation of nucleation and the increased rates of the heat and mass transfer [86]. Ultrasound-assisted crystallization could be explained by two mechanisms namely primary and secondary nucleation. During primary nucleation, large number of cavitation bubbles is generated by ultrasound and can act as nuclei [87, 88], whereas during secondary nucleation, the big preexisting ice dendrite crystals could be broken by power ultrasound into small fragments [89]. Heat and mass transfer are enhanced during ultrasound-assisted freezing, which may improve the freezing rate. This behavior is mainly due to the ultrasonic microstreaming that generates violent agitation. According to

Zamanipoor and Mancera [90], the driving force of ice crystal growth is the heat and mass transfer.

Ultrasound-assisted freezing was studied by several research groups. For instance, the effects of ultrasound-assisted freezing on the freezing time and quality of broccoli during immersion freezing was investigated [91]. The authors showed decreased freezing time when applying specific parameters of ultrasound, better microstructure and firmness of broccoli, as well as a reduced drip loss.

In another recent research work, the impact of ultrasound-assisted freezing was investigated on the textural characteristics of dough and the structural characterization of wheat gluten [92]. The authors demonstrated that the ultrasound treatment was efficient to improve the textural properties of frozen dough, to prevent the secondary structure from destruction by freezing, and to obtain more uniform and smaller gluten network, compared to traditional freezing samples.

Many other research works showed the great potential of ultrasound-assisted freezing to improve the textural properties of the frozen products and reduce the freezing time [9, 93]. The challenge now remains in the scaling-up of the process for industrial applications.

8.5 Conclusions

The different applications provided through this chapter for the implementation of ultrasound in different food industrial sectors showed the versatility if this technology and its great potential to be further developed in many other fields. It was successfully tested for different applications such as extraction, filtration, fermentation, emulsification, freezing and crystallization, among others. However, the main research need for its further industrial exploitation consists of its scaling-up. Some applications such as ultrasound-assisted extraction have been successfully scaled-up, whereas others such as ultrasound-assisted filtration and ultrasound-assisted fermentation require deeper investigations. This technology remains nonetheless one of the most promising in food processing that provide high product quality, along with energy saving.

References

1. F. Chemat, N. Rombaut, A.-G. Sicaire, A. Meullemiestre, A.-S. Fabiano-Tixier, M. Abert-Vian, Ultrasound assisted extraction of food and natural products. Mechanisms, techniques, combinations, protocols and applications. A review, Ultrasonics Sonochemistry. 34 (2017) 540–560.
2. M. Koubaa, H. Mhemdi, J. Fages, Recovery of valuable components and inactivating microorganisms in the agro-food industry with ultrasound-assisted supercritical fluid technology, Journal of Supercritical Fluids. 134 (2018) 71–79.
3. M. Koubaa, F.J. Barba, N. Grimi, H. Mhemdi, W. Koubaa, N. Boussetta, E. Vorobiev, Recovery of colorants from red prickly pear peels and pulps enhanced by pulsed electric field and ultrasound, Innovative Food Science & Emerging Technologies. 37, Part C (2016) 336–344.
4. M. Koubaa, H. Mhemdi, F.J. Barba, S. Roohinejad, R. Greiner, E. Vorobiev, Oilseed treatment by ultrasounds and microwaves to improve oil yield and quality: An overview, Food Research International. 85 (2016) 59–66.
5. K. Peng, M. Koubaa, O. Bals, E. Vorobiev, Recent insights in the impact of emerging technologies on lactic acid bacteria: A review, Food Research International. 137 (2020) 109544. https://doi.org/10.1016/j.foodres.2020.109544.
6. S. Roohinejad, N. Nikmaram, M. Brahim, M. Koubaa, A. Khelfa, R. Greiner, Potential of novel technologies for aqueous extraction of plant bioactives, in: H. Dominguez González, M.J. González Muñoz (Eds.), Water Extraction of Bioactive Compounds, Elsevier, 2017: pp. 399–419.
7. M. Al Daccache, M. Koubaa, D. Salameh, R.G. Maroun, N. Louka, E. Vorobiev, Ultrasound-assisted fermentation for cider production from Lebanese apples, Ultrasonics Sonochemistry. 63 (2020) 104952. https://doi.org/10.1016/j.ultsonch.2019.104952.
8. D. Huang, K. Men, D. Li, T. Wen, Z. Gong, B. Sunden, Z. Wu, Application of ultrasound technology in the drying of food products, Ultrasonics Sonochemistry. 63 (2020) 104950. https://doi.org/10.1016/j.ultsonch.2019.104950.
9. H. Kiani, Z. Zhang, D.-W. Sun, Experimental analysis and modeling of ultrasound assisted freezing of potato spheres, Ultrason, Sonochem. 26 (2015) 321–331.
10. M.J. Mota, R.P. Lopes, M. Koubaa, S. Roohinejad, F.J. Barba, I. Delgadillo, J.A. Saraiva, Fermentation at non-conventional conditions in food- and bio-sciences by the application of advanced processing technologies, Crit. Rev. Biotechnol. 38 (2018) 122–140. https://doi.org/10.1080/07388551.2017.1312272.

11. G. Musielak, D. Mierzwa, J. Kroehnke, Food drying enhancement by ultrasound – A review, Trends in Food Science & Technology. 56 (2016) 126–141.
12. A. RežekJambrak, Z. Herceg, Application of Ultrasonics in Food Preservation and Processing, in: S. Bhattacharya (Ed.), Conventional and Advanced Food Processing Technologies, John Wiley & Sons, Ltd, 2014: pp. 515–536.
13. S. Barnett, Nonthermal issues: Cavitation—Its nature, detection and measurement, Ultrasound in Medicine & Biology. 24 (1998) S11–S21.
14. M.W. Miller, D.L. Miller, A.A. Brayman, A review of *in vitro* bioeffects of inertial ultrasonic cavitation from a mechanistic perspective, Ultrasound in Medicine & Biology. 22 (1996) 1131–1154.
15. W. Obrienjr, Ultrasound–biophysics mechanisms, Progress in Biophysics and Molecular Biology. 93 (2007) 212–255. https://doi.org/10.1016/j.pbiomolbio.2006.07.010.
16. C.X. Deng, Q. Xu, R.E. Apfel, C.K. Holland, *In vitro* measurements of inertial cavitation thresholds in human blood, Ultrasound in Medicine & Biology. 22 (1996) 939–948.
17. C.C. Church, E.L. Carstensen, "Stable" inertial cavitation, Ultrasound in Medicine & Biology. 27 (2001) 1435–1437.
18. S. Wang, R. Feng, X. Mo, The cavitation valley phenomenon of rectangular wave modulation ultrasound, Ultrasonics Sonochemistry. 3 (1996) 69–71.
19. N. Kudo, K. Yamamoto, Impact of bubbles on ultrasound safety, International Congress Series. 1274 (2004) 129–132.
20. I. Rosenthal, J.Z. Sostaric, P. Riesz, Sonodynamic therapy--a review of the synergistic effects of drugs and ultrasound, Ultrason, Sonochem. 11 (2004) 349–363.
21. A.F. Fuciarelli, E.C. Sisk, R.M. Thomas, D.L. Miller, Induction of base damage in DNA solutions by ultrasonic cavitation, Free Radic. Biol. Med. 18 (1995) 231–238.
22. P. Riesz, T. Kondo, Free radical formation induced by ultrasound and its biological implications, Free Radic. Biol. Med. 13 (1992) 247–270.
23. B. Rokita, P. Ulański, Studies on the spatial distribution of polymeric reagents in sonochemical reactions application of competitive kinetics, Polimery. 50 (2005) 29–36.
24. M.A. Margulis, G. Leib, Sonochemistry and cavitation, Gordon and Breach, Australia [etc., 1995.
25. F.B. Blake, Acoustic Research Laboratory Memorandum No. 12, Harvard University, 1949.
26. T.J. Mason, Practical sonochemistry: user's guide to applications in chemistry and chemical engineering., E. Horwood, New York, 2002.
27. Z.-L. Wu, Sonochemistry of selected hydrocarbons, sulfur-containing and nitrogen-containing organic compounds in aqueous solutions and nonaqueous liquids, Friedrich Schiller University, 2005.

28. T.J. Mason, Sonochemistry, Oxford University Press, Oxford, New York, 1999.
29. T.J. Mason, J.P. Lorimer, Sonochemistry: theory, applications and uses of ultrasound in chemistry, Ellis Horwood, Chister, 1988.
30. S.I. Nikitenko, C. Le Naour, P. Moisy, Comparative study of sonochemical reactors with different geometry using thermal and chemical probes, Ultrasonics Sonochemistry. 14 (2007) 330–336.
31. S. de L. Rochebrochardd'Auzay, J.-F. Blais, E. Naffrechoux, Comparison of characterization methods in high frequency sonochemical reactors of differing configurations, Ultrasonics Sonochemistry. 17 (2010) 547–554.
32. Cleaning technologies group, Ultrasonics - Transducers - Piezoelectric Hardware, CTG Technical Blog. (2012). https://techblog.ctgclean.com/2012/01/ultrasonics-transducers-piezoelectric-hardware/ (accessed March 7, 2021).
33. Z. Wei, J.A. Kosterman, R. Xiao, G.-Y. Pee, M. Cai, L.K. Weavers, Designing and characterizing a multi-stepped ultrasonic horn for enhanced sonochemical performance, Ultrasonics Sonochemistry. 27 (2015) 325–333.
34. F.J. Barba, P. Putnik, D. BursaćKovačević, M.M. Poojary, S. Roohinejad, J.M. Lorenzo, M. Koubaa, Impact of conventional and non-conventional processing on prickly pear (*Opuntia* spp.) and their derived products: From preservation of beverages to valorization of by-products, Trends in Food Science & Technology. 67 (2017) 260–270.
35. F.J. Barba, Z. Zhu, M. Koubaa, A.S. Sant'Ana, V. Orlien, Green alternative methods for the extraction of antioxidant bioactive compounds from winery wastes and by-products: A review, Trends in Food Science & Technology. 49 (2016) 96–109.
36. F. Chemat, Zill-e-Huma, M.K. Khan, Applications of ultrasound in food technology: Processing, preservation and extraction, Ultrasonics Sonochemistry. 18 (2011) 813–835.
37. M. Koubaa, E. Roselló-Soto, J. ŠicŽlabur, A. RežekJambrak, M. Brnčić, N. Grimi, N. Boussetta, F.J. Barba, Current and new insights in the sustainable and green recovery of nutritionally valuable compounds from *Stevia rebaudiana*Bertoni, J. Agric. Food Chem. 63 (2015) 6835–6846. https://doi.org/10.1021/acs.jafc.5b01994.
38. S.R. Shirsath, S.H. Sonawane, P.R. Gogate, Intensification of extraction of natural products using ultrasonic irradiations—A review of current status, Chemical Engineering and Processing: Process Intensification. 53 (2012) 10–23.
39. B.K. Tiwari, Ultrasound: A clean, green extraction technology, TrAC Trends in Analytical Chemistry. 71 (2015) 100–109.
40. F. Chemat, N. Rombaut, A. Meullemiestre, M. Turk, S. Perino, A.-S. Fabiano-Tixier, M. Abert-Vian, Review of green food processing techniques.

Preservation, transformation, and extraction, Innovative Food Science & Emerging Technologies. 41 (2017) 357–377.
41. G. Cravotto, Innovative techniques and equipments for flavours extraction: Application and effectiveness of ultrasound and microwaves, Fragrances - Supplement to Household and Personal Care Today. (2010) 30–32.
42. M. Singla, N. Sit, Application of Ultrasound in Combination with Other Technologies in Food Processing: A Review, Ultrasonics Sonochemistry. (2021) 105506. https://doi.org/10.1016/j.ultsonch.2021.105506.
43. P.R. Gogate, A.M. Kabadi, A review of applications of cavitation in biochemical engineering/biotechnology, Biochemical Engineering Journal. 44 (2009) 60–72.
44. Y. Chisti, Sonobioreactors: using ultrasound for enhanced microbial productivity, Trends Biotechnol. 21 (2003) 89–93.
45. N. SainzHerrán, J.L. Casas López, J.A. Sánchez Pérez, Y. Chisti, Effects of ultrasound on culture of *Aspergillus terreus*, J. Chem. Technol. Biotechnol. 83 (2008) 593–600.
46. H. Hu, Y. Yonezawa, A. Matsuda, N. Ishida, A. Noguchi, Influences of ultrasound and Ohmic heating on growth of Sake yeast, Food Sci. Technol. Res. 18 (2012) 611–616.
47. T.S. Awad, H.A. Moharram, O.E. Shaltout, D. Asker, M.M. Youssef, Applications of ultrasound in analysis, processing and quality control of food: A review, Food Research International. 48 (2012) 410–427.
48. B. Kwiatkowska, J. Bennett, J. Akunna, G.M. Walker, D.H. Bremner, Stimulation of bioprocesses by ultrasound, Biotechnology Advances. 29 (2011) 768–780.
49. K.S. Ojha, T.J. Mason, C.P. O'Donnell, J.P. Kerry, B.K. Tiwari, Ultrasound technology for food fermentation applications, Ultrasonics Sonochemistry. 34 (2017) 410–417.
50. G. Huang, S. Chen, C. Dai, L. Sun, W. Sun, Y. Tang, F. Xiong, R. He, H. Ma, Effects of ultrasound on microbial growth and enzyme activity, Ultrasonics Sonochemistry. 37 (2017) 144–149.
51. H. Ma, L. Huang, J. Jia, R. He, L. Luo, W. Zhu, Effect of energy-gathered ultrasound on Alcalase, Ultrasonics Sonochemistry. 18 (2011) 419–424.
52. W. FeiFei, C. Ling, L. HuaQiang, Z. LiangZhong, X. YongPing, L. YanZhen, C. Qiong, Z. ZaoMing, Research progress of applications of ultrasonic technology in food industry., Journal of Food Safety and Quality. 8 (2017) 2670–2677.
53. A.S. Polonorum, J. Stadnik, D. Stasiak, Applications of ultrasound in food technology, Acta Scientiarum Polonorum Technologia Alimentaria. 6 (2007) 89–99.
54. A. Vercet, R. Oria, P. Marquina, S. Crelier, P. Lopez-Buesa, Rheological properties of yoghurt made with milk submitted to manothermosonication., Journal of Agricultural and Food Chemistry. 50 (2002) 6165–6171.

55. T.M.P. Nguyen, Y.K. Lee, W. Zhou, Stimulating fermentative activities of bifidobacteria in milk by highintensity ultrasound, International Dairy Journal. 19 (2009) 410–416.
56. S.-K. Yeo, M.-T. Liong, Effect of ultrasound on the growth of probiotics and bioconversion of isoflavones in prebiotic-supplemented soymilk, J. Agric. Food Chem. 59 (2011) 885–897.
57. L. Huezo, A. Shah, F.C. Michel Jr., Effects of ultrasound on fermentation of glucose to ethanol by *Saccharomyces cerevisiae*, Fermentation. 5 (2019) 16. https://doi.org/10.3390/fermentation5010016.
58. K. Matsuura, M. Hirotsune, Y. Nunokawa, M. Satoh, K. Honda, Acceleration of cell growth and ester formation by ultrasonic wave irradiation, Journal of Fermentation and Bioengineering. 77 (1994) 36–40.
59. S. Lanchun, W. Bochu, L. Zhiming, D. Chuanren, D. Chuanyun, A. Sakanishi, The research into the influence of low-intensity ultrasonic on the growth of *S. cerevisiaes*, Colloids and Surfaces B: Biointerfaces. 30 (2003) 43–49.
60. N. Bhargava, R.S. Mor, K. Kumar, V.S. Sharanagat, Advances in application of ultrasound in food processing: A review, Ultrasonics Sonochemistry. 70 (2021) 105293. https://doi.org/10.1016/j.ultsonch.2020.105293.
61. H.M. Kyllönen, P. Pirkonen, M. Nyström, Membrane filtration enhanced by ultrasound: a review, Desalination. 181 (2005) 319–335.
62. M.T. Grossner, J.M. Belovich, D.L. Feke, Transport analysis and model for the performance of an ultrasonically enhanced filtration process, Chemical Engineering Science. 60 (2005) 3233–3238.
63. H.W.D. Camara, H. Doan, A. Lohi, In-situ ultrasound-assisted control of polymeric membrane fouling, Ultrasonics. 108 (2020) 106206. https://doi.org/10.1016/j.ultras.2020.106206.
64. K. Li, L. Fu, Y.-Y. Zhao, S.-W. Xue, P. Wang, X.-L. Xu, Y.-H. Bai, Use of high-intensity ultrasound to improve emulsifying properties of chicken myofibrillar protein and enhance the rheological properties and stability of the emulsion, Food Hydrocolloids. 98 (2020) 105275. https://doi.org/10.1016/j.foodhyd.2019.105275.
65. H. Wu, G.J. Hulbert, J.R. Mount, Effects of ultrasound on milk homogenization and fermentation with yogurt starter, Innovative Food Science & Emerging Technologies. 1 (2000) 211–218.
66. Y. Yamini, A. Saleh, M. Rezaee, L. Ranjbar, M. Moradi, Ultrasound-assisted emulsification microextraction of various preservatives from cosmetics, beverages, and water samples, Journal of Liquid Chromatography & Related Technologies. 35 (2012) 2623–2642.
67. M. Bahmaei, H.A. Mashayekhi, F. Khalilian, M. Bahmaei, H.A. Mashayekhi, F. Khalilian, Application of ultrasound-assisted emulsification microextraction

based on applying low density organic solvents to the extraction and determination of mononitrotoluenes in water samples, Journal of the Brazilian Chemical Society. 26 (2015) 1475–1481.
68. M.M. Delgado-Povedano, M.D. Luque de Castro, Ultrasound-assisted analytical emulsification-extraction, TrAC Trends in Analytical Chemistry. 45 (2013) 1–13. https://doi.org/10.1016/j.trac.2012.12.011.
69. M. Kalhor, N. Hamzehpour, E. Ghashghaei, M. Faridi, H. Kalhor, Ultrasound-assisted emulsification-microextraction coupled to ion mobility spectrometry for monitoring of atrazine and simazine in environmental water, Int. J. Ion Mobil. Spec. 19 (2016) 139–144. https://doi.org/10.1007/s12127-016-0193-4.
70. J. Regueiro, M. Llompart, C. Garcia-Jares, J.C. Garcia-Monteagudo, R. Cela, Ultrasound-assisted emulsification-microextraction of emergent contaminants and pesticides in environmental waters, J Chromatogr A. 1190 (2008) 27–38. https://doi.org/10.1016/j.chroma.2008.02.091.
71. S.J. Kowalski, D. Mierzwa, M. Stasiak, Ultrasound-assisted convective drying of apples at different process conditions, Drying Technology. 35 (2017) 939–947.
72. S.U. Kadam, B.K. Tiwari, C.P. O'Donnell, Effect of ultrasound pre-treatment on the drying kinetics of brown seaweed *Ascophyllum nodosum*, Ultrasonics Sonochemistry. 23 (2015) 302–307.
73. E. Souza da Silva, S.C. Rupert Brandão, A. Lopes da Silva, J.H. Fernandes da Silva, A.C. Duarte Coêlho, P.M. Azoubel, Ultrasound-assisted vacuum drying of nectarine, Journal of Food Engineering. 246 (2019) 119–124.
74. W. Lv, H. Lv, X. Jin, Z. Cui, D. Su, Effects of ultrasound-assisted methods on the drying processes and quality of apple slices in microwave drying, Drying Technology. 38 (2020) 1806–1816.
75. Z.H. Tekin, M. Baslar, The effect of ultrasound-assisted vacuum drying on the drying rate and quality of red peppers, J Therm Anal Calorim. 132 (2018) 1131–1143. https://doi.org/10.1007/s10973-018-6991-7.
76. A. Mulet, J.A. Cárcel, J.V. García-Pérez, E. Rivera, Ultrasound-assisted hot air drying of foods, in: H. Feng (Ed.), Ultrasound Technologies for Food and Bioprocessing, Springer Science+Business Media, LLC, 2011: pp. 511–534. https://doi.org/10.1007/978-1-4419-7472-3_19.
77. E. Alizadeh, N. Chapleau, M. de-Lamballerie, A. Le-Bail, Impact of freezing process on salt diffusivity of seafood: Application to salmon (*Salmo salar*) using conventional and pressure shift freezing, Food Bioprocess Technol. 2 (2009) 257–262.
78. M.D. Alvarez, C. Fernández, W. Canet, Oscillatory rheological properties of fresh and frozen/thawed mashed potatoes as modified by

different cryoprotectants, Food Bioprocess Technol. 3 (2008) 55. https://doi.org/10.1007/s11947-007-0051-9.
79. R. de Paula, R. Colet, D. de Oliveira, E. Valduga, H. Treichel, Assessment of different packaging structures in the stability of frozen fresh Brazilian toscana sausage, Food Bioprocess Technol. 4 (2011) 481–485. https://doi.org/10.1007/s11947-010-0350-4.
80. A. Le-Bail, C. Nicolitch, C. Vuillod, Fermented frozen dough: Impact of pre-fermentation time and of freezing rate for a pre-fermented frozen dough on final volume of the bread, Food Bioprocess Technol. 3 (2010) 197–203. https://doi.org/10.1007/s11947-008-0142-2.
81. M.E. Steffolani, P.D. Ribotta, G.T. Perez, M.C. Puppo, A.E. León, Use of enzymes to minimize dough freezing damage, Food Bioprocess Technol. 5 (2012) 2242–2255. https://doi.org/10.1007/s11947-011-0538-2.
82. B. Xu, M. Zhang, B. Bhandari, X. Cheng, Influence of power ultrasound on ice nucleation of radish cylinders during ultrasound-assisted immersion freezing, International Journal of Refrigeration. 46 (2014) 1–8. https://doi.org/10.1016/j.ijrefrig.2014.07.009.
83. Z. Zhang, D.-W. Sun, Z. Zhu, L. Cheng, Enhancement of crystallization processes by power ultrasound: Current state-of-the-art and research advances, Comprehensive Reviews in Food Science and Food Safety. 14 (2015) 303–316. https://doi.org/10.1111/1541-4337.12132.
84. L. Otero, M. Martino, N. Zaritzky, M. Solas, P.D. Sanz, Preservation of microstructure in peach and mango during high-pressure-shift freezing, Journal of Food Science. 65 (2000) 466–470. https://doi.org/10.1111/j.1365-2621.2000.tb16029.x.
85. A.E. Delgado, D.-W. Sun, Heat and mass transfer models for predicting freezing processes – a review, Journal of Food Engineering. 47 (2001) 157–174. https://doi.org/10.1016/S0260-8774(00)00112-6.
86. Y. Tao, D.-W. Sun, Enhancement of food processes by ultrasound: a review, Crit Rev Food Sci Nutr. 55 (2015) 570–594.
87. R. Chow, R. Blindt, R. Chivers, M. Povey, The sonocrystallisation of ice in sucrose solutions: primary and secondary nucleation, Ultrasonics. 41 (2003) 595–604.
88. M. Saclier, R. Peczalski, J. Andrieu, A theoretical model for ice primary nucleation induced by acoustic cavitation, Ultrasonics Sonochemistry. 17 (2010) 98–105.
89. R. Chow, R. Blindt, R. Chivers, M. Povey, A study on the primary and secondary nucleation of ice by power ultrasound, Ultrasonics. 43 (2005) 227–230.
90. M.H. Zamanipoor, R.L. Mancera, The emerging application of ultrasound in lactose crystallisation, Trends in Food Science & Technology. 38 (2014) 47–59.

91. Y. Xin, M. Zhang, B. Adhikari, The effects of ultrasound-assisted freezing on the freezing time and quality of broccoli (*Brassica oleracea* L. var. botrytis L.) during immersion freezing, International Journal of Refrigeration. 41 (2014) 82–91.
92. Y. Li, Y. Zhang, X. Liu, H. Wang, H. Zhang, Effect of ultrasound-assisted freezing on the textural characteristics of dough and the structural characterization of wheat gluten, J Food Sci Technol. 56 (2019) 3380–3390.
93. G. Song, S.Q. Hu, L. Li, P. Chen, X. Shen, Structural and physical changes in ultrasound-assisted frozen wet gluten, Cereal Chemistry. 86 (2009) 333–338.

9
Irradiation of Food

Monalisa Sahoo[1]*, Pramod Aradwad[2], Chirasmita Panigrahi[3], Vivek Kumar[1] and S. N. Naik[1]

[1]*Centre for Rural Development and Technology, Indian Institute of Technology Delhi, New Delhi, India*
[2]*Division of Agricultural Engineering, Indian Agricultural Research Institute, New Delhi, India*
[3]*Agricultural and Food Engineering Department, Indian Institute of Technology Kharagpur, West Bengal, India*

Abstract

Application of ionizing irradiations to various food commodities for improving microbial safety and product quality is one of the emerging technologies in industrial food sector. Irradiation of food is considered as a technique or process for food preservation and microbial mitigation. Food irradiation has very diverse application, from inhibition of sprouting to extending shelf life of meat, fish, egg, seafood, fruits & vegetables, milk & dairy products etc and commercially produced sterile food products. For consumer safety and health (toxicological and microbiological safety and nutritional suitability of irradiated foodstuffs) International standards and legislations are being formulated and recommended by World Health Organization (WHO), Food and Agricultural Organization (FAO), International Atomic Energy Agency (IAEA) and other scientific committees. National legislations of more than 60 countries worldwide are approved specific applications of food irradiation. Ionizing radiation includes high energy gamma rays, accelerated electron beams, and X-rays. Electron spin resonance (ESR) spectroscopy methods for food containing bone and food containing cellulose, thermoluminescence (TL) for spices and herbs, and chemical methods based on isolation and detection of hydrocarbon and cyclobutanones in food containing fat are five detection method adopted by European Committee for Standardization (CEN). The process and technique details for industrial application of irradiation and commercialization of food products, safety, limitations and hurdles during irradiation and their effects in food stuffs are discussed in detail in this chapter.

**Corresponding author*: saidurga48@gmail.com

Navnidhi Chhikara, Anil Panghal and Gaurav Chaudhary (eds.) Novel Technologies in Food Science, (333–374) © 2023 Scrivener Publishing LLC

***Keywords*:** X-ray irradiator, electron-beam irradiation, gamma rays, wholesomeness, microbial inactivation, sanitation

9.1 Irradiation

Food irradiation is the application of ionizing radiation to food commodities to achieve the desired purpose by ionizing the food molecules or atoms. It is a non-thermal technology also known as cold pasteurization. Interaction between food and ionizing radiation is a complex process and depends upon size, physical state and type of food. Application of radiations is food specifics and cannot be altered as to maintain the quality and organoleptic characteristics of food [1].

More than 55 countries use irradiation technique for different applications such as disinfection, decontamination and preservation of food and food products, control of insects, pests and plant diseases [2–4]. Irradiation is effective in delaying of fruit and vegetable ripening and reducing the germination of tubers and bulbs [5]. Food irradiation is a technique which has the ability to destroying and reduction of insects, parasites and non-spore forming microorganisms. Thus, prevention of foodborne diseases.

However, irradiation of food is decreasing significantly in European Union, increasing considerably in Asia and Latin American countries and increasing gradually in United Nations and Australia [6, 7].

The main aim of this book chapter is to provide basic fundamental knowledge and overview of irradiation technology and its several applications to the food commodity. Moreover, this chapter also provides the recent development and advancement in food irradiation and future scope.

9.1.1 Sources of Radiation

X-rays, gamma rays, UV rays (ultra violet), alpha particles, neutrons and electrons are different types of ionizing radiations are available [8]. However, X-rays, gamma rays and high energy e-beams are limited to food processing applications. Gamma rays and X-rays are short wave length, high penetrating power (high frequency) and exhibits same properties and effects with different source of origin [9].

Cobalt-60 (^{60}Co) and Cesium-137 (^{137}Cs) radio isotopes are the main sources of gamma irradiation, while Cobalt-60 is the most commonly used in food industries because of its high energy and long life (5.27 years half-life) [1]. On the contrary application of Cesium-137 is limited to insect sterilization and irradiation of blood due to its high solubility in water.

The interaction between food and high energy electrons is illustrated in the Figure 9.1. These electrons can pass directly through the samples

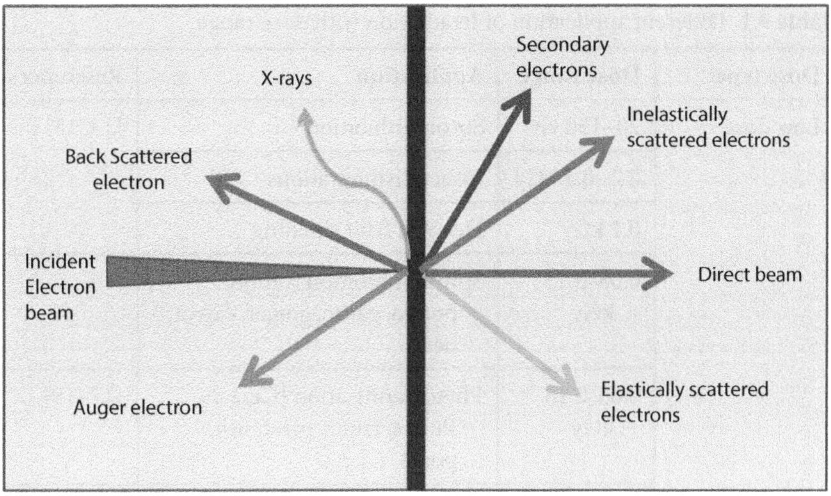

Figure 9.1 Interaction between electrons and matter.

without any interaction (elastic interaction) or they may interact too produce lower energy particles, going through different directions. This happens due to inelastic scattering and energy absorption. Backscattering electrons and augur electrons are produced by secondary electrons and produce further electrons unless the energy is completely dissipated.

9.1.2 Dose Range & Dose Mapping

Gray (Gy) is the unit of radiation dose and one gray is 1 Joule of absorbed energy per kg of radiation units (International Commission on Radiation Units). Different level of irradiation doses are applied to food and agricultural produces and are categorized as follows:

Low dose irradiation of food: Dose < 1kGy
Medium dose irradiation of food: Dose range between 1–10 kGy
High dose irradiation of food: Dose > 10 kGy

Different applications of irradiation includes inhibition of sprouts, insect disinfestations, ripening delay, sanitation of eggs, reduction and inactivation of pathogens [10] and the dose range is mentioned in Table 9.1. Moreover, the doses recommended by Food and Drug Administration (FDA) for various food commodities are mentioned in Figure 9.2.

Table 9.1 Different application of Irradiation with dose range.

Dose type	Dose range	Application	References
Low dose	20–150 Gy	Sprout inhibition	[14, 15]
	0.2 -0.4 kGy	Insect disinfestations	
	0.1 kGy	Delay in fruit ripening	
	0.05-0.15 kGy	Sprout inhibition (onion, potato, garlic, ginger, carrot, beet)	[16]
	0.05-0.15 kGy	Phyto-sanitization (Cereals, Pulses, fruits, meat, fish, pork)	[17–19]
	0.25-1 kGy	Ripening delay (fruits & vegetables)	[20–23]
Medium dose	1–7 kGy	Pathogen control & shelf life extension of meat, fish and dairy products	[8, 15]
	1 kGy	Pathogen control in fresh produces	
	1-3 kGy	Shelf life extension of meat, fish, celery, corn	[24–27]
	1-7 kGy	Disinfection of frozen products (meat, poultry and seafood)	[28]
	2-7 kGy	Dehydration of produces	[29]
High dose	44 kGy	Sterilization of astronaut food	[30]
	25–45 kGy	Sterilization of vacuum packed muscle foods	[8, 31]
	10–30 kGy	Microbial decontamination of dried herbs and spices	[14, 32]
	20-40 kGy	Industrial sterilization of poultry, meat and seafood	[33]
	30 kGy	Decontamination of spices, gums	[34]

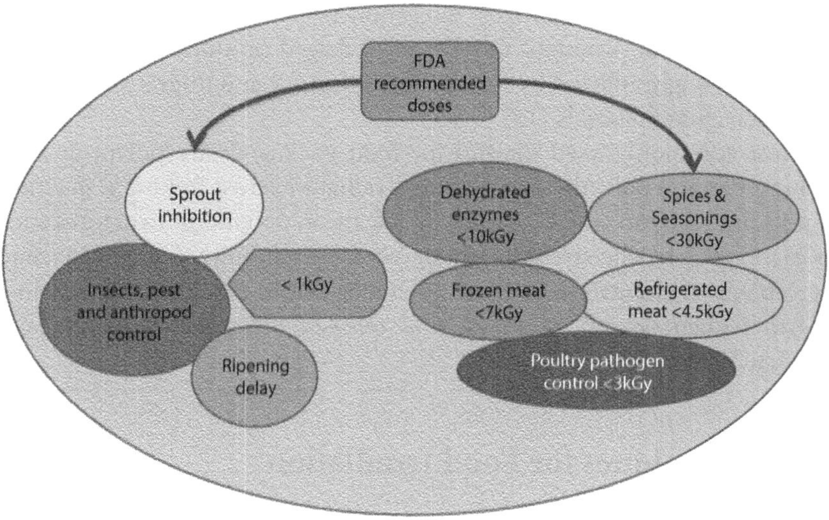

Figure 9.2 FDA recommended doses for irradiation.

9.1.3 Packaging Material for Irradiation

Radiation technology is broadly used for sterilization medical products and irradiation of food. Irradiation of packaged material is more effective and advantageous as compared to the chemical treatments. The vital applications of irradiation on polymeric materials are: (i) food and food packaging irradiation and (ii) sterilization of medical syringes, disposables and tubing [11]. In addition, irradiation has turn out to be a standard treatment in aseptic processing for sterilizing the packages for food and pharmaceuticals. In contrast to medical device and pharmaceutical industries, dairy industries (producing butter, cream etc.), processed food industries (salad dressing, sauces, fruit gels and processed meat) and beverages industries (fruits juices and wine) irradiate packaging materials separately before filling out them.

Natural or synthetic packaging materials are used as packaging where plastic is most commonly used packaging type. Thus, the effect of irradiation on packaging material is key factor in packaging engineering [12]. Free radicals or ions are generally formed after irradiation of packaging material. This ultimately results either in chain scission (decrease in molecular weight) or cross linking (increase in molecular weight). Volatile radiolysis products are released in chain scission which may induce off smells in the polymers by changing the migration functionalities of packaging materials. Polymer additives are also affected by the irradiation, which changes the polymer additives' migration behavior and additive related decomposition of products. Quality of packaged products and consumer safety is strongly affected by the change is sensory and migration of sterilized packaging materials [13].

Most commonly used plastics for food packaging are polypropylene, polystyrene and polyethylene. During irradiation, cross linking is the most significant reaction which has several effects on polymeric structures such as increase in tensile strength and decrease in solubility, crystallinity, and percentage of elongation (%E) of polymers. But in case of chain scission, polymer chain length decreases and releases methane, hydrogen, and hydrogen chloride for chlorine.

9.2 Techniques for Food Irradiation

Gamma rays irradiators, electron beam accelerators and X-rays irradiators are the three different radiation technologies which are used in food irradiation [35]. Only these ionizing rays have been authorized to be used for irradiation applications in food [35]. Irradiation technique is selected on the basis process time, quantity of dose, cost of operation and irradiation purpose (eg. disinfestation, inactivation, etc.) and different attributes of food such as density, size, shape and thickness. For example, e-beam is suitable for treatment of surface contamination and thin surfaces such as meat and vegetable thin slices while gamma, x-ray are suitable for thick surfaces and processed foods such as caresses and pallets [8, 36]. The advantages and disadvantages of various irradiation technologies are presented in Table 9.2.

9.2.1 Gamma Rays Irradiators

Gamma rays are produced with the breakdown of radionuclides Cobalt-60 (^{60}Co) or Cesium-137 (^{137}Cs) sources into radioactive substances (called radioisotopes). They have energy levels of 1.33 MeV and 0.662 MeV respectively. Irradiation of ^{59}Co targeted with fast neutron to produce radioisotopes in the nuclear power plant under the properly protected dry or underwater environment [37–39]. Cobalt-60 is artificial radioisotope with high energy and half-life (5.27 years) and most commonly used source in irradiation of foods while Cesium-137 has been restricted to use only in sterilization of insect and blood irradiation [40]. A irradiation facility includes an irradiation source, source panel, a shielded storage room (dry or wet), a radiation shield around the irradiation room, a control

Table 9.2 Advantages and disadvantages of different irradiation techniques.

Techniques	Advantages	Disadvantages	References
Gamma rays irradiators	▪ Reliable irradiation facilities and simple operation. ▪ Gamma-rays have a high penetration depth and dose uniformity ▪ Can be used to treat food in large packaging units.	▪ Shielding can be expensive ▪ The process is relatively slow and has relatively low dose rate ▪ The radionuclide source (^{60}Co "pencils") require frequent replenishment ▪ A relatively complex and highly regulated process with high acquisition costs	[41, 50]
Electron beam accelerators	▪ More reliable and environmentally acceptable than the gamma rays' facilities ▪ Low cost and limited space needed ▪ Fast process time ▪ Variability of electron-beam current and energy to provide flexibility in the choice of surface and depth treatments for a variety of food items, and ▪ Machine can be turned off or on	▪ Lower penetration capability of e-beams and the need for more shielding when using higher energy e-beams. ▪ Higher requirements for power and cooling. ▪ More sensitive to breakdown and the need for specialized personnel for regular maintenance.	[8, 36, 50, 51]

room, product containers, a product transport and source transition system, a control and safety interlock system, warehouse, areas for loading and unloading of products, automatic control system and equipment services [9, 41]. Four types of product transport systems are of four types: tote box, carrier, pallet carrier, and pallet conveyor systems [42]. Radioactive cobalt-60 is made up of small cobalt pellets and encapsulated into zircaloy or stainless steel sheath tubes. This encapsulated pencil array positioned in concentric rings to from symmetrical beam distributed uniform doses in irradiation chamber [37, 38]. This gamma radiation energy is transferred from the tubes to the food as a result of collisions between the food atoms and the radiation. Here γ-radiation source (^{60}Co) is stored in water reservoir and when it travels on overhead rail into the chamber, these sources are raised above the water reservoir for irradiation of product. Thick steel or concrete roofs and walls protect the workers while the irradiation source is picked up from the reservoir. After completion of irradiation process, sources are dropped down into reservoir. Exposure of product to less than 10 kGy γ-irradiation is considered an effective means in extending the shelf life of food thereby increasing the food safety through killing pathogenic and spoilage microbes [39]. These doses considered to be safe as it does not transmit radioactivity or residual effect in food. Material thickness is an important parameter for complete penetration of radiation and can be achieved. This process not only saves energy but chemical or heat free treatment to the product.

Nordion irradiator, GrayStar Genesis irradiator and Sterigenics irradiators are successfully designed and commercially available food irradiators. Comprehensive information and safety standards of irradiation are given in different literatures [9, 41, 43, 44].

9.2.2 Electron Beam Accelerators

E-beam accelerators along with a scanning and exhaust system are used for removal of ozone from the environment and a material handling system (conveyor) is a part of an accelerator irradiation facility [36]. In E-beam irradiation, Food should be kept vertical in direction on conveyor belt for better stability of operation process. These facilities require product handling and shielding equipment similar to ^{60}Co γ-ray likewise but some covering requirement may be reduced due to orientation of irradiation fields produce by machine source. The companies like Scan Tech Sciences patented electronic cold pasteurization and IBA Industrial solution built continuous wave Rhodotron® e-beam accelerator.

Different acceleration methods are used to generate electron beams of high-power and energy. Accelerator is selected for specific application is based on the required process for beam power and energy of electron.

9.2.2.1 *Direct Methods*

In direct method, accelerating field is excited by generating high potential gradient across a graded insulated column also serves as vacuum interface. This is used for producing electron beam of high kinetic energies as a million eV at a speed almost equal to that of light and quantum energy should not exceed 10 MeV [8]. Direct acceleration methods include dynamitrons, resonant transformers, transformer-rectifier units, and various electrostatic approaches including the Pelletron and the Van de Graaff accelerator.

9.2.2.2 *Induction Methods*

The approach of induction method is to pass a beam via reasonably smaller accelerating sections many times. These accelerating increments are induced by varying magnetic field with time. Betatron and linear accelerator are two practical cases of induction acceleration principle [8]. Betatron accelerators have certain benefits like low cost, simple operation, reliable and lack of high-frequency accelerating systems over linear accelerator [45].

9.2.2.3 *Microwave or Radio-Frequency Methods*

The acceleration originates from electromagnetic fields crated in resonant cavities structure [8]. Accelerating structure consists of single or multiple associated cavities and the beam can pass either one time or multiple times. In Microwave accelerator more common approach is beam passed through series of linear accelerating electromagnetically coupled cavities.

9.2.3 X-Rays (Bremsstrahlung) Irradiators

X-ray is an electromagnetic radiation of short wavelength and high frequency. They are instigated by transition of atomic electron, and less energetic compare to γ-rays. γ and X rays have identical properties and similar effects on food material but their origin is the key dissimilarity between them. Like γ-rays, X-rays can penetrate through dense foods and need heavy shielding for safety. X-ray generator produces varying degrees of X-ray energies and their generation efficiency is somewhat lower. The similarity with e-beams is that non-radioactive substances are involved and the machine can be switched on and off.

The most vital method of making X-rays is based on a process known as bremsstrahlung ("braking radiation"). X-rays are produced when charged particles (or electrons) moving with a very high velocity, are slowed down rapidly by striking a target. A beam of high-velocity electrons is directed at a thin plate of heavy metal (like tungsten or tantalum), producing a stream of X-rays. The fraction of kinetic energy of the electrons that is converted into X-rays, called 'conversion efficiency' is higher for absorbers with a higher atomic number which are used as X-ray converters. The conversion efficiency increases with increasing electron energy and maximum quantum energy should not exceed 5 MeV. In tungsten, at 5 MeV to 10 MeV, it increases approximately 5% to nearly 12.5% [46]. Unlike radionuclide sources, which emit monoenergetic photons, X-ray sources radiate a wide spectrum of photons ranging from the maximum energy posed by the electrons to zero energy. For example, an X-ray beam produced by 5 MeV electrons has around the same penetration characteristics as ^{60}Co radiation. Although the penetration ability is almost same for both X-rays and γ-rays, the production of X-rays is not at present economical [47]. Other features of the X-ray beam, such as pulsing and scanning of the beam result from the properties of the e-beam that created the X-rays [48].

For treatment of food, X-ray machines must work at an energy level of 5 MeV or below. This restriction is because of preventing the induced radioactivity. X-ray production is comparatively inefficient, but it can be competitive with γ-radiation for high-capacity plants. The option of using e-beams and X-rays in the same irradiation facility is even attractive [42, 49]. X-rays or decelerating rays, can be similarly produced in accelerators. In Rhodotron accelerators, fraction of energy can converted into X-ray. In X-ray irradiator, energy is horizontally directed so the arrangement of packages in such way to full fills the horizontal depth and scanning width. The desired dose is achieved by operating the conveyor at the appropriate speed [36]. Compared to other techniques available for food as discussed, gamma ray is most extensively used followed by e-beam. Moreover, e-beams can be converted to X-rays in order to penetrate thick products. In practice, it is desirable to irradiate maximum number of foods using the electron beams [36]. While there is a rising demand for X-rays, still most of the food items are irradiated with high-energy γ-rays (^{60}Co).

9.3 Wholesomeness of Irradiated Foods

The safety assessment of wholesomeness irradiated food can be broken down into four safety aspects as:

- Radiation safety
- Toxicology safety
- Microbiological safety
- Nutritional acceptability

All the scientific results and data regarding the wholesomeness of irradiated food products is governed by the experts of different countries, World health Organization, International atomic energy agency, and food and agricultural organization. According to the joint expert committee (FAO/WHO/IAEA), irradiation of food with a maximum dose of 10 kGy doesn't possess any toxicological hazards including no nutritional and microbial problems [52]. Therefore, the maximum limit of overall dose of irradiation adopted by Codex Alimentarius Commission is 10 kGy [53].

9.4 Application of Irradiation on Different Food Commodities

Food irradiation is known to be among the few technologies that address both food quality and safety by virtue of its capability to destroy spoilage and foodborne pathogenic microorganisms without considerably affecting the food attributes. Application of ionizing radiations to food is the most emerging research domain of twentieth century. It has several potential applications and critical advantages as no chemical residues, cold treatment, and ease in processing packaged food, reproducible and controlled process. Foods are irradiated to provide analogous outcomes and advantages as after being processed by heat, chemical treatment, freezing or drying. Although irradiation is more beneficial than conventional physical and chemical treatments, still its commercial application is limited [5]. However, the fast growth in use of irradiated foods in recent years in could be accounted that nearly 1 million tons per annum of irradiated foods are produced globally [5, 54].

9.4.1 Sanitation and Decontamination

Minimal processing of fruits and vegetables to reduce decontamination and food borne illness by reducing pathogens by food irradiation is the most promising commercial techniques [55]. Irradiation is more advantageous to traditional sanitization processes by reaching internal pores and crevices where pathogens attack. A tolerable dose is sufficient to reduce 3-5 log cycle of commonly found pathogens (*E. coli, Salmonella spp., S. aureus, L. monocytogenes*) with slight impact on nutritional quality [56].

Minimal processing of fruits and vegetables by irradiation is the most promising commercial approach aiming to reduce decontamination and food borne pathogens [56]. Radiation doses in meat, fish and poultry (1–10 kGy), in dairy products (1–7 kGy) and fresh produce (1 kGy) are applied to control pathogenic microbes and extension of shelf life of products [15]. Irradiation is proved more advantageous to traditional sanitization methods as it reaches the internal pores and crevices where pathogens usually attack. An allowable dose is sufficient to reduce 3-5 log cycle of commonly found pathogens (*E. coli, S. aureus, Salmonella spp., L. monocytogenes*) with minimal impact on nutritional quality [56]. Alternative to the use of ethylene oxide fumigation, irradiation (10–30 kGy) can be used as effective method for microbial decontamination in herbs, spices and vegetable seasonings [14] and some Korean food products [32]. Radiation sterilization in the dose range 25-70 kGy extends the shelf life of precooked or enzyme inactivated food products in hermetically sealed containers. 30-50 kGy dose can be useful for industrial sterilization (radiation in combination with mild heat) of poultry, meat, seafood, prepared or cooked foods and hospital diets [14]. In recent years, many research have been carried out for pathogen reduction in fruits and vegetables applying different doses of radiations are presented in Table 9.3.

9.4.2 Sprout Inhibition and Delay in Ripening

Irradiation dose of <1 kGy, also called phytosanitary irradiation is mainly used in inhibition of sprouting in potato and onions, delay fruit ripening and disinfestation [10]. In phytosanitary irradiation, it is applied as a control measure rather than sterilization. Sprouting can be inhibited by applying irradiation in the dose range 20–150 Gy [15]. Low level of irradiation dose inhibits sprouting in tubers and roots and enhances their shelf life. [16] have stated that, 16-ton garlic and 115 ton onion were treated in India at a dose of 60 Gy and in Japan, 5 kiloton of potato was irradiated for the purpose of sprout inhibition. The delay of physiological processes is

Table 9.3 Application and effects of irradiation on food products.

Product	Irradiation techniques	Doses	Targeted microbes	Log reduction	Significance	References
Almonds dates and wheat	X-ray	0.5 - 11 kGy	*Salmonella Enteritidis* PT30	2.5	Food structure water activity of product and storage period affects the efficacy of X-ray	[58]
RTE sliced ham	X-ray	0.2-0.8 kGy	*S. Typhimurium*, *L. monocytogenes* & *E. coli* O157:H7	5.7, 6.9 & 7.2	Effectively used in RTE based products without altering textural and color properties	[59]
Sliced cheese	X-ray	0.2, 0.4, 0.6 & 0.8 kGy	*E. coli* O157:H7, *S. enterica*, *Salmonella typhimurium*, & *L. monocytogenes*	5 (0.6 kGy)	Act as a antimicrobial intervention for packaged dairy products	[60]
Apple juice	X-ray	1.0 kGy	Non-acid-adapted and acid-adapted *Escherichia coli* O157:H7 and *Salmonella Typhimurium*	8.02 and 5.48 log CFU/mL 3.87 and 3.98 log CFU/mL	Significant lethal effect on acid-adapted cells of *E. coli* O157:H7	[61]

(Continued)

Table 9.3 Application and effects of irradiation on food products. (*Continued*)

Product	Irradiation techniques	Doses	Targeted microbes	Log reduction	Significance	References
lettuce leaves	X-ray and gallic acid	0.3 kGy and 0.5% GA	*E. coli* O157:H7, *S. Typhimurium*, and *L. monocytogenes*	5.41, 2.57, and 1.36 log CFU/cm^2	Combined treatment effective against *E. coli* O157:H7	[62]
Frozen green peppers	Gamma	0.5 kGy	*E. coli, S. enterica* and *L. innocua*	4 l, 3.6, 2.2 log CFU/g	Most effective method compare to edible coating, ozonation	[63]
Ground chicken meat	Gamma	0.061 kGy/min	*E. coli*	D_{10}- 0.18–0.61 kGy	Control ExPEC in poultry meat	[64]
Meat	Gamma	6 kGy and 8 kGy	*Shigella* spp., *Staphylococcus* spp., *Salmonella* spp., *Vibrio* spp., *Pseudomonas* spp., & *Listeria* spp.	4 (6 kGy), and 6 (8 kGy)	Effective against multi drug resistant strains	[65]

(*Continued*)

Table 9.3 Application and effects of irradiation on food products. (*Continued*)

Product	Irradiation techniques	Doses	Targeted microbes	Log reduction	Significance	References
Apple juice	Gamma	200, 400, 600 and 800 Gy	*E. coli* O157: H7	>3 (>600 Gy)	significant reduction in *E. coli* O157:H7 without affecting sensory quality	[66]
Hummus	Gamma and refrigerated storage	2.5 kGy	Lactic acid bacterial and yeast and mould	<1 log_{10} cfu/g and 1.7 log_{10} cfu/g	Extend shelf life upto 35 days	[67]
Cucumber	Electron beam	0.2 to 1.0 kGy	*Salmonella Poona*	3.05 log CFU/g	No negative effect on quality parameters	[68]
Raspberries	Electron beam	3 kGy	Mesphilic bacteria, filamentous fungi and *Salmonella Typhimurium*, *Escherichia coli* and *Listeria monocytogenes*	2, 3 and 3 log CFU/g	Preserve phenolic content and antioxidant activity and decrease in more than 80% ascorbic acid	[69]

(*Continued*)

Table 9.3 Application and effects of irradiation on food products. (*Continued*)

Product	Irradiation techniques	Doses	Targeted microbes	Log reduction	Significance	References
Cherry tomato	Electron beam	3.6 kGy	Mesophilic bacteria	4	Preserved the lycopene and antioxidant activity of extracts.	[70]
Duck meat	Electron beam	3 and 7 kGy	TAB, Y&M, and coliform	3 and 6 log	Improvement of microbiological quality.	[71]
Sausages	Electron beam	2 kGy	*Salmonella typhimurium*	<1 logMNP/gr	reduce bacterial population without affecting sensory properties	[72]

mainly achieved when irradiated at doses of 0.25-1 Gy. Fruit ripening can be delayed even with 0.1 kGy dose [31].

9.4.3 Insects and Pest Control

Disinfestation of insect is generally accomplished by applying doses in between 0.2 - 0.4 kGy [15]. Irradiation dose range of 0.2-1 kGy is aimed at preventing losses caused by insect pests in stored grains, cereals, pulses, spices, coffee beans, dried nuts, dried fruits, dried fishery products and other dried foods [14]. The inactivation of some pathogenic parasites of public health significance such as tapeworm and trichina in meat can be achieved at doses in the range 0.3-1 kGy. In fruits and vegetables, minimum irradiation dose of 150 Gy is required to provide quarantine security from fruit flies, while dose of 300 Gy could prevent entering other insect species in non-infested areas [57]. Maximum 500 Gy is required for some spices [10]. However, these doses can guarantee near-complete mortality of all insects and life stages seem to have adverse effects on the sensory properties of most produce.

9.5 Advantages and Disadvantages of Irradiation of Food

Irradiations of food possess both merits and demerits which are well explained below.

9.5.1 Advantages of Food Irradiation

There are several potential advantages in irradiation of food. These come under the conduct of Good Manufacturing Practices (GMP) [73–78]. Merits and benefits of irradiation technology and irradiated food include:

1) Improved sanitation
Radiation is an effective way to control disease and epidemics by controlling foodborne microorganisms such as Campylobacter, E. coli, Vibrio, Salmonella, L. monocytogenes and Toxoplasma, thus making food safer. Decreasing the risk of food borne diseases to the consumers is the most essential advantages of irradiation processing technology. This is done by inactivation of food borne pathogenic microorganisms. Raw and frozen state of fruits, vegetables, cereal grains, and legumes, ready to eat products, meats and meat products, and seafood are preserved by this method.

Decontamination of food containing insects, molds, yeasts and pathogenic bacteria improves the hygienic quality of food material, thus prevents potential health hazards. Bacteria and parasites in meat and seafood, insects in fruits, legumes, cereals and dried fish, insects and bacteria in vegetables and seasonings, and spices can be decontaminated by applying specific irradiation doses.
2) Wide spectrum quarantine processing
Sensory quality and nutritional values of the food is maintained. Less or insignificant loss of vitamins, changes in flavor, color, and texture occurs as compared with the other conventional or novel preservation techniques.
3) Packaged foods can be treated and both in fresh and frozen state.
4) No radioactivity is produced by treating with specific sources of radiation (X-rays, gamma rays, and e-beam) under prescribed conditions. Many people are unaware of this fact.
5) Enhancement of shelf life and food quality by delay in ripening and deterioration process, killing pests, thus escalating food supply.
6) Fresh-like quality of the agricultural commodities can be maintained as it is a cold processing technique.
7) Irradiation facilities are environmentally friendly and secure for workers and the public.

9.5.2 Disadvantages of Food Irradiation

Related issues regarding irradiation of food are as follows:

1) This radiation technique cannot be applied to all types of food products as the quality of some food commodities are degraded by applying radiations. Some examples are pears, plums, avocados and, cantaloupes.
2) Chemical changes are induced by irradiation in some food products which affects the sensory attributes of the food. Therefore, milk and dairy based products are not suitable for processing through irradiation technique as an off flavor is produced even at a dose of 0.1 kGy.
3) Softening and damaging of wall cells occurs in some fresh vegetables and fruits after radiation treatment.
4) Development of unpleasant color and off flavor is produced by deterioration of sensory properties of meat products after exposure to a high dose of radiation.

5) Irradiation results in the loss of fat and water soluble vitamins such as vitamin C, vitamin B12, riboflavin, thiamin, pyridoxine and vitamin E.
6) Irradiation produces certain products such as 2-alkylcyclobutanones that are considered toxic.
7) Oxidation compounds such as ketones, alcohols, aldehydes etc.
8) Irradiation cannot help in eliminating foodborne viruses and prions.
9) Reluctant perceptions towards radiation processing that a bad or spoiled food product can be look good.
10) High capital cost.
11) Industry, consumer and public resistance to nuclear technology (irradiation).

9.6 Factors Affecting Irradiation of Food

Effectiveness of irradiation treatment on food products depends on temperature, percentage of moisture content and atmosphere (oxygen concentration).

9.6.1 Water Content

Irradiation of foods with excessive moisture content, at higher temperature (>45 °C) and in aerobic environment, is an effective means of inactivation of microbes due to less radio resistance of microorganism [79]. Even if, small concentration of oxygen is present, it increases radio sensitivity of microbes. Irradiation treatment itself generate small amount of oxygen in the environment, especially during e-beam irradiation. During dry environmental processing conditions, pathogens are more resistant and therefore higher radiation doses required. Due to low or absent water radiolysis, it affects indirectly on pathogen DNA [79].

9.6.2 Temperature

Food is treated at given temperature may influence radiation resistance of microorganisms [80]. Irradiation of vegetative cells, at elevated temperature (>45 °C) synergistically increase the lethal effect. This enhancement is caused due to damage in microbial repair system, when it operates at ambient or slight above temperatures. General rule indicates, irradiation at low temperature, microorganism are less radiosensitive due to low water activity and immobilization of free radicals [81]. Alternatively, Irradiation of chilled fresh products (-5°C) packaged in the atmosphere with the presence of a high concentration of oxygen produces ozone, which enhances the killing effectiveness of process by increasing the radiation sensitivity of Salmonella and Listeria species [82, 83]. In irradiation process freezing temperature is affected to increasing resistance to vegetative cells.

In irradiation of frozen foods microbial resistance is 2-3 folds higher as compared with ambient temperature. This is attributed to loss of mobility of free radicals and prohibition of diffusion of frozen medium. Irradiation of food in frozen state prevents indirect damage of DNA by OH- radicals group. This observed change in microbial resistance with temperature indicates the importance of indirect action of radiation in presence of high moisture food [80].

Oxygen

Ionizing radiation, presence of oxygen increase lethal effect on the microbial cells [84]. In wet condition and in absence of oxygen, radiation resistance increases by about 2-4 folds, on the other hand in dry conditions radiation resistance increases by about 8-17 folds. The processing of irradiation in combination with vacuum and modified atmosphere packaging prolongs the quality of product and shelf life due to the absence or low concentration of oxygen in the packaging, limit the aerobic and oxidizing reactions [85, 86]. Combined treatments of these techniques are termed as hurdle approach. Cutting-edge nanotechnologies and edible film technology represent a promising approach to improving the food irradiation process [87–93]. Other methods such as chemicals, heat and cold temperature can be combined with irradiation treatment [94]. The complex nature of foods acts as a substrate for different microorganisms may affect the irradiation sensitivity of targeted microbes because it affect the dose quantity for specific application.

9.7 Interaction of Ionizing Radiation and Food Components

Low and medium range doses results minimal changes to the main components of food after exposure to ionizing radiation [95]. As in ordinary hydrolysis, carbohydrates are breakdown into simple compounds such as sugars, sugar acids and ketones. High dose of radiation results in softening of plant tissue and loss of texture. Lipid oxidation of unsaturated fats occurs when food is treated [96, 97]. Irradiation of vacuum or modified atmospheric packed foods could be an alternative to delay or reduce the oxidation process [98]. Similarly, minor breakdown of protein takes place. In fact, irradiated muscle foods preserves considerably more protein than the steam treated products [95].

Only vitamin B1 (thiamine) is the high sensitive micronutrients. Therefore, it is problem to irradiate high thiamine content foods such as pork meat. Other vitamins like vitamin C (ascorbic acid) lose their potency after interacting with generated radicals during irradiation [8]. But, use of low temperature and absence of oxygen in the irradiating process retains the potency of vitamins and antioxidants intact. Overall, the wholesomeness of food is maintained at an appropriate dose level during irradiation. Effect of irradiation technologies on food components are reviewed extensively by [99].

9.8 Interaction of Ionizing Radiation and Biological Cells

Food irradiation is targeted to eliminate the insects, mites, molds, yeasts, and bacteria in the food [8]. Deoxyribonucleic acid (DNA) molecules gets damage, when exposure to ionizing radiation, thus causing cell injury and death. The interaction of irradiation and cell nucleus is the main effect on biological cell materials [100]. The DNA of microorganism is damaged by two ways either directly hitting the double or single strands or water radiolysis in the food components. Virus is the most resistance to irradiation while, molds and vegetative bacteria are the least resistance. Gram negative bacteria are least resistance than the gram positive bacteria, among the vegetative bacteria. Moreover, yeasts and bacterial spores are more resistance to radiation than vegetative bacteria [101]. The quantity of does varies from 2.0 to 4.8 kGy to inactivate the biological material and kill different microorganisms.

Water molecule in free matter is ionizes into free radicals in secondary mode that is, indirect effect. These free radicals not only produce highly reactive hydroxyl (.OH) and hydrogen (H) radicals but also generated zone (O3) and hydrogen peroxide (H2O2) reactive radicals. Simultaneously, radiation inactivates other biomolecules such as enzymes present in cell. Thus efficacy of the irradiation treatment is enhances by irradiation damage. Comprehensive research work has been carried out on radiolysis of water and food irradiation chemistry and an insight overview presented by [99].

9.9 Interaction of Ionizing Radiation and Food Packaging Materials

To avoid potential food borne hazardous through contamination of perishable foods, irradiation of pre-packaged food is essential. Polymer, papers, coatings and incorporation of additives such as antioxidant are various types of approved packaging material by the Food and Drug Administration and other regulatory bodies. Among all the packaging materials mentioned above, polymer is the most common and widely used in packaging. All material is tested in methodical way to ensure its safety. However, there may be changes in the physical structure and properties of polymeric packaging materials. This may be due to the radiation induced changes such as chain incisions and cross linking when irradiated inappropriate or high doses [102]. Approved doses for packaging material is in between 4.5 and 10 kGy [103]. Therefore, it is necessary to follow the correct safety protocols for irradiating packaging materials. Effect of radiation on packaging material including edible films and active packaging had been comprehensively researched [95, 104].

9.10 Detection and Risk Assessment

9.10.1 Detection of Irradiation

Reliability of the commercial irradiated food can be achieved by identifying the food subjected to ionizing radiations by means of analytical methods. There are several validated methods for food category specific based on physical, chemical and biological changes induced by irradiation. Detailed analytical methods are illustrated in the Figure 9.3.

9.10.2 Risk Assessment of Irradiated Foods

Risk assessment is a whole process of hazard identification, hazard characterization, risk assessment and risk characterization [105]. Risk analysis is simply identifying and analyzing issues that could interfere with a set of particular assets or processes and their impact. It is divided into two

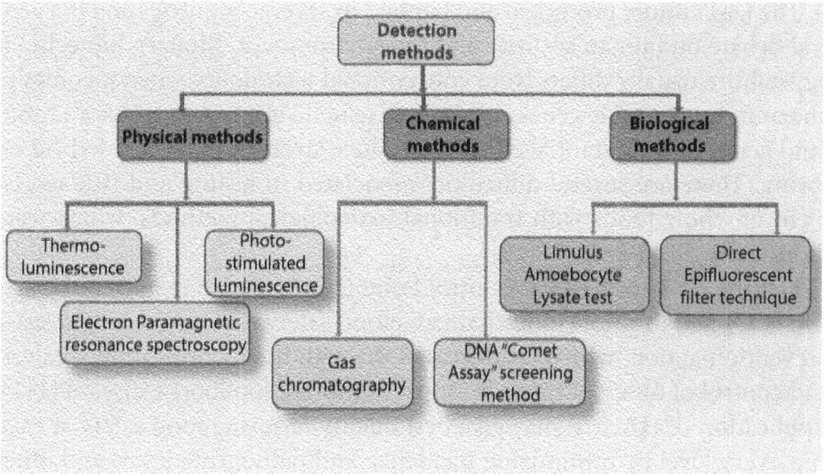

Figure 9.3 Classification of irradiation detection techniques.

primary types qualitative and quantitive. Qualitative risk analysis simply identifies the major threats to which it is exposed. On the other hand, quantitative risk analysis provides the direct correlation to assets values that need protection [106].

Food irradiation is well tested and established technique in accord with good manufacturing practice (GMP) and standard food laws and regulation. Several research studies have concluded that, food product produced through irradiation process can be consider safe and does not have negative effect on human health, as well as food composition cannot be change in terms of toxicological point of view. During irradiation, radiolytic products formed which contains very small quantities may cause health problems if it consumed in higher amount that actually found in irradiated food [52].

The change in linear energy transfer radiation may lead to induced instability in transformed human cells, but not in normal human cells [107]. This contrast should be further investigated due to the implications that induced instability may have a potential role in tumour development. Another form of effect retardation is an adaptive response, which influences the various levels of toxicity following radiation exposure, although the effect is shorter than genomic instability [108]. With respect to toxicity potential, 2-ACBs with Cyto- and genotoxic properties, not only inhibit bacterial growth, but the promoter in tumor development in the colon of rats has been shown to inhibit bacterial growth [109].

The GM culture promoted and banned in several countries and is a general debate on human security and the environment. Modern biotechnology culture usually differs from conventional techniques when it comes to one or more desirable genes. Traditional genetic improvement uses a thousand genes and substantial effort to produce strong and healthy hybrid off spring. There are several difficulties associated in testing and risk assessment of whole foods with traditional toxicological methods, which were thoroughly discussed by [110].

Hazard Analysis Critical Control Point (HACCP) is a management system is applied to whole food chain as a food safety tool food-borne disease prevention action, which is addressed through the identification, analysis and control of different hazards for prevention of foodborne disease [111]. Application of HACCP may give best quality ensuring food safety at each and every level by minimizing incidence and pathogenic levels and other means of threat to food [112].

The evaluation and safety of irradiated foods can be determined on the basis of product history. Monitoring of irradiation of product has commenced from demonstration phase to the commercial phase. Therefore, a monitoring system is required at every step to collect, compile and analyze irradiation data [113].

9.11 Consumer Behavior Towards Irradiated Food

For successful commercialization of a new technology or product it is always necessary to adopt the technology or product at every step of food chain, starting from industry to retailer and market to consumers through purchase it. Consumer uncertainty and food safety concerns of the treated foods have been always faced by the newly developed irradiated or non irradiated food products [114–117]. Consumer attitude and beliefs plays a significant role in manufacturing of food, because it affects the demand in market place, retailers and manufacturing sector [118].

Although it is proven that irradiated food and food products are safe, still there is unwillingness of consumers towards irradiated food [119]. Consumer awareness and education is the key factor for consumer acceptance of irradiated food. Moreover, consumer attitude can influence the food supply chain from farm to market [120]. Preferences and heating habits of consumers changes continuously. Link between health and diet creates the awareness among people regarding safety and public health. Therefore, food safety and health education is the need of the hour [121].

Perception of people, their risk taking and benefits, and perception of consumers and retailers towards buying of a product is directly related to the acceptance of new technologies [122]. International and national organization related to food and health play a vital role in building confidence among consumers about the irradiation technology and safety and benefits of irradiated foods [95]. However, expansion and application of irradiation technology widely to the products faces many challenges due to various beliefs among consumers such as it may contains radioactive substance in food or eating will develop cancer. This beliefs will result carefulness among food producer and retailer [123–125].

Social awareness with complete information and market trial to the consumers increases the commercial scale of irradiated foods. This has been experimented with some irradiated food products e.g., meat and Hawaiian papaya and proved that consumers are well aware and understand the health benefits and safety and gives preference to buy irradiated food over conventional processed food products [95, 126–128]. Furthermore, Food borne illness and public health hygienic regarding food products is raising concerns among consumers and food manufacturer to consider different perspectives of irradiation technology [129]. Even though, there are adequate evidences that at retail store consumer will buy the irradiated foods, while this information should be pass to producers and retailers [1, 124, 125].

Thus scientific information and benefit and safety of irradiated food with effectively communication will produce more awareness among consumers, retailers and producers [130]. Safety aspects and accurate information regarding ionized food products will encourage people to prefer irradiated food over conventional treated food. The acceptance of irradiated food and adoption of this technology will increase when the advantages and benefits will be well explained compared with conventional technologies.

9.12 Standards, Regulations and Legislation on Food Irradiation

All the international organizations have approved irradiated foods as safe, economic benefit and wholesomeness. Food and drug administration (FDA) have issued certain guidelines for irradiation of different food commodities and approved irradiation as a safe practice [131]. FDA recommended doses for specific purposes are illustrated in Figure 9.2. Codex standards and international code of practice have been developed for irradiated food and radiation processing [53, 132].

In Europe, there is a legal record keeping facility for irradiated foods for each batch of treatment which approves and checks dosimetrically. Moreover, radiation treated foods are to be leveled as regulations and guidelines of European parliament and of the council on the provision of food information to consumers, amending Regulations.

9.12.1 International Standards

The international standards may be divided into human health, plant protection, labeling, dose administration, quality assurance and facility management.

9.12.1.1 Human Health

The Codex Alimentarius commission (CAC) is lead organization for human health standards relating to food production and safety. Food irradiation shall be conducted in accordance with good manufacturing and management practices, in accordance following the principles of good hygiene set out in Codex Alimentarius [53]. The basic principles of food irradiation have been established around the world with the General Codex Alimentarius Standard in 1983 [35] and a major review in 2003 [132]. This standard set the guidelines for use of irradiation sources according to hygiene practices, technical requirement and process control. It also deals with post-irradiation, re-irradiation and verification. The Codex general standard [132] defines the minimum and maximum absorbed dose to be achieved by the technological objective and consumer safety including structural, functional, nutritional and sensory properties respectively. For food maximum dose should not be exceed 10kGy except some legitimate technological needs. Food should be treated as per the standards and given dose limit set.

9.12.1.2 Labelling

The General standard requires that irradiated food should be labeled in compliance with codex standard for prepackaged food [134]. This standard requires that label of irradiated food should include a written declaration that the treatment is carried out in close range to the name of the food. The FDA requires that irradiated foods carry the international Radura symbol, the use of the Radura symbol, is voluntary, but when used, it must be present in the immediate vicinity of the food name. Irradiated foods used as ingredients in other foods must be reported in the list of ingredients on the label. Material prepared from irradiated single-ingredient raw materials and the label should contain information indicating the treatment.

9.12.1.3 *Plant Protection*

International Plant Protection Convention (IPPC) is an international treaty to take effective action for preventing and controlling entry and spreading of plant pests. It had issued Phytosanitary Measures of different international standards for phytosanitary use of irradiation. ISPM 18 provides the technical guidance *for the use of irradiation as phytosanitary measure* [135] for application of ionizing radiation in specific procedures for fresh fruit and vegetables and countries should adopt for trading. ISPM 28 sets the guidelines for minimum dose requirement in regulated pests and recognized that 150 Gy is a generic dose for prevention of adult Tephritid fruit flies in host product [136].

9.12.1.4 *Facilities*

Any treatment facility using gamma, electronic and radiographic sources for any purpose must meet the safety standard established by the International Atomic Energy Agency [137]. Food Safety Practice should be followed in irradiation facilities guided by Codex Code of Practice [138]. More information should be included in ISO 14,470 by the International Organization for Standardization. A recent milestone, quality management in radiation installations has been formalized in ISO 9001 [139]. Manual of Good Practice in Food Irradiation is not a global standard but serves as a comprehensive guideline for facility operators [140]. The American Society for Testing and Materials is well known for its standards, some standards are published in conjunction with ISO. A guidance document was published in 2014 for establishments that deal with agricultural products for a phytosanitary purpose [141]. In standards and guidelines a common characteristic is that production, processing and handling of food should be adhered to Good Manufacturing Practice.

9.12.1.5 *Dosimetry*

In food irradiation, Accuracy of dose delivery and its knowledge is essential. Absorbed dose should be measured accurately using dosimeters, which provides chemical or physical well-defined response taking into account the system calibration curve. Adequate calibration and traceability of benchmarks are fundamental principles. Prior to use of dosimeters in commercial application, it confirms that a prescribed dose or dose range can be applied to a particular food and packaged food. In specific configured food packaging in commercial application, dose mapping operation defines the position of maximum and minimum dose application and the routine dosing control system shall be established to verify compliance with the specified dose limits. The precision and accuracy of the system has to be established and well defined. International standards [40, 44, 142–145] are used for good dosimetry practice.

9.12.1.6 Packaging

Foods to be irradiated should be packaged in materials which preserve their integrity during or after the process by preventing diffusion of toxic materials into the food. Mainly based on USA research a list of suitable material for food irradiation is available on website [146, 147]. The listed packaging materials are generally accepted by several countries, but ASTM has separately produced a document of standard guidance on packaging materials [148].

9.12.2 National Regulations

9.12.2.1 Regulations for Human Health

More than 60 countries have endorsed the use of irradiated food as part of their health or food regulations for at least one, and generally more, food type or food class. The IAEA maintains food authorizations database [149]. However, there's no commitment as it were asking for nations to hold up endorsements with international organization and it is far-fetched for updating also. Many countries state that their regulations are based on the Codex standard [132], however, the Standard makes no reference to foods that may or may not be irradiated. All foods may be irradiated at their maximum authorized dose. Very few countries such as Singapore and Brazil have regulations that permit the irradiation of any food subject solely to Codex compliance. Brazil, for illustration, decided to adopt a regulation on "any food, any dose, any purpose", on the condition that the food is tolerant to the treatment. Most countries still choose to give approvals case-by-case basis for a food type or food category in designated purpose. In countries like US, Australia, Canada and New Zealand approval is provided in response to a petition to change existing regulations, but it is more common for an industrial organization to submit such petition.

9.12.2.2 *Regulations for Labeling*

Most countries adopt the Codex General Standard for the Labeling of Prepackaged Foods as part of their regulations [134]. However, there are variations in the interpretation and implementation of the labelling provisions. In EU, Australia and New Zealand, irradiated food should be labelled with text indicating clearly in the vicinity of the food or the individual item. Irradiated ingredients should be labelled with accurate concentration without fail. Whole food or ingredient irradiation must be labelled in restaurant and food catering trade. In the European Union, the labeling of text is specified but not a Radura symbol [133]. In New Zealand and Australia, text wording is suggested but not compulsory and logo use is voluntary as well.

A number of countries permit the irradiation statement on the label stating its purpose or benefit. Indonesia requests that the purpose should be included on the label. Irradiation purpose or benefit is silent in regulation of several countries. For whole foods, U.S. regulations require that the label be marked with the Radura symbol (U.S. version) and the term "radiation treated" or "radiation treated". In U.S. regulations, whole irradiated food should bear Rudra symbol (U.S.A) and "treated with or by radiation" phrase on label. However, special labeling is not required on retailer packaging if ingredient is irradiated and added to non-irradiated food [150]. In Malaysia and Canada, food labeling is not necessary if whole food contain irradiated ingredient amounts to less than 5% and 10% by weight.

9.12.2.3 *Regulations for Plant Health*

Two countries have treatment regulations on phytosanitary to accelerate future benefit. In the United States, a dose of 400 Gy is recognized as a generic dose for sterilization or prevention of emerging adult insects, apart from pupae and adult Lepidoptera, on all host products [151]. Australia has embraced an Interstate Certificate Confirmation (ICA) National Convention for all deliver moving between its States and Domains [152] that perceives 150 Gy as a recommended dose for Tephritid fruit flies and dose of 400 Gy as nonexclusive measurements for all adult insects, apart from pupae and adult Lepidoptera. The ICA operates within the framework of food standards that Australia and New Zealand have endorsed the phytosanitary utilize of radiation for the product.

9.13 Future Perspectives and Conclusions

Irradiation is a nonthermal, nonchemical, environmentally clean, energy-efficient food processing and preservation technique used for improvement and extension the shelf life of fresh or processed foods by preventing tampering or contamination with harmful bacteria and other forms of parasitic life. Gamma rays, electron beams, and x-rays are used for irradiation of foods. In recent years; irradiation is emerging in several countries for insect disinfestations of insects and pests; sprout inhibition; sanitization of agricultural produces and meat, dairy products. Apart from having immense industrial potential, application of ionizing irradiation to food processing is still under estimation. Radiation processor needs to understand the risk and health hazards that the food trade perceives, when extra and unfamiliar technical steps applied into food supply chain. Cooperation between them is important to make irradiation process smooth and safe. Regarding risk assessment of irradiation, few studies have performed in this direction, still in-depth research is required for proper validation. Therefore, there is need to conduct epidemiological studies for the risks associated with the consumption of irradiated feed for animals and humans.

References

1. Moreira, Rosana G., & Castell-Perez, M. Elena (2021). Fundamentals of Food Irradiation. Innovative Food Processing Technologies: A Comprehensive Review, 1-18. https://doi.org/10.1016/B978-0-12-815781-7.00008-1
2. Antonio, A.L., Cabo Verde, S., Ferreira, I.C.F.R., (2018). Introduction. In: Ferreira, I.C.F.R., Antonio, A.L., Cabo Verde, S. (Eds.), Food Irradiation Technologies: Concepts, Applications and Outcomes, Food Chemistry, Functions and Analysis No. 4. The Royal Society of Chemistry, UK, 1–4.
3. Maherani, B., Hossain, F., Criado, P., Ben-Fadhel, Y., Salmieri, S., & Lacroix, M. (2016). World market development and consumer acceptance of irradiation technology. Foods, 5(4), 79.
4. Lung, H. M., Cheng, Y. C., Chang, Y. H., Huang, H. W., Yang, B. B., & Wang, C. Y. (2015). Microbial decontamination of food by electron beam irradiation. Trends in Food Science & Technology, 44(1), 66-78.

5. Zanardi, E., Caligiani, A., &Novelli, E. (2018). New insights to detect irradiated food: an overview. Food analytical methods, 11(1), 224-235.
6. Sharma, A., &Madhusoodanan, P. (2012). Techno-commercial aspects of food irradiation in India. Radiation Physics and Chemistry, 81(8), 1208-1210.
7. Kume, T., Furuta, M., Todoriki, S., Uenoyama, N., & Kobayashi, Y. (2009). Status of food irradiation in the world. Radiation Physics and Chemistry, 78(3), 222-226.
8. Miller, R. B. (2006). Electronic irradiation of foods: an introduction to the technology. Springer Science & Business Media.
9. Szczawińska, M. E. (2017). Application of Ionizing Radiation for Control of Salmonella in Food. Current Topics in Salmonella and Salmonellosis, 253.
10. Hallman, G. J., & Loaharanu, P. (2016). Phytosanitary irradiation–Development and application. Radiation Physics and Chemistry, 129, 39-45.
11. Arvanitoyannis, I. S., & Tsarouhas, P. (2010). Food Packaging Materials for Irradiation, Irradiation of food commodities. Academic Press, 43-64.
12. Haji-Saeid, M., Sampa, M. H. O., & Chmielewski, A. G. (2007). Radiation treatment for sterilization of packaging materials. Radiation Physics and Chemistry, 76(8-9), 1535-1541.
13. Welle, F., Mauer, A., & Franz, R. (2002). Migration and sensory changes of packaging materials caused by ionising radiation. Radiation Physics and Chemistry, 63(3-6), 841-844.
14. Kalyani, B., & Manjula, K. (2014). Food irradiation–technology and application. International Journal of Current Microbiology and Applied Science, 3(4), 549-555.
15. Hallman, G. J. (2011). Phytosanitary applications of irradiation. Comprehensive Reviews in Food Science and Food Safety, 10(2), 143-151.
16. Prakash, A. (2016). Particular applications of food irradiation fresh produce. Radiation physics and chemistry, 129, 50-52.
17. Pillai, S. D., & Shayanfar, S. (2017). Electron beam technology and other irradiation technology applications in the food industry. In Applications of Radiation Chemistry in the Fields of Industry, Biotechnology and Environment, 249-268.
18. Cornwell, P. B. (Ed.). (2013). The entomology of radiation disinfestation of grain: a collection of original research papers. Elsevier.
19. McDonald, H., McCulloch, M., Caporaso, F., Winborne, I., Oubichon, M., Rakovski, C., & Prakash, A. (2012). Commercial scale irradiation for insect disinfestation preserves peach quality. Radiation Physics and Chemistry, 81(6), 697-704.
20. Sea, S., Rakovski, C., & Prakash, A. (2015). Ripening quality of 'Bartlett' pears (Pyrus communis L.) subjected to phytosanitary x-ray irradiation treatment followed by simulated retail display. HortScience, 50(2), 279-287.
21. Hossain, F., Parvez, A. K., Munshi, M. K., Khalil, I., &Huque, R. (2014). Postharvest treatments of radiation and chemical on organoleptic and biochemical properties of mango (Mangifera indica L.) in relation to delay

ripening. American Eurasian Journal of Agricultural and Environmental Science, 14(6), 555-564.

22. Yadav, M. K., & Patel, N. L. (2014). Optimization of irradiation and storage temperature for delaying ripening process and maintaining quality of Alphonso mango fruit (Mangifera indica L.). African Journal of Agricultural Research, 9(5), 562-571.
23. Surendranathan, K. K., & Nair, P. M. (2013). Carbohydrate metabolism in ripening banana and its alteration on gamma irradiation in relation to delay in ripening. Journal of the Indian Institute of Science, 62(8), 63.
24. Fadhel, Y. B., Leroy, V., Dussault, D., St-Yves, F., Lauzon, M., Salmieri, S., ... & Lacroix, M. (2016). Combined effects of marinating and γ-irradiation in ensuring safety, protection of nutritional value and increase in shelf-life of ready-to-cook meat for immunocompromised patients. Meat science, 118, 43-51.
25. Kumar, S., Gautam, S., & Sharma, A. (2015). Hurdle technology including chlorination, blanching, packaging and irradiation to ensure safety and extend shelf life of shelled sweet corn kernels. Journal of Food Processing and Preservation, 39(6), 2340-2347.
26. Yang, Z., Wang, H., Wang, W., Qi, W., Yue, L., & Ye, Q. (2014). Effect of 10 MeV E-beam irradiation combined with vacuum-packaging on the shelf life of Atlantic salmon fillets during storage at 4 C. Food chemistry, 145, 535-541.
27. Prakash, A., Inthajak, P., Huibregtse, H., Caporaso, F., & Foley, D. M. (2000). Effects of low-dose gamma irradiation and conventional treatments on shelf life and quality characteristics of diced celery. Journal of Food Science, 65(6), 1070-1075.
28. Javanmard, M., Rokni, N., Bokaie, S., &Shahhosseini, G. (2006). Effects of gamma irradiation and frozen storage on microbial, chemical and sensory quality of chicken meat in Iran. Food control, 17(6), 469-473.
29. Mongpraneet, S., Abe, T., &Tsurusaki, T. (2002). Accelerated drying of welsh onion by far infrared radiation under vacuum conditions. Journal of food engineering, 55(2), 147-156.
30. Food and Drug Administration, HHS. (2008). Irradiation in the production, processing and handling of food. Final rule. Federal register, 73(164), 49593.
31. Cabo Verde, S., (2018). Food irradiation as sanitary treatment. In: Ferreira, I.C.F.R., Antonio, A.L., Cabo Verde, S. (Eds.), Food Irradiation Technologies: Concepts, Applications and Outcomes, Food Chemistry, Functions and Analysis No. 4. The Royal Society of Chemistry, UK, 183–209
32. Song, B. S., Lee, Y., Park, J. H., Kim, J. K., Park, H. Y., Kim, D. H., ... & Kang, I. J. (2018). Toxicological and radiological safety of chicken meat irradiated with 7.5 MeV X-rays. Radiation Physics and Chemistry, 144, 211-217
33. Dionísio, A. P., Gomes, R. T., & Oetterer, M. (2009). Ionizing radiation effects on food vitamins: a review. Brazilian Archives of Biology and Technology, 52(5), 1267-1278.

34. SádECká, J. (2007). Irradiation of spices–a review. Czech Journal of Food Science, 25(5), 231-242.
35. CAC, (1983). Codex Alimentarius Commission. General Standard for Irradiated Foods (CODEX STAN 106-1983). Codex Alimentarius, FAO/WHO, Rome.
36. Miller, R. B., 2018. Electron beam and x-ray equipment for food irradiation applications. In: Ferreira, I.C.F.R., Antonio, A.L., Cabo Verde, S. (Eds.), Food Irradiation Technologies: Concepts, Applications and Outcomes, Food Chemistry, Functions and Analysis No. 4. The Royal Society of Chemistry, UK, pp. 51–65.
37. McLaughlin, W. L., Boyd, A. W., Chadwick, K. H., McDonald, J. C., Miller, A., 1989a. Dosimetry for Radiation Processing. Taylor and Francis, London (1989).
38. McLaughlin, W. L., Boyd, A. W., Chadwick, K. H., Mc Donald, J. C., Miller, A., 1989b. Physical methods of dosimetry. In: Dosimetry for Radiation Processing. Taylor & Francis, London, pp. 140–141.
39. Aquino, K.A.S., 2012. In: Adrovic, F. (Ed.), Sterilization by Gamma Irradiation, Gamma Radiation. InTech. Available from: http://www.intechopen.com/books/gammaradiation/ sterilization-by-gamma-irradiation.
40. IAEA, 2002. Dosimetry for Food Irradiation. Technical Report Series No. 409. International Atomic Energy Association, Austria, Vienna, p. 161.
41. Craven, E., Schlecht, J., Stein, R., 2018. Gamma irradiation plants. In: Ferreira, I.C.F.R., Antonio, A.L., Cabo Verde, S. (Eds.), Food Irradiation Technologies: Concepts, Applications and Outcomes, Food Chemistry, Functions and Analysis No. 4. The Royal Society of Chemistry, UK, pp. 28–50.
42. Farkas, J. (1988). Irradiation of Dry Food Fngredients (pp. 77–88). Boca Raton, FL: CRC Press.
43. IAEA, 1996. Manual on Panoramic Gamma Irradiators (Categories II and IV). IAEA-PRSM-8, International Atomic Energy Association, Austria, Vienna.
44. ISO/ASTM, (2004). International Organization for Standardization/American Society for Testing and Materials. Practice for Dosimetry in Gamma Irradiation Facilities for Food Processing. ISO/ASTM 51204:2004. ISO, Geneva.
45. Moskalev, V. A., & Sergeev, G. I. (2003). A cyclic induction accelerator. Russian Physics Journal, 46(3), 270–277.
46. Seltzer, S. M., Farrell, J. P., & Silvermann, J. (1983). Bremsstrahlung beams from high-power electron-accelerators for use in radiation processing. IEEE Transactions on Nuclear Science, 30, 1629–1633.
47. Morrison, R. M. (1989). An economic analysis of electron accelerators and cobalt-60 for irradiating food (technical bulletin No. 1762). Washington, DC: Economic Research Service, U.S. Department of Agriculture.

48. Stichelbaut, F., Bol, J.-L., Cleland, M. R., Gregoire, O., Herer, A. S., Jongen, Y., & Mullier, B. (2004). The Palletron: A high-dose uniformity pallet irradiator with X-rays. Radiation Physics and Chemistry, 71, 291–295.
49. Wilkinson, V. M., & Gould, G. W. (1996). Food Irradiation: A Reference Guide. Oxford: Butterworth–Heinemann.
50. IAEA, (2015). International Atomic Energy Agency. Manual of Good Practice in Food Irradiation. Technical Report Series No.481. IAEA, Vienna.
51. Dethier, P., Mullier, B., 2018. Technical and economic considerations in food irradiation. In: Ferreira, I.C.F.R., Antonio, A.L., Cabo Verde, S. (Eds.), Food Irradiation Technologies: Concepts, Applications and Outcomes, Food Chemistry, Functions and Analysis No. 4. The Royal Society of Chemistry, UK, pp. 369–382.
52. World Health Organization. (1994). Safety and nutritional adequacy of irradiated food. World Health Organization, Geneva.
53. CAC, Codex Alimentarius Commission (2003a). General standard for irradiated foods. CODEX STAN 106–1983, Rev 1–2003 Available at: http://www.codexalimentarius.org/standards/list-of-standards/. Accessed on 18 Sept. 2020
54. Kume, T., & Todoriki, S. (2013). Food irradiation in Asia, the European Union, and the United States: A status update. Radioisotopes, 62(5), 291-299.
55. Barkai-Golan, R., & Follett, P. A. (2017). Irradiation for quality improvement, microbial safety and phytosanitation of fresh produce. Academic Press. 302.
56. Niemira, B. A., & Fan, X. (2006). Low-dose irradiation of fresh and fresh-cut produce: safety, sensory and shelf life. Food irradiation research and technology, 32(5), 45-52.
57. ICGFI, 1991. Irradiation as Quarantine Treatment of Fresh Fruits and Vegetables. International Consultative Group on Food Irradiation, Betesda, MD, USA.
58. Steinbrunner, P. J., Limcharoenchat, P., Suehr, Q. J., Ryser, E. T., Marks, B. P., & Jeong, S. (2019). Effect of Food Structure, Water Activity, and Long-Term Storage on X-Ray Irradiation for Inactivating Salmonella Enteritidis PT30 in Low-Moisture Foods. Journal of food protection, 82(8), 1405-1411.
59. Cho, G. L., & Ha, J. W. (2019). Application of X-ray for inactivation of foodborne pathogens in ready-to-eat sliced ham and mechanism of the bactericidal action. Food Control, 96, 343-350.
60. Park, J. S., & Ha, J. W. (2019). X-ray irradiation inactivation of Escherichia coli O157: H7, Salmonella enterica Serovar Typhimurium, and Listeria monocytogenes on sliced cheese and its bactericidal mechanisms. International journal of food microbiology, 289, 127-133
61. Lim, J. S., & Ha, J. W. Effect of acid adaptation on the resistance of Escherichia coli O157: H7 and Salmonella enterica serovar Typhimurium to X-ray irradiation in apple juice. Food Control, 120, 107489.

62. Jeon, M. J., & Ha, J. W. (2020). Bactericidal and synergistic effects of X-ray irradiation and gallic acid against foodborne pathogens on lettuce. Food Microbiology, 92, 103584.
63. Maherani, B., Harich, M., Salmieri, S., & Lacroix, M. (2019). Antibacterial properties of combined non-thermal treatments based on bioactive edible coating, ozonation, and gamma irradiation on ready-to-eat frozen green peppers: Evaluation of their freshness and sensory qualities. European Food Research and Technology, 245(5), 1095-1111.
64. Xu, A., Scullen, O. J., Sheen, S., Johnson, J. R., & Sommers, C. H. (2019). Inactivation of extraintestinal pathogenic E. coli clinical and food isolates suspended in ground chicken meat by gamma radiation. Food microbiology, 84, 103264.
65. Mrityunjoy, A., Israt, I., & Rashed, N. (2019). Effects of gamma irradiation on the propagation of microbial growth in commonly available meat in Bangladesh. International Food Research Journal, 26(4), 1211-1218.
66. Fernandes, D., & Prakash, A. (2020). Use of gamma irradiation as an intervention treatment to inactivate Escherichia coli O157: H7 in freshly extracted apple juice. Radiation Physics and Chemistry, 168, 108531.
67. Al-Qadiri, H., Amr, A., Al-Holy, M. A., & Shahein, M. (2020). Effect of gamma irradiation against microbial spoilage of hummus preserved under refrigerated storage. Food Science and Technology International, 1082013220975891.
68. Joshi, B., Moreira, R. G., Omac, B., & Castell-Perez, M. E. (2018). A process to decontaminate sliced fresh cucumber (Cucumis sativus) using electron beam irradiation. Lwt, 91, 95-101.
69. Elias, M. I., Madureira, J., Santos, P. M. P., Carolino, M. M., Margaça, F. M. A., & Verde, S. C. (2020). Preservation treatment of fresh raspberries by e-beam irradiation. Innovative Food Science & Emerging Technologies, 66, 102487.
70. Madureira, J., Severino, A., Cojocaru, M., Garofalide, S., Santos, P. M. P., Carolino, M. M & Verde, S. C. (2019). E-beam treatment to guarantee the safety and quality of cherry tomatoes. Innovative Food Science & Emerging Technologies, 55, 57-65.
71. An, K. A., Jo, Y., Arshad, M. S., Kim, G. R., Jo, C., & Kwon, J. H. (2017). Assessment of microbial and radioactive contaminations in Korean cold duck meats and electron-beam application for quality improvement. Korean journal for food science of animal resources, 37(2), 297.
72. Bouzarjomehri, F., Dad, V., Hajimohammadi, B., Shirmardi, S. P., & Salimi, A. Y. G. (2020). The effect of electron-beam irradiation on microbiological properties and sensory characteristics of sausages. Radiation Physics and Chemistry, 168, 108524.
73. Urbain, W. M. (1989). Food irradiation: the past fifty years as prologue to tomorrow. Food technology, 43(7), 76-92.
74. Farkas, J. (1998). Irradiation as a method for decontaminating food: a review. International journal of food microbiology, 44(3), 189-204

75. Farkas, J. (2006). Irradiation for better foods. Trends in food science & technology, 17(4), 148-152.
76. Thayer, D. W., & Rajkowski, K. T. (1999). Developments in irradiation of fresh fruits and vegetables. Food technology (Chicago), 53(11), 62-65.
77. Borsa, J., Lacroix, M., Ouattara, B., & Chiasson, F. (2004). Radiosensitization: enhancing the radiation inactivation of foodborne bacteria. Radiation Physics and Chemistry, 71(1-2), 137-141.
78. O'bryan, C. A., Crandall, P. G., Ricke, S. C., & Olson, D. G. (2008). Impact of irradiation on the safety and quality of poultry and meat products: a review. Critical Reviews in Food Science and Nutrition, 48(5), 442-457.
79. da Silva Aquino, K. A. (2012). Sterilization by gamma irradiation. Gamma radiation, 9, 172-202.
80. Joshua Ajibola, O. (2020). An overview of irradiation as a food preservation technique. Novel Research in Microbiology Journal, 4(3), 779-789.
81. Thayer, D. W., & Boyd, G. (2001). Effect of irradiation temperature on inactivation of Escherichia coli O157: H7 and Staphylococcus aureus. Journal of food protection, 64(10), 1624-1626.
82. Gomes, C., Moreira, R. G., Castell-Perez, M. E., Kim, J., Da Silva, P., & Castillo, A. (2008). E-beam irradiation of bagged, ready-to-eat spinach leaves (Spinacea oleracea): An engineering approach. Journal of Food Science, 73(2), E95-E102.
83. Gomes, C., Moreira, R. G., & Castell-Perez, E. (2011a). Radiosensitization of Salmonella spp. and Listeria spp. in ready-to-eat baby spinach leaves. Journal of food science, 76(1), E141-E148.
84. Farkas, J. (2007). Physical methods of food preservation. In Food Microbiology: Fundamentals and Frontiers, Third Edition (pp. 685-712). American Society of Microbiology.
85. Yurttas, Z. S., Moreira, R. G., & Castell-Perez, E. (2013). Combined vacuum impregnation and electron-beam irradiation treatment to extend the storage life of sliced white button mushrooms (Agaricusbisporus). Journal of Food Science, 79(1), E39-E46.
86. Karagöz, I., Moreira, R. G., & Castell-Perez, M. E. (2014). Radiation D10 values for Salmonella Typhimurium LT2 and an Escherichia coli cocktail in pecan nuts (Kanza cultivar) exposed to different atmospheres. Food control, 39, 146-153.
87. Gomes, C., Moreira, R. G., & Castell-Perez, E. (2011b). Microencapsulated antimicrobial compounds as a means to enhance electron beam irradiation treatment for inactivation of pathogens on fresh spinach leaves. Journal of food science, 76(6), E479-E488.
88. Takala, P. N., Salmieri, S., Vu, K. D., & Lacroix, M. (2011). Effects of combined treatments of irradiation and antimicrobial coatings on reduction of food pathogens in broccoli florets. Radiation Physics and Chemistry, 80(12), 1414-1418.

89. Lacroix, M., & Follett, P. (2015). Combination irradiation treatments for food safety and phytosanitary uses. Stewart Postharvest Review, 11(3).
90. Severino, R., Ferrari, G., Vu, K. D., Donsì, F., Salmieri, S., & Lacroix, M. (2015). Antimicrobial effects of modified chitosan based coating containing nanoemulsion of essential oils, modified atmosphere packaging and gamma irradiation against Escherichia coli O157: H7 and Salmonella Typhimurium on green beans. Food control, 50, 215-222.
91. Dias, M. I., Ferreira, I. C., & Barreiro, M. F. (2015). Microencapsulation of bioactives for food applications. Food & function, 6(4), 1035-1052.
92. Ndoti-Nembe, A., Vu, K. D., Doucet, N., & Lacroix, M. (2015). Antimicrobial effects of essential oils, nisin, and irradiation treatments against Listeria monocytogenes on ready-to-eat carrots. Journal of Food Science, 80(4), M795-M799.
93. Tawema, P., Han, J., Vu, K. D., Salmieri, S., & Lacroix, M. (2016). Antimicrobial effects of combined UV-C or gamma radiation with natural antimicrobial formulations against Listeria monocytogenes, Escherichia coli O157: H7, and total yeasts/molds in fresh cut cauliflower. LWT-Food Science and Technology, 65, 451-456.
94. Pinela, J., & Ferreira, I. C. (2017). Nonthermal physical technologies to decontaminate and extend the shelf-life of fruits and vegetables: Trends aiming at quality and safety. Critical reviews in food science and nutrition, 57(10), 2095-2111.
95. Loaharanu, P. (2003). Irradiated foods. American Council on Science and Health, New York.
96. Donnelly, J. K., & Robinson, D. S. (1995). Invited review free radicals in foods. Free radical research, 22(2), 147-176.
97. Silva, A. C. D. O., Mársico, E. T., Jesus, E. F. O. D., Guimarães, C. F., & Cortez, M. A. S. (2011). Effect of gamma radiation on lipids by the TBARS and NMR. Brazilian Archives of Biology and Technology, 54(6), 1343-1348.
98. Kilcast, D. (1995). Food irradiation: Current problems and future potential. International Biodeterioration & Biodegradation, 36(3-4), 279-296.
99. Ferreira, I.C.F.R., Antonio, A.L., Cabo Verde, S. (Eds.), 2018. Food Irradiation Technologies: Concepts, Applications and Outcomes. Food Chemistry, Functions and Analysis No. 4. The Royal Society of Chemistry, UK.
100. Iwanami, S., & Oda, N. (1985). Theory of Survival of Bacteria Exposed to Ionizing Radiation: I. X and γ Rays. Radiation research, 102(1), 46-58.
101. Monk, J. D., Beuchat, L. R., & Doyle, M. P. (1995). Irradiation inactivation of food-borne microorganisms. Journal of Food Protection, 58(2), 197-208.
102. Komolprasert, V. (2016). Packaging food for radiation processing. Radiation physics and chemistry, 129, 35-38.
103. USFDA, 2016. United States Food and Drug Administration. Irradiation in the production. Processing and Handling of Food. Final rule. 21 CFR179.26. Federal Register, 1 April.

104. Jamshidian, M., Lacroix, M., 2018. Packaging for food irradiation. In: Ferreira, I.C.F.R., Antonio, A.L., Cabo Verde, S. (Eds.), Food Irradiation Technologies: Concepts, Applications and Outcomes, Food Chemistry, Functions and Analysis No. 4. The Royal Society of Chemistry, UK, pp. 123–168.
105. European Food Safety Authority (2006). Summary report: EFSA scientific colloquium 6: Risk–benefit analysis of foods: Methods and approaches: 13–14, Tabiano (Province of Parma), Italy. Author. Parma, Italy.
106. Schreider, T. (2003). Risk assessment tools: A primer, Vol. 2. Rolling Meadows, IL: Information Systems Audit and Control Association.
107. Dugan, L. C., & Bedford, J. S. (2003). Are chromosomal instabilities induced by exposure of cultured normal human cells to low-or high-LET radiation?. Radiation research, 159(3), 301-311.
108. Rigaud, O., &Moustacchi, E. (1996). Radioadaptation for gene mutation and the possible molecular mechanisms of the adaptive response. Mutation Research/Fundamental and Molecular Mechanisms of Mutagenesis, 358(2), 127-134.
109. Marchioni, E., Raul, F., Burnouf, D., Miesch, M., Delincee, H., Hartwig, A., & Werner, D. (2004). Toxicological study on 2-alkylcyclobutanones—results of a collaborative study. Radiation Physics and Chemistry, 71(1-2), 147-150.
110. Atherton, K. T. (2002). Safety assessment of genetically modified crops. Toxicology, 181, 421-426.
111. Alimentarius, C. (1997). Joint FAO/WHO Food Standards Programme Codex Alimentarius Commission. Food hygiene, Supplement to, 1.
112. Bauman, H. E. (1995). The origin and concept of HACCP. In HACCP in meat, poultry, and fish processing (pp. 1-7). Springer, Boston, MA.
113. Butterweck, J. S. (1990). Monitoring pathogens from irradiated agriculture products. International Journal of Radiation Applications and Instrumentation. Part C. Radiation Physics and Chemistry, 35(1-3), 345-348.
114. Bruhn, C. M., & Schutz, H. G. (1989). Consumer awareness and outlook for acceptance of food irradiation. Food technology (Chicago), 43(7), 93-97.
115. Conley, S. T. (1992). What do consumers think about irradiated foods. FSIS Food Safety Review, 2, 11–15.
116. Resurreccion, A. V. A., Galvez, F. C. F., Fletcher, S. M., &Misra, S. K. (1994). Consumer attitudes toward irradiated food: results of a new study. Journal of food protection, 58(2), 193-196.
117. Frewer, L. J., Bergmann, K., Brennan, M., Lion, R., Meertens, R., Rowe, G., ... & Vereijken, C. M. J. L. (2011). Consumer response to novel agri-food technologies: Implications for predicting consumer acceptance of emerging food technologies. Trends in Food Science & Technology, 22(8), 442-456.
118. Thomas, P. A. (1990). Food irradiation and the consumer. International Journal of Radiation Applications and Instrumentation. Part C. Radiation Physics and Chemistry, 35(1-3), 342-344.

119. Farkas, J., Ehlermann, D.A.E., Mohácsi-Farkas, C. (2014). Food technologies: food irradiation. In: Motarjemi, Y. (Ed.), Encyclopedia of Food Safety, 3. Academic Press, Waltham, 178–186.
120. Lusk, J. L., & Briggeman, B. C. (2009). Food values. American journal of agricultural economics, 91(1), 184-196.
121. Lusk, J. L., & McCluskey, J. (2018). Understanding the impacts of food consumer choice and food policy outcomes. Applied Economic Perspectives and Policy, 40(1), 5-21.
122. Bearth, A., & Siegrist, M. (2016). Are risk or benefit perceptions more important for public acceptance of innovative food technologies: A meta-analysis. Trends in Food Science & Technology, 49, 14-23.
123. Lyndhurst, B. (2009). An evidence review of public attitudes to emerging food technologies. Social Science Research Unit, Food Standards Agency, Crown, 1-83.
124. Eustice, R. F., & Bruhn, C. M. (2013). Consumer acceptance and marketing of irradiated foods. Food irradiation research and technology, 2, 173-195.
125. Roberts, P. B. (2014). Food irradiation is safe: Half a century of studies. Radiation Physics and Chemistry, 105, 78-82.
126. Johnson, A. M., Reynolds, A. E., Chen, J., & Resurreccion, A. V. A. (2004). Consumer Acceptance of electron-beam irradiated ready-to-eat poultry meats. Journal of food processing and preservation, 28(4), 302-319.
127. Eustice, R. F. (2017). Global status and commercial applications of food irradiation. In Food Irradiation Technologies, 397-424.
128. Frewer, L. J. (2017). Consumer acceptance and rejection of emerging agrifood technologies and their applications. European Review of Agricultural Economics, 44(4), 683-704.
129. Hunter, C. (2000). Changing attitudes to irradiation throughout the food chain. Radiation Physics and Chemistry, 57(3-6), 239-243.
130. Wall, P. G., & Chen, J. (2018). Moving from risk communication to food information communication and consumer engagement. npj Science of Food, 2(1), 1-5.
131. Ravindran, R., & Jaiswal, A. K. (2019). Wholesomeness and safety aspects of irradiated foods. Food chemistry, 285, 363-368.
132. CAC, Codex Alimentarius Commission (2003b). Code of practice for radiation processing of foods. CAC/RCP 19–1979. Available at: http://www.codexalimentarius.org/standards/list-of-standards/. Accessed on 18 sept. 2020.
133. EU, (1999). European Union. Directive 1999/2/EC of the European Parliament and of the Council. Concerning food and food ingredients treated with ionizing radiation. February 22nd.
134. CAC, 2010. Codex Alimentarius Commission. General Standard for the Labelling of Pre-packaged Foods (CODEX STAN 1-1985, Rev. 7-2010). Codex Alimentarius, FAO/WHO, Rome

135. FAO, (2003). Food and Agriculture Organisation of the United Nations. Guidelines for the Use of Irradiation as a Phytosanitary Measure ISPM No. 18. FAO, Rome.
136. FAO, (2009). Food and Agriculture Organisation of the United Nations. Phytosanitary Treatments for Regulated Pests. ISPM No. 28. FAO, Rome.
137. IAEA, (2008). International Atomic Energy Agency. Radiation Safety of Gamma, Electron and X Ray Irradiation Facilities. Iaea Safety Standards Series No. SSG-8, IAEA, Vienna.
138. CAC, (2003c). Codex Alimentarius Commission. Code of Practice for Radiation Processing of Food. CAC/RCP 19-1979, Rev. 2-2003. Codex Alimentarius, FAO/WHO, Rome.
139. ISO, (2008). International Organization for Standardization. Quality Management Systems: Requirements. ISO 9001:2008. ISO, Geneva.
140. IAEA, 2015. Manual of Good Practice in food Irradiation. Sanitary, Phytosanitary and Other Applications. Technical Report Series No. 481 IAEA Radiation Technology Series 4. IAEA, Austria, Vienna, p. 91. Retrieved from. https://www.iaea.org/publications/10801/manual-of-good-practice-in-food-irradiation.
141. ASTM, (2014). American Society for Testing and Materials. Standard Guide for the Irradiation of Fresh Agricultural Produce as a Phytosanitary Treatment. ASTM F1355-06(2014). ASTM, West Conshohocken, PA, USA.
142. ISO/ASTM, (2005). International Organization for Standardization/ American Society for Testing and Materials. Practice for Dosimetry in Electron Beam and X Ray (Brehmsstrahlung) Irradiation Facilities for Food Processing. ISO/ASTM 51431:2005. ISO, Geneva.
143. ISO/ASTM, (2013). International Organization for Standardization/ American Society for Testing and Materials. Practice for Calibration of Routine Dosimetry Systems for Radiation Processing. ISO/ASTM 51261:2013. ISO, Geneva.
144. ISO/ASTM, (2015). International Organization for Standardization/ American Society for Testing and Materials. Guide for Estimation of Measurement Uncertainty in Dosimetry for Radiation Processing. ISO/ ASTM 51707:2015. ISO, Geneva.
145. ISO, (2011). International Organization for Standardization. Food Irradiation: Requirements for the Development, Validation and Routine Control of the Ionizing Radiation Process Used for the Treatment of Food. ISO 14470:2011. ISO, Geneva.
146. USFDA, 2015a. United States Food and Drug Administration. Irradiated Food and Packaging. ⟨http://www.fda.gov/Food/IngredientsPackagingLabeling/IrradiatedFoodPackaging/default.htm⟩ (accessed 03.16).
147. USFDA, 2015b. United States Food and Drug Administration. Packaging Materials for Use during the Irradiation of Pre-packaged Foods. 21 CFR 179.45. US Government Printing Office, Washington

148. ASTM, 2003. American Society for Testing and Materials. Standard Guide for the Packaging Materials for Food to be Irradiated. ASTM F1640-03. ASTM, West Conshohocken, PA, USA.
149. IAEA, 2012. International Atomic Energy Agency, Food Authorization database.
150. USFDA, 2015c. United States Food and Drug Administration. Irradiation in the production, processing, and handling of food; final rule. 21 CFR 179.26. 1 April.
151. USDA-APHIS, 2006. United States Department of Agriculture- Animal and Plant Health Inspection Service. Treatments for fruits and vegetables. Federal Register, 71(18), pp. 4451–64.
152. ICA, (2011). Interstate Certification Assurance National Protocol (Number 55). Irradiation treatment. Australian Subcommittee on Domestic Quarantine and Market Access. 8 March.

10

Active Packaging in Food Industry

Roopa H.[1], Anil Panghal[2], Anju Kumari[1*], Navnidhi Chhikara[3], Ekta Sehgal[1] and Kritika Rawat[1]

[1]Centre of Food Science and Technology, Chaudhary Charan Singh Haryana Agricultural University, Hisar, India
[2]Department of Processing and Food Engineering, AICRP-PHET, Chaudhary Charan Singh Haryana Agricultural University, Hisar, India
[3]Department of Food Technology, Guru Jambheshwar University of Science and Technology, Hisar, India

Abstract

Active packaging is addition of additives which prevent deterioration or degradation of quality and increase in shelf-life of food. To extend the shelf life of food, an active packaging system uses a packaging material that interacts with the internal gas environment. Such new technologies alter the gas environment on a continuous basis by removing or adding gases to the headspace inside a package. Antimicrobial packaging plays a main role in active food packaging in order to prevent or control the microbial growth after processing of food. Primary objective of packaging is to give convenience to carry or transport and to avoid secondary contamination. In order to improve the packaging objective like maintain the quality of food from several environmental factors like moisture, humidity, oxygen, microbes, insects and intrinsic factors like relative humidity, ethylene concentration, temperature,water activity, pH and acidity, nutrient content, redox potential, antimicrobial chemicals, competitive micro flora, and homogeneity etc. the deterioration of food quality by environmental factors can be prevented by incorporation of different active components by different incorporation method, by considering the safety issue modern technology like nanotechnology. The efficacy of combining active packaging with intelligent packaging was good. Year after year, the market value of active and intelligent packaging grows. When compared to other packaging methods, active packaging has both advantages and disadvantages.

**Corresponding author*: anjugaina@hau.ac.in

Navnidhi Chhikara, Anil Panghal and Gaurav Chaudhary (eds.) Novel Technologies in Food Science, (375–404) © 2023 Scrivener Publishing LLC

***Keywords*:** Shelf-life, antimicrobial packaging, nanotechnology, deterioration, food quality

10.1 Introduction

After processing of food, contamination issues have arisen as a result of improper packaging or handling methods, which may result in food borne illness, a major global issue and a financial loss for the food industry. As a result, proper packaging techniques are required to control primary and secondary contamination. Different food products require different packaging. For example, lipid-rich foods require antioxidant or oxygen-scavenging materials in packaging [1], while fresh fruits and vegetables require an ethylene scavenger [2], Low moisture foods necessitate the use of a humidity controller in their packaging [3]. Active packaging perform the extra functions rather than the normal packaging (Figure 10.1).

Antimicrobial packaging is a potential form of active packaging that refers to the incorporation of antimicrobial agents into packaging systems with the aim to extend the product shelf life maintaining its quality and safety [4]. Potential active packaging systems include oxygen, ethylene Scavenging, Co_2-scavengers, emitters, moisture regulators, release or adsorption of flavors, odors, as well as the incorporation of antioxidants to reduce oxidation of food, as well as antimicrobial agents, are used to control the growth of microorganisms on the surface of food

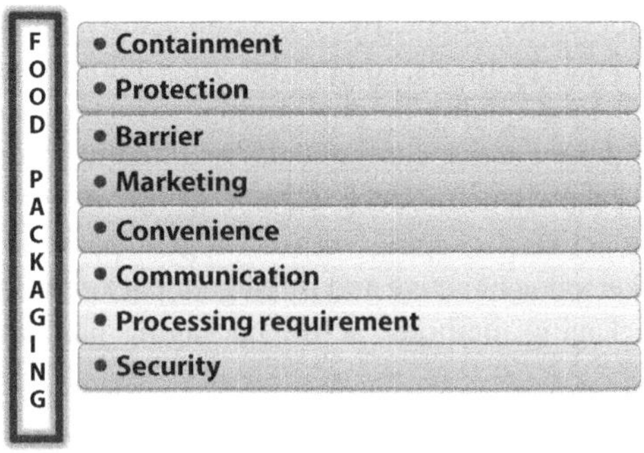

Figure 10.1 Functions of food packaging.

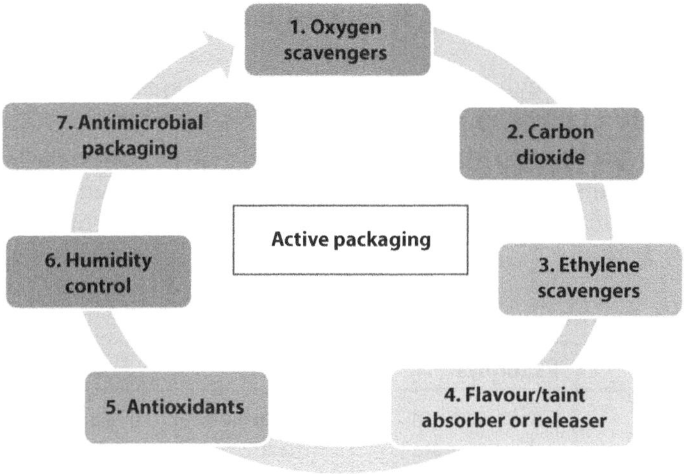

Figure 10.2 Active packaging components.

as mentioned in Figure 10.2 [5, 6]. In antimicrobial packaging, agents may be applied by different methods like coating, incorporating, immobilizing, or modifying the surface of packaging materials. Many compounds have been tried in antimicrobial packaging with varying degrees of success, including organic acids, bacteriocins, enzymes, spices, and polysaccharides (chitosan) [7–9]. Nano-size materials for packing food substances with appropriate properties result in better packaging performance and longer food shelf-life [10].

In addition to standardizing the process of active packaging, it is critical to understand the differences in antibacterial activity of the agents when the packaging film's original activity has been integrated into the packaging film in order to determine the levels that must be incorporated for effective bacterial inhibition. This understanding of antimicrobial activity differences will also aid in the regulation of factors in the production of antimi crobial films. The preservation of packaging films' antimicrobial function during product storage is an even more important factor that needs to be investigated from an industrial standpoint. The sensitivity of the majority of antimicrobial compounds to the high processing temperature used in the manufacture of packaging polymers is a major challenge in commercializing this technology. One way to solve this problem is to develop effective systems for coating antimicrobial compounds onto fully formed polymer materials [11].

10.2 Active Packaging Components

As mentioned by Singh *et al.* [12], active packaging system consists of scavenging and emitting agents, which are listed in Tables 10.1 & 10.2. Unlike traditional packaging, which must be totally inert, active packaging is designed to interact with the contents and/or the surrounding environment (Figure 10.3) [13]. Active Packaging is an innovative concept that can

Table 10.1 List of emitters, examples and application.

Emitters	Examples	Products	References
Carbon dioxide	Citric, sodium bicarbonate (NaHCO3); ferrous carbonate (FeCO3)	Fresh fish and meat	[6, 14]
Ethanol	Sodium stearate	Chinese baby berry	[15]
Antioxidants	Rosemary extract; Green ,black and oolong tea extract; Chitosan	Aqueous and fatty foods; fresh pork; Pork sausage	[16–18]
Antimicrobial agent			
Ethanol	Ethanol emitter	Bakery product; rye bread; Durum wheat bread	[19, 20]
Sulfur dioxide	calcium sulfite	Walnuts, coconuts, peppers, tomatoes, cabbage, gourds, beans, peas, potatoes, and carrots	[21, 88]

Table 10.2 List of scavengers and examples.

Scavengers	Examples	References
Oxygen	Unsaturated hydrocarbons, tocopherol, ascorbic acid, enzyme and microorganism-based scavengers, iron, platinum, and palladium metals.	[22]
Ethylene	Zeolite,ceramics, halloysite nanotubes, Japanese Oya stone, cristobalite, bentonite, coral,Activated charcoal, propylene glycol, potassium permanganate, benzene, pyridines	[2]
Moisture	sorbitol and fructose; cellulose; Calcium oxide (CaO) and silica gel, potassium chloride (KCl), potassium carbonate (K_2CO_3), calcium chloride ($CaCl_2$) and bentonite	[14, 23]
Odour	Photo catalysis with TiO_2 and POM; O_3, O3/UV/H_2O_2, O_3/UV.	[24, 25]
Humidity	sorbitol ($C_6H_{14}O_6$), glycerin (C_3H8O_3), sodium chloride (NaCl), and magnesium chloride ($MgCl_2$), calcium chloride (CaC_{12})	[26, 27]

be defined as a mode of packaging in which the package, the product, and the environment interact to prolong shelf life or enhance safety or sensory properties, while maintaining the quality of the product [14].

10.2.1 Oxygen Scavengers

In a package, the presence of oxygen can trigger or speed up oxidative reactions that result in food deterioration. Oxygen promotes the growth

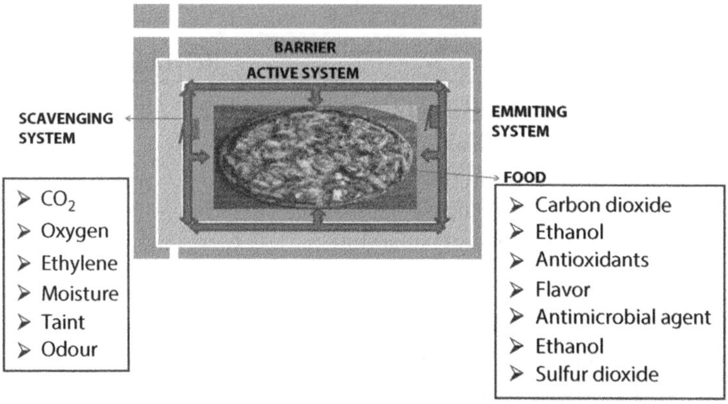

Figure 10.3 Illustration of active packaging system.

of aerobic microbes and moulds. Adverse characteristics such as off-odors, off-flavors, undesirable color changes and decreased nutritional quality result from oxidative reactions. Minimizing the oxidation of foodstuffs will help prevent the degradation of products. Changes in color, nutritional value and flavor can be directly related to degradation due to oxygen combined with food [28]. It is used to control the oxidation mechanism by the action of the galvanic couple between Zn and Fe in recent studies using nanoparticles for the preparation of oxygen scavengers, i.e., Zn-Fe nanoparticles produced by System of hybrid sputtering/gas agglomeration [29]. Oxygen can be removed from modified atmosphere packaging (MAP) and vacuum packaging, but these had disadvantages such as the inability to reduce oxygen by 0.5-2 percent and the inability to control permeated oxygen during storage [22, 30, 31]. Oxygen scavengers not only reduce lipid oxidation, but they are also very effective against insects because they remove the oxygen required for insect survival, such as in rice, wheat, pulses, and other grains. Instead of fumigation, oxygen scavengers play a key role because residues are unlikely to occur [32].

Oxidation of iron and iron salts are the type of oxygen scavenging system that has been used commercially for many years as iron-based sachets. It generally involves a reaction contained in a permeable sachet between iron powders that may also contain a desiccant. Using the following reactions, iron powder reacts with oxygen and the iron powder is stored in small oxygen-permeable bags that prevent food contact [33, 34].

$$4Fe\,(OH)_2 + O_2 + 2H_2O \rightarrow 4Fe(OH)_3$$

Comprehensive information on a range of commercially accessible oxygen scavengers introduced by Suppakul *et al.* [7]. Ageless (Chemical Co., Mitsubishi Gas, Japan) is the most prevalent system for oxygen scavenging based on iron oxidation. These scavengers are designed to reduce the concentration of oxygen to less than 0.1%.

In lipid-containing foods, ascorbic acid and unsaturated fatty acids (oleic, linoleic fatty acids) are oxidized and susceptible to oxidation, resulting in the generation of undesirable flavors and toxic products such as hydro peroxides. (Figure 10.4) L-Ascorbic acid is a water-soluble compound also known as vitamin C. It is used widely as an additive in foods and cosmetics owing to its strong reducing ability [35]. Oxidation of photosensitive coloring matter: increased oxygen levels in food packages may notably reduce the shelf life of meat due to colour changes, the onset of lipid oxidation, microbial growth and nutritional losses. Oxygen absorbing systems provide an alternative to vacuum and gas flushing technologies as a means of improving product quality as well shelf life [36].

Enzymatic oxidation by Glucose oxidase: These techniques can reduce oxygen levels to < 0.01 percent that is much lower than the 0.3-3.0 percent actual level. Achieved with packaging for the residual oxygen-modified atmosphere [37]. Depending on the oxygen permeability of the packaging material, this relatively low level of oxygen; it will be maintained for long periods. Adding sachets is the great drawback in liquid food. Some O_2-scavengers use a combination of two enzymes, glucose oxidase and catalase, that would react with some substrate to scavenge incoming O_2. The glucose oxidase transfers two hydrogens from the -CHOH group of glucose, that can be originally present or added to the product, to O_2 with the formation of glucono-delta-lactone and H_2O_2.

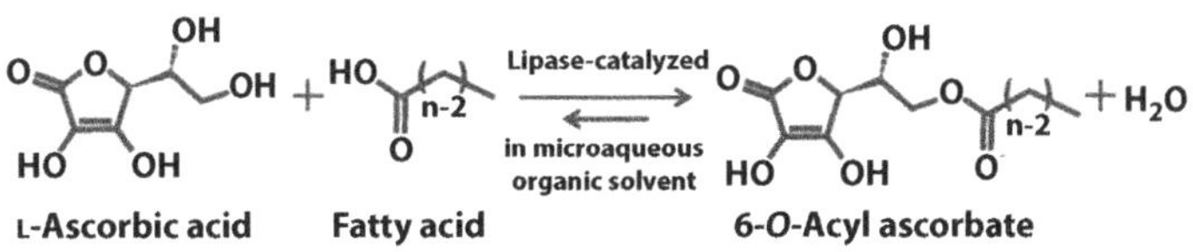

Figure 10.4 Condensation of ascorbic acid with fatty acid in a micro aqueous organic solvent with low water content catalysed by lipase.

Oxygen Interceptors are used to block the negative effect of oxygen in the air on the food, before oxygen can enter the food [38]. Antioxidants are also used to prevent oxidation and rancidity in many foods. Compounds that react with lipid or peroxide radicals or, in light, with singlet oxygen, and that are themselves oxidized to generate what are generally innocuous nontoxic compounds. Butylated Hydroxy Anisole (BHA), Butylated Hydroxy Toluene (BHT) and Propyl Gallate (PG) are synthetic antioxidants that have serious impact on human body, which is why they should be used in limited quantity [39]. Oxidation can be avoided by removing radicals (oxo, hydroxyl, and superoxide) as soon as they form. Fresh meat is preserved using a rosemary-based varnish that improves the stability of myoglobin and fresh meat against oxidation processes [40]. [41] explained that migration of α -tocopherol slows the oxidation of lipids in whole milk powder. Oil, nuts, butter, fresh meat, meat derivatives, bakery products, fruits, and vegetables can all benefit from antioxidants [42].

Oxygen absorbers

Absorbers remove oxygen from the air by physically trapping it rather than by a chemical reaction. Oxygen (O_2) is responsible for the deterioration of many foods either directly or indirectly. For example, direct oxidation reactions result in browning of fruits and rancidity of vegetable oils. Food deterioration by indirect action of O_2 includes food spoilage by aerobic microorganisms. Therefore, the incorporation of O_2 scavengers into food packages can maintain very low O_2 levels, which is useful for several applications, because it will extend the food's shelf life.

UV Absorbing Films: Film based on nanocrystallinetitania is a commonly used UV absorbing material (TiO_2). It has been proven that TiO_2-coated films exposed to sunlight can inactivate faecal coliforms in water. Under UV irradiation, TiO_2 absorbs visible light and increases its photo catalytic activity. Doping TiO_2 with silver has been shown to improve photo catalytic bacterial inactivation. This combination resulted in good antibacterial properties of TiO_2/Ag nanoparticles in a nanocomposite with PVC [43, 44].

10.2.2 Carbondioxide Absorber/Emitter

CO_2 at a high concentration (10-80%) inhibits surface microbial growth and extends shelf life. For extending the shelf life of chicken meat, modified atmosphere packaging containing CO_2 is widely used. Traditional modified atmosphere packaging can be replaced with active packaging by adding CO_2 emitter sachets to meat packages. Microbial growth inhibition was proportional to CO_2 concentration. In comparison to sixty percent CO_2, storage in hundred percent CO_2, both with and without a CO_2 emitter sachet, resulted in a microbiological shelf-life extension of 7 days. At the end of the storage period, the dominant species were *Carnobacteriumdivergens, Carnobacterium sp., and Lactococcus* sp. The carbon dioxide dissolves during storage in pure CO_2, collapse of the packageoccurs due to CO_2 [45]. High concentrated carbon dioxide usually used in non climacteric fruits, meat, poultry, fish, Fresh meat, cheeses and strawberries [46]. Ferrous carbonate, mixture of ascorbic acid and sodium bicarbonate [42].

10.2.3 Ethylene Scavengers

The primary goal of an ethylene scavenger is to absorb ethylene from a product or package in order to postpone the ripening process. It increases the shelf life of the product. Under the right conditions, ethylene biosynthesis from fresh produce could speed up ripening and shorten postharvest life. Ethylene scavenging systems, also known as active packaging, help to extend the shelf life of fresh produce by preventing ethylene-induced deterioration. Ethylene is a gas produced naturally by fresh produce that can have a negative impact on the product's quality, nutrients, and marketability. Ethylene removal systems, on the other hand, can effectively delay the effects of ethylene deterioration during storage [47, 48].

Controlling ethylene concentration during storage is critical for extending fresh produce's postharvest life. An effective scavenging system can be used to control the concentration of this gas [49]. Sadeghi *et al.* [50] explained that the ethylene produced by ripening fruit or vegetables is trapped by a chemical reagent incorporated into the packaging film. The pink colour indicates the extent of the reaction and indicates when the scavenger is depleted. Apples, bananas, mangos, tomatoes, onions, and carrots are all examples of foods that contain this compound. Bentonite, Kieselguhr and crystalline aluminosilicates, as well as zeolites, are examples of activated charcoal alone or after adsorption with bromine.

10.2.4 Flavor & Odor Absorber/Emitter

The addition of essences and odors increases the consumer's desire for the food, improves the aroma of the fresh product, and enhances the flavour of the food when the package is opened. The gradual release of odors can help to compensate for the natural loss of taste or odor in products that have a long shelf life. A few commercial products have been developed to eliminate undesirable compounds or compounds that emit odors selectively. However, removing odors from the undercarriage of food packages can be both time-consuming and costly. In the world of active packaging, the benefits of odor/aroma removal technologies like activated carbon odor absorbers are critical. Concealment odors develop in a variety of foods, including fresh poultry and cereal products. Slight and generally insignificant deterioration odors, such as sulphurous compounds from protein/amino acid breakdown, are detectable. Both advantages and disadvantages are present [51]. Foul smelling volatile amines related to fish protein breakdown, such as trimethylamine, are alkaline and can be neutralized by a number of acidic compounds. Acetylated paper may be used to eliminate large amounts of limonene, accompanied by cellulose acetate gel beads [52].

10.2.5 Humidity Control

Some of the product may face sweating or condensation problems inside packaging material, mainly in fruits and vegetable packages, it can be controlled by a humidity controller; it not only prevents condensation but also maintains the product moisture without drying out the product itself. In flowers and table grapes, mold growth can be reduced by maintaining the moisture content by humidity control [26].

Silica gels, molecular sieves (zeolites), cellulose fibre pads (soaker pads) are placed in the bottom of meat, poultry, and fresh produce trays. Ethanol, carbon dioxide, silver ions, chlorine dioxide, antibiotics, organic acids, essential oils and spices, and other substances are examples.

10.3 Antimicorbial Packaging

Active antimicrobial packaging could be an alternative solution to prevent development of spoilage and pathogenic microorganisms in fresh food and to extend their shelf life. Among the possible strategies used for active packaging, controlled release of antimicrobials from the package itself

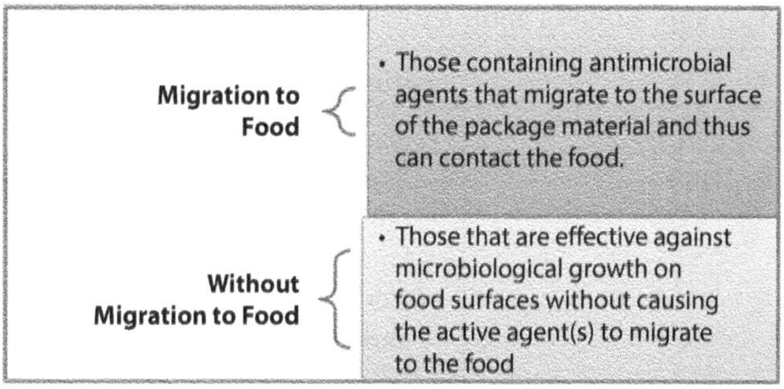

Figure 10.5 Antimicrobial agent migration mechanism.

or from additional release elements placed within the package represent the most common options. As shown in Figure 10.5, the antimicrobial packaging migration mechanism takes two forms: migration to the surface and without migration to the surface. Non-migratory antimicrobial polymers are used in the non-migration system because the antimicrobial agent does not migrate out of the polymer. It has a covalent bond with the polymer backbone [53].

Among the possible strategies used for active packaging, controlled release of antimicrobials from the package itself or from additional release elements placed within the package represent the most common options. A broad range of antimicrobials have received attention, also in consideration of the requirement for the release elements to be non-intrusive, cheap, and compatible with the geometry and physico-chemical properties of food. Whereas volatile antimicrobials (such as essential oils or their active components) have been proposed in the case of foods with a low mass/volume ratio (such as salads), water-soluble antimicrobials have been indicated as most appropriate when direct contact with the food is involved (for instance, meat and meat cuts).

10.3.1 Composition

Antimicrobial packaging system consisting of main two components as shown in the Table 10.3. the selecting of films and antimicrobial agent is a main function in the antimicrobial packaging because that films and antimicrobial agents should be suitable for each particular food. When an antimicrobial agent is added to the packaging material, polymer morphology studies are very useful in predicting possible loss of physical integrity.

Table 10.3 Composition of antimicrobial packaging.

Films	Antimicrobial agents
LDPE (low density polyethylene)	**1. Natural antimicrobial agents** - Plant sources: extracts of spices and oils. - Animal sources: chitosan. -Microorganism sources: bacteriocins and enzymes.
LLDPE (linear low density polyethylene)	**2. Chemical antimicrobial agents**
PET (polyethylene terephthalate)	Benzoic acid
Polyolefin and EVA (ethylene vinyl acetate)	Sorbic acid
Corn zein film	
Methyl cellulose	

Antimicrobial agents of a small size can be mixed together. Polymers and can be placed in amorphous regions of the polymeric structure without causing damage. Interfering significantly with polymer-polymer interconnections [54].

10.3.2 Mechanism of Antimicrobial Agents

To design an antimicrobial package, it is essential to select the right antimicrobial agent for the right package and food, and to achieve a balance between antimicrobial release and microbial growth. Antimicrobial Package, packaged food interactions and environmental conditions are crucial for ensuring the efficacy of a system [55]. Antimicrobials may alter the mechanical and barrier properties of polymers when they are incorporated into a package. Certain antimicrobials, such as enterocin-modified polyvinyl alcohol (PVOH), act as plasticizers, improving the tensile properties, particularly the flexibility, of some polymers. However, increasing concentration has been linked to a decrease in film strength. Changes in crystallinity can be directly linked to the addition of antimicrobials to some polymers, which undoubtedly contribute to the tensile property

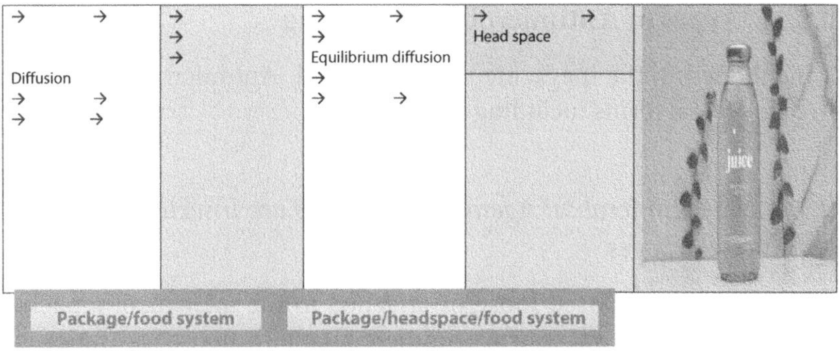

Figure 10.6 Pictorial representation of antimicrobial agent mechanism in package.

changes observed [56]. Antimicrobial packaging materials are used either inside the package space or inside foods. The food products, the headspace atmosphere, and the packaging materials make up the majority of food packaging systems. To improve antimicrobial efficiency, any one of these three components could include an antimicrobial element in Figure 10.6 [54]. Gaseous agents have been used to inhibit the growth of bacteria in addition to the use of antimicrobial packaging materials or antimicrobial inserts in the package headspace [57].

The barrier properties of films are influenced by interfacial compatibility and hydrophobicity of systems. Low interfacial compatibilities result in poor adherence and voids in the material structure, resulting in higher gas transmission rates. The presence of new compounds in the polymer matrix may also change the partition coefficient of gases [58]. Due to the nature of the antimicrobial compound, gases will be more or less soluble in the package material. Physical and chemical factors influence antimicrobial compound diffusion through packaging materials. The presence of additives, as well as interactions with the packaging polymer via hydrogen bonds, hydrophobic or electrostatic interactions, etc., are extremely important such as nano-clays, might decrease diffusion rates due to tortuosity [58].

Diffusional migration dominates the mass transfer of nonvolatile antimicrobials. Nonvolatile agents will first be placed in packaging materials or between the package and the food's surface. If nonvolatile antimicrobial agents are incorporated into a packaging material matrix, the agents' diffusivity is a primary kinetic constant of the matrix to surface transfer [59, 60].

10.3.3 Types of Antimicrobial Packaging

As mentioned in the magazine written by [61]. Antimicrobial packaging can take several forms including (Figure 10.7).

10.3.3.1 Antimicrobial Agent Sachets/Pads are Inserted Into Packages

Sachets that are enclosed loose or attached to the interior of a package have been the most successful commercial application of antimicrobial packaging. Oxygen absorbers, moisture absorbers, and ethanol vapor generators have all been popular. Oxygen and moisture absorbers are commonly used in bakery products and pasta production and packaging for meat to prevent oxidation and water condensation despite the fact that oxygen absorbers aren't designed to be antimicrobial, a reduction in oxygen inhibits the growth of aerobes, particularly moulds. Moisture absorbers can lower water activity (a_w), affecting microbial growth indirectly.

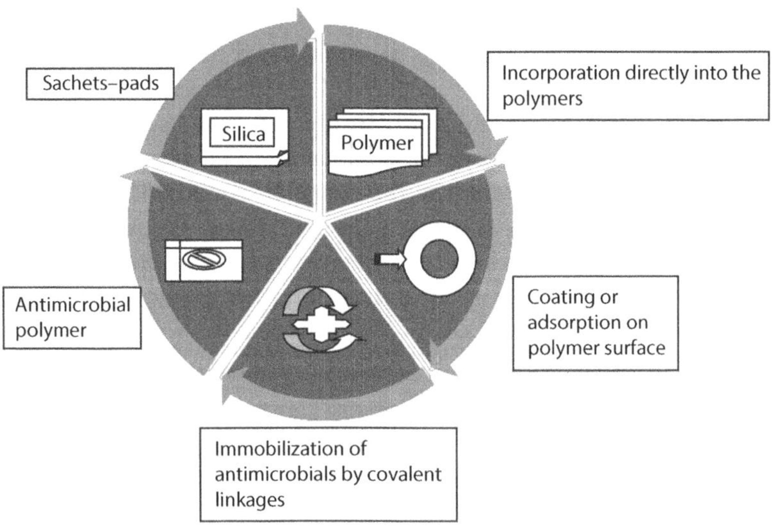

Figure 10.7 Methods of incorporating antimicrobial agents into films.

10.3.3.2 *Antimicrobial Agents are Directly Incorporated Into Polymers*

Drug and pesticide delivery, household goods, textiles, surgical implants, and other biomedical devices have all benefited from the incorporation of bioactive agents such as antimicrobials into polymers. According to the number of recently published articles and patents, research into the incorporation of antimicrobials into food packaging has more than doubled in the last five years. Silver substituted zeolites are the most widely used antimicrobial polymer additives for food applications, particularly in Japan. Silver ions, which are antimicrobial against a wide range of bacteria and moulds, replace sodium ions in zeolites. Zeomic, Apacider, AgIon, Bactekiller and Novaron are commercial examples of silver substituted zeolites.

10.3.3.3 *Coating or Adsorbing Antimicrobials to Polymer Surfaces*

Fungicides were incorporated into waxes to coat fruits and vegetables, and shrink films coated with quaternary ammonium salts to wrap potatoes in the early days of antimicrobial packaging. Other early innovations included absorbic acid-coated wax paper and cellulose casings for wrapping sausages and cheeses. Antimicrobials that are unable to withstand the high temperatures used in polymer processing are frequently used. They're frequently applied to the material after it's been formed, or they're mixed in with cast films.

10.3.3.4 *Immobilization of Antimicrobials by Ionic or Covalent Linkages to Polymers*

Functional groups on both the antimicrobial and the polymer are required for this type of immobilization. Peptides, enzymes, polyamines, and organic acids are examples of antimicrobials with functional groups. Immobilization may require the use of spacer molecules to link the polymer surface to the bioactive agent, in addition to functional antimicrobials and polymer supports. These spacers provide enough movement for the active portion of the agent to come into contact with microorganisms on the food surface. Due to their low toxicity and widespread use in foods, dextrans, polyethylene glycol, ethylenediamine and poly-ethylene-imine could be used as antimicrobial packaging spacers.

10.3.4 Commercial Antimicrobial Packaging Products and Manufactures

In the present era, there are so many manufacturer involved in production of active compound for antimicrobial as well as active packaging, some are listed below Table 10.4.

Table 10.4 Commercial antimicrobial packaging products and manufactures.

Trade name	Active compounds	Manufacturer	Reference
Microban	Triclosan	Microban Products Co. (USA)	[38]
Micro garde (USA)	Clove and others	Rhone- poulenc	[38]
Take guard	Bamboo extract	Takex. Co. (Japan)	[38]
Microfree	Ag, copper oxide, zinc Silicate	DuPont (USA)	[38, 89]
Micro atmosphere	Chlorine dioxide	Southwest (USA), Bernard Technologies (USA) Research Institute	[38]

10.4 Uses of Active Packaging

- To improve food safety
- To maintain the quality of food during its shelf life
- To extend the product's shelf life
- To cut down on food waste
- To cut down on the use of food additives
- As a promotional tool
- To create new products.

10.5 Comparison Between Active and Intelligent Packaging

The active packaging refers to the active functions outside the inert passive containment and protection of the product. Whereas intelligent packaging refers to the ability to measure the attribute of the product; the shipping environment or the inner atmosphere of the product and various other factors.

The main objective of this type of packaging is to provide information; sense and indicate whether or not the product's quality has deteriorated. Intelligent or smart packaging is designed to monitor and communicate information about food quality. Examples include time-temperature indicators (TTIs), ripeness indicators, biosensors, and radio frequency identification. These smart devices may be incorporated in package materials or attached to the inside or outside of a package. intelligent and smart packaging usually involve the ability to sense or measure an attribute of the product, the inner atmosphere of the package, or the shipping environment. This information can be communicated to users or can trigger active packaging functions. Active packaging prevents deterioration, whereas intelligent packaging indicates the deterioration to maintain the active packaging.

10.6 Market Report on Active and Intelligent Packaging

In 2020, the market for active and intelligent packaging was estimated to be worth USD 17.5 billion. It is forecast to have a value of USD 25.16 billion by 2026, with a compound annual growth rate (CAGR) of 6.78 percent from 2021 to 2026. US$20.386 billion in 2019, the active packaging market is expected to grow at a CAGR of 6.74 percent to US$30.145 billion by 2025. From 2010 to 2015, active and intelligent packaging technology grew at the fastest rates, with CAGRs of 10.5 percent and 12.1 percent, respectively. Modified atmosphere technology accounted for the largest share of the total market in advanced packaging technology (approximately 54 percent). From 2010 to 2015, the global advanced packaging technology market is expected to grow at an annual rate of 8.2 %. (https://www.marketsandmarkets.com/Market-Reports/smart-packaging-324.html). According to Market Research Future, 2021 [62] estimation the active and intelligent packaging market was worth USD 15.11 billion in 2016 (https://www.marketresearchfuture.com/reports/active-and-intelligent-packaging-market-5550). The active and intelligent packaging market was worth USD 17.48 billion in 2017 and is expected to grow to USD 28 billion by 2023, with a CAGR of 8.17 percent (2018-2023). (Global Growth, Trends, and Forecasts for the $28 Billion Active and Intelligent Packaging Market) (2018-2023) (ResearchAndMarkets.com; www.businesswire.com) [63].

During the forecast period of 2019-2027, the active, smart, and intelligent packaging market for cosmetics and personal care is expected to grow at a CAGR of 7.8% ([87]; www.transparencymarketresearch.com). The overall expected value with respect to year as shown in the Figure 10.8.

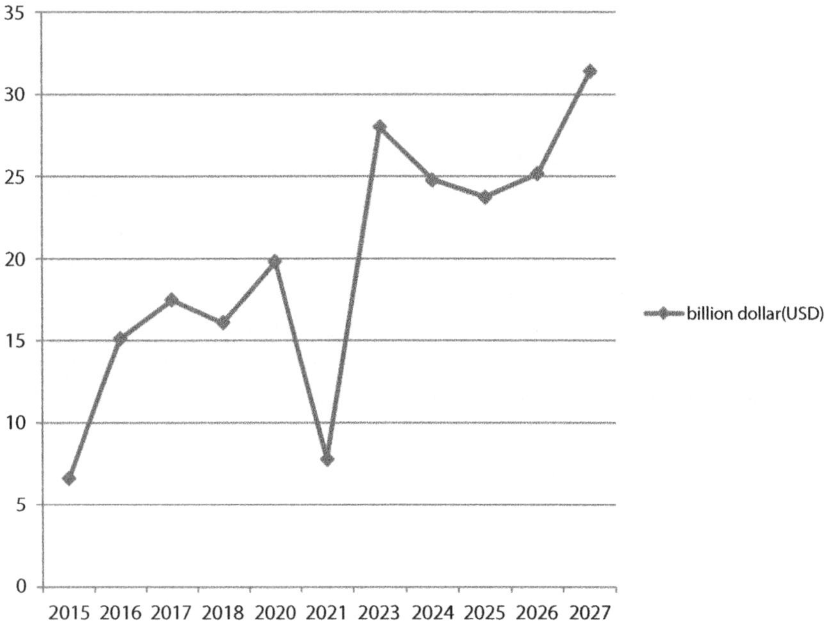

Figure 10.8 Expected market value of active and intelligent packaging.

10.7 Disadvantage

- Cannot be used in liquid foods.
- Cannot be used in packages made of flexible film, as the film will cling to the sachet and prevent it from performing its function.
- Moisture absorbing sachets can alter the sensory properties of packaged foods in some cases, and their absorbing ingredients are unlikely to be consumed unconsciously by customers [2].
- Another limitation is the perforation of outer layers in the pad's structure, which should be spaced properly to prevent film tearing and the release of absorbent material [64].
- Moisture-absorbing material from the film migrates into packaged foods, which is a serious problem. Another challenge to consider is the requirement for large quantities of absorbing materials to uniformly disperse them throughout the packaging films [49].
- Risk of accidental ingestion by consumers

- Materials containing active components
- Incorporating the scavenger into the packaging material itself
- More efficient
- It's important to consider the activation mechanisms, moisture absorption capacity, and safety. Price, regulation, and consumer acceptance are all important and influential factors.

10.8 Advantage

- As a means of improving product quality and shelf life, O2 absorbing systems are an alternative to vacuum and MAP technologies.
- Keeping the oxidation of the edibles to a minimum will help to keep the product from degrading. Color, nutritive value, and flavour changes may be directly linked to degradation caused by oxygen combining with the comestibles.
- Prevent the growth of aerobic microorganisms.
- Reduce or eliminate the need for preservatives and antioxidants
- Slow down metabolism of food.
- Prevent oxidation phenomena: rancidification of fats and oils and consequent emergence of off-odors and off-flavors, loss or change of colors characteristic of food, loss of oxygen-sensitive nutrients

10.9 Safety Issues in Active Packaging

In order to conduct a thorough risk analysis, it's critical to assess the potential for packaging constituents to migrate into food and to assess their potential hazard. Despite this, there have been few studies published on the effects of nanomaterials on ingestion or the potential interaction of nanomaterial-based food contact materials with food components. The migration of silver from three kinds of nanocomposites into food stimulants has been investigated, including an examination of the form of silver migrating either ions or particles migrating. Their findings revealed that silver migrated into food stimulants, with acidic foods showing the most migration. Furthermore, Heat causes migration, and microwave heating causes more migration than a traditional oven [65]. Silver migration could occur via two mechanisms, according to the authors: detachment of silver nanoparticles from composites or oxidative dissolution of silver ions [66].

The percentage of nano-filler in nano-composites, rather than size of particle, temperature, or contact time, was found to be one of the most important factors driving migration in the study [65]. This research also included the development of a model to study particle migration from food packaging. This model was a reliable indicator of nano-silver migration and, to a lesser extent, nano copper migration into food stuff and,when further developed and validated, it has the potential to save industry time and money by reducing the time and costs normally associated with migration studies [67].

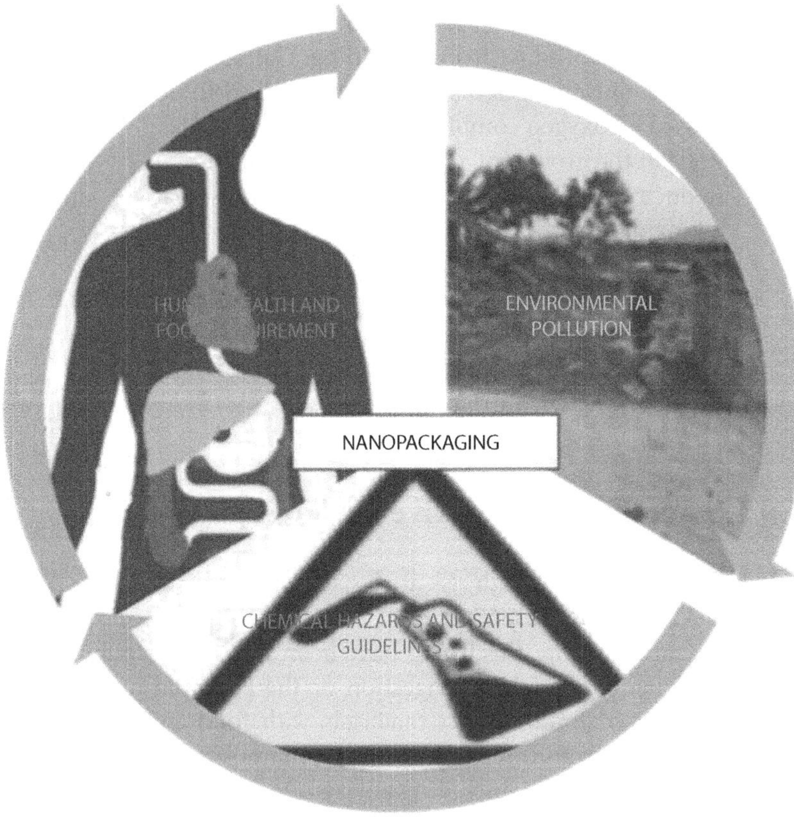

Figure 10.9 Considerable factors in nanopackaging.

A recent study looked at the migration and toxicological profile of an organo-modified clay polylactic acid nanocomposite that could be used as a food contact material. Under the conditions of the experiment, migration studies revealed that less than 10 mg per dm 2 of the nanocomposite migrated in water, and further analysis of the food stimulant revealed that the levels of metals measured were below the permitted values. The authors also tested the potential toxicity of migration extracts *in vitro* and *in vivo* [65, 68].

As mentioned by MarijaJovic, 2017 [69], in Europe, the use of nanoparticles is generally prohibited unless they are specifically authorized in their nano form, and the European Food Safety Authority (EFSA) recommends that nanomaterial risk assessment be done on a case-by-case basis, with only a few exceptions. The EFSA also proposes that the 50% nanoparticle threshold be lowered to 10% only in cases where there are environmental, health, or safety concerns, which pertains to food and medical applications (Figure 10.9).

10.10 Applications in Food Industry

Food contamination and Food spoilage caused by pathogens and microorganisms found in food is a serious issue [70]. As a result, there is a growing demand for antimicrobial drugs in food packaging. Most importantly, antibacterial agents are required to prevent bacteria from growing in food [71]. It is important to store commercial food items for longer periods for remote transport or to conserve them for off-season. The antibacterial agent can be divided into two groups, one natural and the other synthetic antibacterial agent. Antibacterial agents that are synthesized have the advantages of low cost and high activity. Compared with natural antibacterial agents, but their toxicity is typically higher than that of natural antibacterial agents. Finally, potential production of the packaging of antimicrobial foods is suggested. Designing and synthesizing antibacterial drugs with high efficacy and low toxicity is an urgent topic for researchers. Polymeric food packaging films are some of the most commonly used films because they are easy to produce and have excellent performance. Polymeric food packaging has properties such as flexibility, strength, stiffness, and a barrier to oxygen and moisture that can protect food from microbial attack. According to the source of antimicrobial agents, this review introduces polymeric film-based antibacterial packaging and focuses on natural antimicrobial packaging and synthetic organic antimicrobial packaging [72]. Antimicrobialscan be produced by a num-

Table 10.5 Application of natural active agents in food by using different technology.

Technology	Material or metal agent	Scientific name	Application	References
Ultra-nano filtration	Jambolan	*Syzygiumcumini (L.)*	Natural antioxidant	[78]
coating	Carp	*Cyprinuscarpio*	Prevention of oxidation, antimicrobial	[79]
Nanotechnology	Essential oil and vegetable oil (flaxseed oil)	*Linumusitatissimum*	Natural antioxidant	[80]
Nanotechnology	Chitosan	–	Antimicrobial, antioxidant	[81]
Plasticizing	Alginate and yerba mate	*Ilex paraguariensis St. Hil.*	Antioxidant	[82]

ber of compounds, such as organic acid; benzoic acid incorporated in tilapiafilletsstimulantsthat inhibit the growth of bacteria [73]. Sorbates are used as an antifungal agent in cheese culture [74]. Enzymes can also be used as antimicrobials, such as lysozyme, nisin, and EDTA, which are used to inhibit the growth of *E. coli* and *Lactobacillus plantarum* [75]. Bacteriocins are also used in beef culture to stop *Brochothrixthermosphacta* and *Staphylococcus aureus* from growing [76]. Antifungal agents such as benomyl are used in media culture, and imazal is used in bell pepper cheese to prevent mould formation. [77] some of the polymers like chitosan and herb extract used to control the growth of *E. coli, Lb.plantarum, Fusariumoxysporum, S. cerevisiae.*

In pre-cooked chicken meatballs, the aqueous extract and dried leaves of yerba mate were used. Both interventions had a beneficial effect on the formation of secondary lipid oxidation products, particularly when used at 0.10 percent (w/w) concentrations [83].

[84] Natural extracts from grapefruit seed extract used in ground beef to prevent the development of aerobes and coliforms. Clove extracts used in culture media to avoid contamination by *E. coli, Lb. plantarum, Fusariumoxysporum* and *S.cerevisiae.* Eugenol and cinna-maldehyde used in bologna ham to inhibit the growth of *Enterobacter, Lactic acid bacteria, Lb. sakeiserratia*sp. Horseradish extracts were also used in ground beef to inhibit *E. coli.* [85]. Two of the most common pathogens involved in food borne outbreaks are *Escherichia coli* O157:H7 and *Staphylococcus aureus.* Food safety depends on the control of these pathogens in foods. Consumers prefer more natural than synthetic preservative foods, so the use of natural antimicrobial compounds found in edible plants to control food borne pathogens is of great interest. Wasabi (Japanese horseradish) and several other edible plants contain allyl isothiocyanate (AITC), an antimicrobial compound. Some of the application of natural active agents in food by using different technology mentioned in the Table 10.5.

10.11 Recent Advancement in Antimicrobial Packaging Films

The utilization of Nanotechnology in active packaging systems involving polymeric films or coating materials having antibacterial, antifungal and anti-oxidative nature in food products has proved beneficial in improving food quality and thus, enhancing shelf life. Oplon (Israel) along with Reynold group commercialized the antimicrobial packaging basically for perishable foods like: fresh meat, beverages and liquid foods also avoiding the usage of preservatives. It also reduces the cost of refrigeration and chilling, protecting food against antibiotic resistant and heat resistance pathogens. Moreover, the Non-toxic and environment friendly base do not leach or release chemicals from the films thus adding to the safety of the final product. The majority of research has been done on the properties of active packaging agents in the absence of food. It is preferable to investigate the active package agent in conjunction with various foods, as well as the effect on packaging material, active agent release, and active agent loss effectiveness on food.

10.12 Challenges

- Susceptibility to the high processing temperature used to manufacture packaging polymers.
- Regulation issues.
- These concepts, which are successfully utilized in the US and Japan, have seen only limited development in Europe. This could be due to legal restrictions, to a lack of knowledge about both the acceptability of these systems to consumers and their effectiveness in packaging, or to their economic and environmental impact [86].

10.13 Conclusion

To commercialize the application, it is critical to understand how the antibacterial activity of the agents changes when they're incorporated into the packaging film compared to their original activity, so you can figure out what levels to use for effective bacterial inhibition. Antimicrobial packaging's global market is steadily growing, from 6.51 million tons in 2015 to 10.09 million tons by 2024 (11 billion USD). By considering the limitations of active packaging, improved and cost-effective techniques can be used to overcome these limitations. The cost of waste management can be reduced by using natural active agents derived from industrial waste byproducts. In comparison to synthetic active agents, natural active agents have fewer or no side effects. Development of active packaging based on consumer expectations and public health, as well as cost effectiveness. Meat, milk, fruits, and vegetable products can be packed in a variety of ways depending on their characteristics. Active ingredients used in the packaging of a variety of products. The active component made from synthetic chemicals should be non-toxic, comply with regulations, and not pollute the environment.

Active packaging technology passively protects food items, by not only inhibiting pathogenic growth, but by also providing an extended shelf life combating a variety of environmental factors. Since the nano material antimicrobial agents possess a higher surface area-to-volume ratio in comparison to their normal counterparts, they are considerably more efficient against food pathogens. Various antimicrobials based nanocomposite food packaging films incorporated with antimicrobial agents have shown enhanced thermal, physicochemical, mechanical, and optical properties. The active packaging at present day made a main role in food industry to store the different perishable, sensitive food for longer period without altering the food product flavor, colour, odor, and texture. When developing an active component from a chemical, it should be compared to appropriate foods to see if the active packaging is feasible.

References

1. Gomez-Estaca, J., López-de-Dicastillo, C., Hernández-Muñoz, P., Catalá, R., & Gavara, R., Advances in antioxidant active food packaging. *Trends Food Sci. Technol.*, 35(1), 42-51, 2014.
2. Gaikwad, K.K., Singh, S., & Negi, Y.S., Ethylene scavengers for active packaging of fresh food produce. *Environ. Chem. Lett.*, 18(2), 269-284, 2020.
3. Gupta, A., Sahni, P., Sharma, S., & Singh, B., New insight in active packaging of fruits, in: *Emerging Technol. Shelf-Life Enhancement of Fruits*, 129, 2020.
4. Barbiroli, A., Bonomi, F., Capretti, G., Iametti, S., Manzoni, M., Piergiovanni, L. & Rollini, M. Antimicrobial activity of lysozyme and lactoferrin incorporated in cellulose-based food packaging. *Food Control*, 26(2): 387-392, 2015.
5. Corradini, M.G. Shelf life of food products: from open labeling to real-time measurements. *Annu. Rev. Food Sci. Technol.*, 9, 251-269, 2018.
6. Vilela, C., Kurek, M., Hayouka, Z., Röcker, B., Yildirim, S., Antunes, M. D. C., Freire, C.S., A concise guide to active agents for active food packaging. *Trends Food Sci. Technol.*, *80*, 212-222, 2018.
7. Suppakul, P., Miltz, J., Sonneveld, K., & Bigger, S.W., Active packaging technologies with an emphasis on antimicrobial packaging and its applications. *J. Food Sci.*, 68(2), 408-420, 2003.
8. Krasniewska, K., & Gniewosz, M., Substances with antibacterial activity in edible films–a review. *Pol J of Food NutrSci, 62*(4), 199-206, 2012.

9. Ali, A., Noh, N. M., & Mustafa, M. A., Antimicrobial activity of chitosan enriched with lemongrass oil against anthracnose of bell pepper. *Food Package Shelf Life*, 3, 56-61, 2015.
10. Kim, I., Viswanathan, K., Kasi, G., Thanakkasaranee, S., Sadeghi, K. & Seo, J., ZnO Nanostructures in Active Antibacterial Food Packaging: Preparation Methods, Antimicrobial Mechanisms, Safety Issues, Future Prospects, and Challenges. *Food Rev. Int.*, 1-29, 2020.
11. Kerry, J.P., O'grady, M. N., & Hogan, S. A., Past, current and potential utilisation of active and intelligent packaging systems for meat and muscle-based products: A review. *Meat Sci.*, 74(1), 113-130, 2006.
12. Singh, P., Wani, A.A., Saengerlaub, S., Active packaging of food products: recent trends. *Nutr. Food Sci.*, 2011.
13. Pereira de Abreu, D. A., Cruz, J. M., & PaseiroLosada, P., Active and intelligent packaging for the food industry. *Food Rev. Int.*, 28(2), 146-187, 2012.
14. Yildirim, S., Röcker, B., Pettersen, M. K., Nilsen-Nygaard, J., Ayhan, Z., Rutkaite, R., Coma, V., Active packaging applications for food. *Compr Rev Food Sci F, 17*, 1, 165-199, 2018.
15. Mu, H., Gao, H., Chen, H., Fang, X., & Han, Q., A novel controlled release ethanol emitter: preparation and effect on some postharvest quality parameters of Chinese bayberry during storage. *Journal of the Science of Food and Agriculture, 97*(14), 4929-4936, 2017.
16. Qin, Y. Y., Yang, J. Y., Lu, H. B., Wang, S. S., Yang, J., Yang, X. C., & Cao, J. X., Effect of chitosan film incorporated with tea polyphenol on quality and shelf life of pork meat patties. *Int. J. Biol. Macromol., 61*, 312-316, 2013.
17. Yang, H. J., Lee, J. H., Won, M., Song, K. B., Antioxidant activities of distiller dried grains with solubles as protein films containing tea extracts and their application in the packaging of pork meat. *Food Chem, 196,* 174-179, 2016.
18. Piñeros-Hernandez, D., Medina-Jaramillo, C., López-Córdoba, A., & Goyanes, S., Edible cassava starch films carrying rosemary antioxidant extracts for potential use as active food packaging. *Food Hydrocoll. 63,* 488-495, 2017.
19. Salminen, A., Latva-Kala, K., Randell, K., Hurme, E., Linko, P., & Ahvenainen, R. The effect of ethanol and oxygen absorption on the shelf-life of packed sliced rye bread. *PackagTechnolSci, 9*(1), 29-42, 1996.
20. Del Nobile, M. A., Bove, S., La Notte, E., & Sacchi, R., Influence of packaging geometry and material properties on the oxidation kinetic of bottled virgin olive oil. *J. Food Eng.*, 57, 2, 189–197, 2003.
21. Schroeter, L. C., Sulfur Dioxide: Applications in Foods, Beverages and Pharmaceuticals. Oxford and New York, 1966.
22. Dey, A., & Neogi, S., Oxygen scavengers for food packaging applications: A review. *Trends Food Sci.Technol., 90*, 26-34, 2019.
23. Gaikwad, K. K., Singh, S., & Ajji, A., Moisture absorbers for food packaging applications. *Environ. Chem. Lett.*, 17(2), 609–628, 2019.

24. Antonopoulou, M., Evgenidou, E., Lambropoulou, D., & Konstantinou, I., A review on advanced oxidation processes for the removal of taste and odor compounds from aqueous media. *Water Resour. Res. 53, 215-234,* 2014.
25. Fotiou, T., Triantis, T. M., Kaloudis, T., Papaconstantinou, E., & Hiskia, A., Photocatalytic degradation of water taste and odour compounds in the presence of polyoxometalates and TiO2: Intermediates and degradation pathways. *J. Photochem. and Photobiol. A: Chem., 286*, 1-9, 2014.
26. Lopez-Rubio, A., Lagarón, J. M., & Ocio, M. J., Active polymer packaging of non-meat food products. *Smart packaging technologies for fast moving consumer goods*, 19-32, 2008.
27. Quincot, G., Azenha, M., Barros, J., & Faria, R., Use of salt solutions for assuring constant relative humidity conditions in contained environments. *Fundação para a Ciência e a Tecnologia,* 2011.
28. Farrell, C. J., & Tsai, B. C., U.S. Patent No. 4,536,409. Washington, DC: U.S. Patent and Trademark Office, 1985.
29. Castro, A., Carvalho, I., Marques, L., Ferreira, P. J., Cavaleiro, A., & Carvalho, S., Galvanic oxidation of bimetallic Zn-Fe nanoparticles for oxygen scavenging. *Appl. Surf. Sci., 537*, 147896, 2021.
30. Church, I. J., & Parsons, A. L., Modified atmosphere packaging technology: a review. *J. Sci. Food Agric., 67*(2), 143-152, 1995.
31. Mills, A., Oxygen indicators and intelligent inks for packaging food. *Chemical Society Reviews, 34*(12), 1003-1011, 2005.
32. Vermeiren, L., Heirlings, L., Devlieghere, F., Debevere, J., Oxygen, ethylene and other scavengers. *Novel food packaging techniques*, 22-49, 2003.
33. Cooksey, K., Oxygen scavenging packaging systems. *Encyclopedia of polymer sci. and technol.*, 2002.
34. Vermeiren, L., Devlieghere, F., De Kruijf, N., & Debevere, J., Development in the active packaging of foods. *Afr. J. Food Sci.*, 5, 1, 6-13, 2000.
35. Watanabe, Y., Antioxidative properties of acyl ascorbates on lipid oxidation in bulk, disperse and microcapsule systems. *J. Food Sci. Technol., 21*, 5, 639-647, 2015.
36. Ozdemir, M., & Floros, J.D., Active food packaging technologies. *Crit. Rev. Food Sci. Nutr., 44*(3), 185-193,2004.
37. Sen, A., Dhavan, P., Shukla, K. K., Singh, S., & Tejovathi, G., Analysis of IR, NMR and antimicrobial activity of β-sitosterol isolated from Momordicacharantia. *Sci. Secure J. Biotechnol., 1*(1), 9-13, 2012.
38. Brody, A.L., Strupinsky, E.P., & Kline, L.R., *Active packaging for food applications.* CRC Press, 2001.
39. Lobo, V., Patil, A., Phatak, A., & Chandra, N., Free radicals, antioxidants and functional foods: Impact on human health. *Pharmacogn. Rev., 4*(8), 118–126, 2010.
40. Prasad, P. & Kochhar, A., Active packaging in food industry: a review. *JESTFT*, *8*(5), 1-7, 2014.

41. Khan, I.T., Nadeem, M., Imran, M., Ullah, R., Ajmal, M., & Jaspal, M.H., Antioxidant properties of Milk and dairy products: a comprehensive review of the current knowledge. *Lipids Health Dis.*, *18*(1), 1-13, 2019.
42. Nandulal A M., slideshare. https://www.slideshare.net/NandhuLal/active-packaging-64534625,2016.
43. Parejo, P.G., Zayat, M., & Levy, D., Highly efficient UV-absorbing thin-film coatings for protection of organic materials against photodegradation. *J. Mater. Chem. A*, *16*(22), 2165-2169, 2006.
44. Younas, H., Qazi, I. A., Hashmi, I., Awan, M. A., Mahmood, A., Qayyum, H.A., Visible light photocatalytic water disinfection and its kinetics using Ag-doped titania nanoparticles. *Environ SciPollut Res, 21,* 1, 740-752, 2014.
45. Holck, A. L., Pettersen, M. K., Moen, M. H., & Sørheim, O., Prolonged shelf life and reduced drip loss of chicken filets by the use of carbon dioxide emitters and modified atmosphere packaging. *J. Foodprot.*, *77*(7), 1133-1141, 2014.
46. Day, B.P., Active packaging of food. *Smart packaging technologies for fast moving consumer goods*, *1*, 2008.
47. Amit, S.K., Uddin, M.M., Rahman, R., Islam, S.R. and Khan, M.S. A review on mechanisms and commercial aspects of food preservation and processing. *Agric. & Food Secur.*, *6*(1), 1-22, 2017.
48. Sivakumar, D.; Bautista-Baños, S. J. Review on the Use of Essential Oils for Postharvest Decay Control and Maintenance of Fruit Quality during Storage. *J. Crop Prot.*, 64, 27–37, 2014.
49. Gaikwad, K. K., & Lee, Y. S. Current Scenario of Gas Scavenging Systems Used in Active Packaging: A Review. *Korean J. Package Sci. Technol.*, *23*(2), 109–117, 2017.
50. Sadeghi, K., Lee, Y., & Seo, J., Ethylene scavenging systems in packaging of fresh produce: A review. *Food Rev. Int.*, *37*(2), 155-176, 2021.
51. Carchedi, M., Bartolino, R., Gabriele, B., Fazio, A., & Sindona, G., *Synthesis of innovative material for food packaging* (Doctoral dissertation), 2013.
52. Coles, R., McDowell, D., & Kirwan, M. J. (Eds.), *Food Package Technol.*, (Vol. 5). CRC Press, 2003.
53. Steven, M.D., Hotchkiss, J.H., Non-migratory bioactive polymers (NMBP) in food packaging. In: Ahvenainen, R. (Ed.), Novel Food Packaging Techniques. Woodhead Publishing Ltd, Cambridge, UK, 71-102, 2003.
54. Han, Jung H., Plastic films in food packaging. *Antimicrobial Packaging Systems*, 151–180, 2013.
55. Malhotra, B., Keshwani, A. & Kharkwal, H. Antimicrobial food packaging: Potential and pitfalls. *Front. Microbiol.*, *6*, 611, 2015.
56. Marcos, B., Aymerich, T., Monfort, J.M., & Garriga, M., Physical performance of biodegradable films intended for antimicrobial food packaging. *J. Food Sci.*, *75*(8), E502-E507, 2010.
57. Rawdkuen, S., Punbusayakul, N., Lee, D. S., Antimicrobial packaging for meat products. In *Antimicrobial Food Packaging*. Academic Press, 229-241, 2016.

58. Han, J.H., Edible films and coatings: a review. Innovations in food packag., 213-255, 2014.
59. Choi, J.H., Choi, W.Y., Cha, D.S., Chinnan, M.J., Park, H.J., Lee, D.S., Diffusivity of 174 plastic films in food packaging potassium sorbate in kappa-carrageenan based antimicrobial film. *Food Sci. Technol.* 38, 417-423, 2005.
60. Han, J.H., Antimicrobial food packaging. *Food Technol. 54*(3), 56-65, 2000.
61. Katke S D, Deshpande H W and Sawate A R, Antimicrobial is one of many applications of active packaging. *Food Beverage News* (magazine), 2020.
62. Market research future (2021). https://www.marketsandmarkets.com/Market-Reports/smart-packaging-324.html
63. ResearchAndMarkets.com, www.businesswire.com.
64. Otoni, C. G., Espitia, P. J., Avena-Bustillos, R. J., & McHugh, T. H., Trends in anti-microbial food packaging systems: Emitting sachets and absorbent pads. *Food Research International,* 83, 60–73, 2016.
65. Bhunia, K., Sablani, S. S., Tang, J., & Rasco, B., Migration of chemical compounds from packaging polymers during microwave, conventional heat treatment, and storage. *Compr. Rev. Food Sci. F, 12*(5), 523-545, 2013.
66. Echegoyen, Y., & Nerín, C., Nanoparticle release from nano-silver antimicrobial food containers. *Food and Chemi. Toxicol., 62,* 16-22, 2013.
67. Cushen, M., Kerry, J., Morris, M., Cruz-Romero, M., & Cummins, E., Migration and exposure assessment of silver from a PVC nanocomposite. *Food Chem., 139*(1-4), 389–397, 2013.
68. Kuswandi, B., Nanotechnology in food packaging. In *Nanosci. Food Agri. 1* (pp. 151-183). Springer, Cham., 151-183, 2016.
69. MarijaJovic,Challenges of Using Nanomaterials in Packaging, 2017. https://www.prescouter.com/2017/07/using-nanomaterials-packaging
70. Estevinho, L. M., Combarros-Fuertes, P., & Paula, V. B., Recent Advances in Applied Microbiology. *Microorganisms, 8*(9), 1364, 2020.
71. Kamala, K., & Kumar, V. P., Food Products and Food Contamination. In *Microbial Contamination and Food Degradation.* Academic Press, 1-19, 2018.
72. Huang, T., Qian, Y., Wei, J., & Zhou, C., Polymeric antimicrobial food packaging and its applications. *Polymers, 11*(3), 560, 2019.
73. Huang, N., Angeles, E. R., Domingo, J., Magpantay, G., Singh, S., Zhang, G., & Khush, G. S., Pyramiding of bacterial blight resistance genes in rice: marker-assisted selection using RFLP and PCR. *Theor. Appl. Genet., 95*(3), 313-320, 1997.
74. Padgett, T., Han, I. Y., & Dawson, P. L., Incorporation of food-grade antimicrobial compounds into biodegradable packaging films. *J. Food Prot., 61*(10), 1330–1335, 1998.
75. Cooksey, K., Utilization of antimicrobial packaging films for inhibition of selected microorganisms, 17-25, 2000.
76. Halek, G.W., & Garg, A., Fungal inhibition by a fungicide coupled to an ionomeric film. *J. FoodSaf., 9*(3), 215-222, 1988.

77. Yi SS, Yun HS, Yoon, S., Calcareous nannoplankton from the Seoguipo Formation of Cheju Island, Korea and its paleoceanographic implications. *Paleontol. Res.* 2:253e265, 1998.
78. Koop, B. L., Knapp, M. A., Di Luccio, M., Pinto, V. Z., Tormen, L., Valencia, G. A., & Monteiro, A. R., Bioactive Compounds from Jambolan (Syzygiumcumini (L.)) Extract Concentrated by Ultra-and Nanofiltration: a Potential Natural Antioxidant for Food. *Plant Foods Hum. Nutr.*, 1-8, 2021.
79. Tkaczewska, J., Kulawik, P., Jamróz, E., Guzik, P., Zając, M., Szymkowiak, A., & Turek, K., One-and double-layered furcellaran/carp skin gelatin hydrolysate film system with antioxidant peptide as an innovative packaging for perishable foods products. *Food Chem.*, 129347, 2021.
80. Wrona, M., Silva, F., Salafranca, J., Nerín, C., Alfonso, M.J., & Caballero, M. Á., Design of new natural antioxidant active packaging: Screening flowsheet from pure essential oils and vegetable oils to ex vivo testing in meat samples. *Food Control, 120*, 107536, 2021.
81. Kongkaoroptham, P., Piroonpan, T., & Pasanphan, W., Chitosan nanoparticles based on their derivatives as antioxidant and antibacterial additives for active bioplastic packaging. *Carbohydr. Polym., 257*, 117610, 2021.
82. de Farias, N. S., Silva, B., de Oliveira Costa, A. C., & Müller, C. M. O. Alginate based antioxidant films with yerba mate (Ilex paraguariensis St. Hil.): Characterization and kinetics of phenolic compounds release. *Food Packag. Shelf Life, 28*, 100548, 2021.
83. Racanicci, A. M., Danielsen, B., & Skibsted, L. H., Mate (Ilex paraguariensis) as a source of water extractable antioxidant for use in chicken meat. *Eur. Food Res. Technol., 227*(1), 255-260, 2008.
84. Ha, Y., Stevens, D. J., Skehel, J. J., Wiley, D. C. X-ray structures of H5 avian and H9 swine influenza virus hemagglutinins bound to avian and human receptor analogs. *Proc. Natl. Acad. Sci., 98*(20), 11181-11186, 2001.
85. Lu, Z., Dockery, C. R., Crosby, M., Chavarria, K., Patterson, B., & Giedd, M. Antibacterial Activities of Wasabi against Escherichia coli O157: H7 and Staphylococcus aureus. *Front. Microbiol., 7*, 1403, 2016.
86. Véronique, C. O. M. A., Bioactive packaging technologies for extended shelf life of meat-based products. *Meat Sci., 78*(1-2), 90-103, 2008.
87. Market-Snapshot, R. L. G. (2019). Active, smart and intelligent packaging market: Packaging, Transparency Market Research. TMRGL766 232 Page.
88. Kuswandi, B. Active and intelligent packaging, safety, and quality controls. Fresh-cut fruits and vegetables, 243-294, 2020.
89. Vermeiren, L., Devlieghere, F., & Debevere, J. Effectiveness of some recent antimicrobial packaging concepts. Food additives & contaminants, 19(S1), 163-171, 2002.

11

Supercritical Fluid

Cassia Pereira Barros[1], Jonas Toledo Guimarães[1], Tatiana Colombo Pimentel[2], Erick Almeida Esmerino[1], Socorro Josefina Villanueva-Rodríguez[3] and Adriano Gomes da Cruz[3]*

[1]Faculty of Veterinary, Federal Fluminense University (UFF), Niterói, Brazil
[2]Federal Institute of Paraná (IFPR), Paranavaí, Paraná, Brazil
[3]Food Technology Unit, Centro de Investigación y Asistencia en Tecnología y Diseño del Estado de Jalisco A.C., Guadalajara, Jalisco, México

Abstract
Supercritical fluid technology has emerged as a potential non-thermal technology for the safety and the improvement of the quality of processed foods producs. In this chapter, the definition of the possible applications of supecritical fluid at the food industry as well as the impact on the quality parameters of food products.

Keywords: Supercritical fluid, food, preservation

11.1 Introduction

With the advance of food science and knowledge about human nutrition, we could improve our way of eating to live better and longer. Therefore, in the current days, people are more aware of the contribution of foods to human health, increasing the demand for healthier food products. In this way, new technologies for the processing of food aiming microbiological safety are needed, but also those that preserve the nutritional compounds, increase their functional activities, do not produce harmful substances, and present advantages compared to the traditional processing

**Corresponding author*: adriano.cruz@ifrj.edu.br

Navnidhi Chhikara, Anil Panghal and Gaurav Chaudhary (eds.) Novel Technologies in Food Science, (405–450) © 2023 Scrivener Publishing LLC

methods. Studies have been conducted with emerging technologies, mainly non-thermal technologies, such as cold plasma, high hydrostatic pressure, pulsed electric field, ohmic heating, high-intensity ultrasound, ultraviolet light, supercritical fluids, and others. These studies aim to understand the effects of these technologies on foods to produce healthier and innovative products.

The supercritical fluids (SCFs) can be defined as materials presenting pressure and temperature higher than the respective critical values. In this way, they are not in the gaseous, liquid, or solid states, but in the supercritical state. SCFs present specific physicochemical characteristics, as the diffusivity and viscosity are similar to gas, while dielectric constant, density, and solvating properties are similar to liquids [1]. Traditionally, the SCFs were applied as excellent options for bioactive compounds' extraction and fractionation of complex matrices [2]. However, more recently, other food processing applications for SCFs have been studied, such as microbial inactivation, control of the enzymes that impair the foods' quality, and microencapsulation [3].

Most of the studies related to food processing by SCFs use carbon dioxide as a solvent since it is efficient for a wide range of food components, but it is also inexpensive, non-flammable, chemically inert, environmentally friendly, non-toxic, and generally recognized as safe to be used in foods by EFSA and FDA. Furthermore, it can be removed when expanded at ambient pressure values, and its supercritical conditions are easily reached (pressure above 7.38 MPa and temperature above 31 °C) [2, 4]. In addition, carbon dioxide is usually a recyclable solvent, leaving no waste [5]. On the other hand, other fluids like propane, dimethyl ether, and butane have been used in extractions using subcritical fluids.

Therefore, the modern approach for SCFs technology, more specifically supercritical carbon dioxide (SC-CO_2), also called dense phase carbon dioxide (DPCD) or high-pressure carbon dioxide (HPCD), is not only related to the separation and extraction of beneficial and bioactive components [6], but also the manufacturing of liquid (dairy beverages and vegetable and fruit juices) and solid (grains, cheese, oysters, shrimp and cooked ham) food products and solid-liquid mixtures [3, 7–12]. The studies aim to inactivate the endogenous enzymes and pathogenic and spoilage microorganisms (associated with the deterioration of the products) using low processing temperatures. In this way, thermally labile compounds can be preserved. Therefore, a promising effect of this technology is the microbial inactivation of foods to create safe and more stable products besides maintaining nutritional and sensory properties.

Thus, this chapter aims to make and updated and comprehensive overview of SCFs technology, highlighting its fundamentals, advantages, and limitations. Furthermore, the potential utilization in the processing of food, fractionation and extraction of food components, enzymatic and microbial inactivation, and the possible mechanisms of action are discussed. The effects of the SCFs technology in some physicochemical quality parameters and sensory attributes when compared to the conventional processing methods are also discussed.

11.2 Supercritical Carbon Dioxide (SC-CO_2) Technology: General Aspects and Fundamentals

SCFs are pure substances that are pressurized and heated at pressures and temperatures higher than those of the critical values, Pc and Tc, respectively [4]. In a phase diagram of temperature and pressure can be visualized the physical phase of a pure substance (Figure 11.1). In this way, the gas, liquid, and solid states are demarcated by the lines of sublimation, melting, and boiling, respectively. The critical temperature is interpreted as the highest temperature in which a gas can be transformed to a liquid because of an increased pressure. Meanwhile, the critical pressure is interpreted as the highest pressure in which a liquid can be transformed to a gas because of an increased temperature. Thus, these conditions are related to the critical point. The vapor pressure begins at the triple point or triple state, a point where three states coexist, and ends at the critical point (boiling point, liquid, and gas coexist). Above this point it can be visualized the supercritical region, where the component is observed as a single-phase, non-condensable fluid, exhibiting some physicochemical properties, such as density similar to liquid, viscosity similar to gases, and diffusivity twice times higher than that of liquids [4, 13–15]. Therefore, these compounds exist as fluid and not as a liquid or gas, but they are able to dissolve matters as a liquid and diffuse through solids as a gas. Furthermore, changes in pressure and temperature can adjust the properties [4, 11].

The SCFs properties are understood in accordance with the diffusivity, viscosity, and density. In Table 11.1 it is possible to observe some

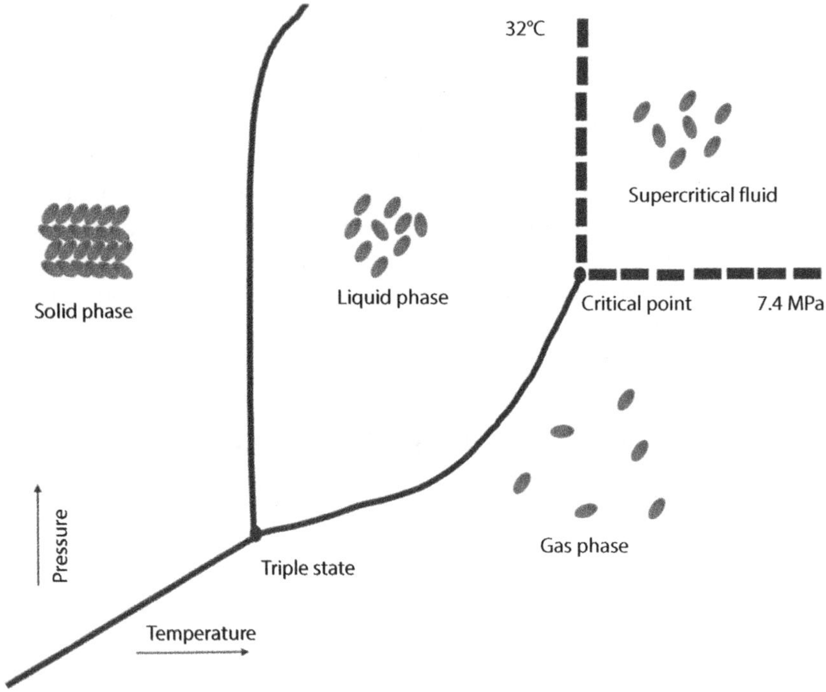

Figure 11.1 Pressure–temperature diagram of phase of a pure substance. Adapted from Lopes [16].

typical values of diffusivity, viscosity, and density for the gaseous, liquid, and supercritical states. The SCFs show high selectivity and solubility due to the pressure-dependent solvent power and high-density values. Furthermore, SCFs show a fast penetration into the samples' pores, resulting in the extraction of the inner material. This is associated with their low viscosity, medium diffusivity, and lack of surface tension. Thus, all these factors combined increase the efficiency of processing using supercritical technology and help to extract compounds from products, such as vegetable oils, beta-carotene, vanilla, and caffeine). Furthermore, they inactivate the vegetative cells. The mentioned advantages have increased the interest of the pharmaceutical, cosmetic, and food industries in the application of a non-thermal methodology allowing to add value to their products. This is associated with the fact that several drugs and cosmetics

Table 11.1 Characteristic values according to the state of aggregation [17].

Fluidstate	Density (g/cm^3)	Diffusivity (cm^2/s)	Viscosity (g/cm/s)
Gas			
1atm 15-30°C	$(0.6\text{-}2.0) \times 10^{-3}$	0.1 – 0.4	$(0.6\text{-}2.0) \times 10^{-4}$
Liquid			
1atm, 15-30°C	0.6-1.6	$(0.2\text{-}2.0) \times 10^{-5}$	$(0.2\text{-}3.0) \times 10^{-2}$
Supercritical fluid			
Pc, Tc	0.2-0.5	0.7×10^{-3}	$(1\text{-}3) \times 10^{-4}$
4Pc, Tc	0.4-0.9	0.2×10^{-3}	$(3\text{-}9) \times 10^{-4}$

are based on natural dyes, and the bioactive compounds associated with the nutritional aspects of food products are sensitive to heat treatments [11, 15].

The solute solubility in the supercritical fluid should be determined and modeled in order to design efficient SCFs, as it is considered the most important thermo-physical property. In this way, the influence of the pressure, density, and temperature of the SCF on the solute solubility should be understood, enabling the specification of unit operations and their operating conditions, such as separators, valves, extractors, transfer lines, and process controllers [18].

The solvent used for processes in the food industry must be chemically inert, non-flammable, stable, odorless, colorless, non-toxic, and non-corrosive [19]. Carbon dioxide (CO_2) is the most used SCF in food-related applications, as it is a by-product of several industrial operations, such as combustion, fermentation, and ammonia synthesis. It can be obtained at a low cost and with a high degree of purity, high density, and low viscosity. Furthermore, it leaves no residue, is relatively recognized as safe and shows a relatively low critical point, 7.38 MPa and 31.1 °C, which is easy to achieve. This last factor favors the use of CO_2 in volatile and heat-labile products since the compounds of interest are not necessarily exposed to high rates that can induce the thermal degradation of the components of the food [2, 4, 20]. Furthermore, due to their relatively low critical pressure, pressures are not high, and investment and manufacturing costs are reduced compared to other supercritical compounds, such as supercritical water (22.064 MPa and 373, 95°C) [3, 4, 19, 20].

Most effects observed by the application of CO_2 in the supercritical state to foods are related to its good solubility in many compounds and penetration of several structures. Evidence indicates a significant correlation between cell death and CO_2 permeability through the cell membrane [21]. The CO_2 solubility is related to its density, which is dependent on the pressure and temperature [22], then, the processing parameters of the SC-CO_2 technology may influence its effects and the food properties. Despite the influence of pressure and temperature on microbial inactivation and other physicochemical parameters of the food, the CO_2 concentration still seems to be the most influential factor [23, 24].

Few studies have investigated the solubility of CO_2 in foods. However, according to Brunner [4], some solvation characteristics of SC-CO_2 when applied to food ingredients, include

(1) dissolution of non-polar or slightly polar compounds because CO_2 is non-polar. However, it is possible to overcome this limitation by using small concentrations of a polar co-solvent or modifiers with SC-CO_2 to improve the extraction efficiency;
(2) the solvent power is dependent on the molecular weight of the compounds, and it decreases with the increase in the molecular weight, such as saponins and xanthophylls [25];
(3) free fatty acids and their glycerides have poor solubility;
(4) SC-CO_2 shows increased affinity with medium molecular weight oxygenated organic compounds;
(5) pigments has low solubility;
(6) liquid water has low solubility in SC-CO_2 (< 0,5% w/w);
(7) the utilization of higher pressures values can increase the solvent power of SC-CO_2, allowing the dissolution of high molecular weight, high polarity, and less volatile compounds [26]; and
(8) polysaccharides, proteins, mineral salts, and sugars are insoluble [27, 28].

11.3 Supercritical Carbon Dioxide (SC-CO_2) Processing

When used for the pasteurization of liquids, systems with SC-CO_2 can be operated in batch, semi-continuous, or pseudo-continuous and continuous, which makes the supercritical technology adaptable to any production chain. Due to the application of high pressure during supercritical carbon dioxide processing, most of the laboratory scale equipment operates in a batch mode, using a stainless-steel vessel to maintain the product pressurized. On the other hand, in continuous and semi-continuous/pseudo-continuous systems a higher contact between CO_2 and food can be found, resulting in a more effective inactivation of enzymes and microorganisms because of the faster saturation with CO_2 [3]. However, the pseudo-continuous operational systems require a longer processing time, and this is associated with a higher diffusion of CO_2 in plant cells compared to the continuous mode, therefore, it is ideal for processing of drinks based on fruits or vegetables [11].

Generally, a basic SC-CO_2 batch system must include a CO_2 cylinder, a pressurization vessel (where the sample is placed), a cooling bath (chiller) to maintain the CO_2 in the gas state, a pneumatic pump to pressurize the vessel with CO_2, a heating bath, air filters, manometers, thermometers, appropriate valves to control the sample and CO_2 flow and an exhaust system (for releasing the pressure after the process is finished), as can be observed in Figure 11.2. The process starts when the CO_2 leaves the cylinder towards the cooling bath (chiller) at a temperature below -10ºC, in which the carbon dioxide is cooled, ensuring that it reaches the pneumatic pump in the liquid phase. In the pneumatic pump, the liquid CO_2 is pressurized with the aid of a compressed air compressor to the desired pressure and transported into the pressurization vessel, where the food is located. The pressurization vessel is coupled inside the reactor. The processing temperature of the product is defined by a heating bath and controlled by thermometers, and a gas flow meter is used to measure the accumulated gas volume and flow rates passing through the apparatus. A stirrer can be incorporated into the system to ensure a better homogenization of the mixture. The pressure can be maintained at around 5 to 40 MPa and the pressurization time depends on food matrix, its characteristics, and the purposes. The temperatures used can usually vary around 35 to 95 °C. After the residence time required for the treatment, the mixture leaves the compression cell and passes to the separator/collection bottle

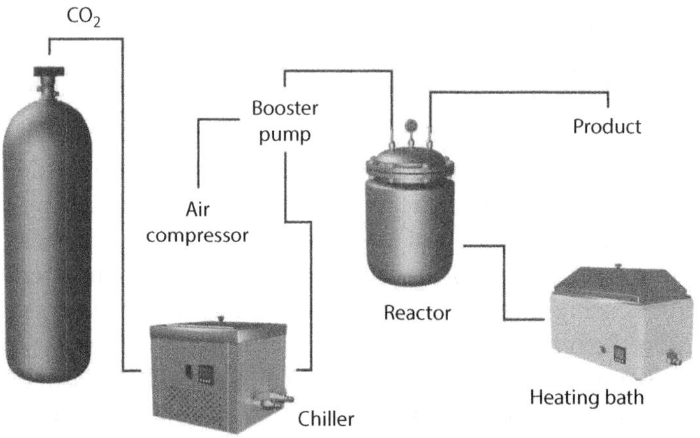

Figure 11.2 A simplified scheme of a basic SC-CO_2 processing batch system.

where the pressure is reduced, causing the vaporization of a large part of the CO_2. Subsequently, the flow rate of CO_2 is controlled, and it is sent to the recompression process for later reuse [13, 29].

In the semi-continuous system, the CO_2 flows in a continuous form. Two or more high-pressure reactors are interconnected in series. In this way, it is possible to mimic the continuous operating mode by inserting the loading, processing, and unloading steps in each reactor. Thus, one product is subjected to a constant pressure, while the others are pressurized or discharged, which reduces the processing time and allows the recovery of energy [3, 30]. However, the industrial application of these systems is limited because of the low processing capacity.

In the continuous system, food and CO_2 are pumped separately and mixed. Then, the mixture passes through the high-pressure pump, where the pressure is increased up to the operating value [3]. Finally, the CO_2 is released from the mixture by using an expansion valve at the end of the process. The processing is operated at flow rates lower than 130 mL/min, temperatures of 25-45 °C, and residence times of 10 min [31–33]. Most of the studies involving the application of supercritical technology in liquid foods were conducted using the continuous system.

The utilization of SC-CO_2 for the treatment of solid foods has been scarcely reported compared to liquids. In this sense, data on equipment is limited. Generally, equipment for food pasteurization can comprise one or several containers for food samples, which are conditioned in a pressure vessel [34–39]. Initially, a pump promoted the passage of CO_2 into the pressure vessel and controls the mass flow rate. To promote CO_2 to the supercritical state, the temperature and pressure are adjusted. Then, when the treatment is finished, a pressure regulator promotes the depressurization of the vessel. There is no published reports of continuous SC-CO_2 systems for solid foods, as the operation if difficult and requires specific skills [33].

11.4 Applications in Food Processing

11.4.1 Extraction and Fractionation of Food Compounds

Classical methods of extraction, such as maceration, solvent extraction, hydro-distillation using organic or inorganic solvents, and cold pressing have been widely applied to isolate chemical compounds and mixtures from complex matrices. However, such approaches have as disadvantages high energy costs, poor selectivity, utilization of large volumes of organic solvents that generate waste and emission of polluting gases responsible for causing environmental impacts, and mainly, the use of high temperatures during the operational process that usually destroy valuable substances such as phenolic compounds. On the contrary, supercritical fluid extraction (SFE) has been considered as an alternative methodology capable of overcoming these disadvantages by providing a product with higher quality and functionality due to the minimized risk of thermal oxidation while meeting the requirements of the concept of green process. The proposal is to avoid or reduce the utilization of organic solvents with toxicity, decrease the process temperature, extraction time, and energy consumption, as well as increase the mass transfer and extraction yields, thus contributing to the optimization of the industrial process [40–42].

The SCFs can be found at a reasonable cost and relatively pure grade compared to liquid solvents of industrial grade. CO_2 is the most used extraction fluid in SFE, as it is leaves no waste and is usually recycled. Therefore, the SC-CO_2 characteristics make this technology highly effective to extract and separate different compounds free of organic solvents, mainly due to the high diffusivity and solubility through materials [43]. This technology has already been used for the extraction of oil and fat in excess from vegetable matter and fried food, lipids, alcohol from beer, wine and other alcoholic beverages, essential oils, off-flavors, spices, and natural compounds, such as vanilla, beta-carotene, caffeine, among others [3, 25, 44–47].

An interesting and modern application for this technology is for extracting foods' bioactive compounds, as CO_2 can reach the supercritical state at a low critical temperature (31.1°C). It is a non-oxidative medium because O_2 is efficiently displaced from the solute matrix, favoring the preservation of the thermally unstable and easily oxidized compounds [42, 45]. Besides, in many cases, this technology allows achieving maximum yields in reduced extraction time. For these reasons, several studies have demonstrated that these compounds extracted with SC-CO_2 showed higher yields, antimicrobial, antioxidant, and anti-proliferative activity than the compounds extracted with organic solvents [42, 48].

Despite the use of this technology to obtain bioactive extracts, another useful application is the fractionation of these extracts into more specific and purified compounds, which is desirable when there are several compounds to be recovered from the food matrix. Therefore, it can be used in the purification and fractionation of aroma constituents, for example, for later addition in processed foods to improve product quality [42, 45, 49]. The concept is based on the separation between compounds of higher and lower molecular weight, such as volatile oil. The fractionation can be carried out through multiple extractions, in which the density is increasing at each step; or using a single-stage extraction involving multiples and consecutive separation phases with different conditions [42, 50].

In general, the success of the SC-CO_2 extraction will depend on the following parameters [42]: (1) pressure and (2) density – the higher the pressure the higher the density of SC-CO_2 and solubility of the solutes, resulting in a higher extraction yield; (3) temperature – the higher the temperature the lower the solvation power of SC-CO_2 (isobaric process), resulting in a lower extraction rate; (4) extraction time – the longer the contact time between the solutes and the solvent in SC-CO_2, the greater the extraction yield; (5) flow rate–the extraction capacity is increased with higher solvent flow rates, with a reduction in the external mass transfer resistance; and (6) co-solvents – increase the extraction yield, adjusting the CO_2 affinity for substances with low fluid solubility. Despite the influence of the supercritical processing parameters on extraction efficiency, the food characteristics are also important. Then, the preparation of the raw material is important for an effective extraction process, such as cutting, milling, and grinding, which increase the surface contact with the CO_2 and reduce the diffusion resistance [1].

The mechanism behind the extraction of compounds from vegetable matrices is mainly related to the solubility and high diffusivity of the supercritical fluid. Therefore, the extraction process may be divided into a solubilization stage, where the compounds to be extracted are solubilized in the SCF and the separation stage. In the solubilization step, the plant cells absorb the SCF into their structures, such as membranes and intracellular channels, causing the solubilization of the compounds of interest. At this moment, it occurs the mass transfer, taking the compounds from inside the cell to the exterior, being solubilized in the supercritical solvent [6]. Then, the extracted compound needs to be separated from the solvent (supercritical fluid). The separation process depends on the composition of the extracted substance and its solubility in the supercritical fluid, which can be manipulated by changing the fluid variables, such as temperature and pressure [1]. As previously mentioned, the application of modifiers as co-solvents may be interesting to enhance extraction efficacy, since they improve the solubility of solutes, mainly polar compounds, extending the variety of extractable compounds. However, the disadvantage is that the modifiers may also need to be separated after extraction, making the process more laborious [51].

Table 11.2 contains several researches that evaluated the main effects resulting from the extraction from food compounds by SC-CO_2 in recent years.

Table 11.2 Potential effects of SC-CO_2n extracting compounds from a variety of matrices.

Compound/sample	Processing parameters	Results/main effects	References
Thermolabile volatile compounds/Basil (*Ocimumbasilicum* L.)	SC-CO_2 (100, 150, 200 and 300 bar; 60°C; extraction time of 4 h and flow rate of 3.225 g CO_2/min	The utilization of SC-CO_2 (150 bar, 60°C) promoted the highest extraction yield. The utilization of SC-CO_2 (100 bar) increased the antioxidant activity (3 times) compared to basil essential oil. SC-CO_2 can be applied for extracting volatile non-polar components and prepare extracts with high antioxidant activity.	[52]

(*Continued*)

Table 11.2 Potential effects of SC-CO_2n extracting compounds from a variety of matrices. (*Continued*)

Compound/sample	Processing parameters	Results/main effects	References
Oils/palm kernel	SC-CO_2(44.6 MPa, 60°C and extraction time of 50 min)	The yield of oil was 49.2%. The utilization of SC-CO_2 kept the nutrition composition in de-oiled palm kernel cake during extraction similar to that of palm oil mill.	[53]
Cholesterol/whole milk powder	SC-CO_2 (8 – 35 MPa; 35 – 80°C; and 10 – 50 mL co-solvent)	The optimized conditions (17 MPa, 48 °C, 31 mL co-solvent, 6 L/min of flow rate of expanded CO_2, dynamic time of extraction of 80 min, static time of 10 min) reduced the cholesterol levels in 46%. There were no changes in the physicochemical properties.	[54]

(*Continued*)

Table 11.2 Potential effects of SC-CO_2n extracting compounds from a variety of matrices. (*Continued*)

Compound/sample	Processing parameters	Results/main effects	References
Essential oils/ spearmint leaves	SC-CO_2 (85–120 bar; 38–50°C; mean particle size - 0.177–2 mm, CO_2 flow rate - 0.059–0.354 g/min; and extraction time - 20–120 min	The increase in the CO_2 flow rate promoted an increase in the extraction yield at the first stage. The increase in the CO_2 flow rate promoted a decrease in the extraction yield at the following stages. The optimized conditions were: 42.6 °C of temperature, 119.7 bar of pressure, 0.248 mm of mean particle size, 0.169 g/min of CO_2 flow rate, and 109.9 min of static time of extraction.	[55]

(*Continued*)

Table 11.2 Potential effects of SC-CO_2n extracting compounds from a variety of matrices. (*Continued*)

Compound/sample	Processing parameters	Results/main effects	References
Flavonoids/brown onion skin	SC-CO_2 (200 and 300 bar; 40, 55 and 70 °C; extraction time - 30–240 min; and modifier quantity - 95, 90 and 85% EtOH %)	The efficiency in the extraction of phenolic compounds in in the SC-CO_2 process is influenced by the composition of solvent modifier. An extraction yield of 16% could be achieved at the optimized conditions (40 °C, EtOH 85%, and 100 bar). A high antioxidant activity was reported.	[56]
Lutein and chlorophyll/ spinach (*Spinacia oleracea*)	SC-CO_2 (10, 30, 50 MPa; 40, 50, and 60 °C; co-solvent percentages -0, 5, and 10%; and extraction time – 1, 3, and 5 h)	The optimum conditions were: 56 °C of temperature, 39 MPa of pressure, 10% ethanol as co-solvent, and extraction time of 3.6 h. 50% of chlorophyll and 72% of lutein were obtained under the optimum conditions.	[44]

(*Continued*)

Table 11.2 Potential effects of SC-CO_2n extracting compounds from a variety of matrices. (*Continued*)

Compound/sample	Processing parameters	Results/main effects	References
Cholesterol/dairy cream powder	SC-CO_2 (100 – 250 bar; 40 – 75 °C; flow rate – 5 L/min; and static time – 30 min	The removal of cholesterol was increased by using higher operating pressures, lower extraction temperatures, and moderate dynamic times. The optimum conditions were: 75 °C of temperature, 204 Bar of pressure, and 3.5 h of dynamic time. A 39% of cholesterol removal and 10.6% of total fat reduction in the cream powder was observed at the optimum conditions.	[57]
Caffeine/black tea	SC-CO_2 (250, 375 and 500 bar; 55, 62.5 and 70 °C, CO_2 flow rate - 1, 2, 3 L/min, extraction time - 60, 180, 300 min; and modifier quantity - 0, 2.5, 5 mol%)	The complete removal of caffeine could be achieved using 62.5 °C of temperature and 300 min of extraction time. The optimized conditions could be used to decaffeinate black tea on industrial scale.	[58]

(*Continued*)

Table 11.2 Potential effects of SC-CO_2n extracting compounds from a variety of matrices. (*Continued*)

Compound/sample	Processing parameters	Results/main effects	References
Lycopene/tomato	SC-CO_2 (40–80 °C; peel to seeds ratio (30/70 to 70/30) and 30–50 MPa).	The optimum conditions for lycopene yield were: 80 °C of temperature, 50 MPa of pressure, and peel/seeds ratio of 70/30. Lycopene yield can be increased with the increase in the external mass transfer coefficient. Lycopene yield is decreased by the increase in the particle diameter and partition coefficient of the solute between the solid and the fluid phase.	[59]

(*Continued*)

Table 11.2 Potential effects of SC-CO_2n extracting compounds from a variety of matrices. (*Continued*)

Compound/sample	Processing parameters	Results/main effects	References
Oils/Brewer's spent grain	SC-CO_2 (40 and 50°C; 20 and 30 MPa; with and without ethanol as co-solvent, 4 and 8%) compared to Soxhlet using hexane as a solvent	Soxhlet showed a yield of 6.1%, recovery of 78.3%, total phenolic contents of 28.3mg GAE/g of sample, and antioxidant activity of 16.7 mg TEA/g of sample. SC-CO_2(30 MPa, 50 °C, and 8% of ethanol) showed a yield of 6.5 %, recovery of 81.3%, total phenolic contents of 26.2mg GAE/g of sample, and antioxidant activity of 14.2mg TEA/g of sample.	[60]
Volatile aroma compounds/pea (*Pisum sativum* L.) flour	SC-CO_2 (28.68, 31.03, 34,47, 37.92 and 40,27 MPa; 33.2, 40, 50, 60 and 66.08 °C; extraction time – 40 min; with ethanol as co-solvent in percentages 0, 10, 25, 40 and 50.2 %	The optimum conditions for flavor modification of pea flour were: 86 °C of temperature, 42.71 MPa of pressure, and 22% of ethanol, resulting in a total volatile compound content of 0.55 µg/g. SC-CO_2 + EtO could be used to remove undesirable flavors and aroma from pea flour.	[25]

In general, it can be observed that the suitable selection of the processing parameters of the SC-CO_2 increases the extraction yield and preserves the nutritional and functional compounds of the product. Furthermore, it reduces the extraction time and avoids economic losses.

11.4.2 Enzymatic and Microbial Inactivation

The SC-CO_2 technology can be used for microbial and enzymatic inactivation, which is remarkably interesting in the vegetable and fruit juices processing. The utilization of SC-CO_2 processing resulted in a completely inactivation of microorganisms and enzymes using lower pressures than HHP (≈10-time lower pressures, typically <20 MPa) and lower temperatures than HIUS [11, 61, 62]. These characteristics are interesting as they avoid an emergence of pressure-resistant vegetative bacteria after pressure treatment and contribute to the retention of the sensory quality, color, aroma, nutritional and physical characteristics of the food. In addition, the HHP technology requires large investment costs for large-scale implementation in food preservation processes, while the application of SC-CO_2 is simple and inexpensive. Furthermore, there are modern methods that allows its post-processing and reusing [45, 61, 63].

Most studies that applied this technology to process foods used liquid matrices, such as milk, coconut water, whey drink, juices, and other beverages, due to the effectiveness of the supercritical CO_2 to penetrate the whole product [24, 64–67]. However, as previously mentioned, the SC-CO_2 can also be used to process solid foods [68], as well as for solid-liquid food mixtures [12], in which, the inactivation potential will act mostly on the product surface, the reason why different inactivation rates have been obtained, for example, in ground meat, sliced ham, fresh-cut fruits, marine food, and vegetables and mango syrups [35, 38, 69, 70].

The mechanism of action behind the inactivation of enzymes and microorganisms by supercritical CO_2 is not causally related to the temperature, since most studies observed high inactivation rates even at temperatures considered non-thermal (<55°C). However, the temperature may affect the supercritical fluid and the microbial characteristics, then, higher temperatures (> 50 °C) usually improve the inactivation rates, since it increases the cell membrane permeability and the CO_2 diffusivity, which facilitates the CO_2 penetration into the cells with subsequent destruction and inactivation of germinated cells [68, 71]. Increased agitation speed or processing time also contributes to an increase in the rate of microbial inactivation. Stirring increases the contact between CO_2 and microorganisms, the CO_2 decomposition rate into carbonic acid, and the pH reduction of the medium. Meanwhile, the increase in pressure causes greater effectiveness of inactivation because it reduces the exposure times necessary for the inactivation of microbial cells [72].

The inhibitory effect that CO_2 has on microbial growth was discovered about 100 years ago [73]. Since then, several studies have shown that vegetative bacterial cells are completely inactivated by SC-CO_2 treatment low pressure and temperature conditions. However, the mechanisms of action associated with the inactivation of microorganisms are not completely clear, since the supercritical CO_2 processing induces many different effects on the microbial cells, reason why even resistant microorganisms are easily inactivated by this technology [74]. Therefore, a set of effects responsible for microbial inactivation have been observed and suggested such as acidification of the cytoplasm, changes in the cell membrane, rupture of the cells due to high pressure, extraction of lipids and other substances from cell structures, and enzyme inactivation, mainly those responsible for cell metabolism [11, 68, 72] as best described below.

Initially, pressurized CO_2 is solubilized in the liquid near the cells of the microorganisms, reducing the extracellular pH, and contributing significantly to reducing bacterial growth. Such CO_2 solubilization is essential because bacterial cells with low water content are not susceptible to inactivation with carbon dioxide [61]. Due to its high lipophilic affinity, aqueous CO_2 diffuses in the cytoplasmic membrane of microbial cells and an accumulation in the phospholipids occurs, causing a disturbance in the lipid chains. Then, the fluidity of the membrane of the cell and its permeability to CO_2 are increased. The HCO_3- ions also can change the membrane permeability, as well as modify the ionic charges of the proteins and phospholipid groups placed on the whole surface. The increased permeability of the membrane allows aqueous CO_2 to penetrate the cytoplasm of microbial cells, and it is transformed into carbonic acid, resulting in a fast decrease in the intracellular pH due to the creation of HCO_3- and + ions. In an attempt of reverse, the variations in the pH values, the microbial cells expel excess protons (H+). However, when an excessive concentration of CO_2 penetrated the cytoplasm, the expelling of all the protons from the cells is not possible, resulting in the decrease of the pH values and inhibition or inactivation of enzymes. Some of these enzymes are important for the regulatory and metabolic processes.

The dissolved CO_2 and generated HCO_3- can unbalance some important metabolic reactions of the cells, mainly carboxylation and decarboxylation reactions responsible for the synthesis of amino acids and nucleic acids. Extreme stimulation of the mentioned metabolic pathways results in excessive loss of ATP. When the pressure of the applied CO_2 is amplified in the cytoplasm of bacterial cells, it can transform HCO_3- into CO_3^{2-} ions, which can precipitate inorganic electrolytes found inside the cells, such as Mg^{2+} andCa^{2+}. These electrolytes can regulate several activities and help to maintain osmotic relationships between the interior of cells and the external environment. Therefore, the precipitation of these electrolytes has a detrimental effect on the entire cell. Finally, SC-CO_2 can remove essential components from cell membranes or bacterial cells because of its high solvating ability. Thus, during extraction, some intracellular components, such as phospholipids, can cause disruption or alteration of the membrane structure. In addition, rapid depressurization can deform or damage the membrane of the cell causing its rupture. In this context, the use of repeated pressurization/depressurization cycles increases the extraction process and, consequently, increases microbial inactivation [61, 72, 75, 76].

The inactivation of enzymes is much important for the conservation of foods, mainly fruits, and vegetables, in which the enzymes remain active after harvest, changing the natural characteristics (e.g. nutritional and sensory) of the fresh foods during storage [11]. Among the enzymes associated with food quality loss some stand out. Pectin esterase (PE) is responsible for the loss of turbidity in fruit juices, while polyphenol oxidase (PPO) drives unwanted browning in vegetables, fruits, some seafood, and juices. Furthermore, lipoxygenase (LOX) damages chlorophyll and develop ff-flavors in frozen vegetables; while peroxidase (POD) contributes to food discoloration and can be used to attest the effectiveness of heat processing of vegetables and fruits [77].

The process of enzyme inactivation is different from microbial inactivation since the enzymes are proteins with active sites to react with their specific substrates. Carbon dioxide under pressure modifies the protein patterns and isoelectric profiles while CO_2 under atmospheric pressure has not effect on these properties. High pressure has also been associated with changes in the conformational of enzymes [77]. The enzymes can be inactivated if they suffer enough changes to their structure to disrupt the active sites. They can be divided into four types of conformational structures, being inactivated when at least the tertiary or secondary structure is affected. It has been reported that each SC-CO_2 processing variable has a different role in inactivation. The high pressure alters mainly the enzymes' tertiary and quaternary structures; in addition, most studies have observed higher enzymatic inactivation under higher temperature conditions [11].

In Tables 11.3 and 11.4 is possible to observe several papers that investigated the potential use of supercritical technology to inactivate microorganisms and enzymes present in liquids, solids, and solid-liquid food mixtures in recent years. Most of these researches confirmed that SC-CO_2

Table 11.3 The effect following SC-CO_2 treatment on microbial reduction in food products.

Microorganism/ sample	**Processing parameters**	**Results/main effects**	**References**
Staphylococcus aureus/ raw salmon	SC-CO_2 (90–250 bar; 33 °C; process time: 2 h depressurization rate - X2 (10 to 100 bar/min) and the salmon: CO_2 mass relation (1:0.2 to 1:1.0, 1:0.6)	Total inactivation of *S. aureus* achieved with pressures of 225 bar, salmon: CO_2 mass relation of 1:0.6, depressurizing rate of 100 bar/min, process time of 2 h, and temperatures of 33 °C. The microbial inactivation increases with increases in the depressurization rate and system pressure.	[78]
Escherichia coli/Milk	SC-CO_2 (80–180 bar; 30–70 °C; CO_2 volume ratio to milk: 0.05–0.45 wt %), process time: 10–30 min	The microbial inactivation rate was 0.09/min at temperatures of 70 °C, pressures of 80 bar, residence time of 30 min, and CO_2 to milk mass ratio of 0.05.	[66]
Total coliforms, mesophilic microorganisms, molds and yeasts/ apples in syrup	SC-CO_2 (8, 10 and 12 MPa; 35, 45 and 55 °C	The microorganisms were totally inactivated at temperatures of 55 °C, pressures of 12 MPa, and processing time of 15 min. No effect on physicochemical properties (color, pH, and total acidity) was observed.	[79]

(*Continued*)

Table 11.3 The effect following SC-CO_2 treatment on microbial reduction in food products. (*Continued*)

Microorganism/ sample	Processing parameters	Results/main effects	References
-/pork	SC-CO_2 (13.8 MPa; 35 °C; process time: 2h; compared to SC-CO_2 +2.5 g (SCO_2 + R 2.5%) or 5 g (SCO_2 + R 5%) dried rosemary powder/100 g meat	A synergistic effect of SC-CO_2 and rosemary powder (5 g rosemary/100 g meat) was observed for microbial growth inhibition.	[80]
Alicyclobacillus acidoterrestris spores/apple juice	SC-CO_2 (60 MPa; CO_2 volume ratio: 30%, 50–75 °C; 20–40 min.	A higher microbial inactivation (3.5 log) was observed at thermal conditions (75°C) compared to non-thermal conditions (50 °C, 1 log).	[71]
Molds and bacteria/ Tempeh	SC-CO_2 (6.3 and 7.6 MPa; process time: 5, 10, 15 and 20 min;	The inactivation of molds and bacteria reduction (2.4 log) were achieved with processing times of 20 min with SC-CO_2 or sub-supercritical. Tempeh composition protected the microorganisms from SC-CO_2. Gram-negative bacteria were less resistant than Gram-positive.	[81]

(*Continued*)

Table 11.3 The effect following SC-CO_2 treatment on microbial reduction in food products. (*Continued*)

Microorganism/ sample	Processing parameters	Results/main effects	References
Vibrio parahaemolyticus/ oysters (*Crassostrea gigas*)	SC-CO_2 (8–20 MPa; 33°C; 2h; rate of 10.0 MPa/min)	The inactivation rate of *V. parahaemolyticus*is dependent on the environmental conditions and initial microbial counts. SC-$CO_2$was able to inactivate *V. parahaemolyticus*in oysters.	[82]
Lactobacillus casei/ apple juice	SC-CO_2 (10–20 MPa; process time: 10–30 min; 35–55 °C; CO_2 volume ratio: 10–70%	The highest inactivation (6.93 log) was obtained at temperatures of 55 °C, pressures of 15 MPa, processing times of 30 min, and CO_2 volume ratio of 70%.	[23]
Escherichia coli O 157:H7, *Listeria monocytogenes*, *Salmonella*, and natural flora (molds and yeasts, spores, mesophilic bacteria, Enterobacteriaceae)/ apple's slices	SC-CO_2 (125 bar; 50°C; 15 h); Freeze drying (0.2 mbar and 40 °C – desorption; 0.05 mbar and 4 °C – sublimation; 24 h), Air drying (60 °C, 8 h), and	The total inactivation of pathogenic bacteria, and reductions of vegetative bacteria, molds, and yeasts were achieved using 10 min of pressurization and 20 min of depressurization.	[83]

(*Continued*)

Table 11.3 The effect following SC-CO_2 treatment on microbial reduction in food products. (*Continued*)

Microorganism/ sample	Processing parameters	Results/main effects	References
Escherichia coli and natural flora (yeasts, total mesophilic bacteria, and molds)/raw chicken meat	SC-CO_2 (80 and 140 bar; 40°C; and 15–45 min)	Processing time had a positive correlation with microbial inactivation (*E. coli* reduction of 1.4 log CFU/g after 15 min and 5 log CFU/g after 45 min). Mesophilic bacteria were reduced (>2.6 log CFU/g) after 45 min. Molds and yeasts were totally inactivated in most cases. Pressure (80-140 bar) had no significant influence. No synergistic effect was observed between fresh herbs and SC-CO_2.	[84]
E. coli and *S. aureus*/ mango in syrup	SC-CO_2 (10–50 MPa; 45–65 °C; 10–30 min) compared to conventional pasteurization (100 °C/15 min)	The SC-CO_2 processing at temperatures of 60 °C, pressures of 20 MPa, and processing times of 30 min reduced in *S. aureus* and *E. coli* counts (>8-log CFU/mL)in mango syrup.	[12]

Table 11.4 Recent studies on applications and operational parameters of SC-CO_2 with the respective effects on enzyme inactivation in different food products.

Enzymes/sample	Processing parameters	Results/main effects	References
Alkaline phosphatase/milk	SC-CO_2 (80–180 bar; 30–70 °C; CO_2 volume ratio to milk: 0.05–0.45 wt%, process time: 10–30 min;)	The SC-CO_2 processing at temperatures of 70 °C, pressures of 80 bar, processing times of 30 min, and CO_2 to milk mass ratio of 0.05inactivated 94.5% of the enzyme.	[66]
Polyphenol oxidase/ apples in syrup	SC-CO_2 (8, 10 and 12 MPa; 35, 45 and 55 °C	The polyphenol oxidase enzyme was totally inactivated at temperatures of 55 °C, pressures of 12 MPa, and processing time of 15 min. Physicochemical properties (color, pH, and total acidity) were slightly changed.	[79]
Polyphenol oxidases and peroxidases/ apple juice	SC-CO_2 (10, 30 and 60 MPa; 35, 45 and 65 °C; 10, 20 and 30 min)	The enzyme inactivation was dependent on the enzyme origin and process parameters. Higher temperatures and pressures resulted in increased k-values and decreased D-values. POD was more resistant to heating and pressure than PPO.	[67]
Polyphenol oxidase/ quince juice	SC-CO_2 (20 MPa; Process time: 20 min, 25–65 °C)	PPO was totally inactivated at 65 °C, while 35% of residual activity was reported at 55 °C.	[85]

(*Continued*)

Table 11.4 Recent studies on applications and operational parameters of SC-CO_2 with the respective effects on enzyme inactivation in different food products. (*Continued*)

Enzymes/sample	Processing parameters	Results/main effects	References
Polyphenol oxidase and peroxidase/ apple juice	SC-CO_2 (10, 30, and 60 MPa; 45 °C; 30 min)	No alterations in the total polyphenols and sugars. Phenolic compounds stability was dependent on the enzyme activity. A synergistic effect between POD and PPO and catechol and chlorogenic acid was observed.	[62]
Polyphenoloxidase/ Srimps *(Litopenaeusv annamei)*	SC-CO_2 (20 MPa; 35–50 °C) compared to conventional pasteurization (35 – 50 °C)	The enzyme inactivation was increased with increases in temperatures (35-50 °C). HPCD resulted in higher PPO inactivation of shrimps *in vivo*.	[70]
Alkaline phosphatase/raw bovine milk	SC-CO_2 (20 MPa; 20–50 °C; 20–70 min)	Complete inactivation of alkaline phosphatase activity after exposure to SC-CO_2 treatment (20 MPa; 50 °C; 50 min) was observed	[63]
Polyphenol oxidase and peroxidase/ mushroom and horseradish	SC-CO_2 (10–65 MPa; 10–30 min; 35–65 °C) compared to high pressure (HPP) (200–900 MPa; 5–45 °C; 1–15 min)	Pressure had a positive impact on the inactivation of PPO and POD. PPO was more resistant to pressure and temperature than POD.	[86]

(*Continued*)

Table 11.4 Recent studies on applications and operational parameters of SC-CO_2 with the respective effects on enzyme inactivation in different food products. (*Continued*)

Enzymes/sample	Processing parameters	Results/main effects	References
Polyphenol oxidase/ apple juice	SC-CO_2 (10 and 25 MPa; Process time : 20 min; 20 and 50 °C)	The increase in the temperature (40-55 °C) increased the inactivation of PPO from 80 tp 95% (at pressured of 25 MPa). The increase in pressure and temperature caused loss of α-helix conformation.	[87]
Polyphenol oxidase and peroxidase/ apple juice	SC-CO_2 (25–65 °C/20 MPa) compared to conventional pasteurization (25-75 °C)	PPO was completely inactivated using SC-CO_2 at 65 °C, while it was active using conventional pasteurization at 75 °C. POD activity reduced 71% using SC-CO_2 at 65 °C, while conventional pasteurization was less effective. SC-CO_2 decreased browning reactions associated with POD and PPO enzymes and increased the phenolic compounds (catechin and epicatechin) (at 35 °C and 9 MPa).	[88]

(*Continued*)

Table 11.4 Recent studies on applications and operational parameters of SC-CO_2 with the respective effects on enzyme inactivation in different food products. (*Continued*)

Enzymes/sample	Processing parameters	Results/main effects	References
Polyphenol oxidase/ Graham flour	SC-CO_2 (300 and 200 bar; 35 °C; 3 and 24h)	The PPO inactivation was reduced at higher levels under higher pressure values. The flour quality characteristics were kept.	[89]
Polyphenol oxidase and peroxidase/ mango in syrup	SC-CO_2 (10–50 MPa; 45–65 °C; 10–30 min) compared to conventional pasteurization (100 °C/15 min)	The PPO and POD enzymes were completely inactivated using SC-CO_2 processing (60 °C, 20 MPa, and 30 min).	[12]

is an efficient non-thermal pasteurization methodology for eliminating pathogenic or deteriorating microorganisms and inactivating food enzymes without quality deterioration when compared to traditional thermal pasteurization treatment, as it uses temperatures, generally, below 50 °C and pressures below 50 MPa.

11.4.3 Effects on Physicochemical Parameters

Despite the promising effects on microorganisms, to preserve foods and ensure the development of safe products for human consumptions, the supercritical CO_2 processing applied to foods may not impair their chemical and physical characteristics such as nutritional compounds and texture/rheological properties.

In general, this technology preserves the nutritional compounds in foods due to the utilization of milder temperatures compared to conventional pasteurization and decreased pressures compared to high-pressure processing. Indeed, the combination of CO_2, mild temperatures and mild pressures make this technology very efficient to process foods containing thermos sensitive compounds, such as products with fruits and vegetables [11].

Tang *et al.* [12] evaluated the efficiency of SC-CO_2 processing as a pasteurization technology for solid-liquid foods (mango syrup) and its impact on the quality of the products. The SC-CO_2 processing was compared to a conventional pasteurization technology. SC-CO_2 processing resulted in products with a better nutritional and physicochemical properties, with color and pH values neared to the unprocessed products. Furthermore, SC-CO_2 processing kept higher concentrations of vitamin C (48.9%) and total phenol (19.7%) compared to conventional pasteurization. The authors conclude that SC-CO_2 processing can be an alternative pasteurization technology for solid-food mixtures, with a lower impact on the quality parameters.

Silva, Guimarães, *et al.* [24] evaluated the effect of SC-CO_2 processing variables (CO_2, temperature and pressure) on a soursop whey beverage. They observed no impact on color, total soluble solids, and pH, but a slight modification was reported in microstructure and rheology, which were suggested to be related to protein denaturation due to the momentary acidification during the process caused by the CO_2. In another study of the same group, Silva, *et al.* [90] evaluated the impact of SC-CO_2 to process a prebiotic-enriched apple juice. They observed a decreased particle size with higher pressures and most of the functional and nutritional compounds were preserved, such as organic acids, phenolic compounds, antioxidant activity and sugars content.

Liao *et al.* [63] investigated the impact of SC-CO_2 treatment on the alkaline phosphatase (ALP) activity and quality parameters of raw bovine milk. The utilization of SC-CO_2 treatment at temperatures of 50 °C, pressures of 20 MPa, and treatment times of 50 min could totally inactivate the ALP activity, while the lactose and protein contents were kept. SC-CO_2 treatment caused a decrease in the viscosity of the products, but the Newtonian flow behavior was kept. Ceni *et al.* [66] investigated the effect of SC-CO_2 processing on the physicochemical parameters of whole milk. No differences between SC-CO_2-treated milks and untreated milks were observed. The authors conclude the SC-CO_2 processing is a promising technology as it could keep the quality parameters of milk and dairy products.

A research group made an extensive evaluation of SC-CO_2 processing on a whey-grape juice drink [64, 91]. They observed no alterations on many physicochemical characteristics such as titratable acidity, pH, soluble solids, total anthocyanin, and antioxidant activity. On the other hand, some slight changes in particle sizes and rheology occurred. Despite the changes, supercritical processing did not modify the sensory properties of the products.

Huang *et al.* [80] evaluated impact of combined treatment with rosemary powder (2.5% and 5.0% (w/w)) and with SC-CO_2 (13.8 MPa, 35 °C for 2 h) on the physicochemical and microbiological properties of ground pork. A synergistic effect of SC-$CO_2$and rosemary powder (5 g rosemary/100 g meat) was observed for microbial growth inhibition. Furthermore, the utilization of rosemary powder reduced the lipid oxidation during refrigerated storage and altered the color parameters (L* and b*). No impact on a*, TVB-N value, and pH values were observed after rosemary powder addition compared to the products processed with SC-CO_2 individually.

Marszałek *et al.* [92] evaluated the strawberry juice SC-CO_2 processing aiming preservation during storage. The processing conditions (45°C, 30 and 60 MPa, 30 min) inactivated polyphenol oxidases and decreased by ~85% the peroxidases. In addition, the SCCD treatment hydrolyzed the sucrose and degraded ~30% of the ascorbic acid; however, the anthocyanins were not affected. In another study of the same authors, Marszałek *et al.* [62] investigated the effect of SC-CO_2 (45°C, 30 and 60 MPa, 30 min) followed by 10 weeks of storage (4 °C) on PPO and POD activities and quality parameters (vitamin C, physicochemical properties, phenolic profile, and sugars) of cloudy juice. No alterations in the total polyphenols and sugars were observed, whereas the vitamin C and individual polyphenols contents were reduced by 28 and 18%, respectively. A synergistic effect between POD and PPO and catechol and chlorogenic acid was observed.

Ferrentino *et al.* [7] evaluated the optimal SC-CO_2 conditions to process cooked ham. They concluded that this technology was capable of ensuring significant microbial inactivation with minimum impact on the physicochemical properties. The pH values and texture parameters were kept stable. Only slight changes in the color were observed.

In conclusion, the physicochemical changes after SC-CO_2 processing seem to be minimal in several food products, however, only a few studies have been carried out in this sense. The maintenance of the physicochemical properties may be associated with the mild conditions employed (low temperatures and pressures), in addition, the CO_2 concentration has a major role in these effects, mainly in dairy products, in which the proteins may be affected.

11.4.4 Effects on Sensory Properties

The most decisive quality control considers the consumer perception through their senses, therefore, any action that impacts the sensory characteristics or influences the consumer behavior, habits, and beliefs needs to be validated. Supercritical fluids as an emerging technology can significantly modify the characteristics of food and beverages, and such modifications can be perceptible as negative or positive. In this way, sensory measurements, and the evaluation of the impact of the process on the overall liking, acceptance, preferences, and beliefs of consumers is decisive in the success of the advances and adjustments that such technology may require.

Some works have been published considering the impact of SCF technology on the sensory properties of food products, and they provide information of great interest that guides adjustments and advances in the improvement of the technology. A bibliographic analysis of 9 works was carried out, which included sensory or hedonic measurements and the level of acceptance or preference (Table 11.5). Different sensory analysis methodologies were applied, such as discriminative tests, descriptive tests, qualitative and quantitative tests. It could be observed that not all the works used trained judges, and sometimes-quick training were performed. From the studies analyzed, no distinctive pattern could be observed about the effect of SCF on food and beverages, therefore, the impact seems to be specific to each product. It is important to note that there are many points of improvement in the application of sensory or consumer behavior methodology to evaluate the SCFs effects. The correlation of physical, chemical, and physicochemical properties with sensory perception can be of great help to understand in depth and more precisely the positive or negative effect of technology on food and beverages, for this, proper training and calibration of sensory judges are decisive, as well as the statistical and mathematical tools for the correlation of these two types of parameters. In this topic, the use of partial least squares or neural networks has been very useful [93, 94]. In the case of the evaluation of hedonic aspects, measurements of preferences or acceptance, the size of the sample and the correct segmentation of the participating consumers are of great importance for the reliability of the results. For example, for tests of overall liking, the minimum number of participants is 110 consumers [95].

11.5 Advantages and Limitations of Supercritical Carbon Dioxide (SC-CO_2)

CO_2 in a supercritical phase or at high pressures (from 7 to 30 MPa) could be used to pasteurize or sterilize food products, and it could totally or partially replace the thermal processes currently applied, offering better retention of the sensory and nutritional attributes in treated foods in comparison with traditional thermal processes. The SC-CO_2 has some important characteristics, such as non-flammability,

Table 11.5 Impact of supercritical fluids on sensory characteristics of food and beverages.

Product	Applied process	Sensory characteristic impacted	Reference
Apple juice	SC-CO_2/N_2O pasteurization, 100 bar, 36°C/10 min	The judges trained for this study were able to detect differences between the fresh apple juice and the samples treated with both CO_2 and supercritical N_2O. However, great variability was detected between the batches of each of the two treatments, which leads to conclude that it is essential to make adjustments in the process to guarantee the reproducibility of both processes.	[96]
Peach and kiwi juice	SC-CO_2/N_2O, 10 MPa, 35 °C/15 min	Both the color, the taste, and the odor of the samples of both juices treated with the SCFs technology were compared against the same characteristics of the untreated samples. Peach juice treated with SC-CO_2, was perceived slightly lighter and clouded, with a hazel flavor regarding control.	[97]
Oysters	SC-CO_2, 100 bar, 37 °C/20 min; 100 bar, 37 °C/50 min; 200 bar, 37 °C/50 min and 200 bar, 37 °C/20 min compared to untreated oysters	The SC-CO_2 treatments maintained the freshness of the oysters. The processing using pressured of 100 bar, temperatures of 37 °C and treatment times of 20 min resulted in the highest freshness index.	[98]

(Continued)

Table 11.5 Impact of supercritical fluids on sensory characteristics of food and beverages. (*Continued*)

Product	Applied process	Sensory characteristic impacted	Reference
Dry-cured ham	SC-CO_2, 12 MPa, 45 °C/30 min and 12 MPa, 50 °C/15 min compared to the untreated product	No significant differences were observed in aroma and texture compared to the untreated product. The samples processed with SC-CO_2 were perceived to be significantly less shiny than the untreated ham. The sample processed at 12 MPa, 50 °C/15 min has a less intense and homogeneous color compared to the untreated product.	[99]
Orange juice	SC-CO_2, 25MPa, 40 °C/90 min compared to heat-treated orange juice at 90 °C/60 s and untreated juice. All samples were stored at 4 °C for 8 weeks during which time the sensory characteristics were evaluated every week.	No differences were observed in the evaluated attributes (smell, taste, and visual appearance) between the treated juice and freshly extracted juice. The heat-treated juice showed significant differences with respect to the fresh juice. There were also no significant differences due to the storage period, that is, the applied process allows the odor, flavor and appearance characteristics to be maintained for 8 weeks at 4 °C.	[100]

(*Continued*)

Table 11.5 Impact of supercritical fluids on sensory characteristics of food and beverages. (*Continued*)

Product	Applied process	Sensory characteristic impacted	Reference
Coconut pulp	SC-CO_2, 12 MPa, 40 °C/30 min, and 12 MPa, 45 °C/15 min	The level of preference of treated samples was compared to that of untreated fresh coconut samples. Panelists were asked to consider appearance, texture, odor, and taste. The results did not show differences in terms of preference, the three samples are at the same level of preference, so the study concludes that with both treatments acceptable samples are obtained for consumers.	[35]
Water coconut	SC-CO_2, 12 MPa, 40 °C/30 min, and Heat-treating 90 °C/1min, compared to untreated water	The judges evaluated and compared coconut water samples in two stages, a first stage using a nonspecific discriminatory test that showed no perceptible differences between water processed by SC-CO_2 and reference water without process. If differences were observed between thermally processed water, both with reference water and with water treated by SCFs. The profile revealed three descriptors significantly modified by the applied processes, with respect to the profile of untreated coconut water. The process that most modified the smell of the water was the thermal pasteurization that generated intense notes of cooked, pantostado and to a lesser extent hazelnut. The treatment with SC-CO_2 modified the notes of toasted and baked bread slightly.	[101]

(*Continued*)

Table 11.5 Impact of supercritical fluids on sensory characteristics of food and beverages. (*Continued*)

Product	Applied process	Sensory characteristic impacted	Reference
Whey-grape juice drink	SC-CO_2, 14, 16, and 18 MPa, 35 ± 2 °C/10 min), compared to a product pasteurized by conventional thermal HTST at 72 °C/15 s	No significant differences were observed in liking (appearance, flavor, aroma, textural, and overall impression) and purchase intention after SC-CO_2 processing. HTST- treated sample had the lowest scores. The sample treated with 16 showed the highest scores in all attributes, the samples tart at 18 were distinguished by their best flavor and the samples treated with 14 were highlighted in his overall impression. Regarding the purchase intention, there was no difference between the samples treated with SC-CO_2, even though HTST treated samples had a better purchase intention due to their better consistency.	[91]

(*Continued*)

Table 11.5 Impact of supercritical fluids on sensory characteristics of food and beverages. (*Continued*)

Product	Applied process	Sensory characteristic impacted	Reference
Pomegranate juice	SC-CO_2, (12.7±0.5 MPa at 45 °C, 15±1 min); SC-CO_2 + ultrasound (30±5 Wand frequency, a piezoelectric transducer consisting of two ceramic rings: internal diameter of 12,5 mm, external diameter of 35 mm, thickness of 5 mm; High-pressure process (600 Mpa for 3 min); and Heat treatment (90 °C for 1 min)	The judges evaluated and compared pomegranate juice samples treated by the 4 different methods or processing technologies as well as reference samples (fresh juice without any treatment). Two evaluation methods CATA and Nappin were used. 11 odor descriptors, 3 taste and 9 flavor descriptors were generated to classify the juices treated by the different types of process and the reference. It was observed that the juices treated by high pressure were similar to the reference samples and were characterized by floral, green, citric, acid and red fruits notes, while the samples treated by SC-CO_2, CO_2 – Ultrasonic and thermal treatment had common descriptors the earthy notes predominate.	[102]
Volatile aroma compounds /pea (*Pisum sativum* L.) flour	SC-CO_2 (28.68, 31.03, 34,47, 37.92 and 40,27 MPa; 33.2, 40, 50, 60 and 66.08 °C; extraction time – 40 min; with ethanol as co-solvent in percentages 0, 10, 25, 40 and 50.2)	The optimum conditions for flavor modification of pea flour were: 86 °C of temperature, 42.71 MPa of pressure, and 22% of ethanol, resulting in a total volatile compound content of 0.55 μg/g. The SC-CO_2changed the four color. A correlation between total volatile content, color parameters, and sensory attributes was observed.	[25]

non-toxicity, non-corrosiveness, and chemically inertia. Furthermore, it shows critical properties that are relatively easily achieved (above 31°C and 7.3 MPa). It has intermediate diffusivity values, low viscosity, and no surface tension, which results in fast permeation into the matrix pores, increasing the extraction rate of compounds. In addition, CO_2 is usually a recyclable solvent, leaving no waste, prevents oxidation process, and it can be restored from the sample by decreasing the pressure, which avoids the evaporation step to remove solvent with additional energy consumption and significantly reduces environmental emissions [5, 44, 74]. Therefore, SC-CO_2 has many advantages which make this technology more viable for food processing, mainly at non-thermal conditions, and meets the consumers demand for products with higher quality, which are commonly not obtained with conventional processing techniques.

On the other hand, we can cite possible limitations of SC-CO_2 technology, such as: impurities can be carried with modifiers or co-solvents (ethers, methanol, and crown-ethers) used for the polar solutes solubilization in CO_2, the initial capital may be high due to the compressors cost, which are needed for implementing the process [45]; operational handling requires care due to the risk of suffocation, mainly, when used in high concentrations; complete spore inactivation requires long processing times, high temperatures and pressure, which affects the sensory properties of the products. However, the utilization of pressure cycling reduces treatment time under the same temperature and pressure conditions [103]. Besides the decompression of the discharge in the system until ambient pressure can result in softening of solid products. However, an immersion of the products in a liquid medium could overcome this disadvantage [79].

References

1. Knez, Ž., Pantić, M., Cör, D., Novak, Z., Knez Hrnčič, M., Are supercritical fluids solvents for the future? *Chem. Eng. Process.,* 141, 107532. 2019. doi: https://doi.org/10.1016/j.cep.2019.107532.
2. Palmer, M. V., Ting, S. S. T., Applications for supercritical fluid technology in food processing. *Food Chem., 52*(4), 345-352. 1995.
3. Amaral, G.V., Silva, E.K., Cavalcanti, R.N., Cappato, L.P., Guimaraes, J.T., Alvarenga, V.O., Esmerino, E.A., Portela, J.B., Sant'Ana, A.S., Freitas, M.Q., Silva, M.C., Raices, R.S.l., Meireles, A.A., Cruz, A.G., Dairy processing using supercritical carbon dioxide technology: Theoretical fundamentals, quality and safety aspects. *Trends Food Sci.Technol.,* 64, 94-101. 2017.

4. Brunner, G., Supercritical fluids: technology and application to food processing. *J. Food Engineering,* 67, 21-33. 2005.
5. Chemat, F., Rombaut, N., Meullemiestre, A., Turk, M., Perino, S., Fabiano-Tixier, A.S., Albert-Vian, M., Review of Green Food Processing techniques. Preservation, transformation, and extraction. *Innov. Food Sci. Emerg. Technol.,* 41, 357-377. 2017.
6. da Silva, R.P.F.F., Rocha-Santos, T.A.P., Duarte, A. C., Supercritical fluid extraction of bioactive compounds. *TrAC Trends Anal. Chem., 76,* 40-51. 2016.
7. Ferrentino, G., Balzan, S., Spilimbergo, S., Optimization of supercritical carbon dioxide treatment for the inactivation of the natural microbial flora in cubed cooked ham. *Int. J. Food Microbiol.,* 161, 189-196. 2013.
8. Ji, H., Zhang, L., Liu, S., Qu. C., Zhang, C., Gao, J., Optimization of microbial inactivation of shrimp by dense phase carbon dioxide. *Int. J. Food Microbiol.,* 156, 44–49. 2012.
9. Park, H.S., Lee, Y.H., Kim, W., Choi, H.J., Kim, K.H., Disinfection of wheat grains contaminated with Penicillium oxalicum spores by a supercritical carbon dioxide-water cosolvent system. *Int. J. Food Microbiol.,* 156, 239–244. 2012.
10. Sikin, A.M., Walkling-Ribeiro, M., Rizvi, S.S.H., Synergistic effect of supercritical carbon dioxide and peracetic acid on microbial inactivation in shredded Mozzarella-type cheese and its storage stability at ambient temperature. *Food Control,* 70, 174–182. 2016.
11. Silva, E.K., Meireles, M.A.A., Saldaña, M.D.A., Supercritical carbon dioxide technology: A promising technique for the non-thermal processing of freshly fruit and vegetable juices. *Trends Food Sci. Technol.,* 97, 381-390. 2020.
12. Tang, Y., Jiang, Y., Jing, P., Jiao, S., Dense phase carbon dioxide treatment of mango in syrup: Microbial and enzyme inactivation, and associated quality change. *Innov. Food Sci. Emerg. Technol.,* 70, 102688. 2021. Doi: https://doi.org/10.1016/j.ifset.2021.102688
13. Brunner,G., *Gas extraction: an introduction to fundamentals of supercritical fluids ant the application to separation processes.* 1st. Ed. Darmstadl: Steinkopff, New York: Springer, 387 p. 1994.
14. Luque de Castro, M.D., Valcárcel-Cases, M., Tena, M.T., *Analytical Supercritical Fluid Extraction.* Heidelberg: Springer, 321 p. 1994.
15. Taylor, L.T. *Supercritical fluid extraction.* New York: John Wiley & Sons Inc., 181 p. 1996.
16. Lopes, A. Extração supercritica. *O Universo da Engenharia de Alimentos.* Available from: http://uniengenhariadealimentos.blogspot.com. 2018. Acessed on 20 april, 2021.
17. Brunner, G., Stofftrennung mit überkritischen Gasen (Gasextraktion) Chemie-Technik Ingenieur- , 59, 12 – 22p. 1987.
18. Bruno, T.J., *Thermophysical property data for supercritical fluid extraction design.* J.F. Ely (Eds.), Supercritical Fluid Technology, Reviews in Modern Theory and Applications, CRC Press Inc., Boca Raton, Florida. 1991.

19. McHugh, M.A., Krukonis, V.J., *Supercritical fluid extraction: principles and practice*. 2nd Ed. Boston, MA: Butterworth-Heinemann, 512 p. 1994.
20. Rozzi, N.L., Singh, R.K., Supercritical Fluids and the Food Industry. Compr. *Rev. Food Sci. Food Saf.*, 1, 33-44. 2002.
21. Spilimbergo, S., Mantoan, D., Quaranta, A., Mea, G.D., Real-time monitoring of cell membrane modification during supercritical CO2 pasteurization. *J. Supercrit. Fluid., 48*(1), 93-97. 2009.
22. Sanli, D., Bozbag, S. E., Erkey, C., Synthesis of nanostructured materials using supercritical CO2: Part I. Physical transformations. *J. Mater. Sci.*, 47(7), 2995-3025. 2012.
23. Silva, E. K., Alvarenga, V. O., Bargas, M. A., Sant'Ana, A. S., Meireles, M. A. A.,. Non-thermal microbial inactivation by using supercritical carbon dioxide: Synergic effect of process parameters. *The Journal of Supercritical Fluids,* 139, 97-104. 2018.
24. Silva, E.K., Guimarães, J.T., Costa, A.L.R., Cruz, A.G., Meireles, M.A.A., Non-thermal processing of inulin-enriched soursop whey beverage using supercritical carbon dioxide technology. *J. Supercrit., Fluid.,* 154, 104635. 2019. doi: https://doi.org/10.1016/j.supflu.2019.104635
25. Vatansever, S., Hall. C., Flavor modification of yellow pea flour using supercritical carbon dioxide + ethanol extraction and response surface methodology. *J. Supercrit. Fluid.*, 156, 104659. 2020. Doi: https://doi.org/10.1016/j.supflu.2019.104659
26. Zuknik, M.H., Norulaini, N.A.N., Dalila, W.S.W.N., Ali, N.R., &Mohd Omar, A.K.M., Solubility of virgin coconut oil in supercritical carbon dioxide. *J. Food Eng.,* 168, 240-244. 2016.
27. Ferrentino, G., Barletta, D., Donsi, F., Ferrari, G., Poletto, M., Experimental measurements and thermodynamic modeling of CO2 solubility at high pressure in model apple juices. *Industr. Eng. Chem. Res.*, 49, 2992-3000. 2010.
28. Illera, A.E., Sanz, M.T., Beltrán, S., Melgosa, R., High pressure CO2 solubility in food model solutions and fruit juices. *J. Supercrit. Fluid.*, 143, 120-125. 2019.
29. Uddin, S., Sarker, Z.I., Ferdosh, S., Akanda, J.H., Easmin, S., Shamsudin, S.H.B., Yunus, K.B., Phytosterols and their extraction from various plant matrices using supercritical carbon dioxide: a review.*J. Sci. Food Agric.,* 95, 1385-1394. 2014.
30. Moraes, M.N., Zabot, G.L., Meireles, M.A.A., Extraction of tocotrienols from annatto seeds by a pseudo continuously operated SFE process integrated with low-pressure solvent extraction for bixin production. *J. Supercrit. Fluid.*, 96, 262-271. 2015.
31. Ferrentino, G., Plaza, M., Ramirez, R.M., Ferrari, G. Balaban, M., Effects of dense phase carbon dioxide pasteurization on the physical and quality attributes of a red grapefruit juice. *J. Food Sci.*, 74,E333-E341. 2009.
32. Gunes, G., Blum, L.K.,Hotchkiss, J.H., Inactivation of yeasts in grape juice using a continuous dense phase carbon dioxide processing system. *J. Sci. Food Agric.*, 85, 2362-2368. 2005.

33. Smigic, N., Djekic, I., Tomic, N., Udovicki, B., Rajkovic, A., The potential of foods treated with supercritical carbon dioxide (sc-CO2) as novel foods. *Brit. Food J.*, 121. 2019. doi: https://doi.org/10.1108/BFJ-03-2018-0168
34. Ferrentino, G., Balzan, S., Dorigato, A., Pegoretti, A. Spilimbergo, S., Effect of supercritical carbon dioxide pasteurization on natural microbiota, texture, and micro structure of fresh-cut coco. *Journal of Food Science*, 77, E137-E143. 2012.
35. Ferrentino, G., Belscak-Cvitanovic, A., Komes, D., Spilimbergo, S., Quality Attributes of Fresh-Cut Coconut after Supercritical Carbon Dioxide Pasteurization. *J. Chem.* 2013, Article ID 703057. 2013. https://doi.org/10.1155/2013/703057
36. Ferrentino, G., Calliari, N., Bertucco, A. Spilimbergo, S., Validation of a mathematical model for predicting high pressure carbon dioxide inactivation kinetics of escherichia coli spiked on fresh cut carrot. *J. Supercrit. Fluid.*, 85, 17-23.2014.
37. Spilimbergo, S., Komes, D., Vojvodic, A., Levaj, B., Ferrentino, G., High pressure carbon dioxide pasteurization of fresh-cut carrot. *J. Supercrit. Fluid.*, 79, 92-100. 2013.
38. Valverde, M., Marín-Iniesta, F. Calvo, L., Inactivation of *Saccharomyces cerevisiae* in conference pear with high pressure carbon dioxide and effects on pear quality.*J. Food Eng.*, 98, 421-428. 2010.
39. Zhong, Q., Black, G., Davidson, M. Golden, D., Nonthermal inactivation of *Escherichia coli* K-12 on spinach leaves, using dense phase carbon dioxide. *J. Food Prot.*,71, 1015-1017. 2008.
40. Chemat, F., Abert Vian, M., Fabiano-Tixier, A.-S., Nutrizio, M., Režek Jambrak, A., Munekata, P.E.S., Lorenzo, J.M., Barba, F.J., Binello, A., Cravotto, G., A review of sustainable and intensified techniques for extraction of food and natural products. *Green Chem.*, 22, 2325–2353. 2020.
41. Dias, A.l.B., Aguiar, A.C., Rostagno, M., Extraction of natural products using supercritical fluids and pressurized liquids assisted by ultrasound: current status and trends. *Ultrason. Chem.* 2021. doi: https://doi.org/10.1016/j.ultsonch.2021.105584
42. Essien, S. O., Young, B., Baroutian, S., Recent advances in subcritical water and supercritical carbon dioxide extraction of bioactive compounds from plant materials. *Trends Food Sci. Technol., 97*, 156-169. 2020.
43. Xiong, K., Chen, Y., Supercritical carbon dioxide extraction of essential oil from tangerine peel: Experimental optimization and kinetics modelling. *Chem. Eng. Res. Des.*, 164, 412-423. 2020.
44. Derrien, M., Aghababararnejad, M., Gosselin, A., Desjardins, Y., Angers, P., Boumghar, Y. Optimization of supercritical carbon dioxide extraction of lutein andchlorophyll from spinach by-products using response surface methodology. *LWT Food Sci. Technol.*, 93, 79-87. 2018.
45. Khosravi-Darani, K., Research Activities on Supercritical Fluid Science in Food Biotechnology. Crit. Rev.Food Sci. Nutr., 50(6), 479-488. 2010.

46. Manjare, S. D.,Dhingra, K.,Supercritical fluids in separation and purification: A review. *Mat. Sci.Energ.Technol., 2*, 463-484. 2019.
47. Truong, T., Palmer, M., Bansal, N., Bhandari, B., Effect of solubilised carbon dioxide at low partial pressure on crystallisation behaviour, microstructure and texture of anhydrous milk fat. *Food Res. Int.*, 95, 82-90. 2017.
48. Zizovic, I., Senerovic, L., Moric, I., Adamovic, T., Jovanovic, M., Krusic, M. K., Misic, D., Stojanovic, D., Milovanovic, S., Utilization of supercritical carbon dioxide in fabrication of cellulose acetate films with anti-biofilm effects against *Pseudomonas aeruginosa* and *Staphylococcus aureus. J. Supercrit. Fluid., 140*, 11-20. 2018
49. dos Santos, L.C., Johner, J.C.F., Scopel, E., Pontes, P.V.A., Ribeiro, A.P.B., Zabot, G.L., Batista, E.A.C., Meireles, M.A.A., Martínez, J., Integrated supercritical CO2 extraction and fractionation of passion fruit (Passiflora edulis Sims) by-products. *J. Supercrit. Fluid., 168*, 105093. 2021. doi: https://doi.org/10.1016/j.supflu.2020.105093
50. Rovetto, L.J., Aieta, N.V., Supercritical carbon dioxide extraction of cannabinoids from *Cannabis sativa* L. *J. Supercrit. Fluid.*, 129, 16-27. 2017.
51. Yousefi, M., Rahimi-Nasrabadi, M., Pourmortazavi, S. M., Wysokowski, M., Jesionowski, T., Ehrlich, H., Mirsadeghi, S., Supercritical fluid extraction of essential oils. *TrAC Trends in Analytical Chemistry,* 118, 182-193. 2019.
52. Filip, S., Vidović, S., Vladić, J., Pavlić, B., Adamović, D., Zeković, Z.,Chemical composition and antioxidant properties of Ocimum basilicum L. extracts obtained by supercritical carbon dioxide extraction: Drug exhausting method. *J. Supercrit. Fluid.*, 109, 20–25. 2016.
53. Hossain, M.S., Norulaini, N.A.N., Naim, A.Y.A., Zulkhairi, A.R.M., Bennama, M.M., Omar, A.K.M., Utilization of the supercritical carbon dioxide extraction technology for the production of deoiled palm kernel cake. *Journal of CO2 Utilization*, 16, 121–129. 2016.
54. Paul, I.D., Jayakumar, C., Mishra, H.N.,Optimization of process parameters for supercritical fluid extraction of cholesterol from whole milk powder using ethanol as co-solvent. *J. Sci. Food Agric.*, 96, 4885-4895. 2016.
55. Shahsavarpour, M., Lashkarbolooki, M., Eftekhari, M.J., Esmaeilzadeh, F., Extraction of essential oils from *Mentha spicata* L. *(Labiatae)* viaoptimized supercritical carbon dioxide process. *J. Supercrit. Fluid.*, 130, 253-260. 2017.
56. Campone, L., Celano, R., Piccinelli, A.L., Pagano, I., Carabetta, S., Sanzo, R., Russo. M., Ibañez, E., Cifuendes, A., Rastrelli, L., Response surface methodology to optimize supercritical carbon dioxide/co-solvent extraction of brown onion skin by-product as source of nutraceutical compounds. *Food Chem.*, 269, 495-502. 2018.
57. Gosh, M., Shubhangi, C.J.S., Mishra, H.N., Advent of clean and green technology for preparation of low-cholesterol dairy cream powder: supercritical fluid extraction process. *Food Qual. Saf.*, 2, 205-211. 2018.
58. Ilgaz, S., Sat, I.G., Polat, A., Effects of processing parameters on the caffeine extraction yieldduring decaffeination of black tea using pilot-scale

supercriticalcarbon dioxide extraction technique. *J.Food Technol.*, 55, 1407-1415. 2018.
59. Hatami, T., Meireles, M.A.A., Ciftci, O.N.,Supercritical carbon dioxide extraction of lycopene from tomato processing by-products: Mathematical modeling and optimization. *J. Food Eng.*, 241, 18-25. 2019.
60. Ferentino, G., Ndayishimiye, J., Haman, N., Scampicchio, M., Functional Activity of Oils from Brewer's Spent Grain Extracted by Supercritical Carbon Dioxide. *Food Bioproc.Technol.*, 12, 789-798. 2019.
61. Buszewski, B., Wrona, O., Mayya, R.P., Zakharenko, A.M., Kalenik, T.K., Golokhvast, K.S., Piekoszewski, W., Rafińska, K., The potential application of supercritical CO2 inmicrobial inactivation of food raw materials and products, *Crit.Rev. Food Sci. Nutr.*, 2021. doi: 10.1080/10408398.2021.1902939
62. Marszałek, K., Woźniak, Ł.,Barba, F.J., Skąpska, S., Lorenzo, J. M., Zambon, A., Spilimbergo, S., Enzymatic, physicochemical, nutritional and phytochemical profile changesof apple (*Golden Delicious* L.) juice under supercritical carbon dioxide andlong-term cold storage. *Food Chem.*, 268, 279-286. 2018.
63. Liao, H., Zhong, K., Hu, X., Liao, X., Effect of high pressure carbon dioxide on alkaline phosphatase activity and quality characteristics of raw bovine milk. *Innov. Food Sci. Emerg. Technol.*, 52, 457-462. 2019.
64. Amaral, G.V., Silva, E.K., Cavalcanti, R.N., Martins, C.P.C., Andrade, L.G.Z.S., Moraes, J., Alvarenga, V.O., Guimarães, J.T., Esmerino, E.A., Freitas, M.Q., Silva, M.C., Raices, R.S.l., Sant'Ana, A.S., Meireles, A.A., Cruz, A.G.,. Whey-grape juice drink processed by supercritical carbon dioxide technology: Physicochemical characteristics, bioactive compounds and volatile profile. *Food Chem., 239*, 697-703. 2018.
65. Cappelletti, M., Ferrentino, G., Endrizzi, I., Aprea, E., Betta, E., Corollaro, M.L., Charles, M., Gasperi, F., Spilimbergo, S., High pressure carbon dioxide pasteurization of coconut water: a sport drink with high nutritional and sensory quality. *J. Food Eng.*, 145, 73-81. 2015.
66. Ceni, G., Fernandes Silva, M., Valério, C., Cansian, R.L., Oliveira, J.V., Dalla Rosa, C., Mazutti, M.A. Continuous inactivation of alkaline phosphatase and *Escherichia coli* in milk using compressed carbon dioxide as inactivating agent. *J. CO2 Util.*, 13, 24-28. 2016.
67. Marszałek, K., Kruszewski, B., Woźniak, Ł., Skąpska, S., The application of supercritical carbon dioxide for the stabilization of native and commercial polyphenol oxidases and peroxidases in cloudy apple juice (cv. Golden Delicious). *Innov. Food Sci. Emerg.Technol.*, 39, 42-48. 2017.
68. Galvanin, F., De Luca, R., Ferrentino, G., Barolo, M., Spilimbergo, S., Bezzo, F.,Bacterial inactivation on solid food matrices through supercritical CO2: A correlative study. *J. Food En., 120*, 146-157. 2014.
69. Bi, X.F., Wu, J.H., Zhang, Y., Xu, Z.H., Liao, X.J., High pressure carbon dioxide treatment for fresh-cut carrot slicesInnovative.*Food Sci. Emerg.Technol.*, 12, 298-304. 2011.

70. Illera, A.E., Sanz, M.T., Beltrán, S., Benito-Roman, O., Effect of high pressure carbon dioxide on polyphenoloxidase from *Litopenaeus vannamei*. *LWT.*, 109, 359-365. 2019.
71. Porębska, I., Sokolowska, B., Skapsaka, S., Rzoska, S.J., Treatment with high hydrostatic pressure and supercritical carbon dioxide to control *Alicyclobacillus acidoterrestris* spores in apple juice. *Food Control*, 73, 24-30. 2017.
72. Garcia-Gonzalez, L., Geeraerd, A.H., Spilimbergo, S., Elst, K., Ginneken, L.V., Debevere, J., Impe, J.F.V., Devlieghere, F., High pressure carbon dioxide inactivation of microorganisms in foods: The past, the present and the future. *Int. J. Food Microbiol.*, 117, 1–28. 2007.
73. Valley, G., Rettger, L.F.,. The influence of carbon dioxide on bacteria. *J. Bacteriol.*, 14 (2),101–37. 1927.
74. Soares, G.C., Learmonth, D.A., Vallejo, M.C., Davila, S.P., González, P., Sousa, R.A., Oliveira, A.L., Supercritical CO2 technology: The next standard sterilization technique? *Mater. Sci. Eng.C.*, *99*, 520-540. 2019.
75. Spilimbergo, S., Bertucco, A., Non-thermal bacteria inactivation with dense CO_2. *Biotechnol. Bioeng.*, 84, 627–638. 2003.
76. Garcia-Gonzalez, L., Geeraerd, A.H., Elst, K. Ginneken, L.V., Impe, J.F.V., Devlieghere, F.,Influence of type ofmicroorganism, food ingredients and food properties on high-pressure carbon dioxide inactivation of microorganisms. *Int. J. Food Microbiol.*, 129, 253–263. 2009.
77. Spilimbergo, S., Matthews, M.A., Zambon, A., *Supercritical Fluid Pasteurization and Food Safety.* Alternatives to Conventional Food Processing. 2nd Ed. Cambridge, UK. The Royal Society of Chemistry, 170p. 2018.
78. Cuppini, M., Zeni, J., Franceschi, E., Toniazzo, G., Cansian, R.L., Inactivation of *Staphylococcus aureus* in raw salmon with supercritical CO2 using experimental design. *Food Sci. Technol.*, 36, 8-11. 2016
79. Ferrentino, G., Spilimbergo, S., Non-thermal pasteurization of apples in syrup with dense phase carbon dioxide. *J. Food Eng.*, 207,18-23. 2017.
80. Huang, S.R., Liu, B., Ge, D., Dai, J.H., Effect of combined treatment with supercritical CO2 and rosemary on microbiological and physicochemical properties of ground pork stored at 4°C. *Meat Sci.*, 125, 114-120. 2017.
81. Kustyawati, M.E., Pratama, F., Saputra, D., Wijaya, A., Viability of Molds and Bacteria in Tempeh Processed with Supercritical Carbon Dioxides during Storage. Int. J. Food Sci., 2018, Article ID 8591015. 2018. Doi: https://doi.org/10.1155/2018/8591015
82. De Matos, K.H.O., Lerin, L.A., Soares, D., Soares, L.S., de Lima, M., Monteiro, A.R., Vladimir Oliveira, J., Effect of supercritical carbon dioxide processing on Vibrio parahaemolyticus in nutrient broth and in oysters (*Crassostrea gigas*). *J. Food Sci. Technol.*,55(10), 4090–40982018. doi:10.1007/s13197-018-3335-3
83. Zambon, A., Bourdoux, S., Pantanoc, M. F., Pugno, N. M., Boldrin, F., Hofland, G., Rajkovic, A., Devlieghere, F., Spilimbergo, S., Supercritical CO2

for the drying and microbial inactivation ofapple's slices. *Dry. Technol.*, 39 (2), 259–267. 2019.

84. González-Alonso, V., Cappelletti, M., Bertolini, F.M., Lomolino, G., Zambon, A., Spilimbergo, S., Microbial inactivation of raw chicken meat by supercritical carbon dioxide treatment alone and in combination with fresh culinary herbs. *Poul. Sci.*, 99 (1), 536-545.2020.
85. Iqbal, A., Murtaza, A., Muhammad, Z., Elkhedir, A., Tao, M., Xu, X., Inactivation, Aggregation and Conformational Changes of Polyphenol Oxidase from Quince (Cydonia oblonga Miller) Juice Subjected to Thermal and High-Pressure Carbon Dioxide Treatment. *Molecules*, 23 (7). 1743; doi: https://doi.org/10.3390/molecules23071743
86. Marszałek, K., Doesburg, P., Starzonek, S., Szczepańska, J., Woźniak, Ł., Lorenzo, J. M., Skąpska, S., Rzoska, S., Barba, F.J., Comparative effect of supercritical carbon dioxide and high pressure processing on structural changes and activity loss of oxidoreductive enzymes. *J. CO2 Util.*, 29, 46–56. 2019.
87. Murtaza, A., Iqbal, A., Linhu, Z., Liu, Y., Xu, X., Pan, S., Hu, W., Effect of high-pressure carbon dioxide on the aggregation and conformational changes of polyphenol oxidase from apple (*Malus domestica*) juice. Innov. Food Sci. Emerg.Technol., 54, 43-50. 2019.
88. Murtaza, A., Iqbal, A., Marszałek, K., Iqbal, M.A., Waseem Ali, S., Xu, X., Pan, S., Hu, W., Enzymatic, Phyto-, and Physicochemical Evaluation of Apple Juice under High-Pressure Carbon Dioxide and Thermal Processing. *Foods*, 9 (2), 243. 2020. doi:10.3390/foods9020243
89. Podrepšek, G.H., Knez, Z., Leitgeb, M., The Influence of Supercritical Carbon Dioxide on Graham Flour Enzyme Polyphenol Oxidase Activity. *Molecules*, 25(24), 5981. 2020. doi: 10.3390/molecules25245981
90. Silva, E.K., Arruda, H.S., Eberlin, M.N., Pastore, G.M., Meireles, M.A.A., Effects of supercritical carbon dioxide and thermal treatment on the inulin chemical stability and functional properties of prebiotic-enriched apple juice. *Food Res. Int.*, 125, 108561. 2019. doi: https://doi.org/10.1016/j.foodres.2019.108561
91. Amaral, G. V., Silva, E. K., Costa, A. L. R., Alvarenga, V. O., Cavalcanti, R. N., Esmerino, E. A., Guimarães, J.T., Freitas, M.Q., Sant'Ana, A.S., Cunha, R.L., Moraes, J., Silva, M.C., Meireles, A.A., Cruz, A. G., Whey-grape juice drink processed by supercritical carbon dioxide technology: Physical properties and sensory acceptance. *LWT, 92*, 80-86. 2018.
92. Marszałek, K., Skąpska, S., Woźniak, Ł., Sokołowska, B., Application of supercritical carbon dioxide for the preservation of strawberry juice: Microbial and physicochemical quality, enzymatic activity and the degradation kinetics of anthocyanins during storage. *Innov. Food Sci. Emerg.Technol.*, 32, 101-109. 2015.
93. García-Barrón, S.E., Herrera-López, E.J., Valentin, D., Villanueva-Rodríguez, S. J.,*Classifying different mezcal spirits through sensory descriptors and artificial neural networks.* 10th Pangborn Sensory Sciene Symposium. 2017.

94. Villanueva, S., Escalona, H., Estarrón, M., Martín del Campo, S., Cantor Esperanza, A.K., *Classification of tequilas from different categories and regions using gas chromatography-mass spectrometry and sensory evaluation.* ACS National Meeting Book of Abstracts 229:101-AGFD. 2005.
95. ISO 11136:2014/AMD 1:2020 Sensory analysis — Methodology — General guidance for conducting hedonic tests with consumers in a controlled area—Amendment 1
96. Gasperia, F., Aprea, E., Biasioli, F., Carlin, S., Endrizzi, I., Pirretti, G., Spilimbergo, S., Effects of supercritical CO2 and N2O pasteurisation on the quality of fresh apple juice. *Food Chem.*, 115, 129–136. 2009.
97. Spilimbergo, S., Ciola, L., Supercritical CO_2 and N_2O pasteurisation of peach and kiwi juice. *Int. J. Food Sci. Technol.*, 45, 1619–1625. 2010.
98. Meujo, D.A.F., Kevin, D.A., Peng, J., Bowling, J.J., Liu, J., Hamann, M.T., Reducing oyster-associated bacteria levels using supercritical fluid CO_2 as an agent of warm pasteurization. *Int. J. Food Microbiol.*, 138, 63–70. 2010.
99. Ferrentino, G., Balzan, S., Spilimbergo, S., Supercritical Carbon Dioxide Processing of Dry Cured Ham Spiked with Listeria monocytogenes: Inactivation Kinetics, Color, and Sensory Evaluations. *Food Bioproc. Technol.*, 6, 1164–1174. 2013.
100. Oulé, K.M., Dickman, M., Arul, J.,Properties of Orange Juice with Supercritical Carbon Dioxide Treatment. *Int. J. Food Prop.*, 16 (8), 1693–1710. 2013.
101. De Marchi, F., Aprea, E., Endrizzi, I., Charles, M., Betta, E., Corollaro, M.L., Cappelletti, M., Ferrentino, G., Spilimbergo, S., Gasperi, F., Effects of Pasteurization on Volatile Compounds and Sensory Properties of Coconut (*Cocos nucifera L.)* Water: Thermal vs. High-Pressure Carbon Dioxide Pasteurization. *Food Bioproc. Technol.,* 8, 1393–1404. 2015.
102. Mosca, A.C., Menghi, L., Aprea, E., Mazzucotelli, M., Benedito, J., Zambon, A., Spilimbergo, S., Gasperi, F.,Effect of CO2 Preservation Treatments on the Sensory Quality of Pomegranate Juice. *Molecules*, 25 (23), 5598. 2020. Doi: https:// doi:10.3390/molecules25235598
103. Zhang, J., Davis, T.A., Mathews, M.A., Drews, M.J., LaBerge, M., A, Y.H., Sterilization using high-pressure carbon dioxide. *J. Supercrit. Fluid.*, 38(3), 354–372. 2006.

12

Image Processing for Food Safety and Quality

Krishna Kumar Patel[1]*, S. K. Goyal[2] and Yashwant Kumar Patel[3]

[1]Department of Agricultural Engineering, Post Graduate College, Ghazipur (U.P), India
[2]Farm Engineering (Agricultural Engg.) Institute of Agricultural Sciences, BHU, Rajiv Gandhi South Campus, Barkachha, Mirzapur (UP), India
[3]Department of FPT, Atal Bihari Vajpayee Vishwavidyalaya, Bilaspur (C.G.), India

Abstract
With the worldwide increase in population and awareness, the concern about the food safety and quality assurance has been increased. Numbers of novel potential techniques have been emerged in the last decade seeking to full-fill the requirements of food safety standards and the current food demand. The developments in application of image processing techniques among them for food quality evaluation and safety assurance have gained more popularity and momentum. The main principle of image processing technique is based on the imaging system and processing techniques (such as low-level, mid-level, high-level, etc.). The low-level image processing includes capturing of images, noise removing and enhancing of the contrast as per requirements for better segmentation of the objects from the background. Image segmentation, image representation and description, similarly, carried out under the mid-level image processing technique, followed by the object measurements and classification i.e. high-level image processing. The main aim of this chapter is, therefore, to introduce the fundamental concept and basic principle of image processing technique used in computer vision as well as to discuss in details the application of image processing for quality analysis and safety assurance of agricultural produce, dairy products, fish, meat and meat products, etc. This chapter reports the various aspect of image processing technique which would be beneficial not only for students, teachers, researchers, scientist but also fruits and vegetable growers and suppliers of the processed food products such as cheese, breads, pizza, etc.

**Corresponding author*: k_krishna_374@yahoo.co.in

Navnidhi Chhikara, Anil Panghal and Gaurav Chaudhary (eds.) Novel Technologies in Food Science, (451–478) © 2023 Scrivener Publishing LLC

Keywords: Image, image processing, computer vision, noninvasive techniques, food, safety, quality

12.1 Introduction

Today we are living in the concrete world, which is enlightened by artificial light and surrounded by electromagnetic waves. Humans in this era are full of stress, more prone to artificial intelligence and worry about their healthy lives. Consequently, humans are more aware about their food safety and quality. Quality, self defined as the degree of excellence, means it is the sum of all those attributes, which can lead to the production of products acceptable to the consumer when they are combined. As the high quality, food products are the basis for success in this high competitive market. Quality has always been a key factor for the food industry in this modern era. Quality of any product generally depends on the grade standards and internal and external factors. The external factor includes texture, flavor and appearance such as size shape, colour, gloss and consistency [4]. Similarly, internal factors depend on physical, chemical and microbial qualities. Evaluation of these qualities of food products using destructive or laboratory or chemical analysis and traditional methods are not only time consuming, tedious, laborious, costly and inconsistent but also more prone to human error. Most of the industries in Asian countries are still dependent on manual inspection for quality analysis of food products. According to Du and Sun [17] and Patel *et al.* [45, 47, 51], traditional methods of quality inspection are easily influenced by physiological factors that lead to the inconsistent and subjective results.

In addition, as per Ministry of Food Processing Industries report (2018-19), India is the second largest producer of food grain (284.95 Million Tons), fruits and vegetables (3.11.7 Million Tons) and marine/fish (12.60 Million Metric Tons) and largest producer of meat (about 7.7 Million Tons produced annually), milk (176.35 Million Metric Tons) and eggs (about 95.2 billion numbers) globally. However, due to the lack of automation in Indian Industries, India processes only 10% of its total agri-output (2% of fruits and vegetables, 6% of poultry, 21% of meat, 23% of marine and 35% of milk). Most of the processing that is done in our country belongs to the primary processing offers lower level of value addition, in comparison to the secondary processing, through rice, sugar, edible oil and flour mills, etc [18]. Consequently, the MoFPI estimated that the wastages of food in India are as high as US$1.5 billion (Rs 92,000 crores) annually (and the exports of enhance food and processed food currently limited to 2.31 and 10.69%, respectively (MoFPI report 2018-19).

Therefore, the automation of Indian food industries and popularization of image processing techniques is very important to avoid such huge wastages in India. Consequently, the awareness sophistication and expectation of consumers can also be satisfied by the enhancement of the quality of food products [6, 45, 47, 50, 51]. Image processing techniques are found more suitable for the quick quality assessment of foods. The characterization of this technique is based on the information extracted using the image data. Image data, generally, is the form of images of microscopic, X-rays, angiography, tomographic, etc. These techniques, however, are more common in medical fields and sometimes known as medical image processing techniques.

Further, image acquired by these techniques (i.e. infrared imaging: IRI; magnetic resonance imaging: MRI; x-ray imaging and CCD: charged coupled device cameras) have defects of low contrast and high noise. The images with low contrast have implicit boundaries between the objects or the details of the objects are not clear. Similarly, the image with elevated noise level does not reflect correct measuring parameters or characteristics of the object. Image can, thus, be processed by means of optical techniques or by photographic methods and/or by electronically to remove or to reduce the defects in the images in order to achieve the unaffected succeeding outputs [71]. In this method, the application of digital computer based image processing tools and techniques are more prevalent due to rapid, consistent, soft in application and methodical characteristics. Image modified considering the brightness/contrast enhancement and restored in desired format, first. The extraction of information (i.e. analysis and recognition) which followed by the change in structure (i.e. composition and image editing) as per requirement is, then, carried out [3].

Several techniques for the processing of images have now been emerged potentially since the last decade for the purpose of non-destructive or non-invasive, non-contact, real-time/quick, consistent and objective evaluation of food quality and safety. Application of image processing for food quality assessment has been discussed by the several researchers such Patel *et al.* [45, 47, 48, 50, 51], Sun [64], Zheng *et al.* [72, 73], Youssef *et al.* [69], Du and Sun [14], etc.

Considering above data the main objectives of this is to report the fundamentals of some valuable and prevalent non-destructive technologies in the food-processing sector, to introduce the basic concept of image, imaging and image processing and their application in food quality and safety.

Image Acquisition Techniques

Non-destructive techniques can be defined as the techniques, which are not causing destruction of food materials being investigated and require no or minimum sample preparation. In other words, in non-destructive methods, samples or bulk of materials remain untouched and samples are not destroyed during the quality assessment means it remains intact for consumer and consumer can use it even after testing. In contrast to the non-destructive method, in destructive methods the food products are destroyed first and then analyzed in the laboratory using chemical analysis methods. During the chemical analysis in laboratories, small quantity of samples is utilized. Thus, utilized samples are often not usable again for the consumption purposes.

There are several non-destructive imaging techniques for external and internal quality assessment that have now been developed. These technologies have bright future in the food and agriculture sector [45]. Image acquisition technologies such as Computer Vision (technique for external quality assessment), Tomographic Imaging (Nuclear Tomography: X-ray and Computed Tomography, Magnetic Resonance Imaging (MRI)/Nuclear Magnetic resonance (NMR)]; Electrical Tomography, etc.), Infrared/Thermo-graphic imaging, Ultrasound, etc. are prominent and most prevalent in food industries. The x-ray, CT-scan, MRI, ultrasound, etc. are very common image acquisition techniques for internal quality assessment in the food sector.

(1) Image acquisition Technique for External Quality Assessment

Although, the computer vision system is primarily used for the extraction of outer information of food and agricultural produce. The internal quality parameters can also be evaluated, if these parameters are correlated with the external characteristic. The internal and external quality characteristics of agricultural products including degree of maturity, defects, moisture content and nutrients can, thus, be evaluated using computer vision based image processing [70]. The basic concept of computer vision system and image processing techniques has been discussed in this section.

Computer Vision

When different devices such as mechanical, optical, sensing (electromagnetic), etc. are used to look, identify and acquire the images, and thus captured digital videos/images and image processing technologies are combined and work together as an integrated mechanical-optical-electronic software system, it is known as computer vision system. Concept of computer vision was started in the 1960s at MIT (USA) when AI (artificial intelligence) became an academic discipline. Before this, in 1959, David Hubel and Torsten Wiesel had published computer vision related paper entitled "receptive fields of single neurons in the cat's striate cortex" described core response properties of visual cortical neurons as well how a cat's visual experience shapes its cortical architecture [13]. Rapid duplication of the effect of human vision by electronically perceiving and understanding an image is the main objective of the computer vision (Figure 12.1). In other words, computer vision is a technology for optical, non-contact sensing to automatically receive and interpret the image of a real scene in order to obtain the information and/or to control machines or process image [47, 75]. In addition, computer vision offers a high level of flexibility and repeatability at a relatively low cost. Consequently, the application of computer vision has now been expanding rapidly in the various sectors such as autonomous vehicles, automatic manufacturing and surveillance, medical

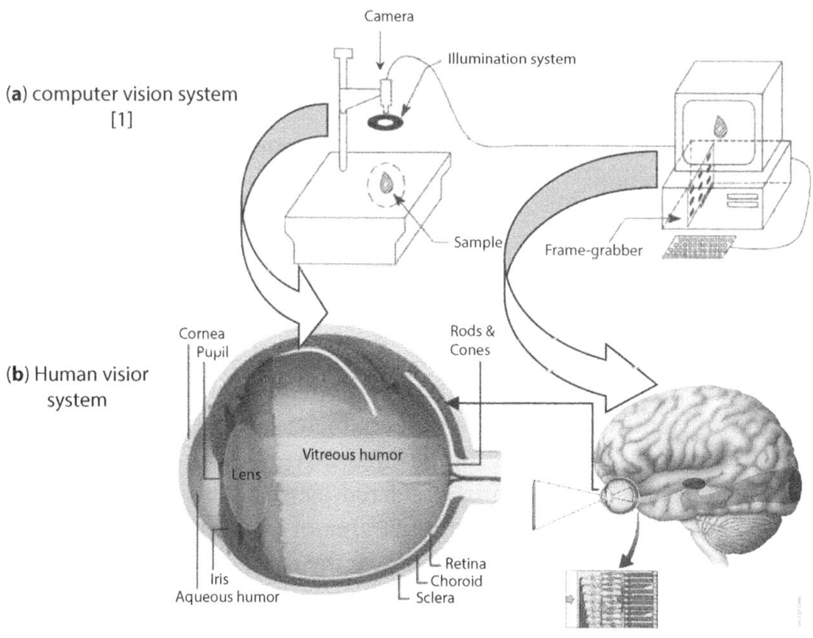

Figure 12.1 Computer vision vs. human vision system [53].

diagnostics, remote sensing, robot guidance, technical diagnostics and in agricultural and allied sectors.

Principle of Computer Vision and Its Basic Components

Computer vision systems comprises five main components such as (i) a camera, (ii) an illumination, (iii) image capture board (frame grabber or digitizer), and (iv) computer hardware and (v) computer software. According to Zuech [78], a fundamental computer vision system has following main functions:

- *Arrangements of lighting devices:* devices dedicated for illumination.
- *Optics*: to couple the image to a sensor.
- *Sensor*: to convert optical image to a sensor.
- *Analog-to-digital (A/D) converter*: to quantize the analog signal.
- *Image processor/vision engine*: s/w or h/w dedicated to reduce the noise and to enhance, process, and analyze and image.
- *Computer*: a decision maker and controller.
- *Operator interface*: devices (such as light pen, terminal, touch panel display, etc.) used by the operator to interface with the system.
- *Input-output:* communication channels to the system and to process it
- *Display:* television or computer monitor to make visual observations

Image acquisition is generally done in a confined chamber, which is known as illumination chamber. This chamber is designed in such a way so that the interaction of natural/unwanted artificial light can be avoided. The image processing among these above stated functions is a combination of various actions and activities, which lead to desired results. Creation of representation of visual characteristics of an object/or real scene/or image is known as digital image. A digital image is generally made of number of small areas referred as pixels or picture elements or pels, etc. The term pixel (a minute area of illumination on a display screen) is most widely used and denotes the elements of a digital image [21].

Image Processing

The term image processing is generally used for the covering of both image processing and analysis of images. The processing of images, however, begins by converting images into digital form followed by preprocessing operation. The main objective of preprocessing of images is to rectify the image and to achieve the dimension identical to the corresponding original image. In addition, the partitioning of the digital images into disjoint and non-overlapping is done using the image segmentation techniques. Similarly, the object measurement and classification operations are also used to measure the characteristics of objects and to identify the objects by separating them into different groups. The image processing techniques (Figure 12.2) can thus be grouped into (a) image acquisition, (b) preprocessing, (c) image segmentation, (d) object measurement and (e) Classification [14, 47].

(a) Image acquisition

Image capturing is considered as being of equal importance as image processing. Generally, image acquisition system consists of a confined chamber (to avid the interference of natural and unwanted artificial light), lighting arrangement, high speed camera, frame grabber, etc. is the first step of image processing and very important as discussed above. Image and imaging process are affected by several factors such application field, illumination specification, quantum efficiency (with reference to technological characteristics), spectral range of camera, etc [38]. The role of a camera in computer vision systems is very important which captures quality images and thus is responsible for quality image processing and analysis of the objects. High-pixel resolution cameras such as CCD (Charge-Coupled Device) and CMOS (Complementary Metal Oxide Semiconductor) are more common for the acquisition of good quality images of the objects. Using these cameras, the detection of photons are organized as arrays of photo-detectors and these photo-detectors then delivered an electrical signal related to the amount of photons that falls on the pixel surface during the integration time. Either in photo-gate or a photodiode detector, photoelectric effect in silicon is used by both sensors [34]. These cameras/sensors

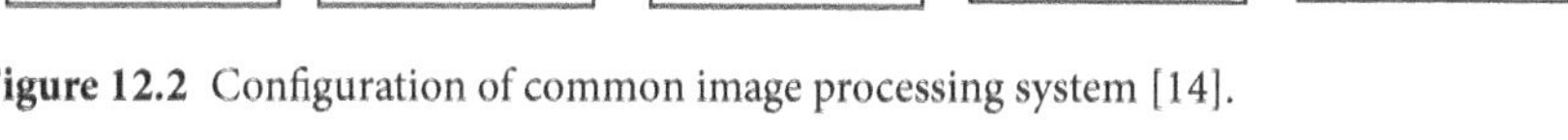

Figure 12.2 Configuration of common image processing system [14].

based on CCD technique using thermionic tube devices have some other applications are more popular than the CMOS. The main contribution of CMOS image sensors, on the other hand, is its capability of addressing, transistors providing buffering, conversion of charge–to-voltage and the combination of inside pixels of the detectors. In addition, CMOS sensors have several benefits regarding power consumption (low), cost (low), capable in integration, imaging speed (i.e. high), radiation-hard, read-out rate, noise, etc. These qualities make CMOS sensors the first choice technology not only for very high end application but also several other applications such as in space applications, high end digital end photography, HD TV (high-definition television) and low-cost photographing markets.

Further, the digitization (i.e. the conversion of image into numerical form or digit) is another important step in computer before processing of images. Digitizer or frame grabber board, which is a vision processor board, is generally used for the digitization of images. Generally, frame grabber board is selected based on results obtained from camera. Camera for image acquisition can be color (RGB) camera, monochrome camera, UV camera, Infrared camera, etc. as per need.

The image is segmented/ partitioned into two-dimensional (2-D) grid (horizontal and vertical lines). The 2-D function $f(x, y)$ where x and y are the spatial (plane) coordinates of the image and f is called the intensity or gray level of image. Thus the smallest unit i.e. one discrete picture element, of an image is defined as pixel or pel. Only two dimensional data is sufficient for agricultural grading, classification and analysis of most of the agricultural produce images while for information and added details the three dimensional grid (3-D) is needed [44, 45]. There are various types of analogue to digital converters (ADC) available but for real time analysis a special type, i.e. known as flesh ADC, analogue to digital converters (ADC) is required. Flash ADC has the ability to process 50-200 mega samples per second.

(b) Preprocessing

Preprocessing of images includes noise removal and contrast enhancing operations. When we capture digital images, there may be various types of noises. Among these noises electronic-noise, read-out-noise, wiring-noise and extraneous noise such as periodic strips, etc. are more common in the images due to imaging systems which required removing before digitizing and storing the images in the computer. Averaging and Gaussian filters are often applied for removing the noises. Averaging by linear filter is the most efficient and feasible approach which includes Gradient and Laplacian filter (i.e. a high–pass filter) and smoothing and Gaussian filter (a low-pass filter). The adjustment of the filter's weight and size is usually based on the type noise required remove [74]. In addition, median filter which is a non-linear filter is very useful in reducing the noise, preserving discontinuities in a step function and smoothing the edge [27].

Contrast of any image is very important characteristics generally defined as the difference in color and brightness (i.e. lightness and darkness) or luminance or grey values between the objects. In other words, the contrast is the difference between maximum and minimum pixel intensity in an image. Images captured with low contrast, have pixels with different intensities, and are unable to separate from each other. The enhancing of contrast means, increasing the differences in pixel intensity values for recognition by computer and/or by human vision. There are two different popular contrast enhancing techniques: (1) histogram scaling i.e. most utilized technique in the food industry and (2) equalization. In histogram equalization, the intensity values of pixels in input image (original image) reassigned so that output image contains uniform density intensities. However, the transformation of original histogram from one scale (smaller) to another higher scale is defined as histogram scaling. Formation (of histogram), calculation (of value new intensities for each intensity level) and replacement (of previous intensity values with the new intensity values) are the three main steps of this technique of contrast enhancement. The scaling of histogram may be linear or non-linear and may be one-to-one or multiple-to-one. But, these scaling functions for transformation of histograms are limited only to some proposed application. Accuracy of processing and analysis of images, thus, can be enhanced by improving the contrast of the images

(c) Image Segmentation

Threshold-based, region-based, gradient-based and classification-based techniques among the several segmentation techniques are commonly used for the segmentation, representation and description of the images in food industries. Using these image segmentation techniques, the food images are partitioned into regions having strong correlation with the area of interest. Thresholding converts the image from a grayscale image, with pixel values ranging from 0 to 255, to a binary image, with pixel values of 0 or 1 (Figure 12.3). However, the fundamental of thresholding is based on image characteristics. There are two types of thresholding: bi-level (background and objects) and multi-level (one background and several objects). Variance-based [43] and entropy-based [57] are the 2 kinds of objective functions that are used mostly. Among these, Otsu's method is referred to as one of the most powerful methods for bi-level thresholding application. Otsu method can provide satisfactory results even in the case of unimodal

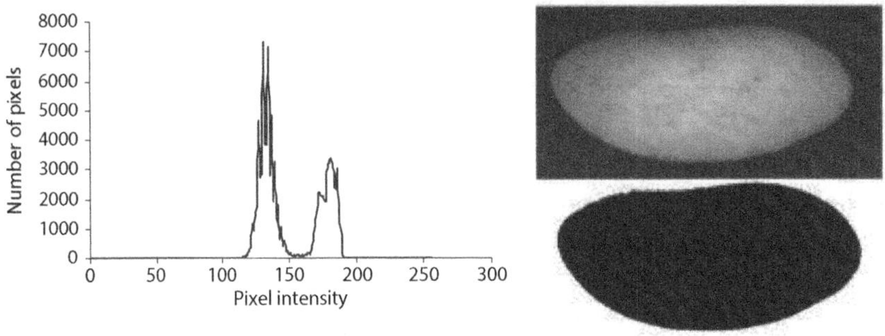

Figure 12.3 Bi-level histogram shows distribution of pixels in object and background.

histogram images which do not have two obvious peaks and can be used as a classical method in real thresholding application [8].

The second most popular segmentation technique used in the food sector, after threshold-based, is classification-based segmentation. Classification-based technique is pixel-oriented technique where every pels are treated as free observer. These observers are classified into the object and background, or defect and non-defect, etc. according to their variables. Image parameters (size/shape, color, texture, etc.) are the main source of these variables. Input of the classification is a matrix which has each pels like as an observer. This technique assumes that the segmentation is able to accurately extract the object of interest from the background image autonomously. Classification-based segmentation methods, in spite of above, lack a structured way for coping with variations in rotation and scale. This drawback limits the applicability of classification based segmentation [14].

Region-based segmentation is more expensive than the threshold based segmentation. However, this segmentation technique is capable of using of various parameters of the image directly and the location of final boundary thus can be determined simultaneously by this novel thresholding method. The region–based technique of two types (1) growing-and-merging (GM), and (2) splitting-and-merging (SM) is generally applied in the food industry.

In contrast to region-based threshold, the gradient-based segmentation technique, also known as edge detection, approaches to detect the edges of object image directly, using their high gradient measures. Gradient operators such as Sobel, Prewitt, Roberts, Krisch, Laplace, Canny, etc., are widely used for the detection of edges of the images. The ACM (Active Contour Model) method, known as Snakes, is another gradient-based segmentation. Edge-identification issues, in this technique, are transformed into an energy optimization cases [74]. Canny edge detector among these is more popular in food sector for boundary extraction of the food products. The application of gradient-based segmentation, however, is limited due to segregations mainly in color, texture and grey level, etc.

Watershed and hybrid based segmentation techniques are also getting popularity so far. Watershed concept originally comes from topography. Different scales of image such as grayscale [66], binary image [9] and gradient [17] can be used for the construction of watersheds. Although, several techniques for segmentation of images have now been developed but no method is complete for efficient and accurate segmentation. By nature, issues regarding the thresholding of the image are still not well-defined. So, several techniques can be combined in order to improve the segmentation process.

(d) Object Measurement and Classification

Object measurement is the calculation of characteristics of the object just after the segmentation. During the image segmentation, objects are discriminated from the background and then the information related to the size, shape, color, texture, etc. of the object is extracted. This information is very important for quality evaluation and inspection tasks or image understanding and interpretation for the classification of the object.

The identification of the objects by separating them into different finite sets or classes is known as object classification. In this technique, the measured features of the new object are compared with those of a known object or other known criteria. In other words, the classification algorithms usually have two phases (training and testing) of processing. In the training phase, first the characteristic properties of typical image features are isolated and then on the basis of these unique descriptions a training class/or classification category is created. Subsequently in the testing phase, these feature-space partitions are used to classify image features. The description of the training class is a very important step in the classification process. Statistical process (i.e. based on an a priori knowledge of probability distribution function) of distribution free process can be used to extract class descriptors in supervised classification while the unsupervised classification relies on clustering algorithms to automatically segment the training data into prototype classes. Although, using image processing techniques the image of food products is quantitatively characterized by a set of features such as size, shape, color and texture, which are the objective data used to represent the food products. These features can, thus, be used to form the training set. Once the training set is having been obtained, the classification algorithm extracts the information/knowledge base necessary to make decisions on unknown cases [16].

Classification is a broad range of decision-theoretic approaches for image identification. For this task, several methods have been employed in the food quality evaluation. Artificial neural network (ANN), fuzzy logic, SVM (support vector machine including binary classification and multi-classification), statistical approaches (Bayesian classification, discriminant analysis and nearest neighbor) and genetic algorithms are some of them, are employed for building knowledge bases into computer structures. In food industries, computer vision system equipped with the ANN (artificial neural networks) and FL (fuzzy logic) has now been applied, effectively [67].

Application of Image Processing

Food quality (the degree of excellence) is defined in terms of characteristics of food, acceptable to the consumers. Quality includes various external (chemical, physical and microbial factors) and internal factors such as texture, flavor, and appearance (colour, consistency, gloss, size, shape). Physical properties (such as mass, volume, perimeter, area, etc.) are most important in determining sizing systems. Although, several studies related to online estimation of mass, volume, size, shape, etc using colour image processing are present in literature. The rapid assessment of quality of fruits and vegetables is still the most interesting and concerning area. Image processing techniques can be used in various applications such as size, shape, color, texture defect detection, quality grading/inspection, variety classification, etc. of the agricultural produce.

Sorting and Grading of Fruits and Vegetables

Sorting and grading of agricultural produce using the traditional method is very tedious, laborious, laborious, and time-consuming and sometimes wrong which causes huge loss not only in terms of economy and quality but also reputation of producers. To overcome these issues image processing can be used for sorting and grading of agricultural produce.

Size and shape on one side are two very important geometric properties of fruits and vegetables. On the other hand, color and texture reflect the appearance and surface properties of a product. For instance, Choi *et al.* [12] used color CCD cameras instead of black and white cameras and extracted features from the HIS color space and achieved a much better classification rate of 77.5% for a total of 120 tomatoes. Significant correlation between color features obtained from the CIEL*a*b* color space and tomato maturity was found [25]. Color (RGB) [50], monochrome [49] and NIR [46] cameras were used in another study for estimating mangoes major, intermediate and minor diameters and found more than 92%, 95% and 90% accuracy, respectively. In addition, various morphological features such as area, diameters (Max Feret and Waddell Disc), Factors (elongation, compactness, Heywood circularity and type), perimeter etc, of mangoes were also analyzed using image processing techniques. A strong correlation was also found between color and the storage temperature, between color and storage time and a model was developed to describe the relationship by Lana *et al.* [28]. Models found to be worked with the accuracy of 84%, 94% and 89% on red, green and blue channels, respectively.

Geometrical characteristics like size and shape of fruits and vegetables (Figure 12.4), thus, contain information important for the quality assessment and for purchase decision by the consumers [29]. According to Jerimposed and Jaisin [26] sweet tamarinds can be sorted as per their size (small, medium, and large), shape (curved, slightly curved and straight) and defects. Using size and shape parameters obtained from the image processing technique various models can be developed for the evaluation of weight and volume of mangoes [49–50]. Similarly, texture features were also used to classify the various fruits and vegetables successfully. Image processing technique, thus, can perform better than the traditional method

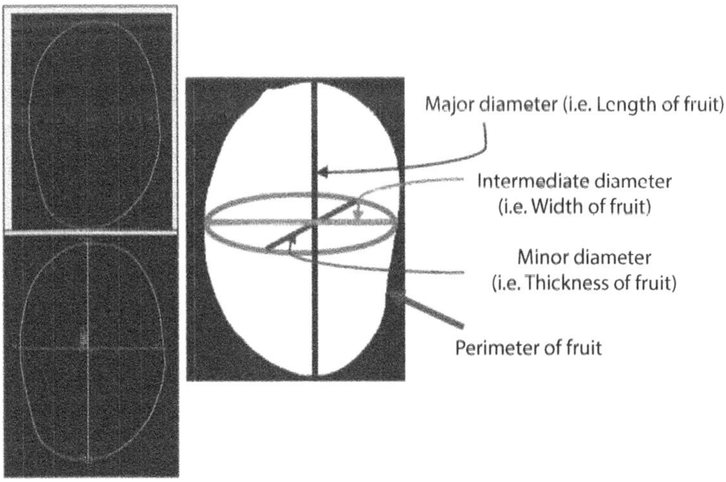

Figure 12.4 Geometric attributes evaluated by image processing technique.

because of the quantitative and objective measurement of color features of agricultural produce.

Defect Detection of Fruits and Vegetables

Fruits and vegetables are more exposed to the impact and vibration than to the compression forces during packing. Impacts are the transitory movements caused by sudden acceleration causing great dissipation of energy and consequent damage to the fruit. Mechanical damage defined as plastic deformation, superficial rupture and tissue destruction due to external forces, however, is more common during fruit handling [51, 58]. Several techniques for image processing have been developed for the detection of defects of fruits such as apples, banana, citrus fruits, plum, mangoes, etc. and vegetables such as tomato, potato, onions, beans, lettuce, etc. Bean weevil (*Acanthosecelides obtectus* Say) damage on beans was determined using image processing technology by capturing images of beans under UV light. Similarly, reflected UV images can also be used to detect the defects on the surface of mangoes and also for the recognition of hidden defects/ruptured tissues, etc. just below the peel [51]. Color [48] and monochrome [45] camera based image processing techniques are also very useful for detecting bruises on the surface area of mangoes with the accuracy of 88.6% and 88.8%, respectively following the same image processing process as discussed above. This technique was also used by Lin *et al.* [32] for the classification of the lemons and tomatoes according to color, defects and volume.

Cereals/Grains Assessment

Image processing has been found very helpful for the identification of classes of cereal/grains. This technique was successfully applied for the classification of the dockage components of wheat (Canada Western Red spring) and other cereals with higher accuracies. Models based on morphology, color and morphology/color were developed and used for the classification of the dockage components [40]. Pourreza *et al.* [54] identified nine Iranian wheat seed varieties by textural analysis using image processing technique. The highest average classification accuracy of 92.2 was achieved by Choudhary *et al.* [10], while identifying five cereal grains using only textural features. In another study by Manickavasagan *et al.* [37] demonstrated the capability of a machine vision system with monochrome camera for the identification of western Canadian wheat classes by bulk sample analysis. They reported that the gray value of the wheat samples were significantly different for different wheat classes.

Further, the measurement of degree of milling is very important in rice industry. Nondestructive rapid measurement of the degree of milling of rice has been developed by the researchers using digital image analysis technique. Ng *et al.* [41] measured mechanical and mould damage of corn kernels using this technique. Similarly, for the classification of maize, the two-dimensional image analysis concept was applied by Steenhoek and Precetti [62]. The image processing technique can, therefore, be applied successfully for the automation of seed classification, extraction of morphological features, grain check, inspection of whole and broken corn kernel, and study of head rice yield during milling, etc. after image acquisition, processing and analysis of image.

Processed Food

The image processing tools and techniques are very helpful in the assessing of the processed products such as breads, biscuits, cake, chips, cheese, chocolates, pizza, etc. accurately and to avoid any type of misguidance to the consumer during the visual based human perception. As colour, size, shape and the other visual features indicate the quality of many foods should be assessed accurately for fair pricing in the market. For instance, the percentage and distribution of pizza topping can be evaluated by the processing and analysis of the captured pizza image [63]. Support vector machine (SVM) based classification model was used for the pizza toppings. The Pizza sauce color was quantified into 256 values and fed into the SVM. The SVM discriminated the pizza sauce into five classes. According to Du and Sun [15] the overall accuracy of SVM for the classification of the 120 samples was 90%.

Ni and Gunasekaran [42] have developed an image processing algorithm to recognize individual cheese shred and to measure the shred length, automatically. Similarly, the image processing technique can be applied effectively to characterize the white wheat products before the manufacturing of the noodle products. Hatcher and Symons [22] have analyzed the impact of hexaploid wheat with a red seed coat in the appearance of Asian noodles using this technique. Further, the color, textural features, etc. of potato chips and French fries are very important quality parameters for the consumer and can also be analyzed quickly using these techniques. Generally, texture feature of image has better information than the color feature of image for the discrimination of the quality parameters of chips and of the consumer preferences [14]. Most approaches of textural analysis are statistical [35, 36] and co-occurrence matrix based method [59]. Several description methods are based on Fourier spectrum; wavelet transform and fractal dimension [55]. Fourier transform was used by Whittaker *et al.* [68] to analyze the texture property of an image obtained from ultrasonic image acquisition to predict the marbling score of beef. Similarly, wavelet method was applied for studying of fresh-beef-images. In this study, first the texture features of the images were extracted and then these features were used for the classification of the steaks into groups (such as tough and tender) of cooked beef tenderness [30, 31]. Therefore, it can be concluded that the technologies for determination of texture feature are as important as texture image feature for the extraction of better information of food products. Fractal technology is one such technology can extract texture features of the image by analyzing the intensity of the image surface using fractal dimensions (i.e. the fractal textures).

Color however another image feature which is used most frequently for quality assessment in food industries. Color (from L*a*b* color space) features can be extracted using statistical measurements and correlated to the lipid rates for fish sorting [33] and it was reported that the computer vision based image processing technique was better than the traditional method.

(2) Image Acquisition Technique for Internal Quality Assessment

(A) Tomographic Imaging

Tomographic imaging is a noninvasive type of imaging of internal structure of the solid objects. Computer vision technology is generally used for the external imaging while the tomography is a method of producing a 3-D (three-dimensional) image of the internal structures of a solid object by the observation and recording of the differences in the effects on the passage of waves of energy impinging on those structures. The solid object may be the human body, agricultural produce or other biological objects or the earth. A significant development has been noticed in the emerging various types of tomographic imaging systems (such as electromagnetic, magneto-acoustic, electrical impedance tomography, ultrasound tomography, nuclear tomography, etc.). There are basic and clear areas of overlap in the development of these imaging technologies. The development of most important tomographic imaging was started in 1972 when a CT machine was developed by Geoffrey Hounsfield (the Nobel Prize winner) and the terms computer assisted tomography (CAT) and computer tomography were emerged [61].

However, the word tomography is made of two ancient Greek words *tomos* meaning "slice" or "section" and *grapho* meaning "to write" or "to describe". Therefore, the imaging by sections or sectioning through the use of any kind of penetrating wave is known as tomography and the device used in tomography is called a tomograph while the image produced is a tomogram. Considering the importance of CT scan in food quality assessment, the x-ray and computed tomographic imaging and magnetic resonance imaging have been discussed in this section.

(a) X-ray and computed tomography

X-ray is a type of electromagnetic radiation has high penetrating power, travel in straight line, and does not deflect in electric or magnetic field, causes glowing in exposed fluorescent materials and can produce photoelectric emission and ionization of gas. In addition, x-rays behave like particles and as well as waves when interacted with the matter, can blacken a photographic film and thus can be used for the imaging of different parts of biological objects in black and white shade. The wavelength of x-rays is ranged between 0.01 to 10 nm with the corresponding frequencies and energies as 30 to 3×10^4 Petahertz and 120eV to 120KeV, respectively. X-ray imaging, similarly, is a transmission based technique for creating pictures of inner side of the biological material. When x-ray beams from a source are passed through the food materials/or biological products undergoes attenuation. During the attenuation, the intensity of x-ray beam decreases gradually due to absorption and scattering. Thus obtained attenuated x-ray beam, on the opposite side, on a film or ionization chamber have different intensity. This difference in intensity of attenuated x-ray beam is dependent on the absorbing power of inner element. Based on the differences in X-rays absorbing power of different elements, radiography locates the position of elements in a composite material. Consequently, the position where there are elements that strongly absorb the x-ray appear brighter and positions where there are elements that do not absorb the X-rays appear dark on a film placed behind the sample. This emission or absorption spectra depends only on atomic number and not on the physical state of the sample or its chemical composition. X-ray imaging has four basic elements (1) X-ray source (2) X-ray converter (3) Imaging medium and (4) casing for imaging medium.

The quality of x-rays imaging depends on these above stated elements. X-ray source includes radioactive substances and X-ray tubes. X-ray tube is more popular for generating of X-rays with varying intensity for X-ray imaging of agricultural produce. There are two more common types of X-rays converters such as phosphors and semiconductor. However, Semiconductor is used for the converting of X-rays into electric charge directly. Sometimes, semiconductor act both as X-ray converter as well as imaging medium. Various imaging medium such as photographic plate/film, digitizing the radiography film, fast industrial radiography film, etc. are now used for radiography of several agricultural produce. The main function of imaging medium is to capture the images while the function of casing is to protect the imaging medium.

Similarly, computed tomography is used for cross-sectional imaging and the CT word coined in the 1970s. The term computed tomography derives from computed (with computer), tomo (to cut) and graph(y) (pictures). When many X-ray images combine during an X-ray procedure it is known as computed tomography or CT-scan or imaging. The X-rays are coupled with an electronic detector array to record a pattern of densities and to create an image of a slice or cut of tissue. Generally, the X-ray beam rotates around the object within the scanner such that multiple X-ray projections pass through the objects. X-ray computed tomography (CT) is a technique that uses X-ray images to reconstruct the internal microstructure of objects and to monitor the internal quality changes in foods. The CT- scan is, thus, a technique of imaging for creating angiographic images with high resolution [7, 44]. This technique is more popular in medical imaging because of its ability of generation of 2-D and 3-D images from the accumulated data. This phenomenon of CT scan imaging helps in studying of physical and physiological indicators.

(b) Magnetic Resonance Imaging (MRI)

The MRI is also known as the nuclear magnetic resonance (NMR). As the hydrogen nucleus is a single proton, charged positively, spins and generates small magnetic moments (means magnetic fields). These magnetic moments align when placed in a larger magnetic field, as in case of a compass needle in the earth's magnetic field, allowing them to display the phenomenon of nuclear magnetic resonance (NMR). The equipment having a strong magnet, a radio transmitter and receiver is generally used to perform nuclear magnetic resonance (NMR). When NMR is used for imaging it is called magnetic resonance imaging (MRI).

Magnetic field across the sample is intentionally made non-uniform by superimposing additional magnetic field gradients that can be turned on and off rapidly. Activation of these additional magnetic fields results in a net gradient in the strength of the magnetic field across the body, which is necessary for spatial localization and imaging [44, 52]. Following are the essential/or major components of MRI system [64]:

(a) ***A large magnet***: this is the main magnet to generate a strong uniform magnetic field over the desired volume of interest. Strong magnetic field provides better SNR (signal to noise ratio) and better resolution in both frequency and spatial domain [64].

(b) ***Smaller electromagnetic coils***: to generate magnetic field gradients. The gradient coils are used to encode spatially the position of the nuclear spins by varying the value of the local magnetic field. Gradient coils are made up of 3-coils, turned on-and-off rapidly and produce the x, y and z-gradients. The Helmholtz, Maxwell and Golay coils are commonly used, so far, to avoid the bad effect on tissues of biological material in case of gradient coils due to eddy current, higher interaction, etc.

(c) ***RF transmit (radio transmitter) and RF receive (radio receiver) coils***: it is associated with transmitting and receiving antennae or coils. A single coil can be employed for transmitting and receiving of the signals but for better performance and improving the SNR (signal to noise ratio) separate coils are required. The RF coils that excite the magnetization are referred as transmit coils and the coils receive the signals from the relaxing nuclei spins [64].

(d) ***A computer***: to coordinate signal generation, for acquisition and image formation and to display

When a biological object lies in a magnet, it becomes temporarily magnetized due to the alignment of the hydrogen nuclei in the object with the magnetic field. When the biological object magnetized, at a particular frequency (i.e. 'Larmor frequency'), begins to respond to the exposure of radio waves by sending back a radio wave signal known as 'spin echo'. This signal (spin echo) is composed of multiple frequencies, reflecting different positions along the magnetic field gradient. With the help of Fourier Transform technique, the signal is broken into multiple component frequencies. Since the magnitude of the signal at each frequency is proportional to the hydrogen density at that location, an image, thus, can be constructed. In MRI, therefore, it can be concluded that the spatial information is contained in the frequency of the signal [44, 52].

(B) Infrared/Thermographic Imaging

(i) Near Infrared Imaging

Near Infrared (NIR) transmittance and reflectance spectroscopy has been used to evaluate internal qualities of many food and agricultural products.

Near infrared (NIR) is a quick, nondestructive, no contact type, predictive technology which is based on the infrared spectral region. NIR imaging is growing in demand around the globe, typically as a replacement for thermal or far-infrared (FIR) vision for night vision. In recent years NIR spectroscopy and imaging techniques have gained momentum and wide acceptance in many fields such as food safety evaluation (including chemical, microbial and physical hazard detection), new-technology induced food safety concern and food traceability [19]. Thermo graphic photography technique is responsible for generating images with infrared light (600-1000 nm). The thermo-graphic imaging is generally based on the principle that each and every object emits some definite magnitude of radiation. Radiation emitted from the objects is the function of their temperature. Means, if the temperature of the object is higher, it will emit more radiation. Infrared camera similar to ordinary camera for visible light, specially built to detect IR radiation [47]. In addition, NIR camera is able to detect the wavelengths of light directly adjacent to visible light spectrum. On-line/real-time applications of these techniques promise to be a huge growth in the near future.

(ii) Thermal Imaging

Thermal imaging is one of the various emerging nondestructive, noninvasive, noncontact and suitable analytical techniques utilizing in food and agricultural sectors for quality assessment now-a-days. Thermal imaging was introduced initially in the field of military applications. Later on its non-invasive and quick online application was started for quality assessment of food products in food industries and in other sectors. An apparent image of thermal distribution of the surface of an object is formed in this imaging technique. As stated above, the thermal imaging works on the basic principle of that the "all substances emit certain amount of infrared radiation". The wavelength of IR (infrared) radiation ranged between 075 to 100 μm on the electromagnetic spectrum. The wavelength ranged from 0.75 to 2.5 μm is near-infrared, from 1.4 to 3.0 μm is shortwave infrared, from 3.0 to 8.0 μm is mid-wave infrared and the radiation with wavelength more than 8 μm is known as long-wave infrared. In addition to these categories, there is an extreme infrared region also with the wavelength range from 15 to 100 μm. Common thermal imaging systems have centered mainly in two wavelengths ranges i.e. middle wavelength IR (3-5 μm) and long-wavelength (8-14 μm). However, there are several thermal imaging systems have now been developed with capability of detection of infrared radiations from short wave (1.4 to 3.0 μm) to long-wave (>8 μm). Typically, long-wave infrared systems, at room temperature, exhibit optimum sensitivity in contrast to the mid-wave infrared systems that exhibit maximum sensitivity at temperatures much higher than the room temperature. For instance, the preferable temperature in some food industry applications is about 400°C. The measurement of the point temperatures, in a large number, over an area of any targeted surface is called thermography and when the result is processed and a thermal map is made, it is known as thermogram [56].

As the emission of the radiation is the function of temperature and emissivity of the object on which the amount of radiation depends. The emitted infrared energy from any object is measured first then it transformed into electrical signal using infrared detectors of the thermal cameras in this technique. Finally, converted signal is displayed by this technique as a color or monochrome thermal image. In this technique only diffusively reflected radiation rather than the emitted radiation is entertained in spite of the existence of near infrared imaging. The main components of a thermal imaging system comprise with camera setup, optical system arrangement having focusing lens, collimating lenses, etc., detector arrays (e.g. micro-bolometers), signal and image processing, etc. Thermal differential is important and required for active thermography measurements in the thermal imaging system. The thermal imaging system, thus, requires a heating or cooling unit, in place of illumination source as in case of multi and hyperspectral imaging to provide thermal differential. Thermography measurements are also popular with terminologies or names such as pulsed thermography, thermal wave imaging, etc.

Application MRI, X-Ray and CT

Magnetic resonance imaging (MRI), X-rays, near infrared (NIR) spectroscopy, and techniques of thermal imaging are mostly used instead of the ordinary camera for the analysis of the quality of grains. Magnetic resonance imaging is a very important novel non-invasive technology, which analyzes not only the physical state of matter but also provides the information related to spatial and chemical constituent [20]. MRI method in radiology, thus, can be used to obtain images of several internal diseases and their distribution in fruits and vegetables. For instance, watercore (a physiological disorder in apple caused when intercellular spaces are filled with liquid) can be investigated using proton magnetic resonance imaging (^{1}HMRI), and thus measure can be taken to cure this postharvest amelioration in apple save to preserve the normal appearance and taste [11]. MRI can also be used to visualize the loss of moisture during the process of rice seeds drying at various temperatures with a single point mapping imaging technique [24]. This technique can also be used for the study of the corn steeping process and at room temperature, the water mobility and distribution rate can be measured. Similarly, the magnetic resonance images of corn during freezing were captured by the Borompichaichartkul *et al.* [5] characterization of the physical state of water. Where water exists, the area found brighter due to proton mobility. The internal quality assessment of the citrus fruits, kiwifruits, mangoes, strawberry, pear, peach, and pomegranate and vegetables such as potato, tomato, etc. using MRI technique is getting momentum so far.

Further, Singh *et al.* [60] have used NIR spectroscopy for the measuring of viscosity and hardness of the kernels of the grains. Color based classification was performed for the identification of damaged kernel and for the detection of insect and mite infested kernel. In addition to these applications, they inspected the mycotoxins, also using NIR spectra. Similarly, thermal imaging technique can also be applied for the detection of insect infested kernels and for the classification of wheat on the basis of different seed classes.

All these applications are performed following the general process (as discussed above in this chapter) of image acquisition using X-rays (soft), NIR spectroscopy, NMR imaging, thermal imaging, etc.; pre-processing (noise reduction and contrast enhancement); segmentation were performed to identify the grains in images. The features of the images (such as color, texture, shape, histogram, etc.) were measured and finally either statistical classifier or neural networks were used linking the image features and the grains quality parameters. MRI (magnetic resonance imaging) and X-ray microtomography can be used in combination to explore the effect of microstructure of rice kernels during the study of cooking behavior. In this technique, MRI allowed the observation of the ingress of water in the rice kernels during cooking in real-time mode with good spatial resolution in 3D [39]. In another study by Takeuchi *et al.* [65] change of moisture distribution in a rice grain during boiling in real-time was observed by NMR T2 imaging and they found that the rise in moisture profile was unexpected. Similarly, Hwang *et al.* [23] examined water distribution and migration in single kernels during the tempering process in a Bruker 9.4T MRI system and they reported that the MRI technique is useful in investigating the dynamic moisture migration within individual rice kernel.

Conclusion

Non-invasive technologies have more future prospects not only in various medical fields but also in the food sector. Automation of food industries, markets, and farms/fields activities for quality assessment, harvesting of fruits and vegetables, and other crops and commodities would be very important in the coming artificial intelligence era. Image processing technologies based on such modern technologies (computer vision, x-ray, CT-scan, infrared/thermal imaging, MRI, etc.) will provide the basis to the food industries sector for online real-time quality assessment (internal and external) of agricultural produce. Image processing is, thus, a very powerful and analytical technique that is widely applicable for the physiological investigation of biological materials.

References

1. Abdullah, M.Z., Image acquisition system. In: Sun D-W, edtitor. Computer vision technology for food quality evaluation-book, Elsevier Inc., Ch.1, p3-35, 2008.
2. Andukar, A.G., Deshmukh, R., Bansod B., A Review on Imaging Techniques in Food Safety and Its Advancements. Int. J Advanced Research in Electrical, Electronics and Instrumentation Engineering, 5(2):980-983, 2016.
3. **Anonymous** (1996). Image processing and image enhancement. http://www.the-aps.org/ (accessed date 23.2.11)
4. Azman, H., Marzuki, M.M., Hussain, A., Hawari, G.K., Automatic detection of starch gelatinization in enzyme 'dodol' processing using computer vision. Proceedings of the International Conference on Electrical Engineering and Informatics Institut Teknologi Bandung, Indonesia June 17-19, B-33:546-549, 2007.
5. Borompichaichartkul, C., Moran, G., Srzednicki, G., Price, W.S., Nuclear magnetic resonance (NMR) and magnetic resonance imaging (MRI) studies of corn at subzero temperatures. Journal of Food Engineering 69:199–205, 2005.
6. Brosnan, T., Sun, D-W., Improving quality inspection of food products by computer vision--a review. Journal of Food Engineering 61:3-16, 2004.
7. Caldemeyer, K.S., Buckwalter, K.A., The basic principles of computed tomography and magnetic resonance imaging. J American Academy of Dermatology, 41(5):768-771, 1999.
8. Cao, L., Shi, Z.K., Cheng, E.K.W., Fast automatic multilevel thresholding method. Electron. Lett. 38:868–870, 2002.

9. Casasent, D., Talukder, A., Keagy, P., Schatzki, T., Detection and segmentation of items in X-ray imagery. Transactions of the ASAE 44:337–345, 2001.
10. Choudhary, R., Paliwal, J., Jayas, D.S., Classification of cereal grains using wavelet, morphological, colour, and textural features of non-touching kernel images. Biosystems Engineering 99, 330–337, 2008.
11. Clark, C.J., MacFall, J.S., Bieleski, R., Loss of watercore from 'Fuji' apple observed by magnetic resonance imaging. Scientia Horticulturae 73(4):213–227, 1998.
12. Choi, K., Lee, G., Han, Y.J., Bunn, J.M., Tomato maturity evaluation using color image analysis. Transactions of the ASAE 38:171-176, 1995.
13. Demush, R. A Brief History of Computer Vision (and Convolutional Neural Networks). https://hackernoon.com/a-brief-history-of-computer-vision-and-convolutional-neural-networks-8fe8aacc79f3. (Access date 07.03.2021), 2019.
14. Du, C-J., Sun, D-W., Recent developments in the applications of image processing techniques for food quality evaluation. Trends in Food Sci Technol 15:230–249, 2004.
15. Du C-J, Sun D-W (2005) Pizza sauce spread classification using color vision and support vector machines. Journal of Food Engineering, 66 (2), 137–145.
16. Du, C-J., Sun, D-W., Jackman, P., Allen, P., Development of a hybrid image processing algorithm for automatic evaluation of intramuscular fat content in beef m. longissimus dorsi. Meat Science 80:1231–1237, 2008.
17. Du, C-J., Sun, D-W., Learning techniques used in computer vision for food quality evaluation: a review. Journal of Food Engineering 72:39–55, 2006
18. FACE, Indian food processing sector: rends and opportunities, 13th edition. CII - Jubilant Bhartia Food and Agriculture, Centre of Excellence Confederation of Indian Industry report, p1-48, 2019.
19. Fu, X., Ying, Y., Food Safety Evaluation Based on Near-Infrared Spectroscopy and Imaging: A Review. Critical Reviews in Food Science and Nutrition, 1913-1924, 2016.
20. Gambhir, P.N., Choi, Y.J., McCarthy, M.J., Development of rapid and non-invasive nuclear magnetic resonance method for identifying freeze damaged citrus fruits. IFT Annual Meeting, Las Vegas, USA, 2004.
21. Gonzalez, R.C., Woods, R.E., Digital image processing. Introduction, ch1, 2nd edition, Ch. 1. Prentice-Hall, Inc. Upper Saddle River, New Jersey, 2002.
22. Hatcher, D.W., Symons, S.J., Image Analysis of Asian noodle appearance: impact of hexaploid wheat with a red seed coat. Cereal Chem 77(3):388-391, 2000.
23. Hwang, S.S., Cheng, Y-C., Chang, C., Lur, H-S., Lin, Ta-Te, Magnetic resonance imaging and analyses of tempering processes in rice kernels. J Cereal Sci 50(1):36–42,2009
24. Ishidaa, N., Naitoa, S., Kanob, H., Loss of moisture from harvested rice seeds on MRI. Magnetic Resonance Imaging 22:871–875, 2004.

25. Jahns, G., Nielsen, H.M., Paul, W., Measuring image analysis attributes and fuzzy consumer aspects for tomato quality grading. Computer and Electronics in Agriculture 31:17-29, 2001.
26. Jarimopas, B., Jaisin, N., An experimental machine vision system for sorting sweet tamarind. J Food Eng 89:291–297, 2008.
27. Koschan, A., Abidi, M. A., Comparison of Median Filter Techniques For Noise Removal in Color Images. Proc. 7th German Workshop on Color Image Processing, D. Paulus, J. Denzler (Eds.), Erlangen, Germany, Institute of Computer Science, October 2001, 34(15):69-79. 2001.
28. Lana, M.M., Tijskens, L.M.M., van Kooten, O., Effects of storage temperature and fruit ripening on firmness of fresh cut potatoes. Postharvest Biology and Technology 35:87-95, 2005.
29. Leemans, V., Destain, M.F., A real-time grading method of apple based on features extracted from defects. Journal of Food Engineering 61:83–89. 2004.
30. Li, J., Tan, J., Martz, F.A., Predicting beef tenderness from image texture features. In: 1997 ASAE Annual International Meeting Technical Papers, Paper No. 973124, ASAE, 2950 Niles Road, St. Joseph, MI 49085-9659, USA, 1997.
31. Li, J., Tan, J., Shatadal, P., Classification of tough and tender beef by image texture analysis. Meat Science 57:341–346, 2001.
32. Lin, L.A.C., Sanches, J., Dal, F.M.I., Image processing techniques for lemons and tomatoes classification. Bragantia Campinas 67:785-789, 2008.
33. Louka, N., Juhel, F., Fazillieau, V., Loonis, A., novel colorymetry analysis used to compare different fish drying processes. Food Control 15:327-334, 2004.
34. Magnan, P., Detection of visible photons in CCD and CMOS: A comparative view. Nuclear Instruments and Methods in Physics Research A, 504: 199–212, 2003.
35. Majumdar, S., Jayas, D.S., Classification of cereal grains using machine vision: II- Color models. ASAE 43(6):1677-1680, 2000.
36. Majumdar, S., Jayas, D.S., Classification of cereal grains using machine vision: III. Texture models. Transactions of the ASAE 43:1681–1687, 2000b.
37. Manickavasagana, A., Sathya, G., Jayasa, D.S., White NDG Wheat class identification using monochrome images. Journal of Cereal Science 47: 518–527, 2008.
38. Martin, D., A practical guide to machine vision lighting, Midwest Sales and Support Manager, Advanced Illumination. http://www.graftek.com/pdf/MachineVision, 2007.
39. Mohoric, A., Vergeldt, F., Gerkema, E., van Dalen, G., van den Doel, L.R., van Vliet, L.J., Van, As, H., van Duynhoven, J., The effect of rice kernel microstructure on cooking behaviour: A combined μ-CT and MRI study. Food Chemistry 115:1491–1499, 2009.
40. Nair, M., Jayas, D.S., Bulley, N.R., Dockage identification in wheat using machine vision. ASAE Annual International Meeting Technical Papers,

Paper No. 973043, ASAE, 2950 Niles Road, St. Joseph, Michigan 49085–9659, USA, 1997.

41. Ng, H.F., Wilcke, W.F., Morey, R.V., Lang, J.P., Machine vision evaluation of corn kernel mechanical and mould damage. In: 1997 ASAE Annual International Meeting Technical Papers, Paper No. 973047, ASAE, 2950 Niles Road, St. Joseph, Michigan 49085–9659, USA, 1997.
42. Ni, H., Gunasekaran, S.A., computer vision method for determining the length of cheese shreds. Artif Intell Rev 12(1–3):27–37, 1998.
43. Otsu, N.A., threshold selection method from gray-level histogram. IEEE Transactions on Systems Man and Cybernetics 9:62–66, 1979.
44. Patel, K.K., Kar, A., Khan, M.A. Nondestructive Food Quality Evaluation Techniques: Principle and Potential, Agricultural Engineering Today, Vol. 36(4):29-34, 2012b.
45. Patel, K.K., Kar, A., Khan, M.A., Monochrome computer vision for detecting common external defects of mango. J Food Sci Technol. https://doi.org/10.1007/s13197-020-04939-9, 2020a.
46. Patel, K.K., Kar, A., Khan, M.A., NIR computer vision for rapid measurements of some physical properties of mangoes. Journal of AgriSearch 7(4):211-219, 2020c.
47. Patel, K.K., Kar, A., Jha, S.N., Khan, M.A., Machine vision system: a tool for on-line sorting and grading of agricultural foods and produces. J Food Sci Technol 49(2):123–141, 2012a.
48. Patel, K.K., Kar, A., Khan, M.A., Common External Defect Detection of Mangoes using Color Computer Vision. Journal of The Institution of Engineers (India): Series A. DOI 10.1007/s40030-019-00396-6, 2019b.
49. Patel, K.K., Kar, A., Khan, M.A., Rapid Assessment of Some Physical Parameters of Mangoes Using Monochrome Computer Vision. Agric Res. https://doi.org/10.1007/s40003-020-00517-9, 2021b.
50. Patel, K.K., Kar, A., Khan, M.A., Development and an Application of Computer Vision System for Nondestructive Physical Characterization of Mangoes. Agric Res 9:109-124, 2020b.
51. Patel, K.K., Khan, M.A., Kar, A., Potential of reflected UV imaging technique for detection of defects on the surface area of mango. J Food Science Technology, 56(3):1295–1301, 2019a.
52. Patel, K.K., Khan, M.A., Kar, A., Recent developments in applications of MRI techniques for foods and agricultural produce – an overview". J Food Science Technology, 52(1): 1-26, 2015.
53. Patel, K.K., Khan, M.A., Kumar, Y. and Yadav, A.K., Novel techniques in post-harvest management of mango-An overview. South Asian Journal of Food Technology and Environment, 5(2):821-835, 2019c.
54. Pourreza, A., Pourreza, H., Abbaspour-Fard, M-H., Sadrnia, H., Identification of nine Iranian wheat seed varieties by textural analysis with image processing. Computer and Electronics in Agriculture 83:102-108, 2012.

55. Quevedo, R., Carlos, L.G., Aguilera, J.M., Cadoche, L., Description of food surfaces and micro-structural changes using fractal image texture analysis. Journal of Food Engineering 53:361–371, 2002.
56. Rahkonen, J. Jokela, H., Infrared radiometry for measuring plant leaf temperature during thermal weed control treatment. Biosystems Engineering, 86(3):257–266, 2003.
57. Sahoo, P., Wilkins, C., Yeager, J., Threshold selection using Renyi's entropy. Pattern Recognition, 30(1):71-84, 1997.
58. Sanches, J., Durigan, J.F., Durigan, M.F.B., Application of mechanical damage in avocados and its effects on fruit quality. Braz Assoc Agric Eng 28(1):164–175, 2008.
59. Shiranita, K., Miyajima, T., Takiyama, R., Determination of meat quality by texture analysis. Pattern Recognition Letters 19:1319–1324, 1998.
60. Singh, C.B., Paliwal, J., Jayas D.S., White NDG Near-infrared spectroscopy application in grain industry. CSBE Paper No. 06-189, Canadian society for Bioengineers, Winnipeg, MB., 2006
61. Soleimani, M., Bayford, R.H., New and emerging tomographic imaging techniques in medical and industrial applications. Phil. Trans. R. Soc. A, 367: 3017–3019, 2009.
62. Steenhoek, L., Precetti, C., Vision sizing of seed corn, In: 2000 ASAE Annual International Meeting, Paper No. 003095 ASAE, 2950 Niles Road, St. Joseph, MI 49085–9659, USA, 2000.
63. Sun, D.W., Inspecting pizza topping percentage and distribution by a computer vision method. J Food Eng 44:245–249, 2000.
64. Sun, D-W., Computer vision technology for food quality evaluation. Food Science and Technology, International Series, Ist edn., Elsevier Inc, 2008.
65. Takeuchi, S., Fukuoka, M., Gomi, Y., Maeda, M., Watanabe, H., An Application of magnetic resonance imaging to the real time measurement of the change of moisture profile in a rice grain during boiling. Journal of Food Engineering, 33(1–2):181–192, 1997.
66. Vincent, L., Soille, P., Watershed in digital space: An efficient algorithm based on immersion simulations. IEEE Transactions on Pattern Analysis and Machine Intelligence 13(6):583-593, 1991.
67. Ying, Y., Jing, H., Tao, Y., Zhang, N., Detecting stem and shape of pears using Fourier transformation and an artificial neural network. Transactions of the ASAE 46(1):157–162, 2003.
68. Whittaker, A.D., Park, B., Thane, B.R., Miller, R.K., Savell, J.W., Principles of ultrasound and measurement of intramuscular fat. J Animal Science 70:942–952, 1992.
69. Youssef, M.M., Abou-Garibia, H.A., El-Haddad, N., Applications of digital image analysis to food quality analysis assessment: an overview. Alex Food Science and Technology 2(1):41-50, 2005.

70. Yud, R.C., Kuanglin, C., Moon, K.S., Machine vision technology for agricultural applications. Computer and Electronics in Agriculture, 28(1):173–191, 2002.
71. Zheng, C., Sun, D-W., Computer vision for quality control. Ch.2 in optical monitoring of fresh and processed agricultural crops edited by Zude M. CRC Press, p127, 2009.
72. Zheng, C., Sun, D-W., Zheng, L., Recent development of image texture for evaluation of food qualities-a review. Trends in Food Science and Technology 17:113-128, 2006a.
73. Zheng, C., Sun, D.W., Zheng L. Recent developments and applications of image features for food quality evaluation and inspection-a review. Trends in Food Science and Technology 17:642-655, 2006b.
74. Zheng, C., Sun, D-W., Image segmentation techniques. In: Sun D-W, editor. Computer vision technology for food quality evaluation, 1st edn. Elsevier, Inc, pp 37-56, 2008.
75. Zuech, N., Understanding and Applying Machine Vision, 2nd eds, Vision Systems International Yardley, Pennsylvania, Marcel Dekker, Inc., p1–336, 2000.

13

High Pressure Processing: An Overview

Yashwant Kumar Patel[1*] **and Krishna Kumar Patel**[2]

[1]*Department of Food Processing and Technology, UTD, Atal Bihari Vajpayee Vishwavidyalaya, Bilaspur, Chhattisgarh, India*
[2]*Department of Agricultural Engineering, PGC, Ghazipur, VBSPU, Jaunpur, Uttar Pradesh, India*

Abstract

The target of processing of foods is not only the transformation of original physical properties of raw products into such products, which has more commercial value and more suitable to consumption but also it is the process of value addition. The valued through preservation, drying, addition of various food additives involve capital cost/or machines, power, employees, etc. During these operations, there are several undesirable changes also take place. The food processing method which results no or minimum effect on smell, color and flavor, texture, and nutritional quality of the end products at economical rates in little process time treated as good food processing method. The HPP technology is such a novel technique that can be applied as the best alternative to heat treatments. Several studies have proven apparent advantages of HPP in different areas of food sector. As this process required very high pressure (vessel pressure ranges from 300 to 800 MPa) for little time to kill the microorganisms in foods, this technology is best suited to the liquid foods and highly perishable food products. This hydrostatic pressure processing is, therefore, very much applicable modern preservation technique to the food processing industries where the dependency of high temperature processing is not required for preservation of foods.

***Keywords*:** HPP, HHP, pascalization, food preservation and bridgmanization, etc.

Corresponding author: profykpatel@gmail.com

Navnidhi Chhikara, Anil Panghal and Gaurav Chaudhary (eds.) Novel Technologies in Food Science, (479–510) © 2023 Scrivener Publishing LLC

13.1 Introduction

Food processing is one of the prime importance to the human being for the processing as well as preservation of agricultural produce, dairy products, meat and meat products, poultry, fish, etc. There is a close relationship between the development of any country and its advancement in processing technologies. After China, India is the leading in the production of fruits and vegetables in the world. India is second largest producer of fruits and vegetables after china. In spite of second largest producer, there are only 2% fruits and vegetables, 21% meat, 23% marine, 35% Milk and 6% poultry are processed in India. Less than 10% of the total food outputs are processed in India in spite of a large production base. The value of India's food industry is more than 258 billion in US$ dollars, which was expected to grow to 482 billion US$ dollars by 2020. According to Dublin (business wire) reports, the global market of food processing by 2024 is expected to touch the 4.1 Trillion US$ (which is an estimate value) with a CAGR of 4.3% from 2019 to 2024).

The transformation of raw plant or animal materials into usable products for consumer considering the stability of food products by reducing/ or preventing the any type of degradation of food quality. Food processing provide the opportunity to the food products store from time of plenty to time of need for transportation of food materials over long distances [1]. Sensory qualities (internal, external) of food materials such as aroma, colour, flavor, smell, size, shape, texture, etc. are very importance for the consumer's acceptability. Thus, to sustain these properties in food products during processing, there is a continuous effort of the scientists, researchers, manufactures/or developers, etc. how to reduce the unwanted changes in foods during processing.

Considering the above goal, the application of traditional technologies/ methods for heating, freezing, drying/dehydration, packaging, pasteurization, sterilization, etc. of foods, are now replaced by some alternative, novel, more efficient and effective tools and technologies. The measure objectives of the developments of these novel methods were not only to reduce the microbial growth, preserve the valuable sensory qualities but also to prevent the undesirable chemical changes and to provide the safe, fresher tasting and nutritive foods without using thermal treatments or any type of chemical preservatives. High pressure processing is one among such technology, which is now getting more popularity in food industries for the non-thermal processing of food product especially liquid foods and highly perishable foods. Application of HPP technology inactivates the microbes, denatures the proteins and finally extends the shelf life of foods. In this processing method, the extension in shelf life occurs without any undesirable effect on flavor or nutrients. Also, there is no impact on sensory, nutritional, or functional properties of foods.

It is unique emerging technology to preserve the foods. Volumetric pressure facilitates the destruction of microorganisms in various foods. The Volumetric Pressure food process can be operated in batches as well as continuous mode. The equipment's for continuous manner required high initial cost but at the same time its processing capacity is more as compare the batch type. Hence, it is best suited for industrial scale. While in case of batch processing, less initial investment is required for small-scale production that is why it is limited to laboratory/experiments purpose.

The main objective of this chapter is, thus, to provide the knowledge regarding basic fundamentals, application, and limitation and future prospects of HPP (high/hydrostatic pressure processing).

13.2 What is HPP?

Hydrostatic/High Pressure Processing (HPP) technique is the application of high pressure from 100 – 800 MPa for the assessment and safety of food products. In this process, foods are held for specific amount of time for inactivation of the microorganisms before releasing of the achieved pressure (i.e. target pressure). Although, the HPP techniques are applied at refrigerated condition/temperatures based on the pressure treatment alone but heat may also be used. However, there is some increment in the temperature generated naturally during the pressure processing due to adiabatic heating. As discussed above, the High Pressure Processing or HPP is also known as the high hydrostatic (HH) pressure (P) or ultra-HHP. Further, the high pressure processing is also known as pascalization, which is a non-heat treatment above 100 MPa. This pressure, however, ranged between 100 MPa-700 MPa in food industries.

13.3 Historical Background

In 17th century, a French named Blaise Pascal find out the effect of pressure on the contained fluids. The processing or pascalization under high pressure was named after his (scientist's) name. The pressure treatment effects on microbes were experimented in the 1884 (early). In 1895, Royer gave the concept of HPP (HHP) and applied for the processing of food to kill

Table 13.1 A brief history of HPP development.

Period		Description of study
1894-1914	:	• Studied on effect of hydrostatic pressures (>700 MPa) during the deactivation of microbes found in milk, fruits and vegetables. • By Professor Burt Heit at the WVAES (West Virginia Agricultural Experiment Station). • Pressure treated fruits like acid products could be preserved because of inactivation of vegetative microorganism • But low acid products like milk and vegetables spoiled because spores remained viable, they would not be inactivated.
1982-1988	:	• By Prof. Farkas, Hoover and Knorr at the Univ. of Delaware. • Main objectives of research were to inactivate broad-spectrum of food spoilage and pathogenic microorganism using cold isostatic pressure (50000 psi) donated by Autoclave Engineers. • Study on preservation of food by high pressure was adopted by US and Japan.
1988-1993	:	• Modified isostatic presses. • First 60000 psi pressure treated product was marketed in Japan. • A research consortium was set up by Japanese food industries for commercialization of high pressure preserved foods such as jams, salad dressings and yogurt.
1993-1995	:	• Study on effect of pressure on microbes in food was adopted by Prof. Hoover at Univ. of Delaware with Prof. Farkas at Oregon State University. • Four food products such as (1) spaghetti with meat sauce, (2) Spanish rice, (3) yogurt with peaches and (4) a fruit mix were preserved successfully. • Samples, microbiologically stable for up to 120 days at room temperature and quality of products reported best when they were refrigerated.

(Continued)

Table 13.1 A brief history of HPP development. (*Continued*)

Period		Description of study
1995-1912	:	• US army involved with Oregon State University (OSU) in the research related to the development of pressure preserved foods. • Raspberry flavored yogurt drink and lemon pudding were preserved at Avomex in 1997. • Avomex, Inc. developed avocado puree (guacamole) using high pressure treatment and extended shelf life for food service use. • Use of tilting pressure vessel in high pressure system was started for accelerating the loading and unloading, and for high cycle rate per hour. • Rapid commercialization of HPP products has been started since 2000.

the bacteria. Again, various studied were carried out on HPP from the end of 1980s and reported that this at a food place, HPP consists of an isostatic pressure in a high-pressure vessel. A researcher named as Bert Hite in Agricultural Research Station in Morganstown (West Virginia, USA) explored the effects of HHP on the processing of food in 1899 [2]. A high pressure processing unit was designed and constructed by him for the pasteurization of milk and some other products (i.e. meat and fruits and vegetables). However, commercial application of HPP came in the 1990 in Tokyo, Japan for the processing of fruit juices, fruit topping, jellies, jams and tenderized meat [2]. Later this technology was diversified for the products such as ham of ox, fishes, meats, sake, etc. Table 13.1 represents the summary of developments history of HPP.

13.4 Principle of High Pressure Processing

When the processing of food products, or inactivation of deleterious pathogens and harmful (spoilage) microbes in their vegetative spore, are done with the help of elevated pressure either in the presence or absence of temperature, the thermal impact on the foods is considerably reduced in this technique. The organoleptic properties in such pressure treated materials are noticed better than the thermal treated food materials. The HPP treatment is, therefore, regarded as the technology where no or minimum heat is required during processing of foods. Means, HPP is a non-thermal technology. The basic principle of HPP comprises of two principles first is Le-Chatelier's principle and second is Isostatic rule. According to this principle "when a system at equilibrium is distributed, the system responds in such a way that tends to minimize the disturbance. In other words "whenever a pressure is applied to a system in equilibrium, bio-molecules obey the Le-Chatelier principle and the system will react so as to counteract the applied pressure. The reactions that result in reduced volume will be promoted under high pressure, and such reactions may result in inactivation of microorganism". In addition to these, any phenomena (such as phase transition, changes in the configuration of molecules/molecular configuration, chemical reaction, etc.) that get together with the volume reduction will be enhanced by pressure [3, 4]. In high pressure processing, food processing equipment was used for denaturation of proteins as well as reduction of microbial counts in very high pressure. Pressure affects hydrophobic bonds and electrostatic interactions that's lead to destruction of microbes and all.

Similarly, the second principle Isostatic Rule also state that when a sample is under pressure, pressure is transmitted through out the sample uniformly and instantaneously. The sample either may be in direct contact with pressure medium or may be in the hermetically sealed in a flexible package that transmits pressure. As the transmission of pressure throughout the sample takes place in a uniform and quasi-instantaneous manner, there is no effect of sample size on the necessary high pressure processing

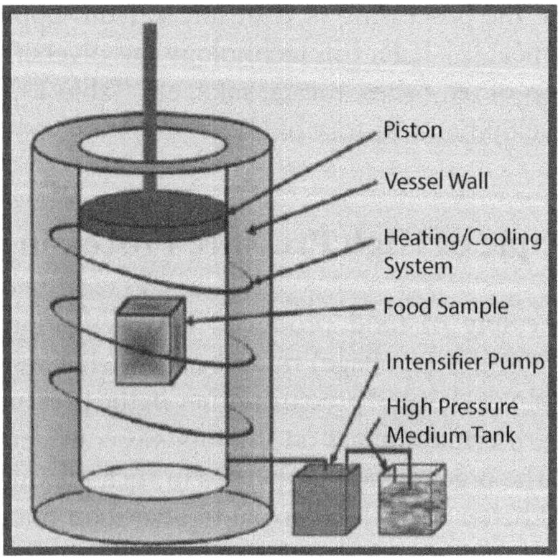

Figure 13.1 Diagram showing main components of HPP [5].

time. In addition, the uniform and simultaneous application of pressure in all direction is the main reason of not crushing of food products during the pressure treatments and product remained uncrushed in contrast to thermal methods. Means, in contrast to thermal processing of food materials where temperature increases gradually, the required time of high pressure processing, is independent of sample size. Figure 13.1 presents a high pressure processing system comprises with piston, a vessel, a pressure-transmitting fluids, a material handling, and supporting units like heating or cooling components, etc. as major components.

First a pressure transmitting fluid is filled in the vessel and the vessel is closed and loaded after removing the air. An automatic de-aeration valve system with a low pressure fast fill-and drain pump is used for the removing of the air from the vessel. The generation of hydrostatic pressure is then take placed either by compression (direct/indirect) or by heating pressure medium [6]. Phase transition along with chemical changes phenomenon leads by a decrease in volume and favored by pressure vice-versa, according to the above stated principles. There are various changes in biological systems, for extant, proteins denaturation, gelation, changes in reactions (hydrophobic), phase changes in the lipids (and therefore in all cell membrane), and increment in the ionization of dissociable molecules due to electrostriction, etc., brought about by the HPP on the reduction of the volume [5, 7]. There are three main parameters *viz.* exposure time (t), temperature (T) and pressure (P) by which high pressure processing technology is compared with thermal preservation [4]. On the other hand, there are only two parameters, temperature (T) and exposure time (t), which are considerable in the thermal based preservation technology. The variability in the designing of three parameter based process, such as HPP (High Pressure Processing), is more than the processes which are based on the two parameters such as thermal food preservation technologies. In addition, a wide range of variability in the enhancement of the shelf-life of foods offered by the high pressure processing (HPP) technology. On the basis of pressure application, exposure time (t) and treatment temperature (T), the HP (high pressure) processing operation can be divided into two groups:

(a) *HP Pasteurization*: To inactivate the vegetative pathogens, foods are treated at pressure varies from 300 to 600 MPa for 1-5 minutes of exposure time and at 5-25 °C initial temperature (of the product).
(b) *HP Sterilization (High T and P):* In this process spores (bacterial) are inactivated using high pressure (400-800 MPa;

sometimes it may be 1000 MPa) when initial temperature of products, process temperature and holding time is 70-90 °C, 110-120 °C and 1 to 10 minutes, respectively.

It can be operated in batches as well as continuous mode. The equipment's for continuous manner required high initial cost but at the same time its processing capacity is more as compare the batch type. Hence it is best suited for industrial scale. While in case of batch processing, less initial investment is required for small scale production that is why it is limited to laboratory/experiments purpose.

13.5 Classification of High Pressure Processing Equipment

High pressure processing system equipments can be classified into following groups (Figure 13.2):

13.5.1 Pressure Application Based HPP Equipments

(i) *Piston direct pressure type*: In this technique piston compresses pressure medium directly in pressure vessel and produced more than 500 MPa

(ii) *External pressurization type:* When pressure medium is pumped in the high pressure vessel externally through the high pressure pump, it is termed as external pressurization.

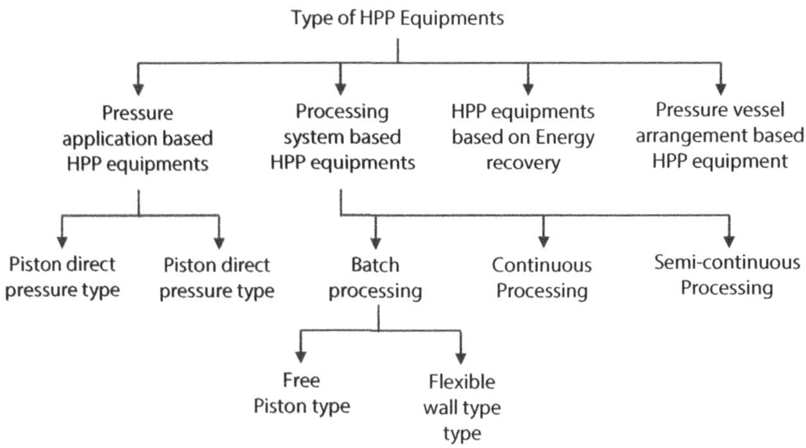

Figure 13.2 Type of HPP equipments.

This technique is used to the system with the pressure of 500 MPa or less

13.5.2 Processing System Based HPP Equipments

High pressure processing is divided into three-groups (i) batch processing, (ii) continuous processing and (iii) semi-continuous processing systems. All these systems are discussed below in brief.

(i) Batch processing

As discussed above the packaged food products are first filled in the pressure vessel and the pressure vessel then closed and sealed carefully. To remove air from the pressure vessel water is pumped and pressurized the water after closing the relief valve till the required pressure is not achieved. Pressure relief valve is opened after the completion of process and pressure medium (water) returned to the atmospheric pressure. Food packets are unloaded and can be used for marketing, so far. Batch type high pressure processing can be divided into two types (a) free piston type and (b) flexible wall type. A brief discussion of both systems is given below:

(a) Free piston type

There is a free piston in pressure vessel, which divides the vessel into two parts. One part is filled with liquid food and other with pressure medium. When pressure is applied on the pressure medium pushes the free piston and transfers the pressure to the liquid food. As there is no pressure difference between two parts, abrasion of packing can be reduced.

(b) Flexible wall type

In this type, the pressure vessel is divided into two parts by a flexible wall such as (1) inner part for liquid food and (2) outer part for pressure medium. High-pressure pump is used to pressurize the pressure medium which pushes the flexible wall and consequently the liquid food.

(ii) Continuous processing

Multiple vessels (each operated automatically) are processed in parallel and obtain continuous treated food in this system. A pressure vessel of 25-littre capacity, with a floating piston and a valve block are generally included in every isolator.

(iii) Semi continuous processing

Unpacked liquid foods in two or three pressure vessel are treated in the semi-continuous systems. By a free-floating system, each vessel is divided into 2-chambers. First chamber contains liquid food and second chamber contains pressure medium. When pressure applied on first chamber resulted compression of liquid food in second chamber. After achieving the holding time, the discharge valve of the product is opened and the contents of food are expelled out in a sterile holding tank. Now, liquid foods can be filled aseptically into pre-sterilized containers and is, thus, ready for shipment [8].

13.5.3 HPP Based on Energy Recovery System

In this HPP system, energy is saved by reusing of the pressure medium, obtained from the pressure vessel in the decompression stage, for pressurizing of other pressure vessel. This system reduced the costs of the equipment and processing time about half of the ordinary HPP system. Energy recovery system comprises with the production equipment, which involve the two 130-liter pressure vessel under 400 MPa pressure, and one set of intensifiers.

13.5.4 HPP System Based on Vessel Arrangement

HPP treatment is given to the pressure vessel operated in different arrangement such as horizontal, vertical or tilting mode in this system. Pressure vessel should be made of stainless steel to avoid any type of corrosion. Simultaneously, pressure vessel can be built with 2-or-more concentric cylinders.

13.6 Effects of HPP on Food Derivatives

13.6.1 Effect of HPP on Color, Texture and Sensory Attributes

Colour, a characteristic of light, is no doubt the most important attribute for acceptance of the food products to the consumer. As color is perceptual phenomenon, it is dependent on the observer and color observation condition and it is measurable in terms of intensity and wavelength. Light of different wavelengths has different color. The wavelength not absorbed but reflected by or transmitted through an object are visible to observers. Thus, the color of foods becomes visible only when light from a luminous object or source illuminates or reflects from the surface [9]. Although, the energy distribution along the visible spectrum which produces a particular sensation in the human eye is normally recognized as color, it depends on the exposure of the pressure and time and on the denaturation of the protein to some extent during the high pressure processing. The pressure induced color modifications, thus, can take place during the HPP treatments and these modifications are dependent on the treatment conditions (pressure, time and temperature). For instance, the color of raw red meat is changed into cooked appearance due to the treatment of high pressure and thus it cannot be sold as raw. Although, some serious color issues are also reported after the HPP treatments of cured or white (or fish) meat, color of meat packed under vacuum with an oxygen scavenger is protected, partly [10]. The fish with red and white muscle when pasteurized with the high pressure, muscle tissue become opaque and shows cooked characteristics rather than a fresh product. This modification in colour may not be required by the consumer. But if the treatment pressure increases, the red and white muscle tissues become lighter and redness decreases in red muscle fish. However, yellow color is not affected in spite of treatments of high pressures [11]. High pressure processing can protect not only the color but other sensory characteristics such as texture, taste, smell, etc.

The textural properties are also considerable during processing of food materials. As there is no sear force generation in HP processing of foods with high moisture contents and thus the structure remains unaffected. However, the gas displacement and liquid infiltration during the HPP treatment of products containing-gas may change the color and texture. Texture information of food products, thus, provides good quality control, sensory evaluation, product development and process equipment's design and evaluation. For instance, the HHP treatment of food products without air-voids/pockets results in minimal or no permanent change in textural properties [12]. However, HPP technique has large potential to modify the texture of food products and as per requirement; this technique can stimulate the changes needed in the texture and structure of new products. For instance, cheese (mozzarella) when treated with high pressure developed required functional qualities on melting [13]. When milk is treated with high pressure at (300-400 MPa), milk yields high cheese about 7% due to high yield of wet curd and reducing the protein loss in whey and loss of whey volume. Similarly, milk when treated at 400 600 MPa matured quickly and developed strong flavor [14]. The hardening of tissue of Alaska Pollock surimi is noticed optimum at 203 MPa pressure for 10 minutes. Tissue becomes softer just after reaching the optimum point (upper limit) which could be an issue for the fish with soft muscle tissues [11].

13.6.2 Effect on Fat

Chemical composition and target pressure are mainly responsible for increasing of the some temperature naturally during HPP treatment of food products. About 3°C temperature of water increases with respect to per 100 MPa increment in the pressure. If there is more compressible ingredient (such as fat) in food, the increment in temperature in can be significantly higher [15]. Time required for crystallization of milk fat is reduced on high pressure treatment because of shifting of transition temperature of milk fat to high value at pressure. At 100 MPa, the crystallization temperature and melting temperature of milk fat is increased by 16.3°C and 15.5°C, respectively. In addition, cream treated with high pressure has higher fat content than the untreated cream. It has also been reported that the ageing time of HPP treated ice cream. Mean diameter of the milk fat globules remain undestroyed even at high pressure up to 800 MPa when treated by HPP.

Further, high pressure treatment speeded up the oxidation rates of lipids in muscle tissue and thus the level peroxide value of oils pressure treated and refrigerated cod muscle is higher in comparison to non-pressurized-refrigerated cod muscle.

13.6.3 Effect on Carbohydrates, Proteins and Molecular Weight of Molecules

The increment in pressure rather than in the temperature can gelatinize the carbohydrates and also can denaturize the proteins during the HPP treatment without increasing the temperature. Figure 13.3 shows the

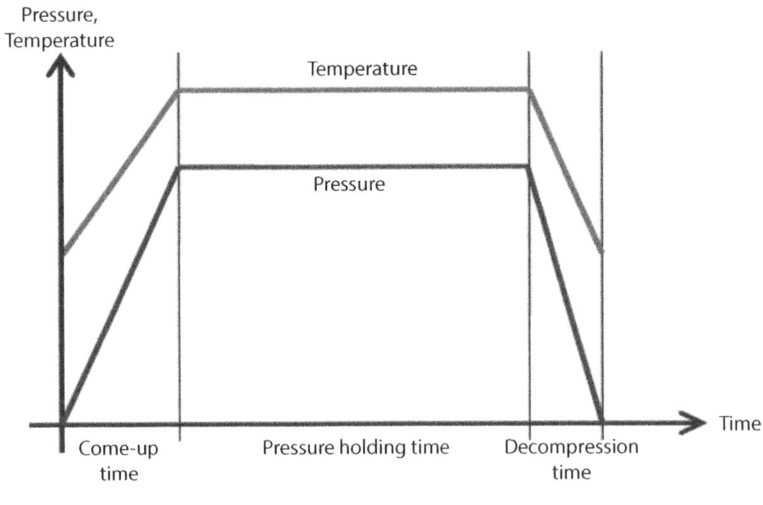

Figure 13.3 Temperature and pressure plot during high pressure processing of foods.

relationship between the pressure and temperature [15]. Treatment of milk by HPP impacts on various milk components especially proteins. However, milk primary proteins which stabilized by covalent bonds are remained almost unchanged by HPP. But, whey proteins denaturize as the pressure increases due to decreasing of the volume of nitrogenous compounds in non-casein protein in milk. Sarcoplasmic proteins, which are fish protein, can be pressurized over 300 MPa for the coagulation of the proteins. The coagulated proteins can then be incorporated into surimi and related products.

The molecules with higher molecular weight are more susceptible, while in case of molecules with lower molecular weight will affect less such as pigments, vitamins and flavoring compounds during HPP treatment. Consequently, the food quality and nutritional value is preserved in HPP treated food products [15].

13.6.4 Effect of HPP on Other Bio-Active Molecules

Vitamins, amino acid, simple sugars and flavour compounds are some other bioactive molecules important for any finished products. These bioactive molecules remain unaffected even after the HPP treatment. For instance, when milk is treated with HPP at 400 MPa, there is no loss in vitamin B complex (especially B_1 and B_6). However the effect on bioactive molecules CLA, sugars and lipids during the HPP is not observed.

There is no temperature induced chemical reaction responsible for flavor changes during the HPP treatment of egg, as in case of the thermal processing. The taste of HPP treated egg is, thus, reported to be closer to the raw egg. This creates more possibilities for the developments of new products.

13.7 Effect on Microorganisms during HPP

The effect of high pressure processing on vegetative microorganism including yeast and lactic acid bacteria (spoilage organisms) and Escherichia coli, Salmonella, and Listeria monocytogenes (Pathogenic microorganisms) is more pronounced and significant and can result in cell death/ or in activated within few minutes if sufficient amount of pressure (about 300 - 400 MPa) is applied. Several researchers have reported that these organisms in different foods can be killed by the HPP and Significant increase in shelf life of foods commodities, thus, can be extended. In spite of above, many studies have also reported that there are some spore-farming organisms which are highly resistant to HPP. For instant, the organism Clostridium botulinum, can germinate, grow and produce a highly potent paralytic neurological toxin in low acid foods [15]. A combined application of heat and pressure of some other antibacterial intervention is required to kill their spore form in the food products [15]. Inactivation of endospores require combination of pressure (>1000 MPa) and temperatures (> 80°C).

Similarly, yeasts and molds ascospores may also be inactivated at high pressures. Food composition and type of microorganisms are the two factors influenced the effectiveness of HPP technique [16]. Means, food nature is very essential parameters to know before processing because it directly affects the microbial population during HPP treatments. Gram negative bacteria, yeasts/molds shows more sensitivity in HPP as compare to gram positive bacteria. The microorganism groups of heat resistant are also pressure resistant; the exponential phase cells are more pressure sensitive than stationery phase cells [17]. Viruses have a wide range of sensitivity towards the pressure application and a careful evaluation is, therefore, needed HPP treatment applied on a specific food. This is the reason why regulatory agencies have specified parameters of critical processing during the high pressure treatments of food [4].

13.7.1 Critical Processing Parameters of HPP

Micro-organisms resistant to the high pressure, as discussed above, depend on various factors such as microbial species, strain, stage of microbial growth, food composition, etc. Therefore, the optimization of high pressure treatment should be optimized to assure the food safety. Considering the factors than can affect the response of microbes including pathogens HPP can be optimized. Some of the critical processing parameters required for optimization/or effective treatment of HPP are listed below:

- Target pressure
- Time interval for achieving the target pressure
- Holding time at target pressure
- Total time of decompression
- Initial temperature of foods
- Water activity (a_w) of the products
- pH of foods
- Initial temperature of pressurized fluid

The above stated specified 'critical processing parameters' of high pressure processing are adopted by the regulatory agencies for the effective treatment of food products to ensure the food safety. Brief descriptions of some these critical processing parameters are given below.

13.7.1.1 Pressure and Time

The pressure required for the treatment of products held in the pressure vessel is defined as the process pressure. In addition, the minimum pressure required by the high pressure technique for the inactivation of microorganisms is generally termed as critical pressure below which no microbial inactivation takes place regardless of process time. In this process, there is a target pressure for each specific product and pressure is increased at certain rate until the target pressure reached which held for specific amount of the time then pressure is released. Time interval for achieving the target pressure and time required for releasing of the pressure are the important processing parameters. The time needed to bring back the food products from the process pressure to atmospheric pressure (0.1 MPa) is termed as decompression time. The depressurization time for many HPP equipments is in the order of seconds. Similarly, the time interval between end of compression and beginning of decompression is the time required by the food to be held at the process pressure and termed as pressure holding time. During this time interval pressure remains constant. Pressure holding time of 3-10 min is often used for economically viable commercial process. If the time for achieving required pressure is long, then this long come-up time will add appreciably to the total process time and it will affect the product flow capacity.

In addition, microbial inactivation kinetics will also be affected. Similarly, the high pressure is not in favor of its commercial application due to high cost of mechanical equipment and increased metal fatigue [18]. Strain gauges and mechanical Bourdon tube type gauges can be used for the measurement of the process temperature. Strain gauges work on the principle of change in resistance under pressure strain (compression-decompression) and displacement transducers [19].

13.7.1.2 Temperature

Different types of microorganism (a heterogeneous group of organisms) are capable to grow in the foods at different temperatures such as below freezing temperatures to boiling temperatures (above 100°C). Extreme psycophiles are able to grow well below freezing temperatures while extreme thermophiles are capable to grow above 100°C. Although, the particular temperature range is best suitable for growth of particular species of microorganisms, but the growth of these species is restricted to those temperature at which cell membranes and cellular enzymes can function [3].

High pressure processing techniques are generally performed at ambient temperature. In general, inactivation rate of microorganisms increased as the pressure, time, or temperature increased during the high pressure processing. The temperature achieved by the food products after reaching the target pressure is known as process temperature. The process temperature, however, depends on the pressure transmitting fluids and their respective heat of compression values, initial temperature of foods and pressure vessel. To avoid the decrement in the process temperature because of loss of heat to the surroundings, thermal lagging of good quality is required [19]. Since, just after the pressurization, mostly the temperature of foods increases slightly due to molecular rearrangement in response to pressure. Hence, for isothermal conditions, there should not be any heat loss during pressure holding time.

The inactivation rate at high pressure (345 MPa) and high temperature (50°C) in combination followed first-order kinetics while at a lower pressure and temperature combination it showed a late tailing effect [20]. The rate of inactivation of pathogens in food products and microbial spoilage increases above the 45°C temperature. The selection of processing temperature, however, depends on the pressure-transmitting media [3]. Several researches have shown that the sensitivity of bacterial cells to high pressure is least at 20 to 25°C. Although because of transition of phase in the membrane above 30°C temperature, the sensitivity of bacterial cells to high pressure increases.

13.7.1.3 pH

The pH (power/or potential of hydrogen) of food products is an important factor responsible for the outgrowing of special microorganisms in some special food products, and also for further survival and growth. Different microorganisms have different pH range and the microorganisms can grow, in general, in such pH range but they grow best at optimum temperature. At neutral pH, spores of bacteria, are more resistant towards the high pressure treatments and are most sensitive to pressure. Spores when killed indirectly follow germination at pressure between 50 to 300 MPa. Bacteria spores, thus, are sensitive to pressure, to temperature and to the low pH [21]. At acidic pH values, however, inactivation of the microorganisms not only can be enhanced but the recovery of injured cells also inhibited [3]. It means that the inactivation of microorganisms in is not only enhanced by the pH, but also it inhibited the outgrowth of the sub-lethally injured cells by heat or high hydrostatic pressure.

Hence, we can say that the effectiveness of HPP treatment is more dependent on the pH of food. If the pH of food shifted, the pressure requirement/ or inactivation rate during HPP treatment may also be shifted as a function of the pH. For instance, the inactivation rate of E.coli present in orange juice increases when pH of orange juice decreases. Similarly, the pH of orange juice changes during the storage and this change in pH affects the survival of microorganisms present in the juice [22]. However, the change in pressure and temperature both causes the shift of pH and pressure causes more pH change during HPP treatment of foods.

13.7.1.4 The Water Activity (a_w)

The ratio of partial vapour pressure of water present in food products and partial vapour pressure of water in its standard state (i.e. partial vapour pressure of pure/distilled water under the ideal condition) is defined as the water activity. In other words, the quantity of water that is available for the growth of microorganisms generally denoted as the water activity of the food systems. Food products usually of have water activity about 0.95 (i.e. 95% of pure water) which provides the suitable moisture condition for the growth of bacteria, yeasts and mold. High pressure processing treatments lower the water activity and consequently protect the microorganism. But, on the contrary, the lower value of a_w prevents the recovery of sub-lethal injured cells.

The effective design of process parameters (temperature, pressure, rate of compression, and multiple compression-decompression cycle) can help to reduce the process time; improve productivity, product quality and economics of the process.

13.8 Kinetics Belongs to Microbial Growth and Inactivation

13.8.1 D Value

It is the time required to achieve one logarithmic reduction at a constant lethal temperature and at constant pressure is termed as decimal reduction time. The

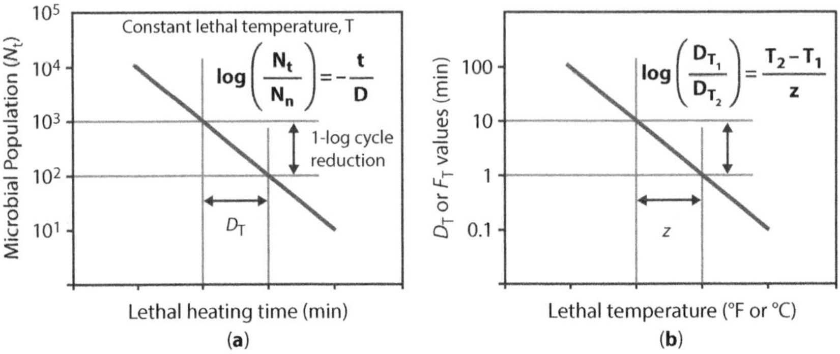

Figure 13.4 First order kinetics based D-z model for microbial inactivation. (a) Decimal reduction time and (b). Thermal resistance [23].

decimal reduction time indicates 90% destruction of microbial population (Figure 13.4a).

The principle of thermo-bacteriology was emerged in 1920s and assumed that the microbial inactivation follows first-order kinetics reaction. It means, remaining microbes is a function of time at constant pressure which is lethal, logarithmic linear behavior is observed. First order kinetic model was developed in 1920s and assumed that the same microbial population has same level of inactivation because it shows the same level of sensitivity to heat and pressure.

$$t/D = \log\frac{(No)}{(Nt)} = \log(No) - \log(Nt)$$

$$\log\left(\frac{No}{Nt}\right) = \frac{t}{D} \quad \text{or}$$

$$\log\left(\frac{Nt}{No}\right) = -\frac{t}{D} \quad \text{or}$$

$$No = Nt\ 10^{t/D} \quad \text{or}$$

$$Nt = No\ 10^{-t/D} \quad \text{or}$$

$$\mathrm{N} = \mathrm{No}\ 10^{-t/D}$$

Where N_0 is the initial microbial population, N_t is the number of remaining microbial populations after the treatments" is the treatment time and D is the decimal reduction time at constant lethal temperature (T). Therefore, the D-value can be defined as the time required for reducing the amounts of viable bacteria 1 $\log_{10}$ or 90% under specified condition.

13.8.2 Z Value (°C)

The thermal resistant constant (z) is defined is the required to achieving one logarithmic reduction (i.e. 90% reduction) in the decimal reduction time. When D values for different temperature are plotted on semi log coordinates and the temperature increases for one log cycle, the changes in values represents the value of z as:

$$z = \frac{T_2 - T_1}{\log D_{T_1} - \log D_{T_2}} = \frac{T_2 - T_1}{\log(D_{T_1}/D_{T_2})}$$

$$\log\left(\frac{D_{T_1}}{D_{T_2}}\right) = \frac{T_2 - T_1}{z}$$

Therefore the thermal resistance constant (z) is the increases in temperature necessary to have a 90% reduction of D or D_T and can be easily determined by plotting log (D) vs. temperature (Figure 12.4b).

13.8.3 F Value (Second)

Thermal death time is the time required to cause a stated reduction in a population of microorganism or spores. This time, generally indicated as 'F', can be expressed as the multiple of 'D' values, the time needed for 90 % reduction. It is customary to represent the 'F' with a subscript indicating the temperature at which it is referred and with a superscript indicating the 'z' value of the microbial considered. Means, the F_T^Z represents thermal death time for a temperature 'T' and a thermal resistant constant 'z'. However, for reference purpose 'z' is taken as 10°C (18°F) and T is as 121 °C (250 °F).

13.8.4 Spoilage Probability

The spoilage probability is used for the estimation of number of spoiled container within a total batch of processed products. However, the probability of surviving of a single microorganism can be calculated for a given heat treatment. If the total exposure of a thermal death time is 'F', then:

$$\log(No) - \log(N) = F/D$$

$$\log\left(\frac{N_0}{N}\right) = F/D$$

If 'r' is the number of food containers processed and N_0 is the initial number of spores, then, the total microbial load at the beginning of the process will be equal to rN_0 (number of container × initial microbial population). Now

$$\log(rN_0) - \log(rN) = F/D$$

If the goal of thermal process is to achieve a probability of one microorganism in the entire batch, then:

$$\log(rN_0) = F/D$$

$$\frac{1}{r} = \frac{N_0}{10^{F/D}} \text{ or } r = \frac{10^{F/D}}{N_0}$$

It means, if the 'F', 'D' and 'N_0' are fixed, there will be only one value of 'r' to satisfy the above equation. Therefore this value of 'r' which is the number of containers should be processed to contain only one microorganism in the entire batch. The reverse of 'r' is generally termed as the spoilage probability, which is the probability to achieve the results.

13.9 Packaging Importance in HPP

The optimum pressure transmission is the primary requirement for the success of HPP technique. For HPP treatment, thus, ideal food products has no gas inclusion and no head/empty spaces in the packaging material but has high moisture content. For best pressure transmission during HPP, the packaging material should be enough flexible to withstand pressure during pressure transmission without changes in any physical or chemical damage. In addition, the package has to allow the reversible deformation after compression of the food due to pressurization. As, tough materials or grasses are unable to withstand the high pressure treatment, the use of these packaging materials are avoided in HPP. In contrast, flexible packaging such as jars, trays, bottles, etc. made of polymers like nylon cast ploy-propylene, poly-propylene, poly-ethylene, polyester tubes and vacuum packed products are noticed ideal for pressure treatment process. When pressure is applied on these packaging materials, these can able to compress with 15% flexibility without any irreversible changes and upon the release of pressure these are able to return to its own shape. Packages made from plastic materials are, therefore, more convenient for high pressure processing due their resiliency, good flexibility and reversible response to compression [15].

13.10 High Pressure Processing Applications

When microorganisms grow in food stuffs, cause spoilage and produce unacceptable flavor, taste, odour, texture and appearance. To eradicate the microorganism present in foods and to enhance the food's safety; an effective preservation technique is required to inactivate the microorganisms in order to increase the shelf life of foods. Physical (Drying/dehydration, heating, freezing and packaging) chemical (reduction of pH or use of preservatives) preservation techniques are commonly used.

These traditional methods prevent the undesirable changes in the foods by reducing the microbial growth and metabolism. Pressure treatment, a novel technique, can be used to process both liquid and high moisture content solid foods effectively. Although lethal to microorganisms, pressure treatment does not break covalent bonds and has a minimum effect on food chemistry. Thus, high pressure processing technique provides a means for retained foods quality while avoiding the need for excessive thermal treatments or chemical preservatives. The novel nature of HPP technology and high equipment costs have been barriers to commercialization, but increased consumer demand for fresher tasting foods containing fewer preservatives is driving increased usage. HPP can preserve food products without heat or chemical preservatives, and its ability to ensure safety and significantly extended the refrigerated shelf life which has opened the new market opportunities particular in the area of "natural" preservatives free meat products.

There are several HPP treated products (jams, fruit juice, soups, oysters, and hams: most recently meat product) available in the markets. Some of them are categorizes in three groups such as (1) vegetables and fruits: juices, salsas, dressings, guacamole, etc., (2) meats: ready to eat meat and poultry, (3) seafood's: shellfish and fish products. However, the range high hydrostatic processing treated product is increasing day by day and spreading world-wide from its origins in Japan.

13.10.1 Fruits, Vegetables and Processed Food Products

Strawberry, apple and kiwi jams represented the first wave of pressure treated commercial products introduced in the Japanese market in 1990. Avocado based products, especially guacamole, were subsequently commercialized in the United States. Fresherized foods (Previously Avomex) began the first industrial production of guacamole in North America in 1997. Quaglia *et al.* [24] observed various positive impacts found in fruits and vegetables qualities after HPP like-wise in case of green pea, 400 to 900 MPa pressure for 5 to 10 min required to highly reduction in the activity of peroxidase enzyme. Pressure range between 400 to 600 MPa is generally used from few seconds to 5 minutes for the treatment of juices and beverages. However, the pressure ranges 500-600 MPa for 2-3 minutes is mostly applied for the preservation of juices [25].

Enriquez *et al.* [26] reported that the shelf life of golden delicious apple juice processed by ultra HPP at 430 MPa pressure for 7 min, and stored in room temperature and in refrigeration (4 °C) for 34 days. There are no effect in physico-chemical properties, nutritional value and sensory qualities of treated apple juice. Table 13.2 represents the effective moisture diffusion value of pineapple at different pressure based on diffusion model. Effect on the texture and drying behavior of pineapple is investigated by Kingsly *et al.* [27]. He found that pineapple slices were dried at 70°C after HP treatment at 50, 100, 300, 500 and 700 Mpa at 25°C. The quality of product increased in terms of reduced hardness, chewiness and springiness.

Table 13.2 Diffusivity of effective moisture of pineapple slices calculated based on diffusion model [27].

Pretreatments	D_{eff} x 09 (m^2/s)	r2
Control	3.60	0.90
50 MPa	3.65	0.93
100 MPa	3.70	0.91
300 MPa	3.90	0.93
500 MPa	4.70	0.96
700 MPa	4.70	0.93
Hot water blanched	4.30	0.93

Landl *et al.* [28] reported that apple puree was processed on 400-600 MPa for 15 min to inactivate aerobic mesophilic bacteria/yeasts/molds; and found 14 to 21 days shelf life extension at 4°C storage.

In case of apple juice, Shahbaz *et al.* [29] processed at 500 MPa for 1 min to inactivate L. monocytogens, S. aureus, E. coli, S. tiphimurium, S. cerevisiae.

Hartyáni *et al.* [30] observed apple orange juices processed at 200-600 MPa for 10 min is sufficient for inactivation of alicyclobacillus acidoterrestris; and found 14 to 28 days shelf life extension at 4°C storage.

Li *et al.* [31] processed the banana smoothie (after degassing by N_2 gas) at 550 MPa for 2–10 min for inactivation of aerobic bacteria, yeasts/molds; and found more than 15 days shelf life extension at 4°C storage.

Cactus juice processed at 600 MPa for 10 min for inactivation of microbial cells yeast/molds & acid tolerant microorganism [32].

Lavinas *et al.* [33] observed the significant reduction in the population of aerobic mesophilic bacteria/yeasts/filamentous fungi/E.coli in cashew apple juice at 250-400 MPa for 3-7 min; and and found 56 days shelf life extension at 4°C storage.

Cucumber juice treated by Zhao *et al.* [34] at 500 MPa for 2 min is sufficient for destruction of the microbial cells, yeasts/molds; and found more than 15 days shelf life extension at 4°C storage.

Fernández-Sestelo [35] reported that in the kiwi puree reduction in viable microbial cells when it is processed at 500 MPa for 3 min; and found 21 days shelf life extension in refrigerated temperature.

Carreño *et al.* [36] found that extended shelf life of mandarin juice when it processed at 150–450 MPa for 0.1–1 min to destroy the L. Plantarum.

Mango nectar and pulp treated at 300-600 MPa for 1 to 20 min for destruction of total aerobic bacteria, yeasts/molds [37, 38].

Orange juice is treated at 100-700 MPa for 1-30 min for destruction of total aerobic bacteria, L. plantarum, L. brevis, E. coli, S. aureus, viable microbial cells, yeasts/molds [39–43].

Fermented cabbage (Sauerkraut) processed by Peñas *et al.* [44] at 300 MPa for 10 min for destruction of total coliforms, aerobic mesophilic bacteria, lactic acid bacteria, faecal coliforms and yeast/molds, which leads the 60 days shelf life.

Marszalek *et al.* [45] observed the extension in shelf life of strawberry puree after processing at 300-500 MPa for 1-5 min. This process destroys the viable microbial cells, yeasts/molds.

Tomato juice is treated at 300-700 MPa pressure for 0.5-10 min for destruction in viable microbial cells, yeasts/molds. Enterobacteria, Lactic Acid Bacteria and B. Stearothermophilus; this leads 28-56 days shelf life [46, 47].

13.10.2 Meat and Sea-Foods

Kraft Foods, Hormel foods, Foster Farms, Perdue and Wellshire Farms are the examples of meat processors that have successfully utilized the technology for a variety of ready to eat, minimally processed meat products [48]. In case of meat and meat products appearance is a strong indicator for sensory parameters [49].

The meat and meat product quality largely affected by the exposure time, processing temperature as well as magnitude of pressure used, which leads denaturation hemi-displacement or release and ferrous atom oxidation [50]. That is why final product is found with cooked like taste [51]. In case of semi cooked and cooked meat products there is no changes found due to their protein content is already denatured during cooking [12]. Several seafood processors such as motivatit seafoods have also employed HPP to improve food safety and shelf life of shellfish with the added benefit of facilitating the removal of flesh from the shell. Meat and seafood products represent a high proportion of the applications of HPP mainly used for decontamination of purposes. Amanatidou *et al.* [52] analyzed the effect of HPP combined with modified atmospheric packaging on shelf life of fresh atlantic salmon.

13.11 Benefits and Drawbacks

Destruction of microorganisms, denaturation of proteins and lowering down the freezing point of water can be used as the major benefits in various sectors of food industries such as dairy sector, sea food, meat industry, etc. In addition, there are various scopes for designing the equipment's for HPP as well as to produce the new products with high sensory appeals in terms of good texture, taste, color, flavor and functional properties. The small molecules responsible for characteristics of flavoring and nutritional components, which are generally crushed during the thermal treatment, remain unchanged in HPP process. The high pressure processing technology, thus, number of benefits foods processed by HPP (such as fresher taste, better appearance, texture and

Table 13.3 Benefits and drawbacks of high pressure processing.

	A. Benefits
1.	HPP is a clean technology.
2.	HPP system is flexible for number of products.
3.	Food products are processed uniformly by this technique.
4.	No/or very less requirement of chemical additives or preservatives.
5.	Low thermal degradation of foods.
6.	Inactivates the microbes, spores and enzymes.
7.	Process time requirement is low.
8.	There is no evidence of toxicity.
9.	Minimal loss of colour, flavour and nutritional value of foods.
10	Maintain the natural antimicrobial system without affecting food quality.
11.	Extend the shelf-life and bioavailability increased.
12.	HPP is almost independence of the shape and size of food products.
13.	HPP processed products have positive consumer appeal.
14.	Uniform and instantaneous application of pressure maintained the food products shape
	B. Drawbacks
1.	As HPP is based on compression, presence of water in food is prime importance.
	For antimicrobial effect, food should have 40% water.
2.	Initial cost is high.
3.	Some enzymes and spores could not be inactivated because they are pressure resistance for some extent.
4.	Delicate & vulnerable foods need more attention.
5.	Option for packaging is limited.
6.	Some microbes can survive.

nutrition, and eliminates cooked off-flavor). However, some of the benefits and drawbacks are listed in Table 13.3.

13.12 Future Prospects of the HPP

In spite of some demerits and drawbacks such as high initial cost, the high pressure technology can be clubbed with other technology to boost up processing sector. For instance, combinations of HPP with moderate temperature, combination of HPP with pulsed electric field, etc. are recent developments. Future development in HPP will be for the production of organic food with supreme quality and to produce locally available fresh fruits. When high pressure processing combined with pulsed electric field leads to notable spore inactivation. Similarly, when HPP is applied in combination to the temperature it inactivates microbes and ensured the food safety. The moderate HPP combined with moderate temperature results higher inactivation (upto 5 log 10 cycles) of different pathogenic microorganism in foods [53]. The HPP technology, therefore, has great future prospect and potential. Since, the current trend for utilizing of the food free from additives or preservatives among the consumers is the big driving force for emerging, further developing and popularizing of such type of non thermal technologies.

The various other researches on HPP definitely impart the several development and innovation in this technique. Pathogenic and spoilage causing microorganisms are pressure and thermal resistant which is function of food matrix, pH and water activity. In addition, reporting on standardization process condition, techniques & effective utilization of equipment's are needed to develop significant data for inactivation pattern. Further more research on the extent mechanism for destruction of bacterial population by using HPP & sterilization is required [4].

13.13 Conclusion

The modern trend among the consumer demands for fresh food materials that have retained its sensory attributes viz. color, flavor, textural properties, minerals & vitamins have increased the value such non-thermal technology. This technology has ability to inactivate the microorganism, spore and enzymes and consequently prolong the keeping quality of food stuffs. As the HPP process is based on the phenomenon of uniform and instant pressure application, the destruction of shape of food material, that occur in thermal methods, completely avoided. Several research already done by various researcher in HPP indicating that there is no adverse effect in nutritional parameters in case of fruits juices, processed products of vegetables (such as tomato puree) or milk and milk products, etc. The HPP method has great opportunities for entrepreneurship in preservation of dairy milk, human milk, fruit juices and colostrums due to in effective nature of HPP on many bioactive components. For efficient application of HPP food processing technology, flexible packaging material is required which withstand high pressure and increased to 15% volume flexibility (able to increase the volume during the process and ability to come into original state) without changing its properties like sealing, integrity and barrier against the environment.

References

1. Lund D (2003) Predicting the impact of food processing on food constituents. Journal of Food Engineering, 56, 113–117.
2. Anonymous. (2011) History of High Pressure Food Processing. https://food-processinghistory.blogspot.com/2011/04/history-of-high-pressure-food.html (access date 05.04.2021).
3. Hogan, E., Kelly, A.L, Sun, Da-Wen (2005) High Pressure Processing of Foods: An Overview. In: Emerging technologies for food processing (Da-Wen Sun, Ed.), p. 2-217, Elsevier Academic Press.
4. Mircea-Valentin Muntean, Ovidiu Marian, Victor Barbieru, Giorgiana M. Cătunescu, Ovidiu Ranta, Ioan Drocas and Sorin Terhes (2016) High Pressure Processing in Food Industry – Characteristics and Applications, Agriculture and Agricultural Science Procedia, Volume 10, Pages 377-383.
5. Naik, L., Sharma, R., Rajput, Y.S., Manu G. (2013) Application of HPP for dairy food preservation - future prospective. J Anim Prod Adv, 3(8):232-241.
6. Mertens B (1995). Hydrostatic pressure treatment of food: Equipment and processing, In: New Methods of Food Preservation (G. W. Gould, Ed.), p. 135: Blackie Academic and Professional, New York.
7. Heremans K (1995). High pressure effects on biomolecules. In High Pressure Processing of Foods, [DA Ledward, DE Johnston, RG Earnshaw and APM Hasting, editors]. pp. 81-97, Nottingham: Nottingham University Press.
8. Balasubramaniam, V. M., Farkas, D., & Turek, E. J. (2008). Preserving foods through high-pressure processing. Food technology, 62(11), 32.
9. Sahin, S., Sumnu, S.G. (2006) Electromagnetic Properties, In: Physical Properties of Foods. Springer Science, p. 162.
10. Cheftel JC, Culioli J (1997) Effects of high pressure on meat: A review. Meat Science, 46 (3), 211–236.

11. Flick Jr. (2003) Novel Applications of High Pressure Processing. Originally appearing in Global Aquaculture Advocate, 6(3):1-4, June.
12. Ting EY, Marshall RG (2002) Production issues related to UHP food. In Engineering and Food for the 21st Century (Welti-Chanes J, Barbosa-Canovas GV, Aguilera JM, eds.). Food Preservation Technology Series, Boca Raton: CRC Press, Ch 44, pp. 727–738.
13. O'Reilly CE, Murphy PM, Kelly AL, Guinee TP, Beresford TP (2002) The effect of high pressure treatment on the functional and rheological properties of Mozzarella cheese. Innovative Food Science and Emerging Technology, 3, 3–9.
14. Huppertz T, Alan L, Kelly Fox PF (2002). Effect of high pressure on constituents and properties of milk. Int. Dairy Journal., 12: 561-572.
15. Ferstl P. (2013) High pressure processing: insights on technology and regulatory requirements. NFL White Paper Series – Volume 10.
16. Abee, T., Wouters, J.A., (1999) Microbial stress response in minimal processing. International Journal of Food Microbiology, 50:65-91.
17. Dring JG (1976). Some aspects of the effects of hydrostatic pressure on microorganisms. In Inhibition and Inactivation of Microorganisms, London: Academic Press, pp. 257-277.
18. Mertens, B., and G. Deplace. (1993). Engineering aspects of high pressure technology in the food industry. Food Technol. 47(6):164-169.
19. Balasubramaniam V.M., Ting E.Y., Stewart C.M. and Robbins J.A. (2004). Recommended guidelines on laboratory practices for conducting high-pressure experiments. Innovative Food Science & Emerging Technologies 5: 299-306.
20. Kalchayanand N, Sikes A, Dunne CP, Ray B (1998) Interaction of hydrostatic pressure, time and temperature of pressurisation and Pediocin AcH on inactivation of food borne bacteria. Journal of Food Protection, 61 (4), 425–431.
21. Smelt JPPM (1998) Recent advances in the microbiology of high pressure processing. Trends in Food Science & Technology, 9, 152–158.
22. Linton M, Patterson MF (2000) High pressure processing of foods for microbiological safety and quality. Acta Microbiologica et Immunologica Hungarica, 47 (2–3), 175–182.
23. Torres, J.A.; Pérez Lamela, C. (2008) Pressure processing of foods: Microbial inactivation and chemical changes in pressure-assisted thermal processing (PATP)-Part 2. Agrofood Ind. Hi-Tech, 19, 34–36.
24. Quaglia GB, Gravina R, Paperi R, Paoletti F (1996) Effect of high pressure treatments on peroxidase activity ascorbic acid content and texture in green peas. Lebensmittel-Wissenschaft-und-Technology, 29, 552–555.
25. Tonello-Samson, C. (2014) Advances in High Pressure Processing: Case Studies. In Proceedings of the IFT Annual Meeting & Food Expo, New Orleans, LA, USA, 21–24 June.
26. Enriquez E.Juarez, Ochoa I.Salmeron, Mendez N.Gutierrez, Ramaswamy H.S., Rivas E.Ortega (2015) Shelf life studies on apple juice pasteurised by

ultrahigh hydrostatic pressure, LWT - Food Science and Technology, 62 (1), 2, pp. 915-919.
27. Kingsly A.R.P., Balasubramaniam V.M. And Rastogi N.K. (2007) Effect of High-Pressure Processing on Texture and Drying Behavior of Pineapple, Journal of Food Process Engineering 32 (2009) 369–381.
28. Landl, A.; Abadias, M.; Sárraga, C.; Viñas, I.; Picouet, P.A. (2010) Effect of high pressure processing on the quality of acidified Granny Smith apple purée product. Innov. Food Sci. Emerg. Technol., 11, 557–564.
29. Shahbaz, H.M.; Yoo, S.; Seo, B.; Ghafoor, K.; Un Kim, Y.; Lee, D.U.; Park, J. (2016) Combination of TiO2-UVPhotocatalysis and High Hydrostatic Pressure to Inactivate Bacterial Pathogens and Yeast in Commercial Apple Juice. Food Bioprocess Technol., 9, 182–190.
30. Hartyáni, P.; Dalmadi, I.; Knorr, D. (2013) Electronic nose investigation of Alicyclobacillusacidoterrestris inoculated apple and orange juice treated by high hydrostatic pressure. Food Control, 32, 262–269.
31. Li, R.; Wang, Y.; Wang, S.; Liao, X. (2015) A Comparative Study of Changes in Microbiological Quality and Physicochemical Properties of N2-Infused and N2-Degassed Banana Smoothies after High Pressure Processing. Food Bioprocess Technol., 8, 333–342.
32. Moussa-Ayoub, T.E.; Jäger, H.; Knorr, D.; El-Samahy, S.K.; Kroh, L.W.; Rohn, S. (2017) Impact of pulsed electric fields, high hydrostatic pressure, and thermal pasteurization on selected characteristics of Opuntiadillenii cactus juice. LWT Food Sci. Technol., 79, 534–542.
33. Lavinas, F.C.; Miguel, M.A.L.; Lopes, M.L.M.; Valentemesquita, V.L. (2008) Effect of High Hydrostatic Pressure on Cashew Apple (Anacardium occidentale L.) Juice Preservation. J. Food Sci., 73, M273–M277.
34. Zhao, L.; Wang, S.; Liu, F.; Dong, P.; Huang, W.; Xiong, L.; Liao, X. (2013) Comparing the effects of high hydrostatic pressure and thermal pasteurization combined with nisin on the quality of cucumber juice drinks. Innov. Food Sci. Emerg. Technol., 17, 27–36.
35. Fernández-Sestelo, A.B.; Sendra de Saa, R.; Pérez Lamela, C.; Torrado-Agrasar, A.; Rúa, M.L.; Pastrana-Castro, L. (2013) Overall quality properties in pressurized kiwi purée: Microbial, physicochemical, nutritive and sensory test during refrigerated storage. Innov. Food Sci. Emerg. Technol., 20, 64–72.
36. Carreño, J.M.; Gurrea, M.C.; Sampedro, F.; Carbonell, J.V. (2001) Effect of high hydrostatic pressure and high-pressure homogenisation on Lactobacillus plantarum inactivation kinetics and quality parameters of mandarin juice. Eur. Food Res. Technol., 232, 265–274.
37. Liu, F.; Wang, Y.; Bi, X.; Guo, X.; Fu, S.; Liao, X. (2013) Comparison of Microbial Inactivation and Rheological Characteristics of Mango Pulp after High Hydrostatic Pressure Treatment and High Temperature Short Time Treatment. Food Bioprocess Technol., 6, 2675–2684.
38. Liu, F.; Li, R.; Wang, Y.; Bi, X.; Liao, X. (2014) Effects of high hydrostatic pressure and high temperature short-time on mango nectars: Changes in

microorganisms, acid invertase, 5 hydroxymethylfurfural, sugars, viscosity, and cloud. Innov. Food Sci. Emerg. Technol., 22, 22–30.

39. Bull, M.K.; Zerdin, K.; Howe, E.; Goicoechea, D.; Paramanandhan, P.; Stockman, R.; Sellahewa, J.; Szabo, E.A.; Johnson, R.L.; Stewart, C.M. (2004) The effect of high pressure processing on the microbial, physical and chemical properties of Valencia and Navel orange juice. Innov. Food Sci. Emerg. Technol., 5, 135–149.
40. Katsaros, G.I.; Tsevdou, M.; Panagiotou, T.; Taoukis, P.S. (2010) Kinetic study of high pressure microbial and enzyme inactivation and selection of pasteurisation conditions for Valencia orange juice. Int. J. Food Sci. Technol., 45, 1119–1129.
41. Timmermans, R.A.H.; Mastwijk, H.C.; Knol, J.J.; Quataert, M.C.J.; Vervoort, L.; Van der Plancken, I.; Hendrickx, M.E.; Matser, A.M. (2011) Comparing equivalent thermal, high pressure and pulsed electric field processes for mild pasteurization of orange juice. Part I: Impact on overall quality attributes. Innov. Food Sci. Emerg. Technol., 12, 235–243.
42. Syed, Q.A.; Buffa, M.; Guamis, B.; Saldo, J. (2014) Effect of Compression and Decompression Rates of High Hydrostatic Pressure on Inactivation of Staphylococcus aureus in Different Matrices. Food Bioprocess Technol., 7, 1202–1207.
43. Yoo, S.; Ghafoor, K.; Un Kim, J.; Kim, S.; Jung, B.; Lee, D.U.; Park, J. (2015) Inactivation of Escherichia coli O157:H7 on orange fruit surfaces and in juice using Photocatalysis and High Hydrostatic Pressure. J. Food Prot., 78, 1098–1105.
44. Peñas, E.; Frias, J.; Gomez, R.; Vidal-Valverde, C. (2010) High hydrostatic pressure can improve the microbial quality of sauerkraut during storage. Food Control, 21, 524–528.
45. Marszałek, K.; Mitek, M.; Ska̧pska, S. (2015) The effect of thermal pasteurization and high pressure processing at cold and mild temperatures on the chemical composition, microbial and enzyme activity in strawberry purée. Innov. Food Sci. Emerg. Technol., 27, 48–56.
46. Hsu, K.C.; Tan, F.J.; Chi, H.Y. (2008) Evaluation of microbial inactivation and physicochemical properties of pressurized tomato juice during refrigerated storage. LWT-Food Sci. Technol., 41, 367–375.
47. Krebbers, B.; Matser, A.M.; Hoogerwerf, S.W.; Moezelaar, R.; Tomassen, M.M.M.; Van Den Berg, R.W. (2003) Combined high-pressure and thermal treatments for processing of tomato puree: Evaluation of microbial inactivation and quality parameters. Innov. Food Sci. Emerg. Technol., 4, 377–385.
48. Saiz, A.H., Mingo, S.T., Balda F.P., Samson C.T. (2008) Advances in design for successful commercial high pressure food processing. Food Australia, 60(4): 154-156. Samson, C. Tonello. 2008. Personal communication. NC.
49. Da-Wen-Sun (2005) Emerging Technologies for Food Processing. DOI: 10.1016/B978-0-12-676757-5.X5000-2, Elsevier Academic Press.

50. Mor-Mor M, Yuste J (2003) High pressure processing applied to cooked sausage manufacture: physical properties and sensory analysis. Meat Science, 65 (3), 1187–1191.
51. Hugas M., Garriga M, Monfort JM (2002) New mild technologies in meat processing: high pressure as a model technology. Meat Science, 62, 359–371.
52. Amanatidou A, Schlüter O, Lemkau K, Gorris LGM, Smid EJ, Knorr D (2000) Effect of combined high pressure treatment and modified atmospheres on the shelf-life of fresh Atlantic salmon. Innov Food Sci Emerg Technol 1:87–98.
53. Woldemariam, W. H., Emire, S.A. (2019) High Pressure Processing of Foods for Microbial and Mycotoxins Control: current trends and future prospects. Cogent Food & Agriculture 5: 1622184.

14

Artificial Intelligence in Food Processing

Manish Tiwari, H. Pandey*, Arunima Mukherjee and R. F. Sutar

College of Food Processing Technology and Bio-Energy, Anand Agricultural University, Anand-Gujarat, India

Abstract

The post-harvest food processing requires wide range of novel emerging techniques for efficient production and handling of food. Among these, Artificial intelligence (AI) is catching the attention of businesses across many disciplines and sectors directly and indirectly. AI driven food processing using mechanized inspection and learning methods for improving the food generation model. It is not only analyzes the data digitally, but also accurately replaces most of the human interventions. The complex assessment and management of food quality and quantity can be easily controlled and assured by using predictable models in AI. In global perspective, this chapter highlights the various scenarios and relationships of AI with dairy processing (identify adulteration, standardization and monitoring cleaning in place systems), cereal processing (sorting and grading), meat and poultry product processing (behaviour of carcass and egg quality), beverage processing (simulation of pulping process, detection of polyphenol contents, permeate flux, viscosity and discrimination of brands), fruits and vegetable processing (prediction of physical properties, internal defects, total soluble solids, pH and texture quality of processed fruits and vegetables), tea processing (assessing quality using E-nose), bakery and confectionary (surface color changes for chocolate). It elucidates requirements, demands and safety towards the food processing industries by means of AI. Also, this chapter explains challenges and recommendations in the setup of AI technologies in real-time post-harvest processing. Finally, the chapter notifies the future possibilities of AI revolution in food processing industries as a key to provide quality food.

***Keywords*:** Novel techniques, artificial intelligence, food processing, food engineering, robotics, software development, quality

**Corresponding author*: hridyesh@aau.in

Navnidhi Chhikara, Anil Panghal and Gaurav Chaudhary (eds.) Novel Technologies in Food Science, (511–550) © 2023 Scrivener Publishing LLC

14.1 Introduction

Worldwide, food is commonly used as a nutrient vehicle for the human body. Proper and efficient processing not only makes the food palatable but also serves to the economy of the country. With the expansion of literature, research and technological advancements, food manufacturers require to design automation processing lines which produces higher quality safe finished produce at an affordable price. Advanced food engineering allied with other streams such as networking, computer science, software development etc. provide sound quality and profitable products at the commercial end successfully. However, raw material quality assessment, inline safety control measures, management during the production process, packaging, product trend forecasting and pricing of product are some of the prime issues during food processing. Also, in recent times, the maximum population tends to shifts towards sound quality healthy foods and thus catalyzes the food processing industries to fight against food insecurity and food safety with better lives and emerging economies throughout the world. To overcome these issues, food-processing industries apply a degree of intelligence and control natural variables to initiate the powerful tools in their processing. Therefore, processing lines requires automated smart machines that mimics the human activities. This makes the process precise and accurate, produce high quality product, minimal level of uncertainty, enhance food safety, profitable sales and sound customer acceptance. This is being successfully upgraded by a fast-growing and promising multidisciplinary research techniques called artificial intelligence (AI). It is a branch of science which deals with the combination of two terminologies 'intelligence' and 'artificial device' and that mimics human intelligence. This human intelligence characteristic is further developed in form of algorithms (machine learning and deep learning) which mimics the human characters like sound recognition, visual perception, flavour analysis and decision making in an artificial way [1]. AI is a rapidly growing modified computer-based technique in the food-processing sector that is used to optimize, control, and model distinct engineering aspects at a single reference point.

Algorithm's knowledge represents the symbol systems to solve the practical aspects. In machine learning (ML), algorithm is used both as an input (unstructured) and output (computable prediction function). ML algorithms having prediction functions, which are closely, resemble the non-linear functions. Therefore, the utilization of output functions of ML algorithms is broader as compared to simple linear least squares tools [2].

The existence of AI in many industrial and governmental sectors is limited to infinite computation and minimize the costs of data storage in the virtual era. It exponentially increases the use of AI to unlock the value trapped in vast volumes of data for many industries. AI is a multi-technological based aspect that enable machines to act with higher levels of intelligence and emulate the human capabilities of sense, comprehend and act as shown in Figure 14.1.

In brief, AI is about to bring a widespread change in our world that allows high-level processes such as learning, decision making and problem solving. AI coupled with other technologies improves the analysis, collection and interpretation of data. Therefore, worldwide processing sectors moves along with AI in order to replace human intelligence with AI. Indian economy is growing exponentially with the second largest population in the world. Recognising AI's importance in national economy, Finance minister (2018-2019) mandated National Institution for Transforming India (NITI) Aayog to set up the framework of National Program on AI. The main aim of this program was to transform existing research and development using AI techniques. Hence, NITI Aayog has collaborated with AI developers in order to execute AI related projects in food, agriculture and health sectors [3].

The main objective of food processors is to maintain the quality of product during the mass production by implementing the designs with standard reliable procedures, boost companies networking, serve the customers in short duration at lower cost. Hence, the food processing sector mainly associated with numerous raw materials, uncertainty working patterns,

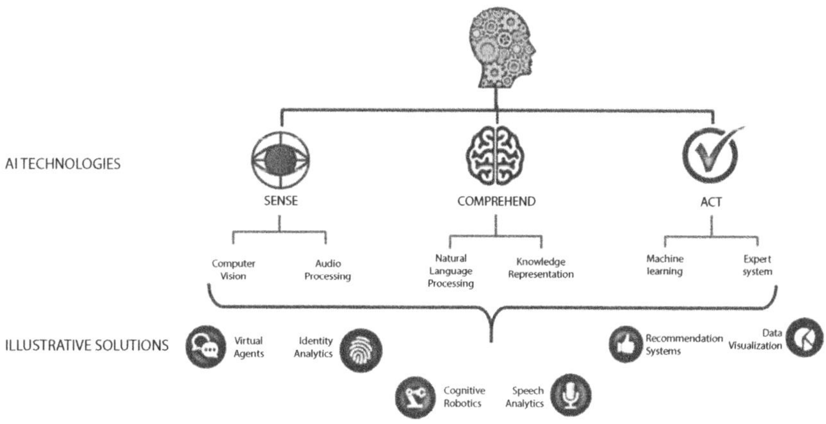

Figure 14.1 Basics of artificial intelligence.

and highly non-linear behaviour in processing [1]. Therefore, this chapter notifies the acceptability of AI in food processing to overcome from the above-mentioned issues and their future possibilities in food processing industries as a key to provide sound safe food.

14.2 Evolution of Artificial Intelligence

Due to industrialization and development of automation, manufacturing sector has potential to exponentially rise the heights of productivity within a short period of time [4]. The historical development of artificial intelligence has been fairly recent, with its birth being traced to the mid-20th century (1940-1956) as depicted in Figure 14.2. Alan turner, who was the pioneered of machine learning during the 1940s and 1950s introduced turning test in 1950. Turning test is a method of inquiry in artificial intelligence (AI) for determining whether or not a computer is capable of thinking like a human being under specific conditions. It was the first serious proposal in the philosophy of artificial intelligence. Though the seemingly recent origin, there still exist some influences that have greatly contributed, albeit indirectly, to the conceptualization of AI. The genesis of Artificial Intelligence unveils the contributions in diverse academic fields, history, art, philosophy logic and mathematics. Following the AI pathway, John McCarthy (father of AI) organized the mythic Dartmouth conference where, in his talk, he first coined the term "artificial intelligence" in 1956.

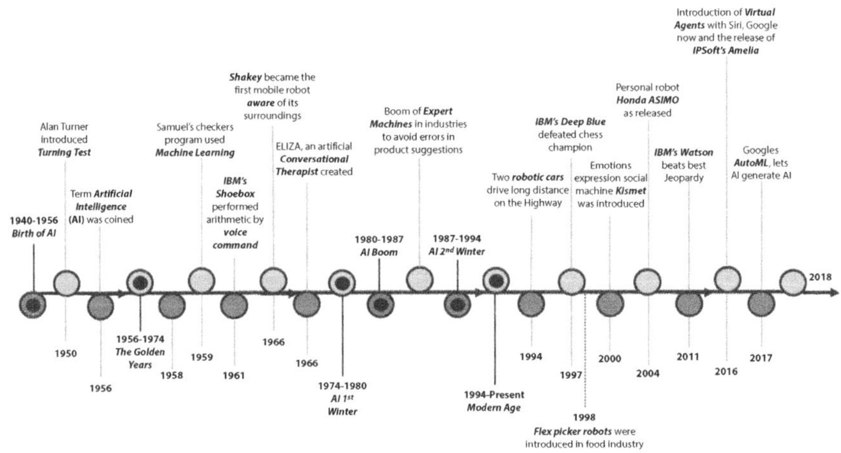

Figure 14.2 Schematic representation of an evolution of artificial intelligence.

According to him AI signifies science and engineering of making intelligent machines.

The period of golden years for AI is considered as 1956-1974, where computers were solving algebra world problems, proving theorems in geometry and learning to speak English. An American pioneer, Arthur Lee Samuel popularized the term "machine learning" in 1959 and programmed world's first successful self-learning programs which was based on fundamental concepts of AI [5, 6]. Late 1961, IBM 's Shoebox performed arithmetic by voice command in which a machine could do simple math calculations using voice commands. Shakey robot project was the first general-purpose mobile robot, which was able to reason about its own actions and started in 1966. While other robots would have to be instructed on each individual step of completing a larger task, Shakey could analyze the commands and break them down into basic chunks by itself. In 1966, Joseph Weizenbaum developed the program ELIZA, which aimed at tricking it users by making them believe that they were having a conversation with a real human being. ELIZA is considered to be the first chatterbot in the history of computer science.

The first Artificial Intelligence (AI) winter was observed in the years of 1974-1980. AI was subject to critiques and financial setbacks in 1970s as researchers had failed to materialize funding for AI. Worldwide, AI expert systems program was implemented by corporations in 1980s which had focus on the mainstream AI research. The rise of expert systems, knowledge revolution, money returns helped once again AI to boom from 1980-1987 in the fifth-generation project.

The second AI winter was taking place in 1987. But, post 1994 favoured the regrowth of AI. Deep blue became the first chess-playing computer system to beat a reigning world chess champion. In 1998, flex picker robots were introduced, revolutionizing the food industry by enabling rapid pick ing and placing of food items. Emotions expression social machine Kismet was introduced in 2000 that can recognize and simulate emotions. The name Kismet comes from a Turkish word meaning "fate" or sometimes "luck". After 2000 to till date lot of modifications were done and every sector increasingly moved with AI.

14.3 Principles of Artificial Intelligence

Artificial intelligence makes use of several techniques individually or in combination to execute several tasks as shown in Figure 14.3. **Case-based reasoning (CBR)** utilizes past problems having similar solution as a

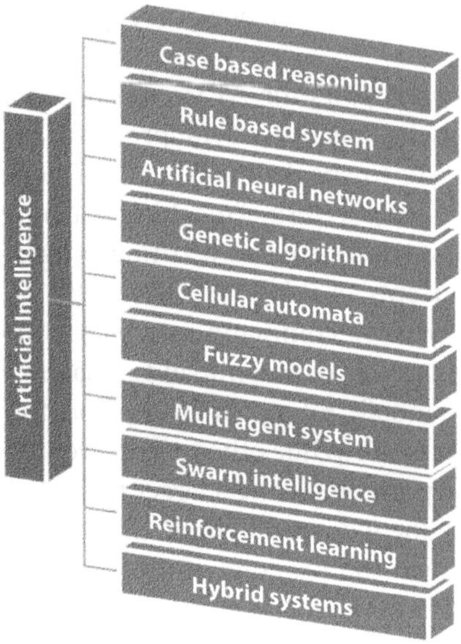

Figure 14.3 Multidiscipilinary use of artificial intelligence technique in different system.

method for problem solving. It recognises the problems which covers four mains steps (1) retrieve (2) reuse (3) revise (4) retain as shown in Figure 14.4 (a).

Firstly, the similar cases are retrieved from the database. The retrieved information is then used to develop a new solution to the current problem. The proposed solution is revised after simulation and successful test executions and the final solution is retained for future use.

Rule-based systems (RBS) solve problems by analysing the new problems on the basis of some pre-fed rules to decide its course of action to develop a solution. Figure 14.4 (b) shows the automation process which utilizes logical operations such as IF (to test the fulfilment of the set of pre-defined rules) and THEN (to decide a course of action for deriving a solution). IF is the condition part whereas THEN is the action decided on the fulfilment of IF condition.

Artificial neural network (ANN) is the simulation of how the human brain processes information. It processes and analyses the data for decision making in a manner similar to that of human brain. The human brain consists of neurons responsible for processing information by carrying information towards (inputs) and away (outputs) from the brain. The ANN is network of artificial neurons also known as processing units. They are highly interconnected by links with weights (w_1, w_2, w_3...w_n) and consists

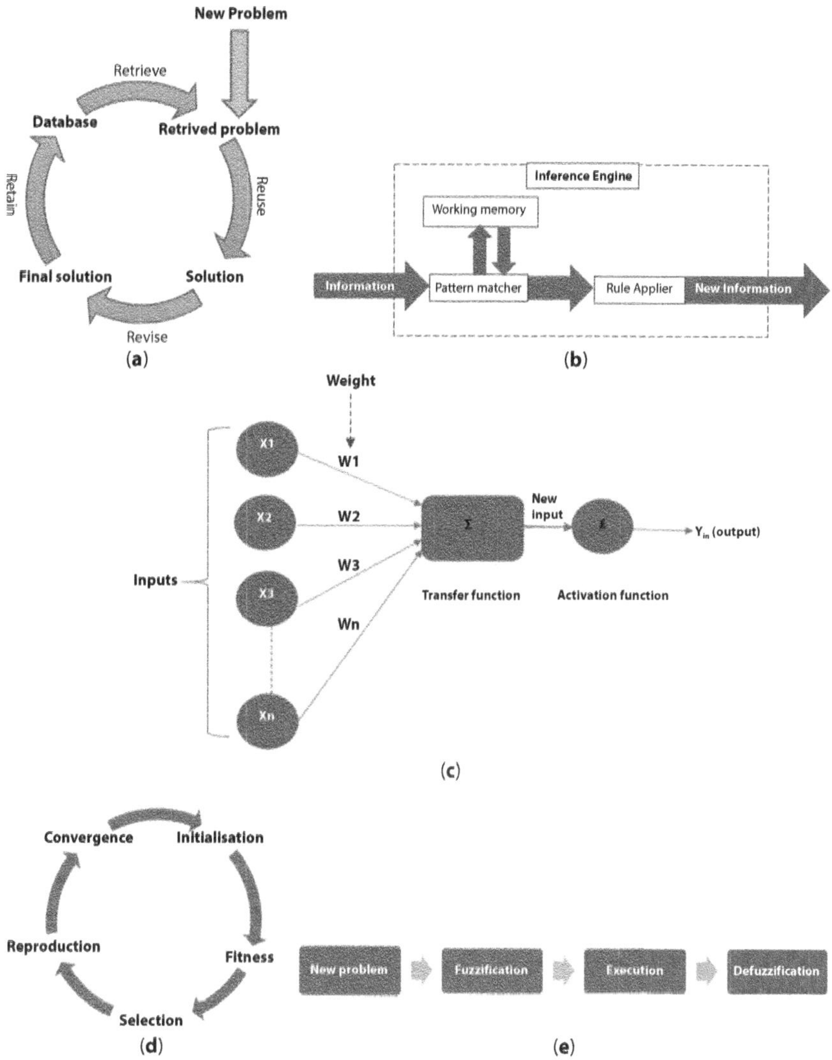

Figure 14.4 Illustration represents the principle of (a) case-based reasoning; (b) rule-based systems; (c) artificial neural network; (d) genetic algorithm and (e) fuzzy system.

inputs (x_1, x_2, $x_3 \ldots x_n$) and output layers as shown in Figure 14.4(c). ANNs can be used to solve different categories of problems such as clustering, pattern classification, prediction, function approximation, optimisation and process control.

Genetic algorithm (GA) is a search technique, which belongs to the class of evolutionary algorithms. It is inspired by the theory of natural selection by Charles Darwin. Genetic algorithm is used for optimization and search problems by taking into use operators such as crossover, mutation and selection as shown in Figure 14.4(d). It then passes its trait to new solution, which replaces the poorer solutions.

Fuzzy logic gives an indication of mathematization of human reasoning and identify those complex data with great accuracy that cannot be simply rectified classical accurate models. It can handle both linguistic knowledge and numerical data at the same time used in food science. Fuzzy logic systems are based on the observation that people make decisions based on imprecise and non-numerical information as shown in Figure 14.4(e). The problem received by fuzzifier translate them into numeral forms followed by linguistic variables. The linguistic variables are reasoning based human interpretation system that results in control logics of the numerical converted problems. Defuzzifier transforms the resulted fuzzy outputs into the required outputs in order to control the system.

Multi-agent system (MAS) is a network of agents interacting in a collective manner to achieve a set of goals. An agent is a software component containing code and data. Individually an agent is incapable of finding solution to a problem. Hence MAS utilizes the method of individual agents interacting with each other on the basis of pre-defined rules. The agents communicate in a high-level agent communication language through which they share information, request services and negotiate with each other. Coordination among the agents is critical, as information is relayed and responses are processed non-synchronously between agents. Coordination depends on resource use, nature of interactions between agents, system's infrastructure and degree of concurrency [7].

Hybrid system is the combination of two or more AI techniques to overcome the weakness. The hybrid system can be categorised into three classes according to how the different AI techniques are combined are sequential hybrid, auxiliary hybrid and embedded hybrid.

The most common type of hybrid which are very widely used are neuro-fuzzy systems. It is a combination of artificial neural networks and fuzzy systems. They are effectively fast, efficient and easily designed, implemented and understood. By combining them, the need to prime fuzzy systems is reduced by learning in ANNs [7].

14.4 Global Developments in Artificial Intelligence

With the development and advancement, AI sets its roots globally in every sector. AI system nowadays focus on bottom to top approach and thus, it is not only widely popular in healthcare sectors but also play an important role as governmental information or service providers, updation (updates occur) regarding natural disasters, conserving biodiversity, improving agricultural practices, crop disease monitoring water management and drought monitoring. Some of these applications are discussed in this section.

Problems involving lower life expectancy and inadequate availability of healthcare resources are serious issues that are required to be tackled at a global level. Several preventable maladies remain undiagnosed and outbreaks of infectious diseases frequently overwhelm the available infrastructures. AI play an important part in health sectors by improving the availability of healthcare workers, enhance human expertise, and increasing the surveillance of diseases [8]. More than 200 million people are afflicted by malaria every year. It is diagnosed by critical analysis of blood samples, which requires technical skills. To overcome this time consuming and complicated task, Maker ere University, Uganda has developed AI-computer vision-based software. It has been trained to detect malaria samples. Systems like this can help in increasing the efficiency of healthcare workers and effectively extend the capabilities to the remote sections of the society [8]. Pakistan has developed a disease response system after the outbreak of dengue epidemic in 2011. The system involves AI to provide early warnings of outbreaks in the future. The statistical algorithms using disease surveillance system are used to assess data from internet news sources, dengue hotline, government agencies, offering hospitals and real-time information on outbreak and hence help to track with a high level of geographic detail [9].

For greater efficiency and enhanced transparency in government sectors such as policy-making, providing services etc., it is important for the citizens to have better access to the information and services. Groups around the world are learning different ways using AI to; help resource-strapped countries; automate e-government assessments in order to make the flexible decision-making structure [10–12].

AI application is not only taken part in planning but also mitigate the chances of natural disasters and conserves the wildlife throughout the globe. Agriculture sector plays a vital role in the surplus of economy. Population in low-medium income countries such as India and Africa depend on agriculture followed by food. Successful harvests and healthy crops are mainly depending on biological factors such as diseases caused by microorganisms, insects, and eventually environmental factors such as flood and drought, which affects the soil fertility. The applications of AI models in agro-processing not only provide critical logics but also gives an indication to improve the quality and efficiency of agricultural practices. Monitoring of crop diseases is a time-consuming endeavour, which requires high level of skills and expertise. Its timely knowledge is necessary for appropriate action in terms of crop protection. Researchers reveal edoptimized methods of a process by capturing crop images for further analysis and classification using AI system. Images of diseased crops are connected with machine learning algorithms, which provide rapid diagnosis and feedback as output. Mapping functions can also be possible to predict diseases over a successive interval of time [13].

AI system was developed by collaborative efforts of Austria and Palestine for water scarcity issues in Palestine. An expert system to oversee mitigation of losses in water systems was developed. The system assessed a range of specific actions-such as active and passive leak control, water pressure management, metering, and public awareness campaigns-against a set of economic, environmental, technical, and socioeconomic evaluation criteria. Makerere University, Uganda has also fabricated an automated AI-based system in order to produce water crises and shortage reports, which summarizes the information to identify relevant trends for the specified parameters such as soil conditions, water and rainfall [8].

14.5 Artificial Intelligence and Food Processing

Food and agriculture organization (FAO) revealed that one out of nine people in 2017 affected from food insecurity [14]. Year 2020 not only stuck with COVID-19 pandemic but also with the outbreaks of desert locust in eastern African and Asian countries, which affected the economic prospects throughout the globe [15]. Also, Asian-African regions are suffering frommal nutrition and lacking of food whereas India and China possess first and second ranks respectively in terms of economical advancements and growing population [16].

Experts found the way to sustain the quality, food security and economy along with the food products advocates the introduction of expert systems in the food manufacturing industries in order to obtain more efficient and automatic control during processing. Modern technologies such as Artificial Intelligence (AI) and computer vision are estimated to break the accessibility of complex data for predictable models and real time intelligent machines in food industries [17]. Robotics is related to the design and implementation of robots, which are programmed and can perform various autonomously and semi-autonomously functions as command them [18].

AI robots are the merger of two technologies called artificial intelligence and robotics where AI programs are installed in Robot systems in the form of developed methods and processes for solving various issues. Robotics contributes with machine-based components such as computers, sensors and effectors whereas AI deals with perception and reasoning it. Korea is known to have the highest robot density while Asian countries reflects the lowest average in 2018. AI based robotic systems have been developed with wide applications for sectors which offered numerous manufacturing solutions in food industry. Among these, less than four percent of robots installed in food industries [19]. Investment in AI sectors raised to four times between 2010-2015 [20]. Food-grade robot which are integrated with AI have been successfully designed and are underutilization in different part of food industry [21]. Specific guidelines for food-handling equipment must be followed while the designing of robots [22, 23]. Ideally, stainless steel have been used for the fabrication of robotic parts having ingress protection (IP) value 67 ('6' represents the protection against dust and '7' against water) along with all visible parts, manual cleaning and accessible for inspection. The hygienic design standards rely on its end use application such as meat and fish products require higher hygienic standards than bakery products.

The AI revolution for the food sector not only used in collecting production data and analytics but also use sensor networks that monitoring the processes, generate smart communication and data sharing among different sections of processing areas and blockchains. The massive complex data can be stored onto the cloud in a simpler form, which can be use in the smooth functioning of various operations in food industry. In addition they can be also used to monitor the processing machines and implement AI driven prognostic algorithms to drive preventive maintenance strategies These installations not only improve the efficiency and security of food supply chain (from farm to fork) but also enables the product traceabil ity. The most important use of digital revolution is that data is not a by-product, it is value addition through automation and should be considered as unique selling point.

14.6 Applications of Artificial Intelligence in Food Processing

The requirement of AI in food processing attract the researchers to focus and modify the available data [24]. Enough data is available for scanning with regard to computer intelligence for predicting food quality.

Massachusetts Institute of Technology (MIT)'s AI system in 2016 was able to identify the nutritional values and ingredients associated with food. After a few months, this technology reached to the public domain in the form of mobile applications [25]. These AI not only takes part during food processing and maintaining quality but also help food sector in advertising for its efficient planning and effective marketing. AI makes the food processing more versatile and its application from distinguishing simple issues of tomato and apples to more complex tasks of identify the type of fatty acids more simplified, in particulars.

14.6.1 Sorting Fresh Produce

One of the biggest challenges faced by food processing plants relates to the irregular availability of feedstock. Food processing plants depend on manual sorting to screen and sort vegetables, leading to loss of efficiency and increased costs. By using artificial intelligence, food-processing companies can achieve significant automation for food cataloguing. It uses a combination of cameras, lasers and machine learning to enable food sorting with enhanced efficiency. For instance, deploying artificial intelligence involving sensor-based optical sorting solutions thus takes less time, produces higher yield with better quality, and less wastage of food. AI is used as a means to better calibrate machines in order to manage several product sizes and reduce waste and costs [1].

14.6.2 Quality Assessment

14.6.2.1 Using AI Methods

AI based computational methodologies such as artificial neural network (ANN), evolutionary computing (EC), fuzzy logic and their integrations are mainly used for those systems, which are not easy to understand, dynamics in progress with uncertain knowledge. ANNs utilization is considered a very important part of food processing for implementing those tasks that are challenging for the human brain on regular basis. ANNs provides faster processing of distorted and noisy data with high variation and lowers the cost of production [26].

EC is simple, domain free, probabilistic and powerful approach and used for generally solving simple optimization problem in food science and technology. Researches revealed that EC algorithms in the beverage industry are successfully derive the optimal or near-optimal solutions of problems with short computational time as shown in Table 14.1 [27].

Table 14.1 Artificial intelligence applications in beverage processing industry.

Beverage	**AI used**	**Aim**	**Network type**	**Observations**	**References**
Hot beverages					
Tea	ANN	Tea classification	Multilayer, back-propagation	R^2 of 0.999, relative error of 0.07	[32]
		Quality evaluation	Back-propagation multilayer Perceptron	Classification rate 50-78%	[33]
		Differentiate tea varieties	Three layers, back-propagation	Discrimination accuracy of 77%	[34]
		Optimization of fermentation time	Recurrent Elman network	Mean percentage error of 0.15%	[35]
		Quality estimation	Multilayer feed-forward, back-propagation	Prediction with no error	[32]
		Total free amino acid	Multilayer perception, back-propagation	Cumulative reliability of 99%	[36]
	Evolutionary Algorithms	Measurement of total anthocyanins content	-	R^2 of 0.859, root mean square error of 0.144	[37]
		Customer preference modelling	-	Mean absolute error of 4.38%	[38]

(*Continued*)

Table 14.1 Artificial intelligence applications in beverage processing industry. (*Continued*)

Beverage	AI used	Aim	Network type	Observations	References
		Identification of samples	-	Accuracy of 97.29%	[39]
Coffee	ANN	Classify genotypic origin and geographical	Radial-basis function network	Performance, genotypically 94% and geographically 100%	[40]
Milk	ANN	Quality control, detection of adulteration	Multilayer perceptron	98% classification rate	[41]
		Crossflow milk ultrafiltration simulation	Feed forward multilayer perceptron	Prediction error less than 3.6%	[42]
		Estimate sensory quality	Multilayer feed-forward	R^2 of 0.97 with root mean square of 4.20%	[43]
		Estimate responses of fat content and lactose crystallinity	General regression neural network, four layers	R^2 value of 0.72	[44]

(*Continued*)

Table 14.1 Artificial intelligence applications in beverage processing industry. (*Continued*)

Beverage	AI used	Aim	Network type	Observations	References
	Evolutionary Algorithms	Optimization of probiotic viability	-	Effective in activity probiotic cultures optimization	[45]
		Conduct supply chain issues	-	No difference in small scale problems, reducing cost of supply chain	[46]
	Fuzzy Logic	Elucidate the ingredients acceptability	-	Final rank as taste >mouth feel >aroma >color	[47]
Alcoholic beverages					
Alcoholic beverages	ANN	Modelling bubble pressure and concentration of Congeners	Back propagation feed forward, two hidden layers	Absolute average deviations of 4.9%	[48]
Beer	ANN	Volatile compounds identification	Kohonen neural network	Successful classification	[49]
		Ester identification	Multilayer feed-forward	Accuracy of prediction 1%	[50]

(*Continued*)

Table 14.1 Artificial intelligence applications in beverage processing industry. (*Continued*)

Beverage	AI used	Aim	Network type	Observations	References
	Evolutionary Algorithms	Optimization of beer game order policy	-	Best performance when sectors have different order policies, optimal (αS, β) around (1,0)	[51]
		Sensorial evaluation	-	Selection of 11 and 15 chromatographic peak areas, for bitterness and grain taste,root mean square error of cross validation of 0.5	[52]
Wine	ANN	Classification of wine Samples	Multilayer perceptron	100% recognition rate	[53]
		Classification of white varietal wines	Multilayer perceptron, quasi-Newton propagation algorithm	97% performance	[54]
		Quality assessment	Kohonen neural network, non-supervised training	100% classification	[55]

(*Continued*)

Table 14.1 Artificial intelligence applications in beverage processing industry. (*Continued*)

Beverage	AI used	Aim	Network type	Observations	References
	Evolutionary Algorithms	Feature selection	-	Correct classification rate of 94.19%	[56]
		Wine Determination of pesticides and metabolites	-	Root mean square error of prediction values below 0.1, variable, Reduction >80%	[51]
		Wine Incident handling		88% average similarity value	[57]
	Fuzzy Logic	Classification of aged wine Distillates	-	Prediction error $\leq$ 10%	[58]
		Wine Aggregation of INDIGO Variables	-	Adapted four indicators	[59]
		Wine Classification of rice wine samples	-	0.986 performance	[60]
Whisky	ANN	Classification of products	Probabilistic neural network, feed-forward	0.999 accurate	[61]

(*Continued*)

Table 14.1 Artificial intelligence applications in beverage processing industry. (*Continued*)

Beverage	AI used	Aim	Network type	Observations	References
Cider	ANN	Classification based on polyphenol content	Multilayer feed-forward, one hidden layer	100% performance	[62]
Non-alcoholic beverages					
Cola	ANN	Recognition of odor	Back-propagation, Levenberg-Marquardt algorithm	86% performance	[63]
Coca-Cola	Evolutionary Algorithms	Generate branded product design based on designer preferences	-	Best solution found after 1000 generations, successful in delivering product shape	[64]
Soft drink	Evolutionary Algorithms	Solve a two-level production problem	-	Production capacity is reduced by 25%, outperform previous evolutionary methods	[65]
Apple juice	ANN	Classification of commercial juice	Multilayer feed-forward, different topologies	Accuracy between 25-100%	[66]

(*Continued*)

Table 14.1 Artificial intelligence applications in beverage processing industry. (*Continued*)

Beverage	AI used	Aim	Network type	Observations	References
Fruit juice	ANN	Prediction of physical Properties	Feed-forward multi-layered Perceptron	*R*2 value of 0.93, root mean square error 0.042	[67]
Orange juice	Evolutionary Algorithms	Reduction of significant Components	-	Success in detecting most significant components, perfect match with three components	[68]
Other beverages					
Vinegar	ANN	Discrimination of vinegars	Three-layer, feed-forward	100% classification rate	[69]
		Identify authenticity	Multilayer, back-propagation	88.6% accuracy	[70]
Soda	ANN	Simulation of pulping Process	Multilayer feed-forward	0.1% mean error	[71]

Where, **ANN**-Artificial neuron network.

The predictive power of regression models may increase using EC as an allied tool for selecting variables subsets [28]. This helps to generate partial least squares and makes the model efficient by lowering down the wavelengths with accuracy, optimization and select the informative models for creating discriminant models [29, 30].

Fuzzy logic accesses the models, which can simultaneously correlate the sensory characteristics and preference of consumers. Higher the number of decision parameters, the fuzzy logic rules become more complex when there is the presence of higher decision parameters. Fuzzy logic plays an important part in food processes and biotechnological functions by successfully using the uncertainty matters [31].

14.6.2.2 Using Integrated Computer Vision-AI System

1. Color

The color is known to as an important parameter, which discloses the food's freshness, wholesomeness and maturity indices of the food and its products. The color parameter is not only used to sort the harvested produce but also check the quality of finished goods using modified technique called integrated computer vision-AI system. So, it is very important to reduce the sorting time on the basis of color and thus required continuous measurement of pixel intensity in red, green, and blue (RGB) using a scanner or digital camera. Sometime, RGB has to convert into L*a*b* values for comparison with colorimeter data.

On the basis of color, meat can be differentiated into two categories named "red meat" which is usually observed in lamb, pork, beef, and "white meat" obtained from fish, turkey, chicken and other birds. As an appropriate indicator, color models are generally used for red meats as shown in Table 14.2. It is done because carcasses of larger animals are required to be cut into pre desired chunks for next unit operations. On the other hand, whole animal is used for processing in order to obtain white meat and therefore the meat shows no muscle color. Salmon have coloured flesh among some of the fishes, and thus it reflects in diet after processing and may affect its quality.

The color of fruit imitates the maturity and quality. The color values may deviate from its standard color value. This difference in the values might be because of the presence of any bruises or scratches or fruit is under ripened or over ripened. Thus, change in color of fruit are something that can be estimated. The average color values were reported in the form of L*a*b* values for various cuts of fruits such as banana, pumpkin, kiwifruit and pear [72–75].

Table 14.2 Computer vision systems application for the evaluation of color in food variants.

Products	Color values	Observations	Reference
Meat			
Beef muscle	Histograms of red, green, and blue (RGB)	Color on its own was ineffective, but it made a valuable contribution to the final models	[81]
Salmon muscle	Mean RGB values	Pigment and fat content strongly correlated ($R^2 = 0.84$)	[82]
Salmon fillets	RGB values	Identified fat linings within the muscles and strong correlated ($R^2 = 0.84$)	[83]
Fish	Hue, saturation, and intensity (HSI) colorspace	96% correct classification	[84]
Trout flesh	L*C*h* (luminance, chromacity, and hue) colorspace	RGB colorspace is not suitable but similar to HSI colorspace	[85]
Fish flesh	L*a*b* (L*-Lightness, a*-Red/Green value, b*-Blue/Yellow value) and L*C*h*	Identified fat and correlated ($R^2 = 0.78$).	[86]

(*Continued*)

Table 14.2 Computer vision systems application for the evaluation of color in food variants. (*Continued*)

Products	**Color values**	**Observations**	**Reference**
Salmon flesh	RGB data into L*a*b*	Data conversion error ≤ 1%	[87]
Salmon flesh	RGB data into L*a*b*	Errors ≤ 2.2 color units	[88]
Fruit and vegetables			
Starfruit	HSI colorspace with hue histogram	100% classifications	[89]
Pomegranate arils	RGB thresholds	90% classification	[90]
Tomato	RGB thresholds	R^2> 0.95	[91]
Potato chips	L*a*b* thresholds	Chemically evaluated acrylamide content with correlations (0.90-0.92)	[76]
Other food products			
Bread	Mean L*a*b* values	Error (<4%)	[92]
Coffee beans	RGB thresholds	(R^2> 0.98)	[77]

In vegetable, potato is widely used in the preparation of snack products in food processing industry. Potato chips are usually manufactured by carrying out various processing operations like washing, peeling, slicing, blanching, frying etc. The color of potato chips varied from golden brown to dark brown during processing. During frying, acrylamide formation (reaction between natural sugars and amino acids) also affect the color of the potato chips. The excessive absorption of fat in chips shows the development of clear areas. It was reported that acrylamide formed after browning of chips lead to lower lightness (L*) and higher (a*) and (b*) values [76].

The monitoring of changes in coffee beans range from green to dark brown color during roasting and is very important. Excessive roasting may cause burning which degrades its acceptability of coffee beans [77]. AI based computer vision techniques uses model setups, which automatically recorded pixels and sense the desired bean color as per optimum requirements. Color for other beans, which is also dependent on processing, were also predicted using AI based computer vision technique. It was possible to obtain more than 90% of specific beans using variance, skewness, and mean of RGB values [78].

It had been reported that moisture content is also correlated with the color properties of various foods. During cereal grains processing, color parameters has been used to predict the moisture content. Algorithms were used correctly to classify 98% cereal grains and estimated moisture content using color parameters [79]. Outer skin of hazelnuts has dark brown skin color than that of white color flesh or pulp (white color). It is difficult to separate pulp from peel because of the difference in color. The outer skin is shredded after roasting followed by peeling which can be differentiated from pulp using RGB values in a computer vision system [80].

2. Texture

Texture is another important sensory property that defines the food quality. During cutting, carcass of red meat, the cut surface shows a flat muscle. AI-computer vision system was used for the estimation of high quality 2-D textural surface, which screened the muscle fibers' geometry and distribution of connective tissues. These results illustrate the flavour, juiciness, tenderness and acceptability of the cooked meat.

The cut in processed meat ham having at least one muscle resembles to the butcher cut of a carcass and thus retains its important textural properties. Overall porosity, roughness and heterogeneity of pork ham has been observed using fractional dimensions and lacunarity functions for its textural properties [93]. These systems have substantial functions to distinguish the quality of ham. The low-quality ham possesses rougher surface as compared to high quality ham which have smoother surface. Fractal dimensions has been also used to observe the changes in micro texture of shrimp as shown in Table 14.3. It is related with the amount of protein present as different proteins shows different characters under different temperature conditions [94]. Strong relations have been observed between the measured protein content by chemical method and the normalized fractal dimension model.

In fruits, fractal dimensions are also used to reflect the textural properties. Browning in banana is directly related to its texture. Fractal dimensions models was used to enlighten this relation [72]. By using these models, researchers found roughness screened on yellow and brown regions of banana flesh. Fractal dimensions model with computer vision system have been used on pears for assessing browning [74]. The change brown spots on fruits is correlated to fractal dimensions and changes accordingly.

The textural characteristics (Heywood shape factor, moisture ratio, and overall color change) of dried pumpkin flesh were observed using wavelet

Table 14.3 Computer vision systems application for the evaluation of texture in food variants.

Products	**Texture expressed**	**Observations**	**Reference**
Meat			
Beef	Wavelet transforms	R^2> 0.8, standard error < 0.333)	[100]
Lamb	Algorithms of grey level histograms	97% accuracy (six grades can be possible)	[101]
Shrimp	Fractal dimensions	Strong correlations	[94]
Fruit and Vegetables			
Banana	Fractal dimensions	Roughness can be well suited	[72]
Dried pumpkin flesh	Wavelet transform	Accurate models (R^2<1)	[73]
Walnut	Local variance and local binary pattern	Accurate classification (98%)	[95]

transform model [73]. In walnut processing industry, the separation of walnut fragments from its pulp is very tedious and important task. Therefore, automation in this sector is really important to retain its quality as labour cost is very high [95].

The formulation of ingredients and baking process affects the texture of bread. These were identified by two-dimensional and three-dimensional images, which were used to assess the fineness and cell size of bread crumb [96, 97]. In chocolate processing industry, surface color and glossiness is correlated to its texture [98]. It has been reported that glossiness and lightness of chocolate surface decreases exponentially with an increase in roughness. Computer vision-based AI are also used in milling industry in order to examine the internal composition of wheat grains and it is reported that whether the grains are vitreous and non-vitreous in nature [99].

3. Shape and Size

Brining of meat is done for preservation of meat muscle, as meat is a perishable commodity and can be spoiled by microorganisms in short period of time. Pore density and area in which pores are distributed is known to have an important quality parameter for ham.

The surface defects in olive can be clearly evaluated by studying morphological properties [102]. Olive can be classified on the basis of their defects as (i) serpeta (mussel scale); (ii) granizo (damaged from hail); (iii) agostado (wrinkled olive); (iv) molestado (undefined damage); (v) purple olive damage. These damages could be differentiated with 100% accuracy using computer vision as shown in Table 14.4.

Computer vision (CV) technique is also used to measure the quality of cooked rice. Rice grain change its morphology and expand during cooking. The expansion rate of rice grains depends on the various parameters like length, breadth, height, area and water absorption. Rice kernels having higher moisture absorption can affects the bearing strength and yield cracked kernels. These cracked grains and irreversibly affects the quality of the finished product. The cooking behaviour of rice for water uptake is modelled using peleg and exponential models [103].

14.6.3 Flavor Identification

Flavor is a perception of both taste as well as aroma. Only skilled and professionals can able to detect the specific sensorial properties. It is time consuming and costly and variable process as the senses can be altered by psychological, intrinsic and external factors. Thus, AI based biosensors

Table 14.4 Computer vision systems application for the evaluation of shape and size of food variants.

Products	Expressed by	Observations	Reference
Meat			
Ham	Magnetic resonance imaging (MRI) with B-spline curves	Identified muscle shrinking with strong correlation ($R^2 = 0.99$).	[104]
Fish	MRI	Classified accuracy (90%)	[105]
Fruits and vegetables			
Olive	Backlighting silhouettes	100% correctly classified	[102]
Date	Near-infrared (NIR) spectroscopy	Classification rates; offline (87%) whereas online (79%)	[106]
Apple	Support vector machine	Calyx and stems classification were 100% and 99% respectively	[107]
Srichompoo pods	Three-dimensional wire frame models	Correct classification (94%)	[108]

such as E-Nose and E-Tongue are used that mimic the human senses. Data obtained from these sensing modules are more logical, consistent, accurate and precise. The signals from receiver are directly read or statistically treated further using artificial neural networks (ANNs) or feed forward neural network (FNN) and fuzzy logic. The quantification of human linguistic expression can be done by using fuzzy logic and thus used in sensorial tests. For particular purpose, fuzzy sets can be applied with currently available technique such as ANN and FNN. ANN supports in the measurement of nonlinear relationships with the fuzzy sets and make it more effective in the analysis of sensory tests. The combining effect of fuzzy sets with FNN and ANN is an effective way to read nonlinear relationships and sensory data having fuzziness.

Electronic noses (E-nose) systems have been not only used in food classification but also equally important in the application of quality assessment of those, which generally have strong smell. AI based biosensors attract the researchers to predict the sensory properties, shelf life, physicochemical and ripeness of various fruits [109–111]. Research reveals that the use of ANNs in E-nose can be more effective and productive for data interpretation as compared to earlier used methods [109, 112].

E-nose system was used to predict the fermentation time of the sausage. The sensory response from these AI based biosensors used both ANN and linear regression methods [113]. The result of these models has been compared with principal component analysis for the scoring of sensory parameters and ANN expression yielded the optimum fermentation time as compare to other methods. The performance of E-nose biosensors can be more effective with the integration of nonlinear ANNs for assessing food quality.

14.6.4 Drying Technology

In drying technologies, genetic algorithms and artificial neural networks are very useful for optimization and assessing the drying process. However, in drying, fuzzy logic has capabilities of both control and classification. The drying of carrot (*Daucus carota*) and narrow-leaved purple coneflower (*Echinacea angustifolia*) are dried using fuzzy logic system approach in order to control the parameters in convective type dryer [114]. Genetic type algorithms are used for computing more complex operations, such as Arrhenius type modeling during drying. A convective drier using fuzzy logic-based control concluded the robustness of the fuzzy control as compared to proportional integral derivative (PID) type control technique. Consequently, AI techniques have various benefits in terms of stability, accuracy and computation time.

14.6.5 Food Safety Compliance

AI enabled cameras are used to ensure safety compliance for individual hygiene food processing industry. AI embedded with object-recognition and facial-recognition software, which recognizes whether workers are complying with good personal hygiene or not as per standard guidelines. If violation is found, it extracts the screen images for review, which can be rectified in the real time. The accuracy of this technology is more than 96% [1].

14.6.6 Cleaning Food Processing Equipment

Cleaning in place systems are programmed to clean equipment in timed cycles. This limits human intervention, which in turn limits the chances of cross contamination from food borne pathogens, but this system operates blindly and are designed for worse case scenarios. Self-optimizing clean in place system (SOCIP) is an AI enabled technology which optimizes the cleaning process using optical fluorescence imaging and ultrasonic sensing that measure microbial debris and foreign residues in the equipment. This leads to savings on water, time and energy. The cleaning time is reduced by 50% using the SOCIP method [1, 115].

14.6.7 Efficient Supply Chain Management

Supply chain management is top priority for all food companies with the increasing need for transparency. The food industry is using AI to improve supply chains through food safety monitoring and testing product at every step of the supply chain to ensure compliance to industry and consumer specifications. More accurate forecasting to manage pricing and inventory [1]. AI-based image recognition technologies enable better and more efficient procurement of produce. AI also helps in efficient and transparent tracking of produce from farm to the consumer, leading to increased confidence amongst consumers.

14.6.8 Anticipating Consumer Preferences

Food manufacturers use artificial intelligence-based solutions that enable them to clearly anticipate and model the flavour preferences of their target consumers, and predicting their response to such new tastes. Such AI-based predictive analytics will help food manufacturers in developing new food products that are closely aligned with consumer tastes and preferences. Kellogg launched AI-enabled technology, which helps the customer to decide with granola should be chosen from 50 ingredients for making a particular customized product [116]. The general data obtained from flavor combinations, information about selections made by consumers, and combinations that consumers are able to create a feedback loop refining the kind of flavours actually liked by them.

14.6.9 Developing New Products

Machine learning enabled with AI technology provide predictive algorithms that modulate the consumer preferences and predict their responses towards new product. The segmented data used by the companies in order to develop new products that correlates with the consumer preferences [1]. With these, manufacturers can know what products will produce before they are available into the market.

14.7 Challenges

Artificial intelligence (AI) is considered as one of the biggest digitalization techniques that change the industrial working process. It has enormous potential and applications but apart from this, AI has been attached with several challenges. Bias in AI system is one of the biggest challenges and are likely to become an increasingly widespread problem in many industries throughout the world. The AI affiliated food processing industries have faced computing power challenges, which is used to carry large number of data. Obtaining and funding such level of data in computing power can be challenging for food entrepreneurs particularly start-ups [117]. The wide volume of data that can be collected in the form of text, sound, images, and videos adds many challenges in finalizing the results. Due to this, AI emerging technology require necessary skills or training in order to develop compatible AI software with varied food processing zones and able to understand the final results. Legal aspects in AI is one of the modest concerns in the food industry. AI collects and store all the confidential data in a cloud, which might be violate the federal or state laws. Organization needs to secure their data with using AI models into their system even if the data is not illegal. Clearly, food manufacturing revolution is just at the beginning and there is still plenty of work to do, but this is a challenge that we all have to take on through a strong, cross-fertilising synergy between end-users, technology providers and innovators.

14.8 Future Aspects

The future of AI reduces the dependency of manual work and enhance the working profile in industrial sectors. AI sectors growing its branches but still it is less than the human intelligence. The future of AI in next forty years is shown in Figure 14.5. Artificial intelligence today is accurately

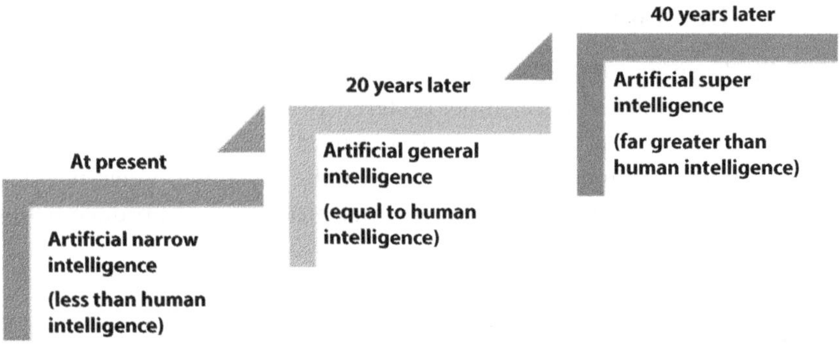

Figure 14.5 Future trends of artificial intelligence.

known as narrow AI (or weak AI) which depends on facial recognition, internet searches etc. However, artificial general intelligence (AGI or strong AI) possesses long-term goal so that machine can able to rectify to any problem and as smart as a human intelligence. The main hypothetical goal in next forty years is to achieve artificial superintelligence (ASI) which is far surpassing than that of human intelligence. Due to recursive self-improvement, superintelligence is expected to be a rapid outcome of creating artificial general intelligence [118].

The trend of AI in food processing positively maximize the efficiency of food quality standards and ensures safety. To apply AI in food processing, the develop model need to be scale up and tested for their effective implementation. The science of food and artificial intelligence require combinations using mathematical tools in order to get acceptable results. Hybrid intelligence system have shown tremendous applications in non-food sectors as compared to single methods. Researchers are working on hybrid intelligence to challenge the drawbacks of existing AI systems in food science and technology. Hence, the application of these systems is to operate the various intelligent system from single point for tracking the best results by utilizing small size data, types of methodologies and limited installed software in short period of time.

Conclusions

This chapter highlights the various applications and future requirement of AI in food sectors. Automatic operating procedures change the processing trends to reflect more reliable automated control during processing of food. Several literatures demonstrate the; advantages of AI, fulfilment of required expert advice in an intelligence form, independent from human expressions and errors, verified results similar to human intelligence, twenty-four hours availability and continuous processing, no need of personnel hygiene requirement, quick access multiple data for optimum solution, simple coded language. AI techniques utilizes non-linear working environment, which are relevant to food processing, provide detailed data and used mainly in smoothening the process performance.

AI has proven its efficacy in various sectors related to food science and technology and plays a significant role in food processing line. Hence, attraction towards the use of AI and AI integrated systems can be set up instantly in food science. Government of developing and developed countries are investing application of AI in various sectors related to agro processing. Already intelligent systems for prediction, optimization, automation, classification and decision-making have been developed at research level and some are already implemented in commercial scale. ANNs has been widely used AI technique among all other techniques as it can be used over a broad range of applications as well as prediction of data in food processing. It involves multiple parallel processing, fault tolerance, flexibility and adaptivity of results during running algorithms, hidden layers and activation functions, which makes it critical tool in processing lines. EC based AI models has ability to convert normalise production process into profitable production process. Therefore, it is widely used in solving optimization problems in many food sectors. Fuzzy logic-based AI models precisely addresses the uncertainties observed in food manufacturing line. The reported applications of fuzzy logic can be categorized into process performance assessment, classification, control and optimization. Robotics, biosensors and computer vision-based AI system allow the fast analysis at low cost and accurate results, which mimics human interpretative and descriptive approach. AI revolution in food sector will not only attract novel ideas from researchers but also become a boon in upcoming industrial revolution.

References

1. Agbai, C.M., Application of artificial intelligence (AI) in food industry. *GSC Biological and Pharmaceutical Sciences*. 13, 1, 171-178, 2020.
2. Goyache, F., Bahamonde, A., Alonso, J., López, S., Del Coz, J.J., Quevedo, J.R., Diez, J., The usefulness of artificial intelligence techniques to assess subjective quality of products in the food industry. *Trends in Food Science & Technology*, 12, 10, 370-381, 2001.

3. Aayog, N.I.T.I., National strategy for artificial intelligence. *Paper*, June, 2019-01, 2018.
4. Volter, M., Stahl, T., Bettin, J., Haase, A., Helsen, S., *Model-driven software development: technology, engineering, management.* John Wiley & Sons, 2013.
5. McCarthy, J. and Feigenbaum, E.A., In memoriam: Arthur samuel: Pioneer in machine learning. *AI Magazine*, 11, 3, 10-11, 1990.
6. Samuel, A.L., Some studies in machine learning using the game of checkers. *IBM Journal of research and development*, 44, 1.2, 206-226, 2000.
7. Chen, S.H., Jakeman, A.J., Norton, J.P., Artificial Intelligence techniques: An introduction to their use for modelling environmental systems. *Mathematics and Computers in Simulation*, 78, 2-3, 379-400, 2008.
8. Quinn, J., Frias-Martinez, V., Subramanian, L., Computational sustainability and artificial intelligence in the developing world. *AI Magazine*, 35, 3, 36-47, 2014.
9. Ahmad, T., Rehman, N.A., Pervaiz, F., Kalyanaraman, S., Safeer, M.B., Chakraborty, S., Subramanian, L., Characterizing dengue spread and severity using internet media sources. In *Proceedings of the 3rd ACM Symposium on Computing for Development*, 1-2,2013.
10. Hossain, M.S., Zander, P.O., Kamal, M.S., Chowdhury, L., Belief-rule-based expert systems for evaluation of e-government: a case study. *Expert Systems*, 32, 5, 563-577, 2015.
11. Yang, H.H., Liu, J.L., Chang, M.C.S., Yang, J.C., Improvement of e-government service process via a grey relation agent mechanism. *Expert Systems with Applications*, 39, 10, 9755-9763, 2012.
12. Magoutas, B., and Mentzas, G., Salt: A semantic adaptive framework for monitoring citizen satisfaction from e-government services. *Expert Systems with Applications*, 37, 6, 4292-4300, 2010.
13. Quinn, J., Computational techniques for crop disease monitoring in the developing world. In *International Symposium on Intelligent Data Analysis*, Springer, Berlin, Heidelberg, 2013.
14. World Health Organization., The state of food security and nutrition in the world 2018: building climate resilience for food security and nutrition. *Food & Agriculture Organization*, 2018.
15. UNICEF., The state of food security and nutrition in the world 2020. *Transforming Food Systems for Affordable Healthy Diets*, 2020.
16. Tiwari, S. and Zaman, H., The impact of economic shocks on global undernourishment. *World Bank Policy Research Working Paper*, 5215, 2010.
17. Kakani, V., Nguyen, V.H., Kumar, B.P., Kim, H., Pasupuleti, V.R., A critical review on computer vision and artificial intelligence in food industry. *Journal of Agriculture and Food Research*, 2, 100033, 2020.
18. Gray, J.O. and Davis, S.T., Robotics in the food industry: an introduction. *In Robotics and Automation in the Food Industry*, Woodhead Publishing, 2013.
19. Kyrylovych, V., Dimitrov, L., Melnychuk, P., Bohdanets, A., Shostachuk, A., Development of a new method for automated selection of robotic

mechanic-assembly technologies based on the technical and economic criteria. *Eastern-European Journal of Enterprise Technologies*, 6, 1, 6-18, 2018.

20. Chen, N., Christensen, L., Gallagher, K., Mate, R., Rafert, G., Global economic impacts associated with artificial intelligence. *Analysis Group*, 1-23, 2016.
21. Masey, R.J.M., Gray, J.O., Dodd, T.J., Caldwell, D.G., Guidelines for the design of low-cost robots for the food industry. *Industrial Robot: An International Journal*, 37, 6, 509-517, 2010.
22. Lelieveld, H. and Keener, L., Global harmonization of food regulations and legislation-the Global Harmonization Initiative. *Trends in Food Science & Technology*, 18, S15-S19, 2007.
23. Lelieveld, H.L.M., Mostert, M.A., Hola, J. and White, B., *Hygiene in Food Processing*. Woodhead Publishing, Cambridge, UK, 2003.
24. Kagaya, H., Aizawa, K., Ogawa, M., Food detection and recognition using convolutional neural network. *InProceedings of the 22nd ACM international conference on Multimedia*, 1085-1088, 2014.
25. Yanai, K. and Kawano, Y., Food image recognition using deep convolutional network with pre-training and fine-tuning. *IEEE International Conference on Multimedia & Expo Workshops (ICMEW)*, 1-6, 2015.
26. Hosu, A., Cristea, V.M., Cimpoiu, C., Analysis of total phenolic, flavonoids, anthocyanins and tannins content in Romanian red wines: Prediction of antioxidant activities and classification of wines using artificial neural networks. *Food chemistry*, 150, 113-118, 2014.
27. Noroozi, A., Mahdavi Mazdeh, M., Rasti-Barzoki, M., Coordinating order acceptance and batch delivery for an integrated supply chain scheduling. *International Journal of Engineering*, 30, 5, 700-709, 2017.
28. Buratti, S., Ballabio, D., Benedetti, S., Cosio, M.S., Prediction of Italian red wine sensorial descriptors from electronic nose, electronic tongue and spectrophotometric measurements by means of Genetic Algorithm regression models. *Food Chemistry*, 100, 1, 211-218, 2007.
29. Galyanin, V., Melenteva, A., Bogomolov, A., Selecting optimal wavelength intervals for an optical sensor: A case study of milk fat and total protein analysis in the region 400–1100 nm. *Sensors and Actuators B: Chemical*, 218, 97-104, 2015.
30. Caredda, M., Addis, M., Ibba, I., Leardi, R., Scintu, M.F., Piredda, G., Sanna, G., Building of prediction models by using mid-infrared spectroscopy and fatty acid profile to discriminate the geographical origin of sheep milk. LWT, 75, 131-136, 2017.
31. Birle, S., Hussein, M.A., Becker, T., Fuzzy logic control and soft sensing applications in food and beverage processes. *Food Control*, 29, 1, 254-269, 2013.
32. Cimpoiu, C., Cristea, V.M., Hosu, A., Sandru, M., Seserman, L., Antioxidant activity prediction and classification of some teas using artificial neural networks. *Food chemistry*, 127, 3, 1323-1328, 2011.

33. Bhattacharyya, R., Tudu, B., Das, S.C., Bhattacharyya, N., Bandyopadhyay, R., Pramanik, P., Classification of black tea liquor using cyclic voltammetry. *Journal of Food Engineering*, 109, 1, 120-126, 2012.
34. Li, X. and He, Y., Discriminating varieties of tea plant based on Vis/NIR spectral characteristics and using artificial neural networks. *Biosystems engineering*, 99, 3, 313-321, 2008.
35. Ghosh, S., Tudu, B., Bhattacharyya, N., Bandyopadhyay, R., A recurrent Elman network in conjunction with an electronic nose for fast prediction of optimum fermentation time of black tea. *Neural Computing and Applications*, 31, 2, 1165-1171, 2019.
36. Guo, Z., Chen, L., Zhao, C., Huang, W., & Chen, Q., Non-destructive estimation of total free amino acid in green tea by near infrared spectroscopy and artificial neural networks. *In International Conference on Computer and Computing Technologies in Agriculture*. Springer, Berlin, Heidelberg, 2011.
37. Xiaowei, H., Xiaobo, Z., Jiewen, Z., Jiyong, S., Xiaolei, Z., Holmes, M., Measurement of total anthocyanins content in flowering tea using near infrared spectroscopy combined with ant colony optimization models. *Food chemistry*, 164, 536-543, 2014.
38. Chan, K.Y., Kwong, C.K., Law, M.C., A fuzzy ordinary regression method for modeling customer preference in tea maker design. *Neurocomputing*, 142, 147-154, 2014.
39. Tudu, B., Ghosh, S., Bag, A. K., Ghosh, D., Bhattacharyya, N., Bandyopadhyay., Incremental FCM technique for black tea quality evaluation using an electronic nose. *Fuzzy Information and Engineering*, 7, 3, 275-289, 2015.
40. Link, J.V., Lemes, A.L.G., Marquetti, I., dos Santos Scholz, M.B., Bona, E., Geographical and genotypic classification of arabica coffee using Fourier transform infrared spectroscopy and radial-basis function networks. *Chemometrics and Intelligent Laboratory Systems*, 135, 150-156, 2014.
41. Moncayo, S., Manzoor, S., Rosales, J.D., Anzano, J., Caceres, J.O., Qualitative and quantitative analysis of milk for the detection of adulteration by laser induced breakdown spectroscopy (LIBS). *Food chemistry*, 232, 322-328, 2017.
42. Razavi, M.A., Mortazavi, A., Mousavi, M., Application of neural networks for crossflow milk ultrafiltration simulation. *International Dairy Journal*, 14, 1, 69-80, 2004.
43. Singh, R.R.B., Ruhil, A.P., Jain, D.K., Patel, A.A., Patil, G.R., Prediction of sensory quality of UHT milk–a comparison of kinetic and neural network approaches. *Journal of Food Engineering*, 92, 2, 146-151, 2009.
44. Koc, A.B., Heinemann, P.H., Ziegler, G.R., Optimization of whole milk powder processing variables with neural networks and genetic algorithms. *Food and bioproducts processing*, 85, 4, 336-343, 2007.
45. Chen, M.J., Chen, K.N., Lin, C.W., Optimization of the viability of probiotics in a new fermented milk drink by the genetic algorithms for response surface modeling. *Journal of food science*, 68, 2, 632-638, 2003.

46. Sadjadi, S.J., Jafari, M., Amini, T., A new mathematical modeling and a genetic algorithm search for milk run problem (an auto industry supply chain case study). *The International Journal of Advanced Manufacturing Technology*, 44, 1, 194-200, 2009.
47. Tahsiri, Z., Niakousari, M., Khoshnoudi-Nia, S., Hosseini, S.M.H., Sensory evaluation of selected formulated milk barberry drinks using the fuzzy approach. *Food science & nutrition*, 5, 3, 739-749, 2017.
48. Faúndez, C.A., Quiero, F.A., Valderrama, J.O., Phase equilibrium modeling in ethanol+ congener mixtures using an artificial neural network. *Fluid Phase Equilibria*, 292, 1-2, 29-35, 2010.
49. da Silva, G.A., Augusto, F., Poppi, R.J., Exploratory analysis of the volatile profile of beers by HS-SPME-GC. *Food Chemistry*, 111, 4, 1057-1063, 2008.
50. Riverol, C. and Cooney, J., Estimation of the ester formation during beer fermentation using neural networks. *Journal of food engineering*, 82, 4, 585-588, 2007.
51. Carneiro, R.L., Braga, J.W., Bottoli, C.B., Poppi, R.J., Application of genetic algorithm for selection of variables for the BLLS method applied to determination of pesticides and metabolites in wine. *Analytica Chimica Acta*, 595, 1-2, 51-58, 2007.
52. da Silva, G.A., Maretto, D.A., Bolini, H.M.A., Teófilo, R.F., Augusto, F., Poppi, R.J., Correlation of quantitative sensorial descriptors and chromatographic signals of beer using multivariate calibration strategies. Food chemistry, 134, 3, 1673-1681, 2012.
53. Penza, M. and Cassano, G., Chemometric characterization of Italian wines by thin-film multisensors array and artificial neural networks. *Food Chemistry*, 86, 2, 283-296, 2004.
54. Kruzlicova, D., Mocak, J., Balla, B., Petka, J., Farkova, M., Havel, J., Classification of slovak white wines using artificial neural networks and discriminant techniques. *Food Chemistry*, 112, 4, 1046-1052, 2009.
55. Santana, J.C.C., Dias, C.G., de Souza, R.R., Tambourgi, E.B., Applying of neural network on the wine sensorial analysis from barbados cherry. *Journal of Food Process Engineering*, 33, 365-378, 2010.
56. Beltrán, N.H., Duarte-Mermoud, M.A., Bustos, M.A., Salah, S.A., Loyola, E.A., Peña-Neira, A.I., Jalocha, J.W., Feature extraction and classification of Chilean wines. *Journal of Food Engineering*, 75, 1, 1-10, 2006.
57. Lam, H.Y., Choy, K.L., Ho, G.T., Kwong, C.K., Lee, C.K. A real-time risk control and monitoring system for incident handling in wine storage. *Expert Systems with Applications*, 40, 9, 3665-3678, 2013.
58. Raptis, C.G., Siettos, C.I., Kiranoudis, C.T., Bafas, G.V., Classification of aged wine distillates using fuzzy and neural network systems. *Journal of food Engineering*, 46, 4, 267-275, 2000.
59. Thiollet-Scholtus, M. and Bockstaller, C., Using indicators to assess the environmental impacts of wine growing activity: The INDIGO® method. *European Journal of Agronomy*, 62, 13-25, 2015.

60. Jha, P., Das, A.J., Dash, K.K., Deka, S.C., Sensory evaluation of black pigmented rice (*Oryza sativa cv. Poireton*) wine fortified with probiotic *Lactobacillus acidophilus* ATCC 4356 and *Lactobacillus sakei* ATCC 15521 using fuzzy logic. *Journal of the Institute of Brewing*, 121, 4, 566-573, 2015.
61. Jack, F.R. and Steele, G.M., Modelling the sensory characteristics of Scotch whisky using neural networks-a novel tool for generic protection. *Food Quality and Preference*, 13, 3, 163-172, 2002.
62. Alonso-Salces, R.M., Guyot, S., Herrero, C., Berrueta, L.A., Drilleau, J.F., Gallo, B., Vicente, F., Chemometric classification of Basque and French ciders based on their total polyphenol contents and CIELab parameters. *Food Chemistry*, 91, 1, 91-98, 2005.
63. Kermani, B.G., Schiffman, S.S., Nagle, H.T. Performance of the Levenberg-Marquardt neural network training method in electronic nose applications. *Sensors and Actuators B: Chemical*, 110, 1, 13-22, 2005.
64. Chau, M.C.A.H.H., Mckay, A., De Pennington, A.L.A.N., Combining evolutionary algorithms and shape grammars to generate branded product design. *In Design computing and cognition'06*. Springer, Dordrecht, 2006.
65. Toledo, C.F.M., de Oliveira, L., de Freitas Pereira, R., Franca, P.M., Morabito, R., A genetic algorithm/mathematical programming approach to solve a two-level soft drink production problem. *Computers & Operations Research*, 48, 40-52, 2014.
66. Gestal, M., Gómez-Carracedo, M.P., Andrade, J.M., Dorado, J., Fernández, E., Prada, D., Pazos, A., Classification of apple beverages using artificial neural networks with previous variable selection. *Analytica Chimica Acta*, 524, 1-2, 225-234, 2004.
67. Chegini, G.R., Khazaei, J., Ghobadian, B., Goudarzi, A.M., Prediction of process and product parameters in an orange juice spray dryer using artificial neural networks. *Journal of food engineering*, 84, 4, 534-543, 2008.
68. Rinaldi, M., Gindro, R., Barbeni, M., Allegrone, G., Pattern recognition and genetic algorithms for discrimination of orange juices and reduction of significant components from headspace solid-phase microextraction. *Phytochemical Analysis*, 20, 5, 402-407, 2009.
69. Gerbi, V., Zeppa, G., Beltramo, R., Carnacini, A., Antonelli, A., Characterisation of white vinegars of different sources with artificial neural networks. *Journal of the Science of Food and Agriculture*, 78, 3, 417-422, 1998.
70. Hajimahmoodi, M., Khanavi, M., Sadeghpour, O., Ardekani, M.R.S., Mazde, F.Z., Khoddami, M.S., Afzalifard, S., Ranjbar, A.M., Application of organic acid based artificial neural network modeling for assessment of commercial vinegar authenticity. *Food Analytical Methods*, 9, 12, 3451-3459, 2016.
71. Labidi, J., Tejado, A., García, A., Jiménez, L., Simulation of tagasaste pulping using soda-anthraquinone. *Bioresource technology*, 99, 15, 7270-7277, 2008.
72. Quevedo, R., Díaz, O., Ronceros, B., Pedreschi, F., Aguilera, J.M., Description of the kinetic enzymatic browning in banana (*Musa cavendish*) slices

using non-uniform color information from digital images. *Food Research International*, 42, 9, 1309-1314, 2009.
73. Zenoozian, M.S. and Devahastin, S., Application of wavelet transform coupled with artificial neural network for predicting physicochemical properties of osmotically dehydrated pumpkin. *Journal of Food Engineering*, 90, 2, 219-227, 2009.
74. Quevedo, R., Díaz, O., Caqueo, A., Ronceros, B., Aguilera, J.M., Quantification of enzymatic browning kinetics in pear slices using non-homogenous L* color information from digital images. *LWT-Food Science and Technology*, 42, 8, 1367-1373, 2009.
75. Fathi, M., Mohebbi, M., Razavi, S.M.A., Application of image analysis and artificial neural network to predict mass transfer kinetics and color changes of osmotically dehydrated kiwifruit. *Food and Bioprocess Technology*, 4, 8, 1357-1366, 2011.
76. Viklund, G., Mendoza, F., Sjöholm, I., Skog, K., An experimental set-up for studying acrylamide formation in potato crisps. *LWT-Food Science and Technology*, 40, 6, 1066-1071, 2007.
77. Hernández, J.A., Heyd, B., Trystram, G., Prediction of brightness and surface area kinetics during coffee roasting. *Journal of Food Engineering*, 89, 2, 156-163, 2008.
78. Kilic, K., Boyaci, I.H., Koksel, H., Kusmenoglu, I., A classification system for beans using computer vision system and artificial neural networks. *Journal of Food Engineering*, 78, 3, 897-904, 2007.
79. Tahir, A.R., Neethirajan, S., Jayas, D.S., Shahin, M.A., Symons, S.J., White, N.D.G., Evaluation of the effect of moisture content on cereal grains by digital image analysis. *Food Research International*, 40, 9, 1140-1145, 2007.
80. Pallottino, F., Menesatti, P., Costa, C., Paglia, G., De Salvador, F.R., Lolletti, D., Image analysis techniques for automated hazelnut peeling determination. *Food and Bioprocess Technology*, 3, 1, 155-159, 2010.
81. Jackman, P., Sun, D.W., Du, C.J., Allen, P., Downey, G., Prediction of beef eating quality from colour, marbling and wavelet texture features. *Meat science*, 80, 4, 1273-1281, 2008.
82. Folkestad, A., Wold, J.P., Rorvik, K.A., Tschudi, J., Haugholt, K.H., Kolstad, K., Morkore, T., Rapid and non-invasive measurements of fat and pigment concentrations in live and slaughtered Atlantic salmon (*Salmo salar* L.). *Aquaculture*, 280, 1-4, 129-135, 2008.
83. Stien, L.H., Kiessling, A., Manne, F., Rapid estimation of fat content in salmon fillets by colour image analysis. *Journal of Food Composition and Analysis*, 20, 2, 73-79, 2007.
84. Zion, B., Alchanatis, V., Ostrovsky, V., Barki, A., Karplus, I., Classification of guppies' (*Poecilia reticulata*) gender by computer vision. *Aquacultural Engineering*, 38, 2, 97-104, 2008.

85. Kause, A., Stien, L.H., Rungruangsak-Torrissen, K., Ritola, O., Ruohonen, K., Kiessling, A., Image analysis as a tool to facilitate selective breeding of quality traits in rainbow trout. *Livestock Science*, 114, 2-3, 315-324, 2008.
86. Stien, L.H., Manne, F., Ruohonene, K., Kause, A., Rungruangsak-Torrissen, K., Kiessling, A., Automated image analysis as a tool to quantify the colour and composition of rainbow trout (*Oncorhynchus mykiss W.*) cutlets. *Aquaculture*, 261, 2, 695-705, 2006.
87. Leon, K., Mery, D., Pedreschi, F., Leon, J., Color measurement in L* a* b* units from RGB digital images. *Food research international*, 39, 10, 1084-1091, 2006.
88. Valous, N.A., Mendoza, F., Sun, D.W., Allen, P., Colour calibration of a laboratory computer vision system for quality evaluation of pre-sliced hams. *Meat science*, 81, 1, 132-141, 2009.
89. Abdullah, M.Z., Mohamad-Saleh, J., Fathinul-Syahir, A.S., Mohd-Azemi, B.M.N., Discrimination and classification of fresh-cut starfruits (*Averrhoa carambola L.*) using automated machine vision system. *Journal of Food Engineering*, 76, 4, 506-523, 2006.
90. Blasco, J., Cubero, S., Gomez-Sanchis, J., Mira, P., Molto, E., Development of a machine for the automatic sorting of pomegranate (*Punica granatum*) arils based on computer vision. *Journal of food engineering*, 90, 1, 27-34, 2009.
91. Lana, M.M., Tijskens, L.M.M., Van Kooten, O., Effects of storage temperature and stage of ripening on RGB colour aspects of fresh-cut tomato pericarp using video image analysis. *Journal of food engineering*, 77, 4, 871-879, 2006.
92. Purlis, E. and Salvadori, V.O., Modelling the browning of bread during baking. *Food research international*, 42, 7, 865-870, 2009.
93. Valous, N.A., Mendoza, F., Sun, D.W., Allen, P., Texture appearance characterization of pre-sliced pork ham images using fractal metrics: Fourier analysis dimension and lacunarity. *Food Research International*, 42, 3, 353-362, 2009.
94. Niamnuy, C., Devahastin, S., Soponronnarit, S., Changes in protein compositions and their effects on physical changes of shrimp during boiling in salt solution. *Food Chemistry*, 108, 1, 165-175, 2008.
95. Jin, F., Qin, L., Jiang, L., Zhu, B., Tao, Y., Novel separation method of black walnut meat from shell using invariant features and a supervised self-organizing map. *Journal of Food Engineering*, 88, 1, 75-85, 2008.
96. Lassoued, N., Babin, P., Della Valle, G., Devaux, M.F., Reguerre, A.L., Granulometry of bread crumb grain: Contributions of 2D and 3D image analysis at different scale. Food research international, 40, 8, 1087-1097, 2007.
97. Brescia, M.A., Sacco, D., Sgaramella, A., Pasqualone, A., Simeone, R., Peri, G., Sacco, A., Characterisation of different typical Italian breads by means of traditional, spectroscopic and image analyses. *Food Chemistry*, 104, 1, 429-438, 2007.

98. Briones, V., Aguilera, J.M., Brown, C., Effect of surface topography on color and gloss of chocolate samples. *Journal of food engineering*, 77, 4, 776-783, 2006.
99. Neethirajan, S., Karunakaran, C., Symons, S., Jayas, D.S., Classification of vitreousness in durum wheat using soft X-rays and transmitted light images. *Computers and electronics in agriculture*, 53, 1, 71-78, 2006.
100. Jackman, P., Sun, D.W., Allen, P., Automatic segmentation of beef longissimus dorsi muscle and marbling by an adaptable algorithm. *Meat Science*, 83, 2, 187-194, 2009.
101. Chandraratne, M.R., Kulasiri, D., Samarasinghe, S., Classification of lamb carcass using machine vision: Comparison of statistical and neural network analyses. *Journal of Food Engineering* 82, 1, 26–34, 2007.
102. Riquelme, M.T., Barreiro, P., Ruiz-Altisent, M., Valero, C., Olive classification according to external damage using image analysis. *Journal of Food Engineering*, 87, 3, 371-379, 2008.
103. Yadav, B.K. and Jindal, V.K., Dimensional changes in milled rice (*Oryza sativa L.*) kernel during cooking in relation to its physicochemical properties by image analysis. *Journal of Food Engineering*, 81, 4, 710-720, 2007.
104. Antequera, T., Caro, A., Rodriguez, P.G., Perez, T., Monitoring the ripening process of Iberian ham by computer vision on magnetic resonance imaging. *Meat science*, 76, 3, 561-567, 2007.
105. Zion, B., Alchanatis, V., Ostrovsky, V., Barki, A., Karplus, I., Classification of guppies' (*Poecilia reticulata*) gender by computer vision. *Aquacultural Engineering*, 38, 2, 97-104, 2008.
106. Lee, D.J., Schoenberger, R., Archibald, J., McCollum, S., Development of a machine vision system for automatic date grading using digital reflective near-infrared imaging. *Journal of Food Engineering*, 86, 3, 388-398, 2008.
107. Unay, D. and Gosselin, B., Stem and calyx recognition on 'Jonagold' apples by pattern recognition. *Journal of food Engineering*, 78, 2, 597-605, 2007.
108. Jarimopas, B. and Jaisin, N., An experimental machine vision system for sorting sweet tamarind. *Journal of food engineering*, 89, 3, 291-297, 2008.
109. Brezmes, J., Llobet, E., Vilanova, X., Orts, J., Saiz, G., Correig, X., Correlation between electronic nose signals and fruit quality indicators on shelf-life measurements with pinklady apples. *Sensors and Actuators B: Chemical*, 80, 1, 41-50, 2001.
110. Natale, D.C., Zude-Sasse, M., Macagnano, A., Paolesse, R., Herold, B., D'Amico, A., Outer product analysis of electronic nose and visible spectra: application to the measurement of peach fruit characteristics. *Analytica Chimica Acta*, 459, 1, 107-117, 2002.
111. Berna, A.Z., Lammertyn, J., Saevels, S., Di Natale, C., Nicolai, B.M., Electronic nose systems to study shelf life and cultivar effect on tomato aroma profile. *Sensors and Actuators B: Chemical*, 97, 2-3, 324-333, 2004.
112. Sinesio, F., Di Natale, C., Quaglia, G.B., Bucarelli, F.M., Moneta, E., Macagnano, A., Paolesse, R., D'Amico, A., Use of electronic nose and trained

sensory panel in the evaluation of tomato quality. *Journal of the Science of Food and Agriculture*, 80, 1, 63-71, 2000.
113. Eklov, T., Johansson, G., Winquist, F., Lundstrom., Monitoring sausage fermentation using an electronic nose. *Journal of the Science of Food and Agriculture*, 76, 4, 525-532, 1998.
114. Erenturk, K., Artificial Intelligence techniques for food drying technology. In *1st International Symposium on Sustainable Development, Sarajevo*, 375-380, 2009.
115. Linko, S., Expert systems-what can they do for the food industry?. *Trends in food science & technology*, 9, 1, 3-12, 1998.
116. Singh, H., How AI is reshaping the food processing business?, Customer think, 2019. Retrieved from https://customerthink.com/how-ai-is-reshaping-the-food-processing-business/
117. Tiempo Development, Artificial intelligence's biggest challenges, Blog post, 2019. Retrieved from https://www.tiempodev.com/blog/artificial-intelligence-challenges/
118. Saleh, Z., Artificial intelligence definition, ethics and standards. Electronics and Communications: Law, Standards and Practice, 2019. Retrieved from https://www.researchgate.net/publication/332548325_Artificial_Intelligence_Definition_Ethics_and_Standards

15

Ohmic Heating

Ramon da Silva Rocha[1,2], Cássia Pereira Barros[1], Tatiana Colombo Pimentel[3], Paola Mutti[4], Massimo Cigarini[4], Matteo Di Rocco[4], Andrea Brutti[4], Cristina Alamprese[5], Marcia Cristina Silva[2], Erick Almeida Esmerino[1] and Adriano Gomes da Cruz[2*]

[1]*Fluminense Federal University, Faculty of Veterinary, Niterói, RJ, Brazil*
[2]*Federal Institute of Science and Technology of Rio de Janeiro, Department of Food, Rio de Janeiro, RJ, Brazil*
[3]*Federal Institute of Paraná, Paraná, PR, Brazil*
[4]*Experimental Station for the Food Preservation Industry (SSICA) – Research Foundation, Parma, Italy*
[5]*Department of Food, Environmental, and Nutritional Sciences (DeFENS), Università degli Studi di Milano, Milan, Italy*

Abstract
Ohmic heating is an electrical technology with clear advantages about conventional processing being able to preserve the nutritional and sensory aspects of the food products. Besides these ones, it is able to guarantee the inactivation of pathogens, contributing for the safety for consumers. In this chapter, the fundamentals, the mechanisms involved in microbial inactivation of ohmic heating and the effects in quality parameters of processed foods are presented in a compreensive manner. Finally, the limitations of this technology are also presented.

***Keywords*:** Ohmic heating, preservation, food products

**Corresponding author*: adriano.cruz@ifrj.edu.br

Navnidhi Chhikara, Anil Panghal and Gaurav Chaudhary (eds.) Novel Technologies in Food Science, (551–610) © 2023 Scrivener Publishing LLC

15.1 Definition

Ohmic Heating (OH) is recognized as a processing in which the electric current passes through the food and heat it. In this way, heat is generated internally of the food, mainly because of the electrical resistance of the food [1]. This is considered the specificity of this technology. The heating can be static or, much more commonly, continuous. The product being heated is in the vast majority of cases a pumpable fluid [2].

When using a continuous heater, the heater can be an axial electric field (i.e. the electric field and the fluid motion are in a parallel direction) or a transverse field (the electric field and the fluid motion in an orthogonal direction) [3]. The first type (normally called tubular ohmic heater) is by far the most used and normally has the appearance of a pipe in which electrically insulating sections (dielectric tubes where the product is heated) alternate with electrically conductive sections (annular electrodes: called earth if connected to the earth potential, high voltage if connected to the high voltage pole of electric generators). The electrical energy applied to the electrodes is supplied by special generators that have the purpose of galvanically isolating the electrode circuit from the power supply network (normally three-phase). The electrodes placed at outlet or inlet of the heater are necessarily connected to the earth potential, being connected to the remaining part of the system. The generators also perform the task of adapting the electrical impedance of the load (the electrical resistance constituted by the product being heated) to the electrical network and of regulating the power provided to the food using an appropriate control signal (normally an analog 0-10 V or 4-20 mA) normally coming from a PLC that manages the heater.

The management of the heater can be of various types, but it is normally based on a logic called "feed-forward" that uses as input data the set point value in temperature to be reached, the temperature of the inlet product, the instantaneous flow rate value of the product (obtained from a flow meter or from the use of a volumetric pump) and the value of the specific heat of the product (often only estimated but generally very close to that of water, i.e. 1 kcal/kg °C). A slow loop on the real temperature measured at the heater outlet corrects the inevitable errors deriving from the open loop calculation (which in any case in applications do not exceed 2-3 °C).

The part of the piping and the electrical panel containing generators and PLC can be assembled in a single base (monobloc ohmic heater) or be physically separated. In this case, the two elements are connected by means of special cables with high electrical insulation. Due to the high voltages and currents involved and the normally used frequency of about 20 kHz, the distance between the two elements cannot in any case exceed a few tens of meters.

The number of generators installed is dependent on the required temperature difference, the product properties, and the total power to be supplied. The power of each individual generator varies, depending on the application, from about 5 to 100 kW. The total power of an industrial ohmic heater can exceed 1 MW, although the most common applications are in the 100-300 kW range. The voltages used in tubular heaters in the industrial field normally range from 1000 to 4000 V.

The electrodes are almost always made of common stainless steel. Exceptions are heaters that still operate at mains frequency (50 or 60 Hz), for which it is necessary to use graphite or glass-carbon electrodes in order to minimize the transfers to the product as a result of electrolytic phenomena (practically absent phenomena in the case of heaters operating at high frequency in which the electrodes in common stainless steel have practically eternal life).

Dielectric tubes can be made of various materials depending on the size, temperature, and pressure: borosilicate glass, ceramic, PTFE, Peek, fiberglass (with or without an internal layer of PTFE) are the most used. In the vast majority of cases the dielectric tubes have a circular internal section of constant diameter. In a few cases (especially in heaters operating at mains frequency) the dielectric tubes have a variable section, and a larger diameter is found in the ends that are in contact with the electrodes, aiming to reduce the density of the electric current on the electrodes [3].

In a tubular Ohmic Heater (OH) the electric current density on the electrodes is said to be the ratio between the current flowing in an electrode (amperes) and the cross section of the tubular electrode (cm^2). This value is of fundamental importance in the design of an ohmic heater, since only if this value is less than a critical value depending on the electric current frequency and the material constituting the electrodes, it is possible to inhibit the electrolyte effects on the electrodes and therefore their dissolution in the product. For heaters operating at a frequency of 20 kHz and using common stainless-steel electrodes, this limit is approximately 0.5 A/cm^2.

Although some authors state that the sterilizing effect of the ohmic heater is not only thermal but is also due to electroporation [4], the design of thermal processes with OH follows the normal rules of traditional heat treatments; the possible reduction in the treatment time could be attributable to the greater uniformity of temperature of the product obtainable with the ohmic heater.

15.2 Microbial Inactivation

The microbial inactivation is an important parameter for several industries because food safety remains a major concern and worldwide foodborne diseases are caused by pathogens or their toxins. Food products must be also protected from the presence of undesired microbes, responsible of spoilage, to prevent deterioration and economic loss.

Ohmic Heater (OH) has long been used as alternative to food pasteurization and sterilization technology in order to eliminate pathogenic and other spoilage-causing microorganisms. The main mechanism involved it the electrical current passage through the food products to heat it. The food product acts as an electrical resistance [5]. High-quality and microbiologically safe food products, and small nutritional, structural, and sensory alterations are observed when short processing times are used [6]. Vegetative pathogenic microorganisms can be inactivated using pasteurization temperatures, while bacterial endospores are inactivated using sterilization temperatures [7]. The microbial inactivation occurs mainly due to thermal effects, such as destruction and denaturation of the structure of the cell membrane and enzyme injuries [8].

As in Conventional Heating (CH) methods, in foods treated with OH, microorganisms are inactivated by increasing the temperature. Microbe inactivation curves obtained with OH are like those obtained with CH, but differences in the slope can be found due to the presence of the electric field.

The kinetics of thermal deaths found in several studies for spoilage and pathogens bacteria can be used to evaluate the application of OH processing, as there are no reports of microorganisms resistant to OH [9]. The values of decimal reduction for the OH microbial inactivation are lower than those found in conventional heating processing, mainly because of the non-thermal nature of the damage to the microbial cell.

Some examples of the microorganism's inactivation through OH, compared to CH, are reported in a study on *E. coli*; D values at 65 °C of 0.86 and 3.5 min were shown for OH and CV, respectively. Furthermore, lower D values (70-80 °C) were obtained for *Bacillus licheniformis* spores in OH rather than CH treatment, with no difference in z values [10]. On the contrary, no differences were found for the yeast *Zygasacchharomyces bailii* when the two processes were compared [11].

The mechanisms of microbial inactivation related to the thermal effects has not been completely understood. High temperatures can impair the physiological and structural properties of non-sporulating and sporulating bacteria. The lethal impact caused by heat proved related to every cellular component (membrane, proteins, outer layers, DNA, RNA); consequently, it is not easy to attribute cellular damage to a single alteration. Moreover, the cell resistance may be dependent on the microorganisms' type and the conditions of the environment. The most important effects of heating on the bacterial cell are:

- Loss of membrane-associated functions
- Loss of intracellular components
- DNA e RNA alterations
- Ribosome conformation loss
- Loss of specific protein functionalities

The cytoplasmic membrane damage and ribosomes denaturation are considered the main reasons of cellular death caused by thermal process. Damages to the cytoplasmic membrane have been also reported by several authors. The role of this membrane is related to the control of solute exchange, so damages caused by heat have a marked effect on bacteria [12].

In Gram negative bacteria, the outer cell layers are affected by heat that can cause a loss of lipopolysaccharide molecules, thus altering the permeability barrier with the release of periplasmic proteins to the treatment medium. In this regard, it has been reported that the cell wall of *S. aureus* was less damaged by high temperatures than the Gram-negative bacteria outer membrane [13, 14].

RNA and ribosomes are structures more thermosensitive than DNA. It has been suggested that one of the most important consequences in heated bacterial cells could be the degradation of RNA, even if it does not seem the main cause of non-spore forming bacterial death. Damages of ribosomes were also reported: the degradation of the 30S ribosomal subunits in cells of *S. aureus* was dependent on the magnesium concentration found in the medium [12, 15].

The main effect of heat on DNA, one of the most thermostable cellular constituents, is the denaturation occurring at sterilization temperatures, even if less intense treatments can also damage the DNA molecule. In vegetative cells exposed to moist heat, single or double strand breaks occur directly or by accelerating the activity of endogenous nucleases [13].

Thermal processes impact on functional and structural proteins, with their denaturation and their coagulation at extremely high temperatures [13].

Recently several studies in literature tried to demonstrate that also non-thermal mechanisms can inactivate microorganisms during the OH processing.

Electroporation is considered the main non-thermal mechanism of microorganism inactivation, and it results in a formation of pores in the membrane of the cell [16–18]. Then, the permeability of the membrane is changed, and cellular materials, such as nucleic acids, protein, and enzymes, may be leaked [19, 20]. A previous study [21] applied an OH treatment using alternating current at frequencies of 50-20000 Hz and electric field strength of 30 V/cm to pomelo juice and to buffered peptone water contaminated with *Salmonella enteritidis* ATCC 13076. The products were treated at the same time-temperature conditions of the traditional treatment (water bath). The OH proved to reduce the concentration of *S. enteritidis* approximatively half to three logs more than the CH in both substrates and the most relevant effect was obtained at 60 Hz and ≥500 Hz frequencies. Another previous study [22] evaluated the effect of heating treatment in water bath (WB) and ohmic heating (10 V/cm) on the membrane permeability, organization of the cells, and leakage of substances in *Escherichia coli* O157:H7 (NCTC 12900, 72 °C). Both technologies (conventional heating and ohmic heating) impair the membrane permeability, alter the cell organization, and promotes leakage of the inner substances. However, a higher leakage of K^+ was observed when OH was used. In this way, the electroporation promoted a pore formation in the cell, altering the cell permeability, and increased the leakage of inner compounds, such as enzymes, nucleic acids, protein, and amino acid. The higher compounds leakage resulted in a higher microbial inactivation of the vegetative cells.

Regarding bacterial endospores, several studies in the literature compared the OH treated endospore inactivation kinetics to the CH treated ones [10, 23]. In one of the studies, the OH (frequencies of 60 Hz and 10 kHz at pulse mode) was compared to CV (temperature of 95, 100, 105, and 110 °C) in the *B. coagulans* spores ATCC 803814 inactivation in tomato juice [24]. Results showed that OH treated endospore inactivation times ($D_{95°C}$ = 7.96 min, $D_{100°C}$ = 1.63 min, $D_{105°C}$ = 0.91 min, $D_{110°C}$ = 0.15 min at 60 Hz and $D_{95°C=}$ 8.81 min, $D_{100°C}$ = 2.27 min, $D_{105°C}$ = 1.03 min, $D_{110°C}$ = 0.13 min at 10 kHz) were lower compared to CH ($D_{95°C}$ = 10.1 min, $D_{100°C}$ = 2.52 min, $D_{105°C}$ = 1.32 min, $D_{110°C}$ = 0.16 min). Electroporation mechanism cannot explain the inactivation effect caused by electric field on endospores. In fact, bacterial spore structure greatly differs from vegetative cells structure. The authors hypothesized that a synergistic effect between temperature and electric current may have occurred, resulting in a higher release of molecules of dipicolinic acid from the spore core, and fragmentation of the enzymes and proteins in the denatured spore. These effects were associated with a faster inactivation. Accelerated inactivation of *B. subtilis* and *B atrophaeus* spores was reported for the same treatment temperature in case of OH in a model system using capillaries, with a reduction by 1 min [25].

The effects of ohmic heating and conventional heating on the inactivation of microorganisms are summarized in Table 15.1.

Table 15.1 Inactivation effects on microorganisms treated with different ohmic (OH) and conventional (CH) heating parameters.

Microorganism	Processed matrix	Processing parameters	Inactivation effects	References
Staphylococcus aureus	Infant formula milk	OH (30 V/cm - 60Hz; T = 57.5, 60.0, 62.5, 65.0 °C) CH (T = 57.5, 60.0, 62.5, 65.0 °C)	Time-temperature combinations necessary for 12 logarithmic cycles reduction: OH 57.5 °C /40.63 min. OH 60.0 °C /17.68 min. OH 62.5 °C /11.47 min. OH 65.0 °C /5.83 min. CH 57.5 °C /184.34 min. CH 60.0 °C /24.06 min. CH 62.5 °C /19.78 min. CH 65.0 °C /17.27 min.	[26]

(*Continued*)

Table 15.1 Inactivation effects on microorganisms treated with different ohmic (OH) and conventional (CH) heating parameters. (*Continued*)

Microorganism	Processed matrix	Processing parameters	Inactivation effects	References
Escherichia coli	Infant formula milk	OH (30 V/cm - 60Hz; T = 57.5, 60.0, 62.5, 65.0 °C) CH (T = 57.5, 60.0, 62.5, 65.0 °C)	Time-temperature combinations necessary for 12 logarithmic cycles reductions: OH 57.5 °C/22.49 min. OH 60.0 °C/8.33 min. OH 62.5 °C/4.99 min. OH 65.0 °C/4.04 min. CH 57.5 °C/17.08 min. CH 60.0 °C/8.71 min. CH 62.5 °C/5.38 min. CH 65.0 °C/3.59 min	[26]
Escherichia coli O157:H7	Apple juice (18 and 72°Bx)	30 V/cm - 20 kHz 60 V/cm - 20 kHz, sine waveform, time = 20 s	Logarithmic reductions: 30 V/cm, 18 °Bx/0.17 log 30 V/cm, 72 °Bx/0.42 log 60 V/cm, 18 °Bx 6.93 log 60 V/cm, 72 °Bx 0.67 log	[27]

(*Continued*)

Table 15.1 Inactivation effects on microorganisms treated with different ohmic (OH) and conventional (CH) heating parameters. (*Continued*)

Microorganism	Processed matrix	Processing parameters	Inactivation effects	References
Escherichia coli O157:H7	Buffered peptone water	OH (30 V/cm; T = 58, 60 °C; time = 30 s) CH (T = 58, 60 °C - time = 30 s)	Logarithmic reductions: OH 58 °C/4.00 OH 60 °C/5.62 CH 58 °C/1.58 CH 60 °C/2.3	[18]
Escherichia coli O157:H7	Pomelo Juice	OH (30 V/cm – 60 Hz; T = 68 °C) CH (T= 68 °C)	OH – Decimal reduction time (s) = 3.6 CH – Decimal reduction time (s) = 8.2	[28]
		OH (30 V/cm – 500 Hz; T = 68 °C) CH (T= 68 °C)	OH – Decimal reduction time (s) = 3.3 CH – Decimal reduction time (s) = 8.2	
	Buffered peptone water	OH (30V/cm – 60 Hz; T = 68 °C) CH (T= 68 °C)	OH – Decimal reduction time (s) = 4.1 CH – Decimal reduction time (s) = 8.5	
		OH (30V/cm – 500 Hz; T = 68 °C) CH (T= 68 °C)	OH – Decimal reduction time (s) = 3.4 CH – Decimal reduction time (s) = 8.5	

(*Continued*)

Table 15.1 Inactivation effects on microorganisms treated with different ohmic (OH) and conventional (CH) heating parameters. (*Continued*)

Microorganism	Processed matrix	Processing parameters	Inactivation effects	References
Salmonella typhimurium	Apple juice - 18, 72°Bx	30 V/cm - 20 kHz 60 V/cm - 20 kHz, sine waveform, time = 20 s	Logarithmic reductions: 30 V/cm, 18 °Bx/0.17 log 30 V/cm, 72 °Bx/0.72 log 60 V/cm, 18 °Bx/6.68 log 60 V/cm, 72 °Bx/0.90 log	[27]
Salmonella typhimurium	Buffered peptone water	OH (30 V/cm; T = 58, 60 °C; time = 30 s) CH (T = 58, 60 °C; time = 30 s)	Logarithmic reductions: OH 58 °C/4.63 OH 60 °C/6.93 CH 58 °C/1.42 CH 60 °C/3.45	[18]
Geobacillus stearothermophilus	Tomato soup	OH (60 Hz, 10 kHz, T = 121, 125, 130 °C) CH (T = 121, 125, 130 °C)	D-values (minutes): OH - 10 kHz, 121 °C/0.88 min OH - 60 Hz, 121 °C/1.17 min CH – 121 °C/2.53 min OH - 10 kHz, 130 °C/0.05 min OH - 60 Hz 130 °C/0.05 min CH 130 °C/0.06 min	[23]

(*Continued*)

Table 15.1 Inactivation effects on microorganisms treated with different ohmic (OH) and conventional (CH) heating parameters. (*Continued*)

Microorganism	Processed matrix	Processing parameters	Inactivation effects	References
Listeria monocytogenes	Apple juice (18 and 72 °Bx)	30 V/cm - 20 kHz 60 V/cm - 20 kHz, sine waveform, time = 20 s	Logarithmic reductions: 30 V/cm, 18 °Bx/0.04 log 30 V/cm, 72 °Bx/0.32 log 60 V/cm, 18 °Bx/5.94 log 60 V/cm, 72 °Bx/0.23 log	[27]
Listeria monocytogenes	Buffered peptone water	OH (30 V/cm; T = 58, 60 °C; time = 30 s) CH (T= 58, 60 °C; time 30 s)	Logarithmic reductions: OH 58 °C/1.11 OH 60 °C/4.37 CH 58 °C/0.41 CH 60 °C/0.97	[18]
Bacillus cereus spores	Tsuyu sauce	26.7 V/cm -25 kHz; T = 95 °C; time = 30 s	log reductions 1.4	[7]
		26.7 V/cm - 25 kHz; T= 125 °C; time = 30 s	log reductions 5.7	
Ms-2 bacteriophage (norovirus surrogate)	Tomato Juice	47.7 V/cm - 0.06, 1 kHz, T = 80 °C	Logarithmic reductions: OH 0.06 KHz/5.80 OH 1.0 KHz/5.13	[29]

(*Continued*)

Table 15.1 Inactivation effects on microorganisms treated with different ohmic (OH) and conventional (CH) heating parameters. (*Continued*)

Microorganism	Processed matrix	Processing parameters	Inactivation effects	References
Alicyclobacillus acidoterrestris spores	Orange juice	OH (30 V/cm; T= 70, 80, 90 °C) CH (T = 70, 80, 90 °C)	D-values (minutes): OH – 70 °C/58.48 CH – 70 °C/83.33 OH – 80 °C/12.24 CH – 80 °C/15.11 OH – 90 °C/5.97 CH – 90 °C/7.84 Z-values: OH 18.80 °C CH 17.89 °C	[30]
		OH (50 V/cm; T 70, 80, 90 °C) CH (T = 70, 80, 90 °C)	D-values (minutes): OH – 70 °C/40.49 CH – 70 °C/83.33 OH – 80 °C/11.43 CH – 80 °C/15.11 OH – 90 °C/4.29 CH – 90 °C/7.84 Z-values: OH 20.00 °C CH 17.89 °C	

(*Continued*)

Table 15.1 Inactivation effects on microorganisms treated with different ohmic (OH) and conventional (CH) heating parameters. (*Continued*)

Microorganism	Processed matrix	Processing parameters	Inactivation effects	References
Alicyclobacillus acidoterrestris spores	Apple juice	26.7 V/cm - 25 kHz; T = 100 °C; time = 30 s	Log reductions 4.8	[31]
Bacillus amyloliquefaciens spores	Carrot puree	50 V/cm; T = 105 °C; time = 5 min; pressure 600 MPa	Log reductions 2.80	[32]
Bacillus amyloliquefaciens spores	Tomato juice	30 V/cm; T = 105 °C; time = 10 min.; pressure 600 MPa	Log reductions 3.05	[32]
Bacillus amyloliquefaciens spores	0.1% NaCl solutions	50 V/cm; T = 105 °C; time = 5 min.; pressure 600 MPa	Log reductions 1.0	[32]
Bacillus amyloliquefaciens spores	0.1% NaCl solutions	50 V/cm; T = 105 °C; time= 10 min.; pressure 0.12 MPa	Log reductions 0.84	[32]
Geobacillus stearothermophilus spores	0.1% NaCl solutions	50 V/cm; T = 105 °C; time = 5 min.; pressure 600 MPa	Log reductions 3.27	[32]

15.3 Applications

OH is considered an emerging technology, however, its application in food is not new. The technique was used for the first time in the late 19th century for the electrical milk pasteurization, which was recognized as "electro-pure process of treating milk" [33]. The wide development of commercial units for food processing was limited by some factors, such as the difficulty of controlling the process parameters, the high electricity cost, and the absence of inert materials used in the fabrication of the electrodes [34]. Subsequently, over the years, several researches were carried out in order to optimize the process, resulting in the development of new technologies for energy supply, and new materials that avoided the effects of electrolysis on the electrode, making possible the efficient and economic use of the method [34]. These technological advances boosted the research in the OH field, resulting in a significant increase of publications. Since 2015, the scientific articles about OH applied to foods doubled with respect to the works published in 2010-2014. Most of publications come from Brazil, South Korea, and USA.

Currently, OH is used in food technology as a thermal method for several purposes, which include defrosting, preheating, bleaching, evaporation, dehydration, extraction, fermentation, sterilization, and pasteurization of meat, dairy, vegetable, and fruit preparations [35–40]. Ohmic Heater or OH application results in a more quickly and evenly heating, shorter heating times, maintenance/retention of nutritional compounds and sensory characteristics, and reduction of the formation of fouling by preventing overheating and unwanted temperature peaks.

In the following paragraphs, OH applications in the field of dairy products, meat and fish, eggs, cereal products, and juices will be reviewed.

15.3.1 Dairy

OH provides advantages for the dairy industry when compared to traditional processes (HTST pasteurization and UHT sterilization) [35]. Figure 15.1 illustrates some examples of the main effects observed with the use of OH in dairy processing.

In recent years, studies involving the effects of OH on the microbiological safety and quality parameters of dairy products have gained increasing attention. In general, the application of OH to dairy products resulted in increased concentrations of bioactive peptides, antioxidant activity [41–43], as well as higher levels of volatile compounds [41, 43, 44]. No negative effects were identified on the nutritional value of whey-based

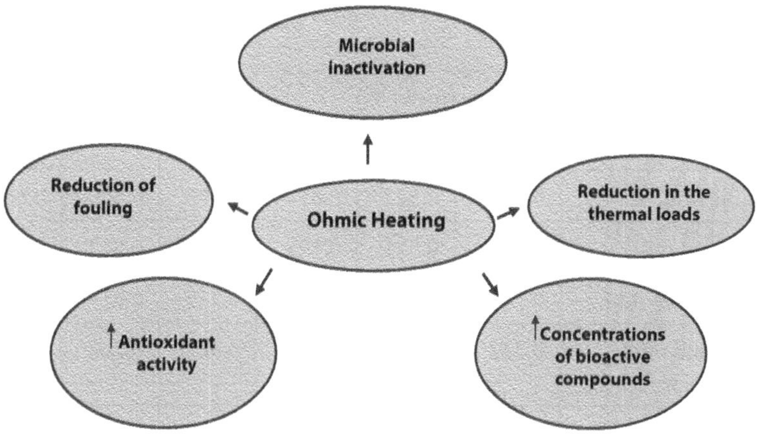

Figure 15.1 Examples of effects of the application of ohmic heating in dairy products.

dairy drinks, dairy desserts, cheeses, and dulce de leche when milk was pasteurized by OH [41–44]. Furthermore, no effect on the nitrogen index of whey protein in infant formulas was observed [45]. It is worth mentioning that dairy products are excellent sources of nutrients, but they can also be sources of foodborne pathogens because they provide adequate conditions for microorganism's growth. OH provides microbiological stability to products similar to those subjected to conventional heat treatment, but with lower energy costs, processing time, and, mainly, with minimal damage to rheological, sensory, or functional properties [38, 41, 44, 46–49]. Therefore, OH technology can be used as an alternative for processing dairy products.

The effectiveness of the technology is dependent on the food matrix characteristics, process parameters (electric field strength, frequency, and electrical conductivity), and equipment design [35]. The food matrix must be electrically conductive, that is, the food must have a high content of water and polar components (minerals and proteins), as well as electrical conductivities in the range of 0.01 and 10 S/m (25 °C) [49]. Milk has ionic components, high water content, and electrical conductivities between 0.3 and 1.0 S/m at temperatures between 4 and 60 °C, therefore, it can be considered an ideal food for OH process [35, 48]. The electrical conductivity can vary significantly according to the temperature, as temperature affects the mobility of ions in the milk. Furthermore, it is dependent, among other factors, on the non-conductive components presence (sugar and fat), which contribute to the proportional increase or decrease in conductivity [35] (Table 15.2).

Table 15.2 Effect of temperature on the values of electrical conductivity (S/m) in dairy products [35].

Dairy food	Temperature (°C)					
	4	22	30	40	50	60
Whole milk	0.357	0.527	0.617	0.683	0.800	0.883
Skimmed milk	0.328	0.511	0.599	0.713	0.832	0.973
Lactose-free milk	0.380	0.497	0.583	0.717	0.817	0.883
Chocolate milk (3% w/w fat)	0.332	0.433	0.483	0.567	0.700	0.800
Chocolate milk (2% w/w fat)	0.420	0.508	0.617	0.700	0.833	1.000
Chocolate milk (skimmed milk)	0.532	0.558	0.663	0.746	0.948	1.089

The effect of fat content of traditional Turkish and Mara ice creams (9.8% and 3.3% fat, respectively) on electrical conductivity was evaluated. Mara ice creams presented a higher electrical conductivity and heating rate in all the voltage gradients used [50]. A similar result was reported by Kim and Kang [51] who found increased electrical conductivities and heating rates in low-fat milk. Furthermore, higher microbial inactivation rates were reported for low fat dairy products (0-3%). Products with a higher fat content can be heated unevenly due to the reduction of electrical conductivity by fat globules, which act as electrical insulators, resulting in lower heating rates in the areas near the globules, which consequently compromises microbiological safety of the product [52]. Therefore, the evaluation of the influence of the factors on the rate of heating is important for understanding the temperature profile distribution in food.

Electrical conductivity is recognized as the main OH parameter. However, other factors also affect the heating rate of the process and, consequently, the physicochemical, rheological, and sensory properties of the products, such as the residence time, electric field intensity, and food flowability. Therefore, the evaluation of the different OH parameters in foods is essential to optimize the process. The thermal process indicators routinely used for conventional treatments may also be used to evaluate OH process parameters impact and define the optimal conditions of the process [52]. In the literature, the following quality indicators stand out: free fatty acids (FFA), Hydroxy-Methyl-Furfural (HMF) and furosine, lactulose, carboxy-methylisine, glycoxidation products, bioactive compounds (ACE inhibitory activity, total phenolics, DPPH, and α-glucosidase and α-amylase inhibitory activities), FAST index (fluorescence of advanced Maillard products and soluble tryptophan), color parameters (L*, a*, b*), denaturation and aggregation of proteins (β-lactoglobulin), pH, among others [35, 41, 53, 54].

Table 15.3 contains several studies that compared the effects of OH to conventional heat treatments in relation to the intrinsic quality of different dairy products. In general, it can be observed that with the proper choice of the operational parameters of the OH it is possible to maintain the nutritional/functional and sensory aspects of the product, in addition to reducing the processing time, avoiding economic losses.

Table 15.3 Impact of OH on the quality parameters of dairy products.

Milk/Dairy product/ objective process	**Processing parameters**	**Main effects**	**References**
Goat milk[1]	Conventional processing (72 °C/ 15 s) and OH (14.4 V/cm, 5–30 A, 72 °C/15 s)	FFA content was similar in products processed by OH or conventional processing.	[55]
Milk[1]	Conventional processing (40, 50, 60, 65, 70, 75 and 80 °C/30 min) and OH (20 kHz; 7.3 and 2 A, 70 and 12 V)	FAST index was similar in products processed by OH or conventional processing. No additional protein denaturation occurred.	[56]
Whey protein film[1]	OH (50 Hz; 10 V/cm, 85 °C/30 min) and conventional processing (85 °C/ 30 min)	OH processing resulted in films with lower permeability to water vapor, slimmer, and with lower protein aggregation and free thiol group.	[57]

(Continued)

Table 15.3 Impact of OH on the quality parameters of dairy products. (*Continued*)

Milk/Dairy product/ objective process	Processing parameters	Main effects	References
Buffalo Milk[1]	OH (50 Hz, 0.6 A, 8.38 V/cm, 72–75 °C/15 s) and conventional processing (72–75 °C/15 s)	The OH and conventional treated products showed similar physicochemical values.	[58]
WPI (Whey Protein Isolate)[1]	OH (6 and 12 V/cm), under different time (10 and 100 s) to target temperature (90 °C) and conventional process (same conditions)	OH promoted a lower protein aggregation and increased the protein solubility. OH at 12 V/cm for 10 s retained a higher protein content, resulting in products with a better quality.	[54]
Liquid infant formula[3]	OH (25 kHz, 15 kW; 300 - 4000 V) and steam injection under similar conditions of pre-heating (50 °C) and heating (120-140 °C)	OH processing resulted in higher K values in the FAST index, higher carboxymethyllysine (CML) and furosine contents, and increased retention of color parameters and vitamin C content.	[53]

(*Continued*)

Table 15.3 Impact of OH on the quality parameters of dairy products. (*Continued*)

Milk/Dairy product/ objective process	Processing parameters	Main effects	References
Sweet whey[1]	Conventional processing (72–75 °C/15 s) and OH (2, 4, 5, 7 and 9 V/cm at 60 Hz, up to 72–75 °C/ 15 s)	Low electrical fields resulted in products with higher color variation (ΔE*), saturation, and luminosity (L*). Higher electrical fields resulted in products with lower antioxidant capacity, and ACE inhibitory activity. OH processing at 4-5 V/cm resulted in products with similar sensory characteristics and higher levels of volatile compounds.	[41]
Whey drink[1]	Conventional processing (65 °C/30 min) and OH (10, 100 and 1000 Hz with 25 V, 65 °C/30 min and 45, 60 and 80 V at 60 Hz)	Low frequencies and voltages (≤100 Hz and 45 V) resulted in products with better characteristics.	[59]

(*Continued*)

Table 15.3 Impact of OH on the quality parameters of dairy products. (*Continued*)

Milk/Dairy product/ objective process	Processing parameters	Main effects	References
Buffalo milk cow milk, and mixed milk (50:50)[4]	OH evaporation, 13.33 V/cm/75 min	OH processing significantly increased instrumental color values (a* and b*), apparent viscosity, HMF content, and FFA. The OH-treated products had lower pH values and whiteness (L*) than those subjected to conventional evaporation.	[38]
Whey Drink[1]	OH (10, 100 and 1000 Hz, 25 V, 65 °C for 30 min and 45, 60 and 80 V at 60 Hz) and conventional processing (65 °C/ 30 min)	Low frequencies resulted in lower concentrations of bioactive compounds (DPPH, total phenolics, ACE inhibitory activity, and FRAP) and lower antioxidant activity. The effects on the fatty acid profile and volatile compounds were dependent on the processing parameters (frequency and voltage).	[60]

(*Continued*)

Table 15.3 Impact of OH on the quality parameters of dairy products. (*Continued*)

Milk/Dairy product/ objective process	**Processing parameters**	**Main effects**	**References**
Whey-raspberry flavored beverage[1]	OH (10, 100 and 1000 Hz, 25 V; 45, 60 and 80 V at 60 Hz, 65 °C for 30 min) and conventional processing (65 °C/30 min)	Heat and OH-treated products showed changes in the color parameters (h*, C*, and ΔE*) during 30 min process. OH-treated products had higher viscosity and increased D[3,2] and D[4,3] values (1000 Hz-25 V and10 Hz-25 V).	[61]
Whey-raspberry flavored beverage[1]	OH (10, 100 and 1000 Hz, 25 V; 45, 60 and 80 V at 60 Hz, 65 °C for 30 min) and conventional processing (65 °C/30 min)	OH processing resulted in products with lower contents of anthocyanins, but higher inhibition of α-glucosidase and α-amylase. Mild conditions (and 45,60-V at 60 Hz and 10,100-Hz at 25 V) resulted in products with higher antioxidant activity.	[62]

(*Continued*)

Table 15.3 Impact of OH on the quality parameters of dairy products. (*Continued*)

Milk/Dairy product/ objective process	Processing parameters	Main effects	References
Minas Frescal cheese[2]	Conventional processing (72–75 °C/15 s) and OH (4, 8 and 12 V/cm, 72 °C/15 s, 60 Hz)	OH processing improved the textural (lower hardness, firmness, and elasticity) and sensory acceptance of the products. OH processing increased the antihypertensive, antioxidant, and antidiabetic properties of improved the fatty acid profile (higher contents of short, medium, and long-chain fatty acids).	[43]
Whey dairy beverage[1]	OH (6 V/cm, 60 Hz, 57.5–60 °C/ 38-4 min and conventional processing (57.5–60 °C/38-4 min)	OH processing decreased the HMF and pH values, improved the sensory properties, and changes the brightness.	[47]
Blueberry flavored dairy desserts[1]	conventional processing (90 °C/ 3 min), and OH (1.82, 3.64, 5.45, 7.30, 9.1 V/cm, 90 °C/3 min, 60 Hz)	Higher electrical fields increased the ACE and α-glucosidase inhibitory activities. No effect on fatty acid profile was observed. OH processing altered the volatile profiling, color parameters, and rheological properties.	[42]

(*Continued*)

Table 15.3 Impact of OH on the quality parameters of dairy products. (*Continued*)

Milk/Dairy product/ objective process	Processing parameters	Main effects	References
Dulce de Leche[2]	Conventional processing (72–75 °C/15 s) and OH (2, 4, 6, 8 and 10 V/cm, 72 °C/ 15 s, 60 Hz	No changes in fatty acid profile, and chemical composition of the products. OH processing had lower energy expenditure. OH-treated products had lower HMF levels and improved homogeneity and volatile compounds.	[44]
Infant formula[1]	Conventional processing (72–75 °C/15 s) and OH (8, 12, 16, 20, and 24 V/cm, 72–75 °C/15 s)	OH-treated products had lower HMF levels (total and free) and viscosity, and higher whiteness index. OH processing improved the volatile compounds profile.	[45]
Infant formula[1]	OH (6 V/cm, 50, 55 and 60 °C/ 45, 24 and 4 min, respectively, at 60 Hz) and conventional processing (50, 55 and 60 °C/45, 24 and 15 min, respectively)	OH processing resulted in products with increased concentration of bioactive compounds (ACE inhibitory activity, DPPH, α-amylase inhibitory activity, α-glucosidase inhibitory activity, and total phenolics) OH processing resulted in products with lower values for parameters related to thermal loads (HMF, whey protein nitrogen index, and HMF).	[63]

[1]Pasteurization, [2]Pasteurization of milk intended for the manufacture of dairy foods, [3]Sterilization, [4]Concentration.

15.3.2 Meat and Fish

In recent years, OH has stood out as an alternative thermal method capable of promoting rapid, volumetric, and homogeneous heating from the inside to the surface of the food by transmitting an alternating electric current through the conductive food matrix. In this way, food matrix acts as a resistance, generating shorter processing times and resulting in reduced surface burn and better nutrient and vitamin retention. Therefore, OH is suitable for processing foods with high protein content and with high susceptibility to aggregation and denaturation when subjected to conventional thermal processing. Therefore, its use has been increasingly common in the meat food processing (fish and meat products), mainly for cooking, pasteurization, and defrosting [64].

15.3.2.1 Meat

Red meat has a great biological value, whose composition is rich in high quality proteins, B_{12}, B_6, D, niacin, and riboflavin vitamins, iron, zinc, phosphorus, selenium, and pantothenic acid. It has also large concentrations of long-chain polyunsaturated fatty acids, mainly omega 3. Furthermore, it presents antioxidants and bioactive compounds, such as carnitine, taurine, glutathione, and ubiquinone [65]. Meat and its derivatives must be cooked before consumption, as the cooking step will ensure the destruction of foodborne pathogens, thus guaranteeing microbiological safety and the achievement of meat quality. However, meat quality characteristics can be affected according to the type and intensity of the thermal method [66]. Different means of heat transfer have been applied for cooking meat, which involve methods of dry, moist, or electric heat, such as microwave, radiofrequency heating, and OH.

CH treatments (sterilization and pasteurization) are commonly used for preserving meat used in the food industry, however, they promote structural changes in proteins, which can cause oxidation, degradation, denaturation, and aggregation. Oxidation associated with protein aggregation and fragmentation can result in loss nutritional and functional quality [67]. Protein denaturation leads to transversal and longitudinal shrinkage of myofibrils and myofibrillar and sarcoplasmic protein gelation [68]. In addition, these changes may impact on the sensory and physical properties of the meat products, and they are directly associated with the palatability of the product and its acceptance by consumer, such as color, texture, juiciness, and flavor [69].

OH cooking of meat products can be carried out on intact parts of muscles or in meat emulsions. Ohmic cooking has several advantages over conventional cooking processes (Figure 15.2), mainly due to the uniform temperature distribution without the need for hot surfaces and the significant reduction in cooking times, resulting in reduced cooking loss with better water retention, and meat with greater juiciness [70, 71]. In addition, the ohmic treatment extends the product's shelf life by inhibiting the growth of pathogens. Furthermore, it reduces the energy costs by around 70%, provides greater heating performance and results in meat/meat product with greater firmness, texture, color preservation and better gelling characteristics [69, 72–74].

Table 15.4 contains studies that evaluated the main effects caused by ohmic and conventional cooking in meat and derivatives.

Ohmic Heating (OH) can also be combined with conventional cooking methods in order to provide a safer and more homogeneous cooking process, since preheating can influence conductivity and generate temperature distribution with higher uniformity when compared to the isolated use of OH [49, 72, 78, 81]. On the other hand, some difficulties have been reported in the use of the ohmic process in solid meat products, as the treatment may be inefficient for obtaining products with desirable changes in the texture and surface color [81, 85]. OH cooking results in products with more stable colors, but this stability can represent a disadvantage, as the consumer prefers a cooked meat with a darker color and showing a surface with a cooked crust layer [81]. Samples of ground beef, whole beef muscle and cooked turkey meat subjected to OH showed a noticeably

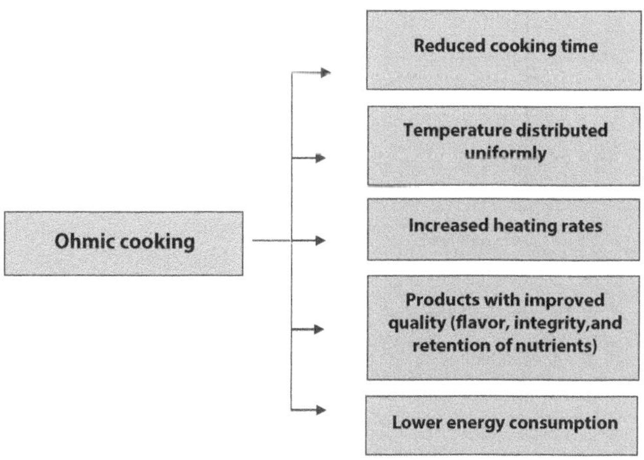

Figure 15.2 OH processing advantages compared to traditional heat treatments.

Table 15.4 Studies that considered OH and conventional heating as cooking technologies in meat and derivatives.

Meat/meat product	Processing parameters	Main effects	References
Meat emulsion batter	OH (230 V; 3, 5 and 7 V/cm; 50 Hz; 73 °C/2 min); CH (73 °C/ 2 min)	No effect on texture and taste, lower elasticity, changes in color, and higher cell damage visualized in the microscope.	[72, 75, 76]
Hamburger	OH (50 V, 0-13 A, 70 °C/15 s); CH (70 °C/15 s)	OH reduced cooking time up to half the time usually required in conventional cooking. No differences on the moisture and oil contents and mechanical properties.	[77]
Beef	OH (25-50 V, 50 Hz, 72 °C/12 min); CH (72 °C/ 2 min)	OH altered the color parameters (less red and lighter). OH decreased the cook loss OH-treated products were tougher.	[78]

(*Continued*)

Table 15.4 Studies that considered OH and conventional heating as cooking technologies in meat and derivatives. (*Continued*)

Meat/meat product	Processing parameters	Main effects	References
Turkey meat	8.33 V/cm; 100 V a) LTLT at 72 °C for 4 min, and 80°C for 3 min. b) HTST at 95 °C for 5 min.	OH processing changed the color (more uniform and lighter). LTLT-treated products showed a higher quality. HTST-treated products showed a better flavor.	[79]
Minced beef	OH (50 Hz, 20, 30 and 40 V/cm, 80 °C)	The electrical conductivity was dependent on the temperature and initial fat content. No effect of voltage gradient on the electrical conductivity.	[80]
Cylindrical ground beef with fat content from 2 to 15%	OH (50 Hz, 20, 30 and 40 V/cm, 80 °C)	No effect of OH on the yield and fat retention. OH-treated products were firmer.	[81]
Pork	OH (10 V/cm, 20–100 °C); CH (20–100 °C)	OH processing decreased the shear force and cooking loss. OH processing increased the pH values, protein solubility, and endothermic transition temperature.	[82]

(*Continued*)

Table 15.4 Studies that considered OH and conventional heating as cooking technologies in meat and derivatives. (*Continued*)

Meat/meat product	Processing parameters	Main effects	References
Beef	OH (12 V/cm, 72 °C) and conventional cooking (72 °C)	OH processing decreased the shear force and cooking loss and promoted a lower protein damage. OH processing increased the redness (a* values)	[69]
Wagyuu beef (WB) and conventional beef (CB)	Only OH (50 V, 50 Hz - 20 kHz, 5-65 °C)	The fat content decreased the electrical conductivity and the heating rate. OH-treater products (20 kHz) had lower drip loss related to shrinkage and cooking.	[83]
Pork	OH (21 ± 1 V/cm, 2, 2.5 and 3 min, 70 °C); CH (70 °C)	There were no significant differences between the cooking methods regarding cooking loss; however, significant differences concerning water holding capacity and change in color (ΔE) were found for OH cooking treatments at 2 min and 2.5 min, respectively.	[84]

lighter color than the conventionally cooked [78, 79, 81, 86]. Such a difference in color occurred due to longer processing times and increased surface temperatures during conventional cooking [79].

The electrical conductivity and the heat rate of OH need to be evaluated. Meat is considered a challenging food matrix due to its heterogeneous structure that affects uniform heat distribution during processing. It is known that meat components, such as fat, moisture, density or direction of muscle fiber and the type of meat (lean or fatty) have a significant influence on electrical conductivity [87]. Fat acts as a thermal insulator because of the extremely low thermal conductivity and lack of water and salt and, therefore, cold spots can be observed, compromising the microbial safety. Therefore, lean cuts of meat are better drivers of electrical conductivity than fat cuts [50, 76, 88]. The composition and structure of meat can be modified during heating, and such changes are attributed to the denaturation of proteins that causes changes in ionic movements, decreasing the electrical conductivity [84, 87, 88].

In addition to the use of OH for cooking meat and derivatives, it can also be used to effectively defrost food products. In the food industry or retail, defrost systems by conduction are commonly used. In these systems, the heat is transferred to the center from the external surfaces of the meat by using water, air, condensation vapor, or contact with a heated surface [89]. Inappropriate defrost can result in excessive weight loss, decrease in the nutrient, and increase in the microorganism counts. Therefore, defrost has a significant impact on the control of the physicochemical properties and the microbiological safety of the product [90]. In this sense, defrost of food should be carried out as quickly as possible and at low temperatures. In general, it is assumed that the development of bacteria, yeasts, and fungi ceases at -10 °C, -12 °C and -18 °C, respectively [91].

One of the major obstacles to the application of OH for meat defrost is the solid, non-foldable nature of frozen foods, since, for the heat to be generated, it is necessary to have direct contact of the product surface with the electrodes [57]. Yun *et al.* [92] combined the OH (60-210 V AC, 60 Hz-60 kHz) with the conventional treatment by immersion in water (10 °C or 20 °C) to defrost pieces of meat. There was no direct contact between the electrodes and the meat under these conditions, but OH reduced the defrost time as the voltage increased, while at lower voltages, the samples had reduced drip loss and a better water retention capacity [92]. In contrast, when OH and immersion in brine or by electrode contact were compared as defrost methods for pork, the defrost was faster in the electrode contact method. The increase in voltage decreased the drip loss, providing increased water holding capacity and decreased in TBARS values. At high voltage levels, it is advisable to use the immersion defrost method, while a low voltage levels the electrode contact should be used [93].

The use of OH for defrosting meat and derivatives with flatter surfaces has also been studied. Pork hamburgers formulated with 0 and 5% salt were defrosted by OH using different power intensities (0-50 V) at 50 Hz. The defrost rate was faster with higher energy intensity (50 V). The addition of salt influenced the defrosting, which was faster in hamburgers added with salt. The increase in the intensity of potency increased the levels of TBARS, and the hamburgers without salt had significantly higher levels of TBARS [94].

In summary, it can be concluded that ohmic defrosting has advantages in reducing meat defrosting time, reducing drip water loss and, consequently, promoting better water retention, which contributes to improvements in meat product quality when compared to defrosting by conventional methods. Therefore, OH is a promising technology for performing meat defrosting by minimizing the changes in meat and enabling the maintenance of its technological properties closer to refrigerated meat.

15.3.2.2 Fish

Fish is a source of unsaturated fatty acids (Omega-3) and protein, and it contains B_2 (riboflavin) and D vitamins, and minerals (phosphorus, calcium, zinc, iron, potassium, magnesium, and iodine. Furthermore, it is an excellent food for the human diet, helping in the maintenance of good health functioning, mainly related to brain and heart. However, its quality can be affected by several factors and it is recognized as a perishable food due to its composition and certain intrinsic characteristics, such as the high moisture content and low acidity [64, 95]. Fish should be subjected to a preservation process (cooling or freezing or any method of processing such as salting, sterilization, and drying) within 12 to 20 h after death, otherwise microbial, chemical, and enzymatic deterioration may occur [64].

The fish shelf life is usually determined by its nutritional value, bacterial activity, and quality parameters, such as appearance, color, flavor, and texture. The quality attributes are affected by freezing, defrosting, and heating/cooking methods [96–98]. Ohmic Heating has been extensively investigated in recent decades, as it can provide faster and more uniform heating, resulting in products with greater quality and minimal thermal deterioration [99, 100]. OH has been used in fish (fish, crustaceans, and seafood) and fish-based products in order to obtain defrosted or cooked products with shorter processing times, lasting color, greater heating performance, and keeping structural, sensory, and nutritional properties [97, 98, 100].

Defrosting

There are different techniques for defrosting food (air, refrigerator, air jet, hot salt water, cold and hot water, and static water), in addition to innovative methodologies, such as high hydrostatic pressure, microwave, radio frequency, infrared, among others). Conventional defrosting generally occurs more slowly than freezing. Long periods of defrosting cause chemical, physical, and microbiological changes with a consequent degradation of product quality [98]. On the contrary, fast freezing and defrosting keep the frozen fish quality close to that of fresh fish [101].

Roberts *et al.* [102] defrosted two blocks of shrimp concurrently using a prototype of an ohmic apparatus. According to the authors, the substantial amount of residual water generated by conventional defrosting by immersion in hot water does not occur in defrosting by OH [102]. Subsequently, the microbiological, sensory, and quality attributes of the defrosted shrimp (conventionally by immersion in water and by OH) were compared. No differences in the sensory or total aerobic microbial counts were observed, the conventionally defrosted shrimp had a higher moisture content [103]. Miao *et al.* [104] defrosted surimi cubes in saline using ohmic and conventional technologies. The authors observed that the highest defrost rate and strongest gels were observed in the products subjected to OH. The temperature distribution was homogeneous when 4% saline solution, 20 V, and 60 Hz were used.

OH was used to defrost tuna fish (200 V, 20 kHz). Fat distribution and presence of membranes in the muscles affected the electrical conductivity of the tuna muscles [99]. In a more recent study [105], the defrosting of tuna cubes was carried out through OH at different voltages (40-60 V) and saline concentrations (0.3-0.5% w/v). The OH methodology was compared to conventional defrost by air (250 °C) and water (27 and 40 °C). OH defrosting shortened the defrosting time and caused lower losses. The OH treatment at 50 V and 0.3% saline solution provided the shortest defrost time. Therefore, OH has great potential for use in the defrosting of fish, resulting in minimal damage to its quality.

Thermal processing

Cooking meat, fish and their derivatives is necessary to improve palatability and ensure microbiological safety, as well as making them more digestible for the consumer through the denaturation of proteins [74, 106]. Immersion in hot water and steam cooking are the processes most used by the food industries, but they have the disadvantage of low penetration rate of heat for the food geometric center and, therefore, they require a longer cooking time. The longer cooking time raises energy costs and leads to overcooking, decreasing yield [74, 107]. The type of methodology used for heating and its characteristics, such as the time/temperature binomial profile, impact the quality and structure of cooked meat products and may cause several undesirable changes [66, 69, 99].

Scientific research involving the use of OH for cooking fish and by-products individually or compared to conventional heat treatments are described in Table 15.5. OH is an efficient alternative for cooking food because it is able to guarantee the quality of the structural, physical-chemical, and sensory properties of cooked fish due to the considerable reduction in processing time.

Table 15.5 Studies on the impact of ohmic (OH) cooking of fish and fishery products.

Fish/Fish product	Processing parameters	Main effects	References
Yellowtail (*Seriola quinqueradiata*) fillets	Only OH (20 V, 50 and 20,000 Hz, 80 °C)	The electric conductivity was dependent on the fat and moisture contents, current direction, temperature, and muscle membrane connecting fiber.	[108]
Shrimp	Only OH (150 to 175 V/cm with salt concentrations 10 - 20 kg/m^3 and 1150 to 1725 V/cm with salt concentrations 13.75 and 25.75 kg/m^3, 72 °C)	No effect on the texture profile. No effect of shrimp size on the heating time. Temperature conduction within the shrimp was dependent on the electrical field strength and salt content.	[109]

(Continued)

Table 15.5 Studies on the impact of ohmic (OH) cooking of fish and fishery products. (*Continued*)

Fish/Fish product	Processing parameters	Main effects	References
Shrimp	OH (15 A, 50 Hz, 0–250 V, 72 °C) and conventional processing (72 °C)	OH-treated products had 50% shorter cooking times. No effect was observed in color, texture, and yield compared to the products steam cooked.	[107]
Scallops	OH (50 V, 20 kHz, 65 °C) and sous-vide (SV) processing (65 °C and V/30 min)	The heating rate increased with temperature and frequency. OH-treated products retained more water within the muscle, and had lower actin denaturation.	[98]
Fish pâté	OH (5 V/cm, 50 Hz, 78 °C) and conventional processing (78 °C)	OH processing reduced the processing time, energy consumption, and energy efficiency.	[100]
Yellowtail (*Seriola quinqueradiata*) fillets	Only OH (20 V ± 0.05 V, 20 kHz, 80 °C)	OH promoted low cooking loss.	[110]

15.3.3 Eggs and Egg Products

Liquid eggs are commonly used as ingredients in food products because of their important technological properties, such as gelling, foaming, and emulsifying properties. Liquid eggs are subjected to pasteurization treatments to reach the sanitary properties and extend their shelf lives. However, the high temperatures may affect the egg components, and protein coagulation is not desired due to its technological properties [111]. Therefore, alternative methods for egg pasteurization are demanded, and they should ensure product safety and result in low product degradation [112]. In this context, OH can be a valuable technology, allowing a rapid temperature reaching with short treatment time, low heat losses, and uniform heating. Moreover, OH can be used in the processing of viscous liquid products because of the reduced shear stresses. This fact is related to the lack of moving parts, as feed pump is excluded [113, 114]. However, heating of protein-rich liquid foods, such as egg products, can be critical because even a slight protein coagulation can change the flow rate of the product, thus affecting temperature distribution and residence times. Therefore, the liquid residence time within the ohmic column is dependent on the pump and piping system characteristics [34].

Recent studies about OH applied to egg products are reported in Table 15.6. Many studies were performed at a lab-scale, on small amounts of sample (5-200 mL), evaluating electrical conductivity changes and color or rheological properties. Only Alamprese *et al.* [112] evaluated the impact of OH on whole egg technological properties considering both a static cell (1 L volume) and a pilot plant (100 L whole egg for each treatment) and measuring gelling and foaming properties besides color and rheological behavior. Electrical conductivity (σ) is an important OH parameter, and egg products are classified as good conductors ($\sigma > 0.05$ S/m). However, the electrical conductivity of food products increases with temperature, water content, frequency, and voltage gradient [114]; therefore, a deep study of these parameters for the considered product and the ohmic system used is especially important. The published papers generally report a dependence of electrical conductivity on temperature, while no or little effect was observed for the electric field frequency. Color and rheological properties are more affected by conventional heating rather than OH, due to a different protein denaturation degree. For the same reason, also technological properties such as foaming and gelling capacities are better preserved with OH. Thus, OH can be considered as a promising novel pasteurization

Table 15.6 OH processing of egg products.

Egg product	Processing parameters	Main results	References
Liquid whole egg	20 V/cm, 50 Hz, 4-60 °C	Temperature dependency of electrical conductivity and rheological consistency index. Sharp increase of viscosity at 60 °C due to protein denaturation. No effect on apparent viscosity. Lower temperature sensitivity of OH-treated product.	[116, 117]
Liquid whole egg, albumen, and yolk	30 V/cm, 60 Hz, 19-60 °C	Linear temperature dependency of electrical conductivity. Higher conductivity of albumen with respect to yolk and whole egg. Significant effect on color parameters.	[118]
Liquid yolk	30 V, 20 kHz, 20-90 °C	Drastic color change from 75 °C on. Correlation of color change with protein denaturation.	[119]

(*Continued*)

Table 15.6 OH processing of egg products. (*Continued*)

Egg product	Processing parameters	Main results	References
Liquid albumen and yolk, shell, and membrane	27.2 V, 10-100,000 Hz, 20-60 °C	Assessment of electrical conductivity of the different egg fractions. Little or no frequency dependence of electrical conductivity. Significant effect of temperature on electrical conductivity.	[117]
Whole egg	20 kHz, different temperature-time combinations with pasteurizing effect.	Similar or even better technological properties of OH-treated whole egg products. Low-temperature treatments could overcome fluidity difficulties caused by protein denaturation.	[112]

technology, but still few studies have been carried out on a large scale and new researches are needed to overcome technical and economic issues and increase the processing of safe foods with better technological properties.

In the future, OH can be considered also for the pasteurization of shell eggs. Indeed, Kasler & Sastry [115] carried out a preliminary study about the electrical conductivity of the different egg fractions (e.g., albumen, yolk, shell, and membrane), just to model heat transfer for a possible application of OH to in-shell pasteurization.

15.3.4 Cereal Products

Cereal products are staple foods at the base of many diets in the world. Many cereal preparations require heat, not only for cooking grains, but also for proofing, baking, dehydration, pasteurization, instant soup rehydration, and so on. However, few studies have been recently performed about OH applied to cereal products (Table 15.7). One of the main applications is for bread baking. A preliminary work by Gally *et al.* [120] focused on the assessment of the dough electrical conductivity as a function of porosity (with or without the presence of yeast) and salt content. Dough formulation can indeed largely affect the conductivity, thus impacting on the OH efficiency. Moreover, Gally *et al.* [120] modeled the dough thermophysical properties and the temperature profile during OH, thus providing useful information for the set-up of the OH baking. The rapid and uniform heating obtained with OH baking positively affects quality and shelf life of the final product [121], especially when a gluten-free dough is used [122]. Actually, in gluten-free products the rapid heating can result in a better structure fixation during the early baking stages, thus allowing a higher gas retention. Moreover, the higher water content of gluten-free batters makes them more suitable for OH, due to the higher electrical conductivity compared to conventional doughs. Further investigations are needed for a better understanding of starch structural modifications and product digestibility, and a better optimization of process variables [122]. Anyhow, the experiments were conducted at a lab-scale using a small ohmic cell, thus the scale-up of the process needs specific optimization.

OH was applied also to bread dough proofing [123], showing that the same expansion rate can be obtained with shorter time, due to the reduced lag-phase observed compared to the conventional fermentation. Reduced times for dough proofing are interesting for industries to decrease the total energy consumption of bakeries. However, the continuous application of OH during the whole proofing process seems to be not beneficial and the process must be optimized based on dough pH, microbial population, and characteristics of the used device (e.g., waveform, electric field strength, frequency). Thus, further studies are encouraged for the industrial implementation of ohmic-assisted fermentation units [124].

Baking by means of OH was studied also for baked goods. In particular, Luyts *et al.* [124] and Deleu *et al.* [125] applied OH baking to pound cakes, demonstrating a better quality of OH-baked products during storage compared to the conventionally baked cakes, due to the limited moisture gradients inside the products. Moreover, since a higher amount of water is necessary in the OH-baked dough, this technology results in reduced-sucrose cakes with protein denaturation and starch gelatinization extents similar to those of a conventionally baked product. OH can be useful also for research purposes, being the heating more homogeneous than

Table 15.7 Recent applications of ohmic heating to cereal products.

Application	Processing parameters	Main results	References
Bread baking	50-120 V	Low temperature gradients, baking time, and final moisture; no significant effect on starch retrogradation during storage.	[122]
Bread baking	50 V, 50 Hz	Effect of salt content and dough porosity on the electrical conductivity. Modeling of dough thermophysical properties and temperature profile during OH.	[120]
Gluten-free bread baking	1000 V, 12 kHz, different heating profiles (0.5-7.5 kW for 0.25-10 min)	Reduced baking time and energy consumption; improved quality characteristics of the final product; increased resistant starch fraction. Variable power in different heating steps for optimized product quality.	[122]
Bread dough proofing	50-150 V, 50 Hz	High heating rates; reduced fermentation lag-phase; reduced energy consumption.	[123]

(*Continued*)

Table 15.7 Recent applications of ohmic heating to cereal products. (*Continued*)

Application	Processing parameters	Main results	References
Cake pound baking	Not reported	Limited moisture gradients inside the products; lower firmness increase during storage.	[124]
Cake pound baking	Not reported	Limited temperature and moisture gradients during baking; possibility to lower sucrose concentration while obtaining similar extents of starch gelatinization and protein denaturation phenomena.	[125]
Cake pre-baking for 3D printing	22, 33, 48 V/cm, 50 Hz	Considerable temperature heterogeneity along the nozzles, affected by nozzle aspect ratio and voltage conditions. Development of a numerical model for experiment simulations.	[127]
Cooking of rice	10.7-11.8 V/cm, 50 Hz	Low energy consumption; different textural properties depending on the rice type.	[128]

(*Continued*)

Table 15.7 Recent applications of ohmic heating to cereal products. (*Continued*)

Application	Processing parameters	Main results	References
Cooking of rice	1.4 kW, 60 Hz	Low energy consumption; great softening rates; negative effects on color.	[130]
Cooking of whole and dehulled pearl millet grains	8-10 V/cm, 60 Hz	Higher effects of dehulling rather than the cooking method.	[129]
Cooking of instant noodles	10-17.5 V/cm, 60 Hz	Good textural properties obtained with the proper combination of and cooking time (i.e., 90 s) and electric field strength (i.e., 15 V/cm)	[131]
Instant rice cake soup cooking	9-18 V/cm, 60 Hz	Good textural properties obtained with the proper combination of and cooking time (i.e., 80 s) and electric field strength (i.e., 15 V/cm)	[132]

in conventional baking. Actually, the use of OH baking was suggested for studying the formation of crumb in gluten-free bread [126].

A very recent study [127] proposed the use of OH for 3D printing of bakery products. OH was considered more suitable then conventional heating because of the batter rheological properties observed during batter-crumb transition. However, a strong temperature heterogeneity was measured along the nozzle axes, leading to clogging at higher temperatures. A numerical model purposely developed allowed the study of the effects of nozzle aspect ratio and voltage conditions on the temperature gradients, thus paving the way for the optimization of the nozzle-ohmic heating configurations able to provide effective 3D printing.

OH has been proposed also as an environmental friendly methodology for cooking cereals due to the short cooking time, high energy efficiency, and reduced energy consumption [128–130]. It can be a valuable alternative to boiling, steaming, or pressure cooking. Cereal grains generally have low electrical conductivity, which must be increased using a salt solution during cooking instead of water. Besides the type of grains in terms of species, variety, and possible pre-treatments (e.g., dehulling, germination), cooking yield as well as cooked products quality characteristics (e.g., color, texture, solubility index, water absorption, and pasting properties) also depend on the achieved temperature, frequency, and electrical field strength. Thus, an optimization of the cooking parameters as a function of the raw material and equipment used should always be performed. For instance, Dias-Martins *et al.* [129] demonstrated that the presence of pericarp in millet grains mainly affects cooking yield and quality characteristics rather than the heating method. Conversely, the OH cooked rice altered textural properties compared to the rice cooked in the electric cooker. The impact was dependent on the rice type [129]. Gavahian *et al.* [130] also demonstrated the higher efficiency of OH compared to microwave cooking of rice and the lack of fouling, although the color parameters were negatively affected.

The good energy efficiency of OH can be applied also to the preparation of instant cereal soups or noodles. Good textural properties can be obtained by optimizing electric field strength and cooking time [131–133].

15.3.5 Juices

The scientific research in the field of OH application to juices is very active and numerous papers have been published in the last 10 years, maybe due to the growing consumption of juices worldwide [133]. Ohmic Heating or OH can be applied in different steps of the production and Table 15.8 shows just few examples of the possible applications. Most studies used OH for promoting microbial inactivation, enzyme inactivation, extension of the shelf life, and improvements in physical stability. OH has been found as more efficient than conventional pasteurization, due to shorter treatment times allowing a higher retention of flavors and bioactive compounds. Moreover, a synergistic thermal and electrical effect has been demonstrated [133–140]. OH can be used to enhance extraction and expressing processes, thanks to the electrical breakdown or electroporation mechanism combined with the tissue softening [133, 141, 142]. These phenomena can be exploited also in the extraction of specific compounds (such as pectins, polyphenols, and so on), as recently reviewed by Gavahian *et al.* [143]. Moving to fruit concentrates production, OH can successfully substitute

Table 15.8 Examples of possible applications of ohmic heating to fruit juices.

Application	Processing parameters	Main results	References
Orange juice pasteurization	30 V/cm, 50 Hz	No significant differences in quality parameters. Significant efficiency gains.	[147]
Sugarcane juice pasteurization	15.2-16.7 V/cm, 60 Hz	Reduction of polyphenol oxidase activity due to the combined thermal and electrical effect; effect of frequency and waveform on juice color.	[146]
Carrot juice extraction	15-25 V/cm	Increased juice recovery by OH coupled with 2-stage pressing. No significant changes in color.	[142]
Grape juice concentration	15-30 V/cm, 50 Hz	Beneficial effect on antioxidant activity; low pH change; no effect on phenol content; minimal damage of sensory properties.	[143]
Sour cherry juice concentration	8.3-13.9 V/cm, 50 Hz	Acidity reduction similar to conventional heating; lower effect on color and turbidity.	[145]

conventional heating in order to make the water evaporation faster, while maintaining good sensory and nutritional quality [133, 143–145].

Juices are multiphase products, thus the electrical conductivity of both liquid and particulates must be considered in the OH implementation to assure the desired thermal effect [133, 146]. The efficiency gain is commonly recognized, but the electricity costs in many Countries are high, thus increasing the unit operative costs [147]. The high acidity of most juices can represent a drawback, because of possible undesirable electrochemical reactions between electrodes and solution interfaces, especially for high electrical field strength [133].

A further step has been taken in this field, combining OH with other innovative technologies. A synergistic effect in microbial inactivation has been observed by combining OH and UV-C irradiation [148]. In combination with sonication, OH kept higher concentrations of the bioactive profile and inactivated polyphenoloxidase at higher levels [149]. As for juice concentration, better quality parameters can be obtained if the evaporation step is carried out by means of OH under vacuum, thus lowering the vitamin C and polyphenol degradation [150]. In particular, OH (> 10 V/cm) could improve the vacuum evaporations system performance [151], while obtaining similar rheological properties of the treated product [152]. Attention must be paid to possible change of food electrical conductivity when low pressures are achieved [153, 154] or as a consequence of the increasing solid concentration [155], thus requiring specific optimizing processes when OH is combined with other technologies.

It has to be highlighted that results of different studies are not always consistent, due to differences in the type of juice considered as well as the equipment used and operative conditions. Thus, further standardized studies are needed to process scale up and meet the food industry needs.

15.4 Commercial Status

Several works have investigated the possible applications of OH technology, optimize the process parameters according to each food matrix, improve the materials of the equipment, and implement it on an industrial scale. Currently, the application of technology in the industry is restricted to only seven countries in the world, which includes: United Kingdom, Canada, Japan, Switzerland, Sweden, United States of America, and Italy. Table 15.9 highlights the companies that develop industrial OH systems and the respective countries where they are located [156].

Table 15.9 Companies manufacturing industrial ohmic heating systems [156].

Companies manufacturing	Countries
APV Baker Ltd.	United Kingdom
C-Tech Innovation	United Kingdom
Agro process	IAI Group, Canada
Yanagiya Machinery Co. Ltd.	Japan
Kasag	Switzerland
Alfa Laval	Sweden
Raztek	United States of America
Emmepiemme SRL	Italy

The first company in the United Kingdom to manufacture commercial systems used for food products was SPX (formerly APV Ltd.). Emmepiemme owns the production of most OH systems for processing fruit and vegetables in Europe. It is known that to date there are 18 commercials OH plants in full operation in the world to pasteurize or sterilize pumpable foods, such as vegetable and fruit products (purees, juices, pulps, etc.), ice cream mix, milk, whey, egg products, soy milk, soups, as well as aseptic packaging [157].

For greater adoption of OH systems in the food industry, technological, economic, social, and environmental issues must be evaluated. There are some obstacles to the widespread application of OH, such as: expensive equipment, complex structures (use and operation require sophisticated skills), and consumer concerns (safety). Therefore, OH needs to be viewed positively by consumers, highlighting its advantages and low complexity for implementation at the industrial level [157, 158]. In addition, approval by regulatory agencies is required for commercial adoption. In the USA, for example, OH for food processing has been approved by the FDA [159].

15.5 Limitations and Advantages

The electrical conductivity is dependent on the temperature, frequency, presence and concentration of electrolytes, and voltage gradient [160].

During the heating process, it is necessary to have a product with constant and homogeneous electrical conductivity.

The differences in electrical resistance and temperature in a two-phase product can impair the heating performance of the system and then a non-thermal uniformity during the treatment can be found.

If in a product there is a zone with lower conductivity, it will be heated less than the whole mass; at the same time if there is a region with a higher conductivity it will be heated more than the rest of the product. Nonpolar constituents, like fats and lipids, have a very low conductivity [161] and they will be heated quite exclusively by conduction.

Applying a right mixing phase for adequate time is possible to equalize the initial condition of the product (initial temperature and conductivity) and it is possible to obtain temperature homogeneity at the exit of ohmic heating phases.

Temperature uniformity is not required in a conventional heat exchanger, because if a part of the product has a lower temperature it receives more energy due to temperature gradient. On the contrary, in OH each part of the product receives the same energy, thus if a portion of the product is colder than the rest of the product, after the heating it will remain colder than the whole product. The temperature differences present at the inlet of the ohmic heater are often even amplified during heating phases. For almost all the products, the higher temperature the higher the electrical conductivity (due to the greater ionic mobility) and a portion of the product with higher electrical conductivity undergoes higher heating [162].

Another critical factor for OH is how the product moves inside the heater. In case of laminar flow, the product advances along the axis of the tube at a speed higher (up to double) than the average speed, which means that the product will be under heated. To overcome this problem, the common solution is to use different heating stages, to minimize the temperature difference for each stage. Each stage is connected to the next using a curved pipe (Figure 15.3) or, in the most critical cases, motorized mixers.

A variable to be kept under control with greater attention than a traditional system is the operating pressure. If an excessive pressure, associated with high temperatures, can damage the dielectric tubes, an insufficient pressure can bring the product to the boiling point. Reaching the product boiling point in a conventional heater does not lead to important consequences. In the case of the ohmic heater even the boiling of a small portion of the product inside the dielectric tubes for a few moments can lead to the rupture of the pipe itself. The operating pressure of the system must be set in such a way as to avoid the product boiling during OH.

Figure 15.3 Ohmic heater. (Courtesy of emmepiemme s.r.l.)

The main difference between an ohmic treatment and a conventional heating technique is the heat transfer mode. In the CH process the heat is transferred by convection, conduction, or radiation from a heat source external to the product. Inside the heating tube there is a very heterogeneous temperature distribution, which results in a deterioration of food product quality [88]. In OH, the passage of alternating electrical current through the food form thermal energy by the Joule's effect and the heating is volumetric because of the internal energy generation [163]. Ohmic processing allows a rapid heating rate of products, uniform heating of biphasic foods, high energetic efficiency, with positive consequences on food quality [49].

In the traditional heating systems, generally the system start-up takes place using water; in ohmic systems the different heating also avoids product loss in the initial start-up phases, with an increase in productivity.

Ohmic Heater (OH) lacks thermal inertia, which is considered an advantage of the methodology. The production flow can be stopped instantly without dangerous temperature rises in the heater. During the shut-down phases the product does not suffer from overheating and the heating resumes with the switch on of the plant.

Compared to traditional heating, the ohmic heater does not suffer from burning food external layers thanks to the absence of hot surfaces, therefore the cleaning costs are lower [35]. If this statement is true for many products, it must be considered that the ohmic heater is very sensitive to the presence of electrically insulating deposits on the electrodes or to conductive deposits on the walls of the dielectric tubes. Electrically insulating deposits on the electrodes do not allow a regular passage of current between the electrode and the product and can give rise to local electrical micro-discharges with consequent burning of the product. In the case of dairy product, OH promotes fouling on the surfaces of the electrodes, due to high mineral content [146, 164, 165]. The electrically conductive deposits, on the other hand, absorbing continuously energy, increase the temperature and produce burning of the food external layers. The use of dielectric tubes and cooled electrodes can reduce these phenomena, but there is a decrease of the heater efficiency.

References

1. De Alwis, A.A.P., Fryer, P.J., A finite element analysis of heat generation and transfer during ohmic heating of food. *Chem. Eng. Sci.*, 45, 1547-1559. 1990.
2. Kaur, N., Singh, A.K., Ohmic Heating: Concept and Applications - A Review. *Crit. Rev. Food Sci. Nutr.*, 56, 2338–2351. 2016.
3. https://www.emmepiemme-srl.com/ohmic-heating
4. Kulshrestha, S., Sastry, S., Frequency and voltage effects on enhanced diffusion during moderate electric field (MEF) treatment. *Innov. Food Sci. Emerg. Technol.*, 4, 189-194. 2003.
5. Reznick, D., Ohmic heating of fluid foods: Ohmic heating for thermal processing of foods: Government, industry, and academic perspectives. *Food Technol.* 50, 250–251. 1996.
6. Rahman, M.S. (eds) Handbook of Food Preservation, 521–532. 1999.

7. Ryang, J., Kim, C., Lee, S., Hwang, I., Rhee, M., Inactivation of *Bacillus cereus* in a tsuyu sauce using continuous ohmic heating with five sequential elbow-type electrodes. *J. Appl. Microbiol.*, 120(1), 175-184. 2016.
8. Kim, S.-S., Sung, H.-J., Kwak, H.-S., Joo, I.-S., Lee, J.-S., Ko, G., Kang, D.-H., Effect of power levels on inactivation of *Escherichia coli* O157: H7, *Salmonella Typhimurium*, and *Listeria monocytogenes* in tomato paste using 915-megahertz microwave and ohmic heating. *J. Food Prot.*, 79, 1616–1622. 2016.
9. FDA-CFSAN, Food and Drug Administration-Center for Food Safety and Applied Nutrition 2000. Kinetics of Microbial Inactivation for Alternative Food Processing Technologies-Ohmic and Inductive Heating. 2000.
10. Pereira, R., Martins, J., Mateus, C., Teixeira, J. A., Vicente, A. A., Death kinetics of *Escherichia coli* in goat milk and *Bacillus licheniformis* in cloudberry jam treated by ohmic heating. *Chem. Pap.*, 6, 121-126. 2007.
11. Palaniappan, S., Sastry, S.K., Richter, E.R., Effects of electroconductive heat treatment and electrical pretreatment on thermal death kinetics of selected microorganisms. *Biotechnol. Bioeng.*, 39, 225-232.1992.
12. Tomlins, R.I., Ordal, Z.J., Thermal injury and inactivaction in vegetative bacteria. In: Skinner F.A., Hugo W.B., editors. Inhibition and Inactivaction of Vegetative Microbes Society for applied Bacteriology Symposium Series No.5. 1976.
13. Russell, A.D., Lethal effects of heat on bacterial physiology and structure. *Sci Prog.*, 86,115-37. 2003.
14. Tsuchido, T., Katsui, N., Takeuchi, A., Takano, M., Shibasaki, I., Destruction of the outer membrane permeability barrier of Escherichia coli by heat treatment. *Appl. Environ. Microbiol.*, 50, 298–303. 1995.
15. Iandolo, J.J., Ordal, Z.J., Repair of thermal injury of *Staphylococcus aureus. J. Bacteriol.*, 91, 134–142. 1996.
16. Uemura, K., Isobe, S., Developing a new apparatus for inactivating *Escherichia coli* in saline water with high electric field AC. *J. Food Eng.*, 53, 203–207. 2002.
17. Lee, S.-Y., Sagong, H.-G., Ryu, S., Kang, D.-H., Effect of continuous ohmic heating to inactivate *Escherichia coli* O157:H7, *Salmonella Typhimurium* and *Listeria monocytogenes* in orange juice and tomato juice. *J. Appl. Microbiol.*, 112, 723–731. 2012.
18. Park, I.-K., Kang, D.H., Effect of electropermeabilization by ohmic heating for inactivation of *Escherichia coli* O157: H7, *Salmonella enterica* serovar *typhimurium*, and *Listeria monocytogenes* in buffered peptone water and apple juice. *Appl. Environ. Microbiol.*, 79, 7122–7129. 2013.
19. Yoon, W.S., Lee, C.Y.J., Kim, K.M., Lee, C.H., Leakage of cellular materials from *Saccharomyces cerevisiae* by ohmic heating. *J. Microbiol. Biotechnol.*, 12, 183–188. 2002.
20. Sun, H., Masuda, F., Kawamura, S., Himoto, J.I., Asano, K., Kimura, T., Effect of electric current of ohmic heating on nonthermal injury to *Streptococcus thermophiles* in milk. *J. Food Proc. Eng.*, 34, 878–892. 2011.

21. Khue, D.N., Tiep, H.T., Dat, L.Q., Kim Phung, L.T., Tam, L.N., Influence of frequency and temperature on the inactivation of *Salmonella enterica* serovar *enteritidis* in Ohmic heating of pomelo juice. LWT – *Food Sci. Technol.*, 129. 2020.
22. Tian, X., Shao, L., Yu, Q., Li, X., Dai, R., Evaluation of structural changes and intracellular substance leakage of *Escherichia coli* O157:H7 induced by ohmic heating. *J. Appl. Microbiol.*, 127(5), 1430-1441. 2019.
23. Somavat, R., Mohamed, H.M.H., Chung, Y.-K., Yousef, A.E., Sastry, S.K., Accelerated inactivation of *Geobacillus stearothermophilus* spores by ohmic heating. *J. Food Eng.*, 108, 69–76. 2012.
24. Somavat, R., Mohamed, H.M.H., Sastry, S.K., Inactivation kinetics of *Bacillus coagulans* spores under ohmic and conventional heating. LWT - *Food Sci. Technol.*, 54, 194-198. 2013.
25. Cho, H.Y., Yousef, A.E., Sastry, S. K., Kinetics of inactivation of *Bacillus subtilis* spores by continuous or intermittent ohmic and conventional heating. Biotechnol. Bioeng., 62(3), 368-372. 1999.
26. Rodrigues, R.Q., Dalmás, M., Muller, D.C., Escobar, D.D., Pizzato, A.C., Mercali, G.D., Tondo, E.C., Evaluation of nonthermal effects of electricity on inactivation kinetics of *Staphylococcus aureus* and *Escherichia coli* during ohmic heating of infant formula. J. Food Saf., 38(1). e12372. 2018.
27. Park, I.-K., Ha, J.-W., Kang, D.-H., Investigation of optimum ohmic heating conditions for inactivation of *Escherichia coli* O157:H7, *Salmonella enterica* serovar *Typhimurium*, and *Listeria monocytogenes* in apple juice. *BMC Microbiol*, 17, 117. 2017.
28. Doan, K.N., Lai, D.Q., Le, P.T.K., Inactivation of *E. coli* O157:H7 by Ohmic Heating at Different Frequencies and Temperatures in Buffer and Pomelo Juice. *Chem. Eng. Trans.*, 78, 475-480. 2020.
29. Kim, S.-S., Choi, W., Kang, D.-H., Application of low frequency pulsed ohmic heating for inactivation of foodborne pathogens and MS-2 phage in buffered peptone water and tomato juice. *Food Microbiol.*, 63, 22–2. 2017.
30. Baysal, A. H., Icier, F., Inactivation kinetics of *Alicyclobacillus acidoterrestris* spores in orange juice by ohmic heating: Effects of voltage gradient and temperature on inactivation. *J. Food Prot.*, 73(2), 299–304. 2010.
31. Kim, N.H., Ryang, J.H., Lee, B.S., Kim, C.T., Rhee, M.S., Continuous ohmic heating of commercially processed apple juice using five sequential electric fields results in rapid inactivation of *Alicyclobacillus acidoterrestris* spores. *Int. J. Food Microbiol.*, 246, 80–84. 2017.
32. Park, S.H., Balasubramaniam, V.M., Sastry, S.K., Lee, J., Pressure-ohmic-thermal sterilization: A feasible approach for the inactivation of *Bacillus amyloliquefaciens* and *Geobacillus stearothermophilus* spore. *Innov. Food Sci. Emerg. Technol.* 19, 115-123. 2013.
33. Anderson, A.K., Finkelstein, R., A Study of the Electro-Pure Process of Treating Milk. *J. Dairy Sci.*, 2, 374-406. 1919.

34. Icier, I. *Ohmic Heating of Fluid Foods*. In: Cullen, P.J., Tiwari, B.K., Valdramidis, V.P. Novel Thermal and Non-Thermal Technologies For Fluid Foods. London: Academic Press, 1st ed. 2012.
35. Cappato, L.P., Ferreira, M.V.S., Guimarães, J.T., Portela, J.B., Costa, A.L.R., Freitas, M.Q., Cunha, R.L., Oliveira, C.A.F., Mercali, G.D., Marzack, L.D.F., Cruz, A.G., Ohmic heating in dairy processing: Relevant aspects for safety and quality. *Trends Food Sci Technol.*, 62,104-1012. 2017.
36. Duygu, B.K., Ümit, G., Use of Ohmic Heating System in Meat Thawing and Its Effects on Microbiological Quality. *MANAS J. Eng.*, 6, 129-142. 2018.
37. Guida, V., Ferrari, G., Pataro, G., Chambery, A., Di maro, A., Parente, A., The effects of ohmic and conventional blanching on the nutritional, bioactive compounds and quality parameters of artichoke heads. *LWT*, 53, 569-579. 2013.
38. Parmar, P., Singh, A.K., Meena, G.S., Borad, S., Raju, P.N., Application of ohmic heating for concentration of milk. *J. Food Sci. Technol.*, 55, 4956-4963. 2018.
39. Gavahian, M., Tiwari, B.K., Moderate electric fields and ohmic heating as promising fermentation tools. *Innov. Food Sci. Emerg.*, 2020, https://doi.org/10.1016/j.ifset.2020.102422.
40. Inmanee, P., Kamonpatana, P., Pirak, T., Ohmic heating effects on Listeria monocytogenes inactivation, and chemical, physical, and sensory characteristic alterations for vacuum packaged sausage during post pasteurization. *LWT.*, 108, 183-189. 2019.
41. Costa, N.R., Cappato, L.P., Ferreira, M.V.S., Pires, R.P.S., Moraes, J., Esmerino, E.A., Silva, R., Cucinelli Neto, R.P., Tavares, M.I.B., Freitas, M.Q., Silveira Júnior, R.N., Rodrigues, F.N., Bisaggio, R.C., Cavalcanti, R.N., Raices, R.S.L., Silva, M.C., Cruz, A.G., Ohmic heating: a potential technology for sweet whey processing. *Food Res. Int.*, 106, 771–779. 2018.
42. Kuriya, S.P., Silva, R., Rocha, R.S., Guimarães, J.T., Balthazar, C.F., Pires, R.P., Tavares Filho, E.R., Pimentel, T.C., Freitas, M.Q., Cappato, L.P., Raices, R.S.L., Cruz, A.G., Silva, M.C., Esmerino, E.A., Assessment the impact of the electric field on the quality parameters of blueberry flavored dairy desserts. *Food Res. Int.*, 134, 109235. 2020.
43. Rocha, R.S., Silva, R., Guimarães, J.T., Balthazar, C.F., Pimentel, T.C., Neto, R.P.C., Tavares, M.I.B., Esmerino, E.A., Freitas, M.Q., Cappato, L.P., Cavalcanti, R.N., Rodrigues, F.N., Raices, R.S.L., Silva, M.C., Cruz, A.G., Possibilities for using ohmic heating in Minas Frescal cheese production. *Food Res. Int.*, 131,109027. 2020.
44. Silva, R., Rocha, R.S., Guimarães, J.T., Balthazar, C.F., Scudino, H., Ramos, G.L.P., Pimentel, T.C., Neto, R.P.C., Tavares, M.I.B., Esmerino, E.A., Duarte, M.C.K.H., Freitas, M.Q., Silva, P.H.F., Cappato, L.P., Raices, R.S.L., Silva, M.C., Cruz, A.G., Advantages of using ohmic heating in Dulce de Leche manufacturing. *Innov. Food Sci. Emerg. Technol.*, 65, 102475. 2020.

45. Pires, R. P. S., Cappato, L. P., Guimarães, J. T., Rocha, R. S., Silva, R., Balthazar, C. F., Cruz, A.G., Ohmic heating for infant formula processing: Evaluating the effect of different voltage gradient. *J. Food Eng.*, 280, 109989. 2020.
46. Coimbra, L.C., Vidal, V.A.S., Silva, R., Rocha, R.S., Guimarães, J.T., Ballthazar, C.F., Pimentel, T.C., Silva, M.C., Granato, D., Freitas, M.Q., Pollonio, M.A.R., Esmerino, E.A., Cruz, A.G., 2020. Are ohmic heating-treated whey dairy beverages an innovation? Insights of the Q methodology. *LWT- Food Sci. Technol.,* 134,110052. 2020.
47. Pereira, M.O., Guimarães, J.T., Ramos, G.L.P.A., Prado-Silva, L., Nascimento, J.S., Sant'Ana, A.S., Franco, R.M., Cruz, A.G., Inactivation kinetics of *Listeria monocytogenes* in whey dairy beverage processed with ohmic heating. *LWT- Food Sci. Technol.*, 127, 109420. 2020.
48. Silva, R., Rocha, R.S., Guimarães, J.T., Balthazar, C.F., Scudino, H., Ramos, G.L.P.A., Pimentel, T.C., Silva, M.C., Silva, P.H.F., Duarte, M.C.K.H., Freitas, M.Q., Cruz, A.G., Esmerino, E.A., Dulce de leche submitted to ohmic heating treatment: Consumer sensory profile using preferred attribute elicitation (PAE) and temporal check-all-that- apply (TCATA). *Food Res. Int.*, 134, 109217. 2020.
49. Jaeger, H., Roth, A., Toepfl, S., Holzhauser, T., Engel, K-H., Knorr, D., Vogel, R.F., Bandick, N., Kulling, S., Heinz, V., Steinberg, P., Opinion on the use of ohmic heating for the treatment of foods. *Trends Food Sci. Technol.*, 55, 84-97. 2016.
50. Icier, F., Tavman, S., Ohmic heating behaviour and rheological properties of ice cream mixes. *Int. J. Food Prop.*, 9, 679-689. 2006.
51. Kim, S.-S., Kang, D.-H., Effect of milk fat content on the performance of ohmic heating for inactivation of Escherichia coli O157:H7, *Salmonella enterica* Serovar *Typhimurium* and *Listeria monocytogenes. J. Appl. Microbiol.,*119, 475-486. 2015.
52. Sastry, S. K., Barach, J.T., Ohmic and inductive heating. *J. Food Sci.*, 65, 42-46. 2000.
53. Roux, S., Courel, M., Birlouez-Aragon, I., Municino, F., Massa, M., Pain, J.-P., Comparative thermal impact of two UHT technologies, continuous ohmic heating and direct steam injection, on the nutritional properties of liquid infant formula. *J. Food Eng.*, 179, 36-43. 2016.
54. Pereira, R., Rodrigues, R. M., Ramos, O. L., Malcata, F. X., Teixeira, J. A., Vicente, A. A., Production of whey protein-based aggregates under ohmic heating. *Food Bioprocess Technol.*, 9, 576-587. 2015.
55. Pereira, R., Martins, R. C., Vicente, A., Goat milk free fatty acid characterization during conventional and ohmic heating pasteurization. *J. Dairy Sci.*, 91, 2925-2937. 2008.
56. Sun, H., Kawamura, S., Himoto, J.-i., Itoh, K., Wada, T., Kimura, T., Effects of ohmic heating on microbial counts and denaturation of proteins in milk. *Food Sci. Technol. Res.*, 14, 117-123. 2008.

57. Pereira, R. N., Souza, B.W.S., Cerqueira, M.A., Teixeira, J. A., Vicente, A.A., Effects of electric fields on protein unfolding and aggregation: Influence on edible films formation. *Biomacromolecules*, 11, 2912-2918. 2010.
58. Shivmurti, S., Harshit, P., Rinkita, P., Smit, P., Comparison of chemical properties of milk when conventionally and ohmically heated. *Int. Food Res. J.*, 21 1425-1428. 2014.
59. Cappato, L.P., Ferreira, M.V., Pires, R.P.S., Cavalcanti, R.N., Bisaggio, R.C., Freitas, M.Q., Silva, M.C., Cruz, A.G., Whey acerola-flavoured drink submitted ohmic heating processing: Is there an optimal combination of the operational parameters? *Food Chem.*, 245, 22-28. 2018.
60. Cappato, L.P., Ferreira, M.V.S., Moraes, J., Pires, R.P.S., Rocha, R.S., Silva, R., Neto, R.P.C., Tavares, M.I.B., Freitas, M.Q., Rodrigues, F.N., Calado, V.M.A., Raices, R.S.L., Silva, M.C., Cruz, A.G., Whey acerola-flavoured drink submitted Ohmic Heating: Bioactive compounds, antioxidant capacity, thermal behavior, water mobility, fatty acid profile and volatile compounds. *Food Chem.*, 263, 81-88. 2018.
61. Ferreira, M.V.S., Cappato, L.P., Silva, R., Rocha, R.S., Neto, R.P.C., Tavares, M.I.B., Esmerino, E.A., Freitas, M.Q., Bissagio, R.C., Ranadheera, S., Raices, R.S.L., Silva, M.C., Cruz, A.G., Processing raspberry-flavored whey drink using ohmic heating: Physical, thermal and microstructural considerations. *Food Res. Int.*, 123, 20-26. 2019.
62. Ferreira, M.V.S., Cappato, L.P., Silva, R., Rocha, R.S., Guimarães, J.T., Balthazar, C. F., Esmerino, E.A., Freitas, M.Q., Rodrigues, F.N., Granato, D., Neto, R.P.C., Tavares, M.I.B., Silva, P.H.F., Raices, R.S.L., Silva, M.C., Cruz, A.G., Ohmic Heating for processing of whey-raspberry flavored beverage. *Food Chem.*, 297, 125018. 2019.
63. Pires, R.P.S., Guimarães, J.T., Barros, C.P., Balthazar, C.F., Chincha, A.I.A., Freitas, M.Q., Duarte, M.C.K.H., Silva, P.H.F., Pimentel, T.C., Abud, Y.K.D., Sant'Anna, C., Sant'Ana, A.S., Silva, M.C., Nascimento, J.S., Cruz, A.G., Ohmic heating increases inactivation and morphological changes of *Salmonella* sp. and the formation of bioactive compounds in infant formula. *Food microbiol.*, 97, 103737. 2021.
64. Richa, R., Shahi, N.C., Singh, A., Lohani, U.C., Omre, P.K., Kumar, A., Bhattacharya, T.K., Ohmic Heating Technology and Its Application in Meaty Food: A Review. *Adv. Res.*, 10, 1-10. 2017.
65. Williams, P.G., Nutritional composition of red meat. *Nutr. Diet.*, S113-S119. 2007.
66. Pathare, P.B., Roskilly, A.P., Quality and Energy Evaluation in Meat Cooking. *Food Eng. Rev.*, 8, 435-447. 2016.
67. Sun, W., Zhao, M., Yang, B., Zhao, H., Cui, C., Oxidation of sarcoplasmic proteins during processing of Cantonese sausage in relation to their aggregation behaviour and in vitro digestibility. *Meat Sci.*, 88, 462-467. 2011.
68. Tornberg, E., Effects of heat on meat proteins — Implications on structure and quality of meat products. Meat Sci., 70, 493-508. 2005.

69. Tian, X., Wu, W., Yu, Q., Hou, M., Jia, F., Li, X., Dai, R., Quality and proteome changes of beef M. longissimus dorsi cooked using a water bath and ohmic heating process. *Innov. Food Sci. Emerg. Technol.*, 34, 259-266. 2016.
70. Sastry, S., Ohmic heating and moderate electric field processing. *Food Sci. Technol. Int.*, 14, 419-422. 2008.
71. Lyng, J.G., McKenna, B.M., Myer, K., *Ohmic Pasteurization of Meat and Meat Products. Handbook of Farm, Dairy and Food Machinery Engineering.* (second ed.), Academic Press, Elsevier Inc., USA, 541-570. 2013.
72. Shirsat, N., Brunton, N.P., Lyng, J.G., McKenna, B., Scannell, A., Texture, colour and sensory evaluation of a conventionally and ohmically cooked meat emulsion batter. *J. Sci. Food Agric., 84,* 1861,1870. 2004.
73. Dai, Y., Lu, Y., Wu, W., Lu, X.-m., Han, Z.-p., Liu, Y., Li, X.-m., Dai, R., Changes in oxidation, color and texture deteriorations during refrigerated storage of ohmically and water bath-cooked pork meat. *Innov. Food Sci. Emerg. Technol.*, 26, 341-346. 2014.
74. Gavahian, M., Tiwari, B.K., Chu, Y.-H., Ting, Y., Farahnaky, A., Food texture as affected by ohmic heating: Mechanisms involved, recent findings, benefits, and limitations. *Trends Food Sci. Technol.*, 86, 328-339. 2019.
75. Shirsat, N., Brunton, N.P., Lyng, J.G., McKenna, B., Water holding capacity, dielectric properties and light microscopy of conventionally and ohmically cooked meat emulsion batter. *Eur. Food Res. Technol.*, 219, 1-5. 2004.
76. Shirsat, N., Lyng, J.G., Brunton, N.P., McKenna, B., Ohmic processing: Electrical conductivities of pork cuts. *Meat Sci.*, 67, 507-514. 2004.
77. Özkan, N., Ho, I., Farid, M., Combined ohmic and plate heating of hamburger patties: quality of cooked patties. *J. Food Eng.*, 63, 141-145. 2004.
78. Zell, M., Lyng, J.G., Denis, A., Cronin, D.A., Morgan, D.J., Ohmic cooking of whole beef muscle—Optimisation of meat preparation. *Meat Sci.*, 81, 693–698. 2009.
79. Zell, M., Lyng, J.G., Cronin, D.A., Morgan, D.J., Ohmic cooking of whole turkey meat — Effect of rapid ohmic heating on selected product parameters. *Food Chem.*, 120, 724-729. 2010.
80. Bozkurt, H., Icier, F., Electrical conductivity changes of minced beef-fat blends during ohmic cooking. *J. Food Eng.*, 96, 86-92. 2010a.
81. Bozkurt, H., Icier, F., Ohmic cooking of ground beef: effects on quality. *J. Food Eng.*, 96, 481–490. 2010b.
82. Dai, Y., Zhang, Q., Wang, L., Liu, Y., Li, X.-m., Dai, R.t., Changes in Shear Parameters, Protein Degradation and Ultrastructure of Pork Following Water Bath and Ohmic Cooking. *Food Bioproc. Technol.*, 7, 1393-1403. 2014.
83. Llave, Y., Udo, T., Fukuoka, M., Sakai, N., Ohmic heating of beef at 20 kHz and analysis of electrical conductivity at low and high frequencies. *J. Food Eng.*, 228, 91-101. 2018.
84. Ángel-Rendón, S.V., Filomena-Ambrosio, A., Condon-Díaz, S., Benítez-Sastoque, E.R., Sotelo-Díaz, L.I., Ohmic cooking: Application of a novel

technology in pork and influences on water holding capacity, cooking loss and colour. *Int. J. Gastron. Food Sci.*, 17, 100164. 2019.
85. Yildiz-Turp, G., Sengun, I.Y., Kendirci, P., Icier, F., Effect of ohmic treatment on quality characteristic of meat: a review. *Meat Sci.*, 93, 441–448, 2013.
86. Zell, M., Lyng, J.G., Cronin, D.A., Morgan, D.J., Ohmic cooking of whole beef muscle — Evaluation of the impact of a novel rapid ohmic cooking method on product quality. *Meat Sci.*, 86, 258-263. 2010.
87. Sarang, S., Sastry, S.K., Knipe, l., Electrical conductivity of fruits and meats during ohmic heating. *J. Food. Eng.*, 87, 351-356. 2008.
88. van der Sman, R.G.M., Model for electrical conductivity of muscle meat during Ohmic heating. *J. Food Eng.*, 208, 37-47. 2017.
89. Lyng, J.G., Clemente, I., McKenna, B.M., *Chapter 23 - Ohmic Pasteurization of Meat and Meat Products.* Handbook of Farm, Dairy and Food Machinery Engineering, (third edition), 593-622. 2019.
90. Külcü, D.B., Gürbüz, U., Use of Ohmic Heating System in Meat Thawing and Its Effects on Microbiological Quality. *Manas J. Eng.*, 6, 129-142. 2018.
91. Öztan A. Et Bilimi ve Teknolojisi. TMMOB Gıda Mühendisleri Odası Yayınları, KitaplarSerisi. 6. Baskı. Yayın No: 1, Filiz Matbaacılık Sanayi, Ankara, 182-191. 2008.
92. Yun, C.G., Lee, D.H., Park, J., Ohmic thawing of a frozen meat chunk. *Korean J. Food Sci. Technol.*, 30, 842-847. 1998.
93. Hong, G.P., Min, S.G., Ko, S.H., Shim, K.B., Seo, E.J., Choi, M.J., Effects of brine immersion and electrode contact type low voltage ohmic thawing on the physico-chemical properties of pork meat. *Korean J. Food Sci. Anim. Resour.*, 27, 416-423. 2007.
94. Kim, J.Y., Hong, G.P., Park, S.H., Spiess, W.E.L., Min, S.G., Effect of ohmic thawing on physico-chemical properties of frozen hamburger patties. *Korean J. Food Sci. Anim. Resour.*, 26, 223-228. 2006.
95. García, M., Vilas, C., Herrera, J.R., Bernárdez, M., Balsa-Canto, E., Alonso, A.A., Quality and shelf-life prediction for retail fresh hake (*Merluccius merluccius*). *Int. J. Food Microbiol.*, 208, 65–74. 2015.
96. Leygonie, C., Britz, T.J., Hoffman, L.C., Impact of freezing and thawing on the quality of meat: Review. *Meat Sci.*, 91, 93–98. 2012.
97. Liu, L., Llave, Y., Jin, Y., Zheng, D., Fukuoka, M., Sakai, N., Electrical conductivity and ohmic thawing of frozen tuna at high frequencies. *J. Food Eng.* 197, 68-77. 2017.
98. Llave, Y., Morinaga, K., Fukuoka, M., Sakai, N., Characterization of ohmic heating and sous-vide treatment of scallops: Analysis of electrical conductivity and the effect of thermal protein denaturation on quality attribute changes. *Innov. Food Sci. Emerg. Technol.*, 50, 112-123. 2018.
99. Gavahian, M., Farahnaky, A., Javidnia, K., Majzoobi, M., Comparison of Ohmic-assisted hydrodistillation with traditional hydrodistillation for the extraction of essential oils from *Thymus vulgaris* L. *Innov. Food Sci. Emerg. Technol.* 14, 85–91. 2012.

100. Aydin, C., Ünal, K., Kaya, Y., Comparison of the Effects of Ohmic and Conventional Heating Methods on Some Quality Parameters of the Hot-smoked Fish Pâté. *J. Aquat. Food Prod. Technol.* 2020. https://doi.org/10.1080/10498850.2020.1741752
101. Duygu, B., Ümit, G.. Application of ohmic heating system in meat thawing. Procedia. *Social Behav. Sci.*, 195, 2822–2828. 2015.
102. Roberts, J.S., Balaban, M.O., Zimmerman, R., Luzuriaga, D., Design and testing of a prototype ohmic thawing unit. *Comput. Electron. Agric.* 19, 211-222. 1998.
103. Roberts, J.S., Balaban, M.O., Luzuriaga, D.A., Comparison of quality attributes of ohmic and water immersion thawed shrimp. *J. Aquat. Food Prod. Technol.* 11, 3-11. 2002.
104. Miao, Y., Chen, J. Y., Noguchi, A., Studies on the Ohmic Thawing of Frozen Surimi. *Food Sci. Technol. Res*, 13, 296–300. 2007.
105. Fattahi, S., Zamindar, N., Effect of immersion ohmic heating on thawing rate and properties of frozen tuna fish. *Food Sci. Technol. Int.*, 26, 453-461. 2020.
106. Uran, H., Gokoglu, N., Effects of cooking methods and temperatures on nutritional and quality characteristics of anchovy (*Engraulis encrasicholus*). *J Food Sci Technol.*, 51, 722–28. 2014.
107. Lascorz, D., Torella, E., Lyng, J.G., Arroyo, C., The potential of ohmic heating as an alternative to steam for heat processing shrimps. *Innov. Food Sci. Emerg. Technol.*, 37, 329-335. 2016.
108. Jin, Y., Chenh, Y.-d., Fukuoka, M., Sakai, N., Electrical Conductivity of Yellowtail (*Seriola quinqueradiata*) Fillets During Ohmic Heating. *Food Bioprocess Technol.*, 8, 1904-1913. 2015.
109. Pedersen, S.J., Feyissa, A.H., Kavli, S.T.B., Frosch, S., An investigation on the application of ohmic heating of cold water shrimp and brine mixtures. *J. Food Eng.*, 179, 28-35. 2016.
110. Jin, Y., Jiang, C., Jiao, Y., Llave, Y., Fukuoka, M., Sakai, N. Analysis of ohmic heating of yellowtail (*Seriola quinqueradiata*) fillets at high frequencies by 3D simulation—Effect of ohmic heating system (batch and pseudo-continuous), sample shape, and size. *Innov. Food Sci. Emerg. Technol.*, 2020.
111. Espina, L., Monfort, S., Álvarez, I., García-Gonzalo, D., Pagán, R., Combination of pulsed electric fields, mild heat and essential oils as an alternative to the ultrapasteurization of liquid whole egg. *Int. J. Food Microbiol.*, 189, 119-125, 2014.
112. Alamprese, C., Cigarini, M., Brutti, A., Effects of ohmic heating on technological properties of whole egg. *Innov. Food Sci. Emerg. Technol.*, 58, 102244, 2019.
113. Lechevalier, V., Croguennec, T., Anton, M., Nau, F., "Processed egg products", in Y. Nys, M. Bain, and F. Van Immerseel, eds., *Improving the Safety and Quality of Eggs and Egg Products. Vol. 1: Egg Chemistry, Production and Consumption,* Cambridge, Woodhead Publishing Limited, Ch. 23, pp. 538-581, 2011.

114. Goullieux, A., Pain, J. P. "Ohmic heating" in D.-W. Sun, ed., *Emerging Technologies in Food Processing*, London, Academic Press, Ch. 22, pp. 399-426, 2014.
115. Kasler, D. R., Sastry, S. K., Effects of frequency on the electrical conductivity of whole shell egg components. *J. Food Process Eng., 42,* e13056, 2019.
116. Icier, F., Bozkurt, H., Ohmic heating of liquid whole egg: rheological behavior and fluid dynamics. *Food Bioproc. Technol., 4,* 1253-1263, 2011.
117. Bozkurt, H., Icier, F., The change of apparent viscosity of liquid whole egg during ohmic and conventional heating. *J. Food Process Eng., 35,* 120-133, 2012.
118. Darvishi, H., Khoshtaghaza, M.H., Zarein, M., Azadbakht, M., Ohmic processing of liquid whole egg, white egg and yolk. *Agric Eng Int: CIGR Journal, 14(4),* 224-230, 2012.
119. Llave, Y., Fukuda, S., Fukuoka, M., Shibata-Ishiwatari, N., Sakai, N., Analysis of color changes in chicken egg yolks and whites based on degree of thermal protein denaturation during ohmic heating and water bath treatment. *J. Food Eng., 222,* 151-161, 2018.
120. Gally, T., Rouaud, O., Jury, V., Le-Bail, A. Bread baking using ohmic heating technology; a comprehensive study based on experiments and modelling. *J. Food Eng., 190,* 176-184, 2016.
121. Derde, L.J., Gomand, S.V., Courtin, C.M., Delcour, J.A., Moisture distribution during conventional or electrical resistance oven baking of bread dough and subsequent storage. *J. Agric. Food Chem., 62,* 6445-6453, 2014.
122. Bender, D., Gratz, M., Vogt, S., Fauster, T., Wicki, B., Pichler, S., Kinner, M., Jäger, H., Schönlechner, R., Ohmic heating - a novel approach for gluten-free bread baking. *Food Bioproc. Technol., 12,* 1603-1613, 2019.
123. Gally, T., Rouaud, O., Jury, V., Havet, M., Ogé, A., Le-Bail, A., Proofing of bread dough assisted by ohmic heating. *Innov. Food Sci. Emerg. Technol., 39,* 55-62, 2017.
124. Luyts, A., Wilderjans, E., Van Haesendonck, I., Brijs, K., Courtin, C.M., Delcour, J.A., Relative importance of moisture migration and amylopectin retrogradation for pound cake crumb firming. *Food Chem., 141,* 3960-3966, 2013.
125. Deleu, L.J., Luyts, A., Wilderjans, E., Van Haesendonck, I., Brijs, K., Delcour, J.A., Ohmic versus conventional heating for studying molecular changes during pound cake baking. *J. Cereal Sci., 89,* 102708, 2019.
126. Masure, H.G., Wouters, A.G.B., Fierens, E., Delcour, J.A. Electrical resistance oven baking as a tool to study crumb structure formation in gluten-free bread. *Food Res. Int., 116,* 925-931, 2019.
127. Khodeir, M., Rouaud, O., Ogé, A., Jury, V., Le-Bail, P., Le-Bail, A., Study of continuous cake pre-baking in a rectangular channel using ohmic heating. *Innov. Food Sci. Emerg. Technol., 67,* 102580, 2021.
128. Jittanit, W., Khuenpet, K., Kaewsri, P., Dumrongpongpaiboon, N., Hayamin, P., Jantarangsri, K., Ohmic heating for cooking rice: Electrical conductivity

measurements, textural quality determination and energy analysis. *Innov. Food Sci. Emerg. Technol., 42,* 16-24, 2017.

129. Dias-Martins, A.M., Cappato, L.P., da Costa Mattos, M., Rodrigues, F.N., Pacheco, S., Carvalho, C.W.P. Impacts of ohmic heating on decorticated and whole pearl millet grains compared to open-pan cooking. *J. Cereal Sci., 85,* 120-129, 2019.
130. Gavahian, M., Chu, Y.H., Farahnaky, A. Effects of ohmic and microwave cooking on textural softening and physical properties of rice. *J. Food Eng., 243,* 114-124, 2019.
131. Jo, Y.J., Park, S.H. Evaluation of energy efficacy and texture of ohmically cooked noodles. *J. Food Eng., 248,* 71-79, 2019.
132. Soisungwan, S., Khampakool, A., You, S.G., Park, S.H. Ohmic cooking of instant rice cake soup: energy efficiency and textural qualities. *Food Sci. Biotechnol., 29,* 641-649, 2020.
133. Jiménez-Sánchez, C., Lozano-Sánchez, J., Segura-Carretero, A., Fernández-Gutiérrez, A. Alternatives to conventional thermal treatments in fruit-juice processing. Part 1: Techniques and applications. *Crit. Rev. Food Sci. Nutr., 57,* 501-523, 2017.
134. Achir, N., Dhuique-Mayer, C., Hadjal, T., Madani, K., Pain, J.P., Dornier, M. Pasteurization of citrus juices with ohmic heating to preserve the carotenoid profile. *Innov. Food Sci. Emerg. Technol., 33,* 397-404, 2016.
135. Ashitha, G.N., Prince, M.V., Sudheer, K.P. Mild thermal processing of cashew apple juice using ohmic heating. *J. Trop. Agric., 58,* 44-52, 2020.
136. Brochier, B., Mercali, G.D., Marczak, L.D.F. Effect of moderate electric field on peroxidase activity, phenolic compounds and color during ohmic heating of sugarcane juice. *J. Food Proc. Preserv.,43,* e14254, 2019.
137. Doan, K.N., Lai, D.Q., Kim Le, P.T., Nhat Le, T. Inactivation of pectin Methylesterase and *Lactobacillus Plantarum* by ohmic heating in pomelo juice. *Int. J. Food Sci. Technol., 56,* 1987–1995, 2021.
138. Funcia, E.S., Gut, J.A.W., Sastry, S.K. Effect of electric field on pectinesterase inactivation during orange juice pasteurization by ohmic heating. *Food Bioproc. Technol., 13,* 1206-1214, 2020.
139. Makroo, H.A., Saxena, J., Rastogi, N.K., Srivastava, B. Ohmic heating assisted polyphenol oxidase inactivation of watermelon juice: Effects of the treatment on pH, lycopene, total phenolic content, and color of the juice. *J. Food Proc. Preserv., 41,* e13271, 2017.
140. Zhong, S., Vendrell-Pacheco, M., Heskitt, B., Chitchumroonchokchai, C., Failla, M., Sastry, S.K., Francis, D.M., Martin-Belloso, O., Elez-Martínez, P., Kopec, R.E. Novel processing technologies as compared to thermal treatment on the bioaccessibility and Caco-2 cell uptake of carotenoids from tomato and kale-based juices. *J. Agric. Food Chem., 67,* 10185-10194, 2019.
141. Gavahian, M., Chu, Y.H., Sastry, S. Extraction from food and natural products by moderate electric field: Mechanisms, benefits, and potential industrial applications. *Compr. Rev. Food Sci. Food Saf., 17,* 1040-1052, 2018.

142. Ranmode, S., Kulshreshtha, M. Enhancement of juice recovery from carrot using 2-stage pressing with ohmic heating. *J. Eng. Sci. Technol., 6(2),* 228-239, 2011.
143. Darvishi, H., Koushesh Saba, M., Behroozi-Khazaei, N., Nourbakhsh, H. Improving quality and quantity attributes of grape juice concentrate (molasses) using ohmic heating. *J. Food Sci. Technol., 57,* 1362–1370, 2020.
144. Darvishi, H., Salami, P., Fadavi, A., Koushesh Saba, M. Processing kinetics, quality and thermodynamic evaluation of mulberry juice concentration process using Ohmic heating. *Food Bioprod. Proc., 123,* 102-110, 2020.
145. Norouzi, S., Fadavi, A., Darvishi, H. The ohmic and conventional heating methods in concentration of sour cherry juice: Quality and engineering factors. *J. Food Engin., 291,* 110242, 2021.
146. Brochier, B., Hertz, P.F., Marczak, L.D.F., Mercali, G.D. Influence of ohmic heating on commercial peroxidase and sugarcane juice peroxidase inactivation. *J. Food Engin., 284,* 1100066, 2020.
147. Atuonwu, J.C., Leadley, C., Bosman, A., Tassou, S.A. High-pressure processing, microwave, ohmic, and conventional thermal pasteurization: Quality aspects and energy economics. *J. Food Process Eng., 43,* 13328, 2020.
148. Kim, S.S., Park, S.H., Kim, S.H., Kang, D.H. Synergistic effect of ohmic heating and UV-C irradiation for inactivation of *Escherichia coli* O157:H7, *Salmonella Typhimurium* and *Listeria monocytogenes* in buffered peptone water and tomato juice. *Food Control, 102,* 69-75, 2019.
149. Abedelmaksoud, T.G., Mohsen, S.M., Duedahl-Olesen, L., Elnikeety, M.M., Feyissa, A.H. Optimization of ohmicsonication for overall quality characteristics of NFC apple juice. *J. Food Process Preserv., 43,* e14087, 2019.
150. Darvishi, H., Mohammadia, P., Fadavi, A., Saba, M.K., Behroozi-Khazaei, N. Quality preservation of orange concentrate by using hybrid ohmic – Vacuum heating. *Food Chem., 289,* 292-298, 2019.
151. Sabanci, S., Icier, F. Enhancement of the performance of sour cherry juice concentration process in vacuum evaporator by assisting ohmic heating source. *Food Bioprod. Proc.*, 122, 269-279, 2020.
152. Sabanci, S., Icier, F. Rheological behavior of sour cherry juices concentrated by ohmic and conventional evaporation processes under vacuum. *J. Food Process. Preserv.*, 44, e14832, 2020.
153. Fadavi, A., Salari, S. Ohmic heating of lemon and grapefruit juices under vacuum pressure - Comparison of electrical conductivity and heating rate. *J. Food Sci., 84,* 2868-2875, 2019.
154. Fadavi, A., Salari, S., Mansouri, A., Hoseini, S. Effects of vacuum and juice concentration on electrical conductivity by the Ohmic method: a case study of sour cherry. *Food Bioproc. Technol.*, 13, 1146-1153, 2020.
155. Icier, F., Cokgezme, Ö.F., Döner, D., Bayana, D., Kaya, O., Çabas, B.M. Mathematical modelling of vacuum ohmic evaporation process. *Innov. Food Sci. Emerg. Technol., 67,* 102560, 2021.

156. Kumar, T., A Review on Ohmic Heating Technology: Principle, Applications and Scope. *Int. J. Agri. Envir. Biotechnol.*, 11, 679-687. 2018.
157. Priyadarshini, A., Rajauria, G., O'Donnell, C.P., Tiwari, B.K., Emerging Food Processing Technologies and Factors Impacting their Industrial Adoption. *Crit. Rev. Food Sci. Nutr.*, 59, 3082-3101. 2019.
158. Salari, S., Jafari, S.M. The Influence of Ohmic Heating on Degradation of Food Bioactive Ingredients. *Food Eng. Rev.* 12, 191-208. 2020.
159. Jambrak, A.R., Donsì, F., Paniwnyk, L., Djekic, I., Impact of novel nonthermal processing on food quality: Sustainability, modelling, and negative aspects *J. Food Qual.*, 2171375. 2019.
160. Assiry, A.M., Sastry, S.K., Samaranayake, C.P., Influence of Temperature, Electrical Conductivity, Power and PH on Ascorbic Acid Degradation Kinetics during Ohmic Heating Using Stainless Steel Electrodes. *Bioelectrochemistry.* 68, 7-13. 2006.
161. Goullieux, A., Pain, J.P., Ohmic heating, p 476 – 479. In Sun DW (ed), Emerging technologies for food processing. Elsevier Academic Press, Inc, San Diego, CA. 2005.
162. Darvishi, H., Khostaghaza, M.H., Najafi, G., Ohmic Heating of Pomegranate Juice: Electrical Conductivity and PH Change. *J. Saudi Soc. Agric. Sci.*, 12, 101-108. 2013.
163. Müller, W.A., Marczack, L.D.F., Sarkis, J.R., Microbial Inactivation by Ohmic Heating: Literature Review and Influence of Different Process Variables. *Trends Food Sci. Technol.*, 99, 650-659. 2020.
164. Ayadi, M.A., Leuliet, J.C., Chopard, F., Berthou, M., Lebouché, M., Experimental Study of Hydrodynamics in a Flat Ohmic Cell-Impact on Fouling by Dairy Products. *J. Food Eng.*, 70, 489-498. 2004.
165. Stancl, J., Zitny, R., Milk Fouling at Direct Ohmic Heating. *J. Food Eng.*, 99, 437-444. 2010.

Index

Also of Interest

Other Books in the series, "Bioprocessing in Food Science"

FUNCTIONAL FOODS, Edited by Navnidhi Chhikara, Anil Panghal, and Gaurav Chaudhary, ISBN: 9781119775560. Presenting cutting-edge information on new and emerging food engineering processes, Functional Foods, the second volume in the groundbreaking new series, "Bioprocessing in Food Science," is an essential reference on the modeling, quality, safety, and technologies associated with food processing operations today. *DUE OUT IN EARLY 2022!*

THERMAL FOOD ENGINEERING OPERATIONS, Edited by Nitin Kumar, Anil Panghal, and M. K. Garg, ISBN: 9781119775591. Presenting cutting-edge information on new and emerging food engineering processes, Thermal Food Engineering Operations, the first volume in the new series, "Bioprocessing in Food Science," is an essential reference on the modeling, quality, safety, and technologies associated with food processing operations today. *DUE OUT IN EARLY 2022!*

Other related titles from Scrivener Publishing

FUNCTIONAL POLYMERS IN FOOD SCIENCE, Part 1: Food Packaging, Edited by Giuseppe Cirillo, Umile Gianfranco Spizzirri, and Francesca Iemma, ISBN: 9781118594896. Provides an extensive overview of the link between polymer properties and food quality. *NOW AVAILABLE!*

FUNCTIONAL POLYMERS IN FOOD SCIENCE, Part 2: Food Processing, Edited by Giuseppe Cirillo, Umile Gianfranco Spizzirri, and Francesca Iemma, ISBN: 9781118595183. Provides an extensive overview of the link between polymer properties and food quality. *NOW AVAILABLE!*

ADVANCES IN FOOD SCIENCE AND TECHNOLOGY Volume 1, Edited by Visakh, P.M., Sabu Thomas, Laura B. Iturriaga, and Pablo Daniel Ribotta,

ISBN: 9781118121023. Written in a systematic and comprehensive manner the book reports recent advances in the development of food science and food technology areas. *NOW AVAILABLE!*

ADVANCES IN FOOD SCIENCE AND NUTRITION Volume 2, Edited by Visakh P.M., Laura B. Iturriaga, and Pablo Daniel Ribotta, ISBN: 9781118137093. This important book comprehensively reviews research on new developments in all areas of food chemistry/science and nutrition. *NOW AVAILABLE!*

Applied Water Science Volume 1: Fundamentals and Applications, Edited by Inamuddin, Mohd Imran Ahamed, Rajender Boddula and Tauseef Ahmad Rangreez, ISBN 9781119724766. Edited by one of the most well-respected and prolific engineers in the world and his team, this is the first volume in a two-volume set that is the most thorough, up-to-date, and comprehensive volume on applied water science available today. *COMING IN SUMMER 2021*

Sustainable Water Purification, by Safiur Rahman and M. R. Islam, ISBN 9781119650997. This is the only book that takes a zero-waste approach to propose 100% sustainable water purification techniques. *NOW AVAILABLE!*

Printed and bound by CPI Group (UK) Ltd, Croydon, CR0 4YY
17/09/2026
14964246-0008